Neural Networks

G. Dreyfus

Neural Networks

Methodology and Applications

With 217 Figures

Gérard Dreyfus
ESPCI, Laboratoire d'Électronique
10 rue Vauquelin
75005 Paris, France
E-mail: Gerard.Dreyfus@espci.fr

Original French edition published by Eyrolles, Paris (1st edn. 2002, 2nd edn. 2004)

ISBN-13 978-3-642-06187-5 e-ISBN-13 978-3-540-28847-3

Springer is a part of Springer Science+Business Media

springeronline.com

Printed in Germany

Cover design: *design & production* GmbH, Heidelberg

Preface

The term artificial neural networks used to generate pointless dreams and fears. Prosaically, neural networks are data-processing techniques that are essentially understood at present; they should be part of the toolbox of all scientists who want to make the most of the data that are available to them, including performing previsions, designing predictive models, recognizing patterns or signals, etc. All curricula oriented toward data processing contain educational programs related to those techniques. However, their industrial impact differs from country to country and, on the whole, is not yet as large as it should be.

The purpose of this book is to help students, scientists and engineers understand and use those techniques whenever necessary. To that effect, clear methodologies are described, which should make the development of applications in industry, finance and banking as easy and rigorous as possible in view of the present state of the art. No recipes will be provided here. It is our firm belief that no significant application can be developed without a basic understanding of the principles and methodology of model design and training.

The following chapters reflect the present state-of-the-art methodologies. Therefore, it may be useful to put it briefly into the perspective of the development of neural networks during the past years. The history of neural networks features an interesting paradox, i.e., the handful of researchers who initiated the modern development of those techniques, at the beginning of the 1980s, may consider that they were successful. However, the reason for their success is not what they expected. The initial motivation of the development of neural networks was neuromimetic. It was speculated that, because the most simple nervous systems, such as those of invertebrates, have abilities that far outperform those of computers for such specific tasks as pattern recognition, trying to build machines that mimic the brain was a promising and viable approach.

Actually, the same idea had also launched the first wave of interest in neural networks, in the 1960s, and those early attempts failed for lack of appropriate mathematical and computational tools. At present, powerful computers are available and the mathematics and statistics of machine learning have made enormous progress. However, a truly neuromimetic approach suffers from the lack of in-depth understanding of how the brain works; the very principles of information coding in the nervous system are largely unknown and open to heated debates. There exist some models of the functioning of specific systems (e.g. sensory), but there is definitely no theory of the brain.

It is thus hardly conceivable that useful machines can be built by imitating systems of which the actual functioning is essentially unknown. Therefore, the success of neural networks and related machine-learning techniques is definitely not due to brain imitation. In the present book, we show that artificial neural networks should be abstracted from the biological context. They should be viewed as mathematical objects that are to be understood with the tools of mathematics and statistics. That is how progress has been made in the area of machine learning and may be expected to continue in future years.

Thus, at present, the biological paradigm is not really helpful for the design and understanding of machine-learning techniques. It is actually quite the reverse, mathematical neural networks contribute more and more frequently to the understanding of biological neural networks because they allow the design of simple, mathematically tractable models of some parts of the nervous system. Such modeling, contributing to a better understanding of the principles of operation of the brain, might finally even benefit the design of machines. That is a fascinating, completely open area of research.

In a joint effort to improve the knowledge and use of neural techniques in their areas of activity, three French agencies, the Commissariat à l'énergie atomique (CEA), the Centre national d'études spatiales (CNES) and the Office national d'études et de recherches aérospatiales (ONERA), organized a spring school on neural networks and their applications to aerospace techniques and to environments. The present book stems from the courses taught during that school. Its authors have extensive experience in neural-network teaching and research and in the development of industrial applications.

Reading Guide

A variety of motivations may lead the reader to make use of the present book; therefore, it was deemed useful to provide a guide for the reading of the book because not all applications require the same mathematical tools.

Chapter 1, entitled "Neural networks: an overview", is intended to provide a general coverage of the topics described in the book and the presentation of a variety of applications. It will be of special interest to readers who require background information on neural networks and wonder whether those techniques are applicable or useful in their own areas of expertise. This chapter will also help define what the reader's actual needs are in terms of mathematical and neural techniques, hence, to lead him to reading the relevant chapters.

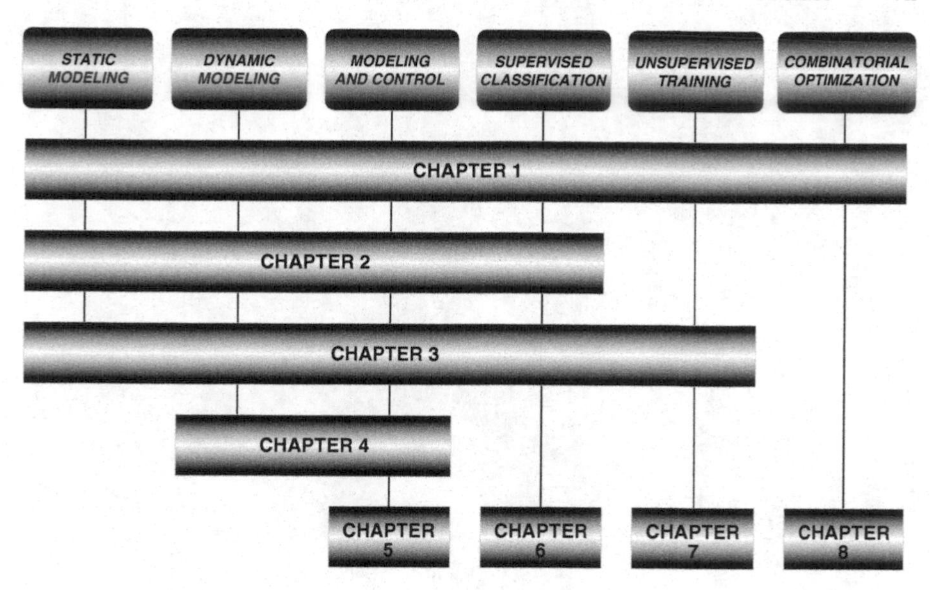

Readers who are interested in static modeling will read Chap. 2, "Modeling with neural networks: principles and model design methodology", up to, and including, the section entitled "Model selection". Then they will turn to Chap. 3, "Modeling methodology: dimension reduction and resampling methods".

Readers who are involved in applications that require dynamic modeling will read the whole of Chaps. 2, 3 and 4, "Neural identification of controlled dynamical systems and recurrent networks". If they want to design a model for use in control applications, they will read Chap. 5, "Closed-loop control learning".

Readers who are interested in supervised training for automatic classification (or discrimination) are advised to read the section "Feedforward neural networks and discrimination (classification)" of Chap. 1, then Chap. 2 up to, and including, the "Model selection" section, and then turn to Chap. 6 and possibly Chap. 3.

For those who are interested in unsupervised training, Chaps. 1, 3 and 7 ("Self-organizing maps and unsupervised classification") are relevant.

Finally, readers who are interested in combinatorial optimization will read Chaps. 1 and 8, "Neural networks without training for optimization".

Paris, September 2004 *Gérard Dreyfus*

[illegible]

[illegible]

[illegible]

[illegible]

[illegible]

Contents

3 Modeling Methodology: Dimension Reduction and Resampling Methods

4 Neural Identification of Controlled Dynamical Systems and Recurrent Networks

List of Contributors

Fouad Badran
Laboratoire Leibniz, IMAG
46 avenue Félix Viallet, 38000 Grenoble, France

Gérard Dreyfus
ESPCI, Laboratoire d'Électronique
10 rue Vauquelin, 75005 Paris, France

Mirta B. Gordon
Laboratoire Leibniz, IMAG
46 avenue Félix Viallet, 38031 Grenoble, France

Laurent Hérault
CEA-LETI, DSIS/SIT, CEA Grenoble
17 rue des Martyrs, 38054 Grenoble Cedex 9, France

Jean-Marc Martinez
DM2S/SFME Centre d'Études de Saclay
91191 Gif sur Yvette, France

Manuel Samuelides[1,2]
[1]École Nationale Supérieure de l'Aéronautique et de l'Espace
Département Mathématiques Appliquées
10 avenue Édouard Belin, BP 4032, 31055 Toulouse Cedex, France

[2]DRFMC/SPSMS/Groupe Théorie, CEA Grenoble
17 rue des Martyrs, 38054 Grenoble Cedex 9, France

Sylvie Thiria
Laboratoire d'Océanographie Dynamique et de Climatologie (LODYC)
Case 100, Université Paris 6
4 place Jussieu, 75252 Paris Cedex 5, France

Méziane Yacoub
CEDRIC, Conservatoire National des Arts et Métiers
292 rue Saint Martin, 75003 Paris, France

1

Neural Networks: An Overview

G. Dreyfus

How useful is that new technology? This is a natural question to ask whenever an emerging technique, such as neural networks, is transferred from research laboratories to industry. In addition, the biological flavor of the term "neural network" may lead to some confusion. For those reasons, this chapter is devoted to a presentation of the mathematical foundations and algorithms that underlie the use of neural networks, together with the description of typical applications; although the latter are quite varied, they are all based on a small number of simple principles.

Putting neural networks to work is quite simple, and good software development tools are available. However, in order to avoid disappointing results, it is important to have an in-depth understanding of what neural networks really do and of what they are really good at. The purpose of the present chapter is to explain under what circumstances neural networks are preferable to other data processing techniques and for what purposes they may be useful.

Basic definitions will be first presented: (formal) neuron, neural networks, neural network training (both supervised and unsupervised), feedforward and feedback (or recurrent) networks.

The basic property of neural networks with supervised training, parsimonious approximation, will subsequently be explained. Due to that property, neural networks are excellent nonlinear modeling tools. In that context, the concept of supervised training will emerge naturally as a nonlinear version of classical statistical modeling methods. Attention will be drawn to the necessary and sufficient conditions for an application of neural networks with supervised training to be successful.

Automatic classification (or discrimination) is an area of application of neural networks that has specific features. A general presentation of automatic classification, from a probabilistic point of view, will be made. It will be shown that not all classification problems can be solved efficiently by neural networks, and we will characterize the class of problems where neural classification is most appropriate. A general methodology for the design of neural classifiers will be explained.

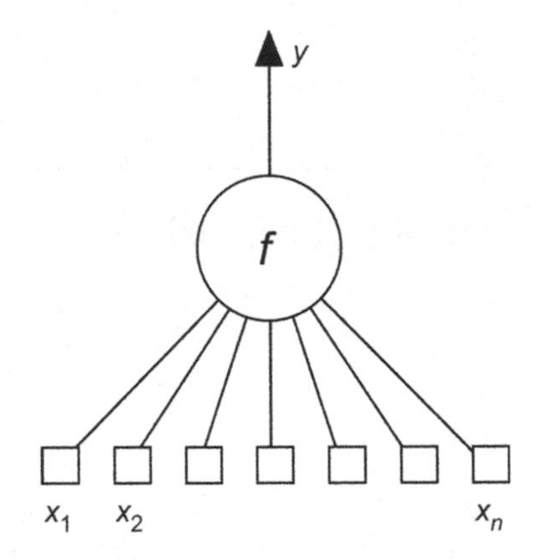

Fig. 1.1. A neuron is a nonlinear bounded function $y = f(x_1, x_2, \ldots x_n; w_1, w_2, \ldots, w_p)$ where the $\{x_i\}$ are the variables and the $\{w_j\}$ are the parameters (or weights) of the neuron

Finally, various applications will be described that illustrate the variety of areas where neural networks can provide efficient and elegant solutions to engineering problems, such that pattern recognition, nondestructive testing, information filtering, bioengineering, material formulation, modeling of industrial processes, environmental control, robotics, etc. Further applications (spectra interpretation, classification of satellite images, classification of sonar signals, process control) will be either mentioned or described in detail in subsequent chapters.

1.1 Neural Networks: Definitions and Properties

A *neuron* is a nonlinear, parameterized, bounded function.

For convenience, a linear parameterized function is often termed a linear neuron.

The variables of the neuron are often called inputs of the neuron and its value is its output. A neuron can be conveniently represented graphically as shown on Fig. 1.1. This representation stems from the biological inspiration that prompted the initial interest in formal neurons, between 1940 and 1970 [McCulloch 1943; Minsky 1969].

Function f can be parameterized in any appropriate fashion. Two types of parameterization are of current use.

- The parameters are assigned to the inputs of the neurons; the output of the neuron is a nonlinear combination of the inputs $\{x_i\}$, weighted by the parameters $\{w_i\}$, which are often termed weights, or, to be reminiscent of the biological inspiration of neural networks, synaptic weights. Following the current terminology, that linear combination will be termed potential in the present book, and, more specifically, linear potential in Chap. 5. The most frequently used potential v is a weighted sum of the inputs, with an additional constant term called "bias",

$$v = w_0 + \sum_{i=1}^{n-1} w_i x_i.$$

Function f is termed activation function. For reasons that will be explained below, it is advisable that function f be a sigmoid function (i.e., an s-shaped function), such as the tanh function or the inverse tangent function. In most applications that will be described in the present chapter, the output y of a neuron with inputs $\{x_i\}$ is given by $y = \tanh[w_0 + \sum_{i=1}^{n-1} w_i x_i]$.

- The parameters are assigned to the neuron nonlinearity, i.e., they belong to the very definition of the activation function such is the case when function f is a radial basis function (RBF) or a wavelet; the former stem from approximation theory [Powell 1987], the latter from signal processing [Mallat 1989].
  For instance, the output of a Gaussian RBF is given by

$$y = \exp\left[-\sum_{i=1}^{n}(x_i - w_i)^2 \bigg/ 2w_{n+1}^2\right],$$

  where the parameters w_i, $i = 1$ to n, are the position of the center of the Gaussian and w_{n+1} is its standard deviation.

Additional examples of neurons are given in the theoretical and algorithmic supplements, at the end of the chapter.

For practical purposes, the main difference between the above two categories of neurons is that RBFs and wavelets are local nonlinearities, which vanish asymptotically in all directions of input space, whereas neurons that have a potential and a sigmoid nonlinearity have an infinite-range influence along the direction defined by $v = 0$.

1.1.1 Neural Networks

It has just been shown that a neuron is a nonlinear, parameterized function of its input variables. Naturally enough, a network of neurons is the composition of the nonlinear functions of two or more neurons.

Neural networks come in two classes: feedforward networks and recurrent (or feedback) networks.

1.1.1.1 Feedforward Neural Networks

General Form

A *feedforward neural network* is a nonlinear function of its inputs, which is the composition of the functions of its neurons.

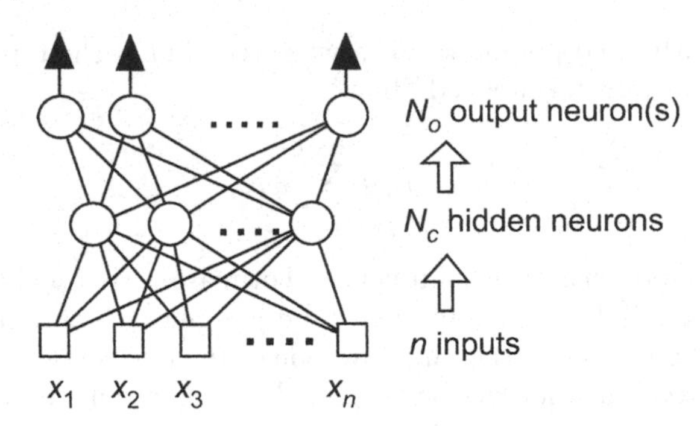

Fig. 1.2. A neural network with n inputs, a layer of N_c hidden neurons, and N_o output neurons

Therefore, a feedforward neural network is represented graphically as a set of neurons connected together, in which the information flows only in the forward direction, from inputs to outputs. In a graph representation, where the vertices are the neurons and the edges are the connections, the graph of a feedforward network is acyclic: no path in the graph, following the connections, can lead back to the starting point. The graph representation of the topology of the network is a useful tool, especially for analyzing recurrent networks, as will be shown in Chap. 2.

The neurons that perform the final computation, i.e., whose outputs are the outputs of the network, are called output neurons; the other neurons, which perform intermediate computations, are termed hidden neurons (see Fig. 1.2).

One should be wary of the term connection, which should be taken metaphorically. In the vast majority of applications, neurons are not physical objects, e.g., implemented electronically in silicon, and connections do not have any actual existence: the computations performed by each neuron are implemented as software programs, written in any convenient language and running on any computer. The term connection stems from the biological origin of neural networks; it is convenient, but it may be definitely misleading. So is the term connectionism.

Multilayer Networks

A great variety of network topologies can be imagined, under the sole constraint that the graph of connections be acyclic. However, for reasons that will be developed in a subsequent section, the vast majority of neural network applications implement multilayer networks, an example of which is shown on Fig. 1.2.

General Form

That network computes N_o functions of the input variables of the network; each output is a nonlinear function (computed by the corresponding output neuron) of the nonlinear functions computed by the hidden neurons.

A *feedforward network* with n inputs, N_c hidden neurons and N_o output neurons computes N_o nonlinear functions of its n input variables as compositions of the N_c functions computed by the hidden neurons.

It should be noted that feedforward networks are static; if the inputs are constant, then so are the outputs. The time necessary for the computation of the function of each neuron is usually negligibly small. Thus, feedforward neural networks are often termed static networks in contrast with recurrent or dynamic networks, which will be described in a specific section below.

Feedforward multilayer networks with sigmoid nonlinearities are often termed *multilayer perceptrons,* or *MLPs.*

In the literature, an input layer and input neurons are frequently mentioned as part of the structure of a multilayer perceptron. That is confusing because the inputs (shown as squares on Fig. 1.2, as opposed to neurons, which are shown as circles) are definitely not neurons: they do not perform any processing on the inputs, which they just pass as variables of the hidden neurons.

Feedforward Neural Networks with a Single Hidden Layer of Sigmoids and a Single Linear Output Neuron

The final part of this presentation of feedforward neural networks will be devoted to a class of feedforward neural networks that is particularly important in practice: networks with a single layer of hidden neurons with a sigmoid activation function, and a linear output neuron (Fig. 1.3).

The output of that network is given by

$$\begin{aligned} g(\boldsymbol{x}, \boldsymbol{w}) &= \sum_{i=1}^{N_c} \left[w_{N_c+1,i} \tanh \left(\sum_{j=1}^{n} w_{ij} x_j + w_{i0} \right) \right] + w_{N_c+1,0} \\ &= \sum_{i=1}^{N_c} \left[w_{N_c+1,i} \tanh \left(\sum_{j=0}^{n} w_{ij} x_j \right) \right] + w_{N_c+1,0}, \end{aligned}$$

where $\boldsymbol{x}$ is the input $(n+1)$-vector, and $\boldsymbol{w}$ is the vector of $(n+1)N_c+(N_c+1)$ parameters. Hidden neurons are numbered from 1 to N_c, and the output neuron is numbered N_c+1. Conventionally, the parameter w_{ij} is assigned to the connection that conveys information from neuron j (or from network input j) to neuron i.

The output $g(\boldsymbol{x}, \boldsymbol{w})$ of the network is a linear function of the parameters of the last connection layer (connections that convey information from the N_c hidden neurons to the output neuron N_c+1), and it is a nonlinear function

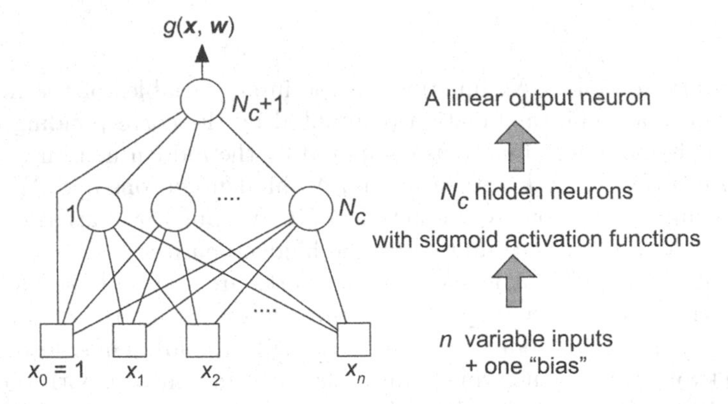

Fig. 1.3. A neural network with $n+1$ inputs, a layer of N_c hidden neurons with sigmoid activation function, and a linear output neuron. Its output $g(\boldsymbol{x}, \boldsymbol{w})$ is a nonlinear function of the input vector $\boldsymbol{x}$, whose components are $1, x_1, x_2, \ldots, x_n$, and of the vector of parameters $\boldsymbol{w}$, whose components are the $(n+1)N_c + N_c + 1$ parameters of the network

of the parameters of the first layer of connections (connections that convey information from the $n+1$ inputs of the network to the N_c hidden neurons). That property has important consequences, which will be described in detail in a subsequent section.

The output of a multilayer perceptron is a nonlinear function of its inputs and of its parameters.

1.1.1.2 What Is a Neural Network with Zero Hidden Neurons?

A feedforward neural network with zero hidden neuron and a linear output neuron is an affine function of its inputs. Hence, any linear system can be regarded as a neural network. That statement, however, does not bring anything new or useful to the well-developed theory of linear systems.

1.1.1.3 Direct Terms

If the function to be computed by the feedforward neural network is thought to have a significant linear component, it may be useful to add linear terms (sometimes called direct terms) to the above structure; they appear as additional connections on the graph representation of the network, which convey information directly from the inputs to the output neuron (Fig. 1.4). For instance, the output of a feedforward neural network with a single layer of activation functions and a linear output function becomes

$$g(\boldsymbol{x}, \boldsymbol{w}) = \sum_{i=1}^{N_c} \left[w_{N_c+1,i} \tanh\left(\sum_{j=0}^{n} w_{ij} x_j \right) \right] + \sum_{j=0}^{n} w_{N_c+1,j} x_j.$$

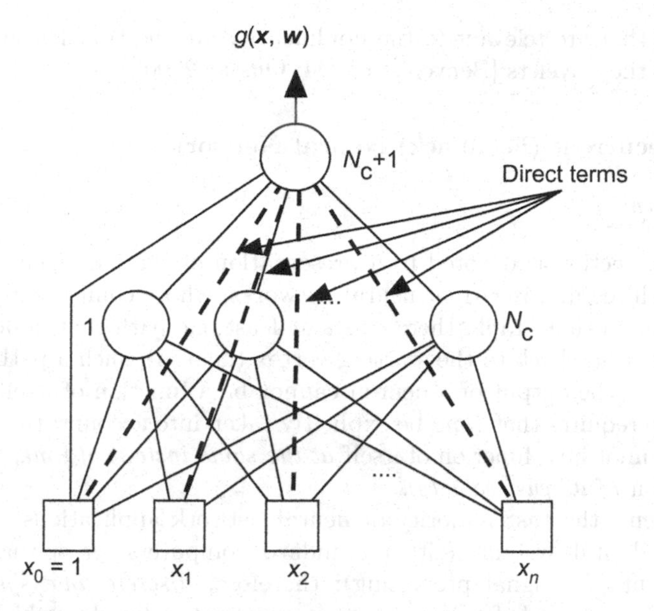

Fig. 1.4. A feedforward neural network with direct terms. Its output $g(\boldsymbol{x}, \boldsymbol{w})$ depends on the input vector $\boldsymbol{x}$, whose components are $1, x_1, x_2, \ldots, x_n$, and on the vector of parameters $\boldsymbol{w}$, whose components are the parameters of the network

RBF (Radial Basis Functions) and Wavelet Networks

The parameters of such networks are assigned to the nonlinear activation function, instead of being assigned to the connections; as in MLP's, the output is a linear combination of the outputs of the hidden RBF's. Therefore, the output of the network (for Gaussian RBF's) is given by

$$g(\boldsymbol{x}, \boldsymbol{w}) = \sum_{i=1}^{N_c} \left[w_{N_c+1,i} \exp\left(-\frac{\sum_{j=1}^{n} (x_j - w_{ij})^2}{2w_i^2} \right) \right],$$

where $\boldsymbol{x}$ is the n-vector of inputs, and $\boldsymbol{w}$ is the vector of $((n+2)N_c)$ parameters [Broomhead 1988; Moody 1989]; hidden neurons are numbered from 1 to N_c, and the output neuron is numbered $N_c + 1$.

The parameters of an RBF network fall into two classes: the parameters of the last layer, which convey information from the N_c RBF (outputs to the output linear neuron), and the parameters of the RBF's (centers and standard deviations for Gaussian RBF's). The connections of the first layer (from inputs to RBF's) are all equal to 1. In such networks, the output is a linear function of the parameters of the last layer and it is a nonlinear function of the parameters of the Gaussians. This has an important consequence that will be examined below.

Wavelet networks have exactly the same structure, except for the fact that the nonlinearities of the neurons are wavelets instead of being Gaussians. The parameters that are relevant to the nonlinearity are the translations and the dilations of the wavelets [Benveniste 1994; Oussar 2000].

1.1.1.4 Recurrent (Feedback) Neural Networks

General Form

The present section is devoted to a presentation of the most general neural network architecture: recurrent neural networks, whose connection graph exhibits *cycles.* In that graph, there exists at least one path that, following the connections, leads back to the starting vertex (neuron); such a path is called a *cycle*. Since the output of a neuron cannot be a function of itself, such an architecture requires that *time* be explicitly taken into account: the output of a neuron cannot be a function of itself *at the same instant of time*, but it can be a function *of its past value(s).*

At present, the vast majority of neural network applications are implemented as digital systems (either standard computers, or special-purpose digital circuits for signal processing): therefore, *discrete-time systems* are the natural framework for investigating recurrent networks, which are described mathematically by recurrent equations (hence the name of those networks). Discrete-time (or recurrent) equations are discrete-time equivalents of continuous-time differential equations.

Therefore, each connection of a recurrent neural network is assigned a *delay* (possibly equal to zero), in addition to being assigned a parameter as in feedforward neural networks. Each delay is an integer multiple of an elementary time that is considered as a time unit. From causality, a quantity, at a given time, cannot be a function of itself at the same time: therefore, the sum of the delays of the edges of a cycle in the graph of connections must be nonzero.

A *discrete-time recurrent neural network* obeys a set of nonlinear discrete-time recurrent equations, through the composition of the functions of its neurons, and through the time delays associated to its connections.

Property. *For causality to hold, each cycle of the connection graph must have at least one connection with a nonzero delay.*

Figure 1.5 shows an example of a recurrent neural network. The digits in the boxes are the delays attached to the connections, expressed as integer multiples of a time unit (or sampling period) T. The network features a cycle, from neuron 3 back to neuron 3 through neuron 4; since the connection from 4 to 3 has a delay of one time unit, the network is causal.

Further Details

At time kT, the inputs of neuron 3 are $u_1(kT), u_2[(k-1)T], y_4[(k-1)T]$ (where k is a positive integer and $y_4(kT)$ is the output of neuron 4 at time kT), and

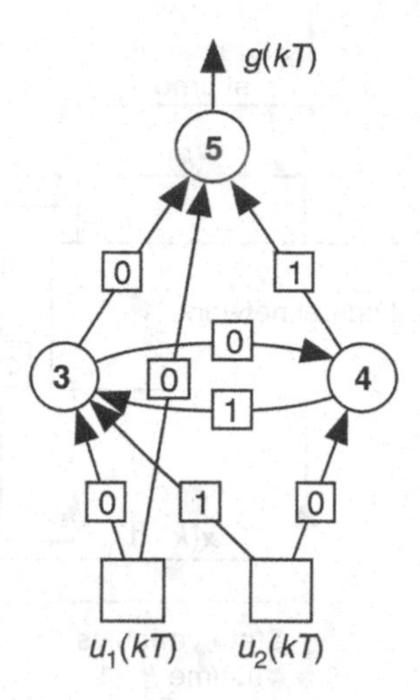

Fig. 1.5. A two-input recurrent neural network. Digits in *square boxes* are the delay assigned to each connection, an integer multiple of the time unit (or sampling period) T. The network features a cycle from 3 to 3 through 4

it computes its output $y_3(kT)$; the inputs of neuron 4 are $u_2(kT)$ and $y_3(kT)$, and it computes its output $y_4(kT)$; the inputs of neuron 5 are $y_3(kT), u_1(kT)$ et $y_4[(k-1)T]$, and it computes its output, which is the output of the network $g(kT)$.

The Canonical Form of Recurrent Neural Networks

Because recurrent neural networks are governed by recurrent discrete-time equations, it is natural to investigate the relations between such nonlinear models and the conventional dynamic linear models, as used in linear modeling and control.

The general mathematical description of a linear system is the state equations,

$$\boldsymbol{x}(k) = A\,\boldsymbol{x}(k-1) + B\,\boldsymbol{u}(k-1)$$
$$\boldsymbol{g}(k) = C\,\boldsymbol{x}(k-1) + D\,\boldsymbol{u}(k-1),$$

where $\boldsymbol{x}(k)$ is the state vector at time $kT, \boldsymbol{u}(k)$ is the input vector at time $kT, \boldsymbol{g}(k)$ is the output vector at time kT and A, B, C, D are matrices. The state variables are the minimal set of variables such that their values at time $(k+1)T$ can be computed if (i) their initial values are known, and if (ii) the values of the inputs are known at all time from 0 to kT. The number of state variables is the *order* of the system.

Similarly the canonical form of a nonlinear system is defined as

$$\boldsymbol{x}(k) = \boldsymbol{\Phi}[\boldsymbol{x}(k-1), \boldsymbol{u}(k-1)]$$
$$\boldsymbol{g}(k) = \boldsymbol{\Psi}[\boldsymbol{x}(k-1), \boldsymbol{u}(k-1)],$$

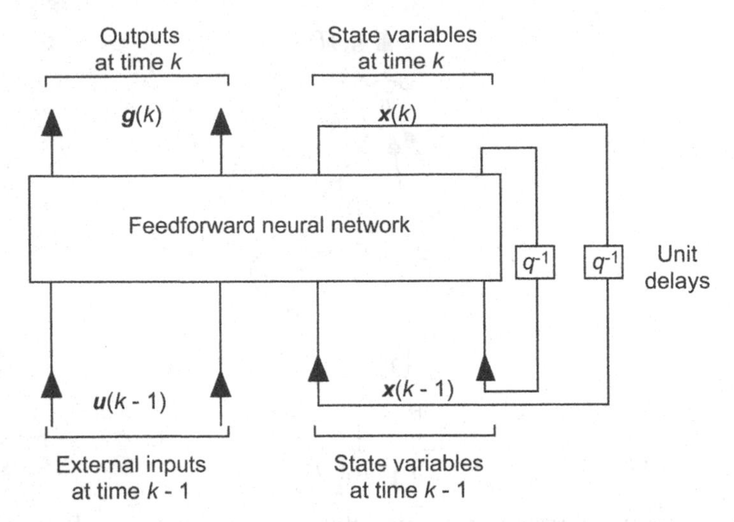

Fig. 1.6. The canonical form of a recurrent neural network. The symbol q^{-1} stands for a unit time delay

where $\boldsymbol{\Phi}$ and $\boldsymbol{\Psi}$ are nonlinear vector functions, e.g., neural networks, and where $\boldsymbol{x}$ is the state vector. As in the linear case, the state variables are the elements of the minimal set of variables such that the model can be described completely at time $k+1$ given the initial values of the state variables, and the inputs from time 0 to time k. It will be shown in Chap. 2 that any recurrent neural network can be cast into a canonical form, as shown on Fig. 1.6, where q^{-1} stands for a unit time delay. This symbol, which is usual in control theory, will be used in throughout this book, especially in Chaps. 2 and 4.

Property. *Any recurrent neural network, however complex, can be cast into a canonical form, made of a feedforward neural network, some outputs of which (termed state outputs) are fed back to the inputs through unit delays [Nerrand 1993].*

For instance, the neural network of Fig. 1.5 can be cast into the canonical form that is shown on Fig. 1.7. That network has a single state variable (hence it is a first-order network): the output of neuron 3. In that example, neuron 3 is a hidden neuron, but it will be shown below that a state neuron can also be an output neuron.

Further Details

At time kT, the inputs of neuron 4 are $u_2[(k-1)T]$ and $x[(k-1)T] = y_3[(k-1)T]$); therefore, its output is $y_4[(k-1)T]$; just as in the original (non-canonical) form, the inputs of neuron 3 are $u_1(kT), u_2[(k-1)T]$, $y_4[(k-1)T]$; therefore, its output is $y_3(kT)$; the inputs of neuron 5 are

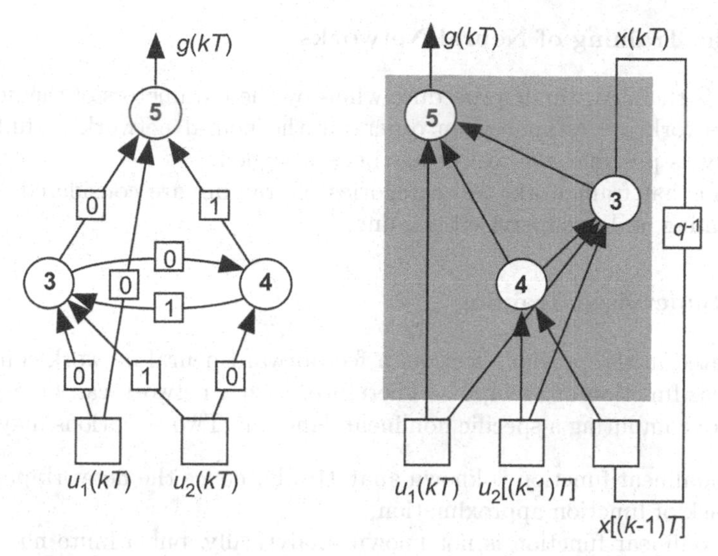

Fig. 1.7. The canonical form (*right-hand side*) of the network shown on Fig. 1.5 (*left-hand side*). That network has a single state variable $x(kT)$ (output of neuron 3): it is a first-order network. The gray part of the canonical form is a feedforward neural network

$y_3(kT), u_1(kT), y_4[(k-1)T]$; therefore, its output is $g(kT)$, which is the output of the network. Hence, both networks are functionally equivalent.

Recurrent neural networks (and their canonical form) will be investigated in detail in Chaps. 2, 4 and 8.

1.1.1.5 Summary

In the present section, we stated the basic definitions that are relevant to the neural networks investigated in the present book. We made specific distinctions between:

- Feedforward (or static) neural networks, which implement nonlinear functions of their inputs,
- Recurrent (or dynamic) neural networks, which are governed by nonlinear discrete-time recurrent equations.

In addition, we showed that any recurrent neural network can be cast into a canonical form, which is made of a feedforward neural network whose outputs are fed back to its inputs with a unit time delay.

Thus, the basic element of any neural network is a feedforward neural network. Therefore, we will first study in detail feedforward neural networks. Before investigating their properties and applications, we will consider the concept of training.

1.1.2 The Training of Neural Networks

Training is the algorithmic procedure whereby the parameters of the neurons of the network are estimated, in order for the neural network to fulfill, as accurately as possible, the task it has been assigned.

Within that framework, two categories of training are considered: supervised training and unsupervised training.

1.1.2.1 Supervised Training

As indicated in the previous section, a feedforward neural network computes a nonlinear function of its inputs. Therefore, such a network can be assigned the task of computing a specific nonlinear function. Two situations may arise:

- The nonlinear function is known analytically: hence the network performs the task of function approximation,
- The nonlinear function is not known analytically, but a finite number of numerical values of the function are known; in most applications, these values are not known exactly because they are obtained through measurements performed on a physical, chemical, financial, economic, biological, etc. process: in such a case, the task that is assigned to the network is that of approximating the regression function of the available data, hence of being a static model of the process.

In the vast majority of their applications, feedforward neural networks with supervised training are used in the second class of situations.

Training can be thought of as "supervised" since the function that the network should implement is known in some or all points: a "teacher" provides "examples" of values of the inputs and of the corresponding values of the output, i.e., of the task that the network should perform. The core of Chap. 2 of the book is devoted to translating the above metaphor into mathematics and algorithms. Chapters 3, 4, 5 and 6 are devoted to the design and applications of neural networks with supervised training for static and dynamic modeling, and for automatic classification (or discrimination).

1.1.2.2 Unsupervised Training

A feedforward neural network can also be assigned a task of data analysis or visualization: a set of data, described by a vector with a large number of components, is available. It may be desired to cluster these data, according to similarity criteria that are not known a priori. Clustering methods are well known in statistics; feedforward neural networks can be assigned a task that is close to clustering: from the high-dimensional data representation, find a representation of much smaller dimension (usually 2-dimensional) that preserves the similarities or neighborhoods. Thus, no teacher is present in that task, since the training of the network should "discover" the similarities between elements of the database, and translate them into vicinities in the new data representation or "map." The most popular feedforward neural networks with unsupervised training are the "self-organizing maps" or "Kohonen maps". Chapter 7 is devoted to self-organizing maps and their applications.

1.1.3 The Fundamental Property of Neural Networks with Supervised Training: Parsimonious Approximation

1.1.3.1 Nonlinear in Their Parameters, Neural Networks Are Universal Approximators

Property. *Any bounded, sufficiently regular function can be approximated uniformly with arbitrary accuracy in a finite region of variable space, by a neural network with a single layer of hidden neurons having the same activation function, and a linear output neuron [Hornik 1989, 1990, 1991].*

That property is just a proof of existence and does not provide any method for finding the number of neurons or the values of the parameters; furthermore, it is not specific to neural networks. The following property is indeed specific to neural networks, and it provides a rationale for the applications of neural networks.

1.1.3.2 Some Neural Networks Are Parsimonious

In order to implement real applications, the number of functions that are required to perform an approximation is an important criterion when a choice must be made between different models. It will be shown in the next section that the model designer ought always to choose the model with the smallest number of parameters, i.e., the most *parsimonious* model.

Fundamental Property

It can be shown [Barron 1993] that, if the model is nonlinear with respect to its parameters, it is more parsimonious than if the model is linear with respect to its parameters.

More specifically, it can be shown that the number of parameters necessary to perform an approximation with a given accuracy varies exponentially with the number of variables for models that are linear with respect to their parameters, whereas it increases linearly with the number of variables if the model is not linear with respect to its parameters.

Therefore, that property is valuable for models that have a "large" number of inputs: for a process with one or two variables only, all nonlinear models are roughly equivalent from the viewpoint of parsimony: a model that is nonlinear with respect to its parameters is equivalent, in that respect, to a model that is linear with respect to its parameters.

In the section devoted to the definitions, we showed that the output of a feedforward neural network with a layer of sigmoid activation functions (multilayer Perceptron) is nonlinear with respect to the parameters of the network, whereas the output of a network of radial basis functions with fixed centers and widths, or of wavelets with fixed translations and dilations, is linear with respect to the parameters. Similarly, a polynomial is linear with respect to the coefficients of the monomials. Thus, neurons with sigmoid activation functions provide more parsimonious approximations than polynomials or radial basis functions with fixed centers and widths, or wavelets with fixed translations and dilations. Conversely, if the centers and widths of Gaussian radial basis functions, or the centers and dilations of wavelets, are considered as adjustable parameters, there is no *mathematically proved* advantage to any one of those models over the others. However, some *practical* considerations may lead to favor one of the models over the others: prior knowledge on the type of nonlinearity that is required, local *vs.* nonlocal function, ease and speed of training (see Chap. 2, section "Parameter initialization"), ease of hardware integration into silicon, etc.

The origin of parsimony can be understood qualitatively as follows. Consider a model that is linear with respect to its parameters, such as a polynomial model, e.g.,

$$g(x) = 4 + 2x + 4x^2 - 0.5x^3.$$

The output $g(x)$ of the model is a linear combination of functions $y = 1, y = x, y = x^2, y = x^3$, with parameters (weights) $w_0 = 4, w_1 = 2, w_2 = 4, w_3 = -0.5$. The shapes of those functions are fixed.

Consider a neural model as shown on Fig. 1.8, for which the equation is

$$g(x) = 0.5 - 2\tanh(10x + 5) + 3\tanh(x + 0.25) - 2\tanh(3x - 0.25).$$

This model is also a linear combination of functions ($y = 1, y = \tanh(10x + 5), y = \tanh(x + 0.25), y = \tanh(3x - 0.25)$), but the shapes of these functions depend on the values of the parameters of the connections between the inputs and the hidden neurons. Thus, instead of combining functions whose shapes are fixed, one combines functions whose shapes are adjustable through the parameters of some connections. That provides extra degrees of freedom, which can be taken advantage of for using a smaller number of functions, hence a smaller number of parameters. That is the very essence of parsimony.

1.1.3.3 An Elementary Example

Consider the function

$$y = 16,71\, x^2 - 0,075.$$

We sample 20 equally spaced points that are used for training a multilayer Perceptron with two hidden neurons whose nonlinearity is $\tan^{-1}$, as shown on Fig. 1.9(a). Training is performed with the Levenberg-Marquardt algorithm

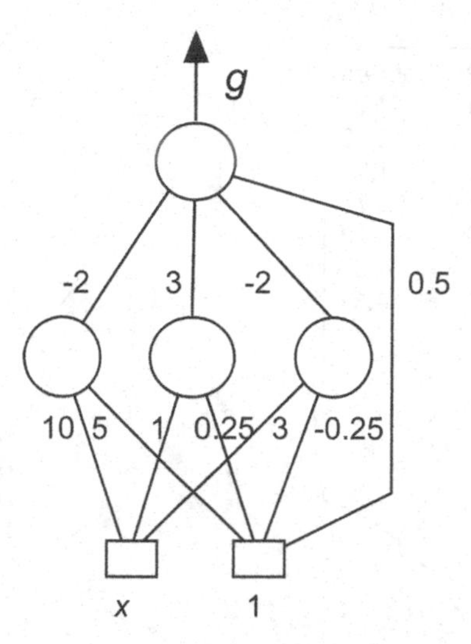

Fig. 1.8. A feedforward neural network with one variable (hence two inputs) and three hidden neurons. The numbers are the values of the parameters

(see Chap. 2), resulting in the parameters shown on Fig. 1.9(a). Figure 1.9(b) shows the points of the training set and the output of the network, which fits the training points with excellent accuracy. Figure 1.9(c) shows the outputs of the hidden neurons, whose linear combination with the bias provides the output of the network. Figure 1.9(d) shows the points of a test set, i.e., a set of points that were not used for training: outside of the domain of variation of the variable x within which training was performed ($[-0.12, +0.12]$), the approximation performed by the network becomes extremely inaccurate, as expected. The striking symmetry in the values of the parameters shows that training has successfully captured the symmetry of the problem (simulation performed with the NeuroOne^{TM} software suite by NETRAL S.A.).

It should be clear that using a neural network to approximate a single-variable parabola is overkill, since the parabola has two parameters whereas the neural network has seven parameters! This example has a didactic character insofar as simple one-dimensional graphical representations can be drawn.

1.1.4 Feedforward Neural Networks with Supervised Training for Static Modeling and Discrimination (Classification)

The mathematical properties described in the previous section are the basis of the applications of feedforward neural networks with supervised training. However, for all practical purposes, neural networks are scarcely ever used for uniformly approximating a *known* function.

In most cases, the engineer is faced with the following problem: a set of measured variables $\{\boldsymbol{x}^k, k = 1 \text{ to } N\}$, and a set of measurements $\{y_p(\boldsymbol{x}^k),$

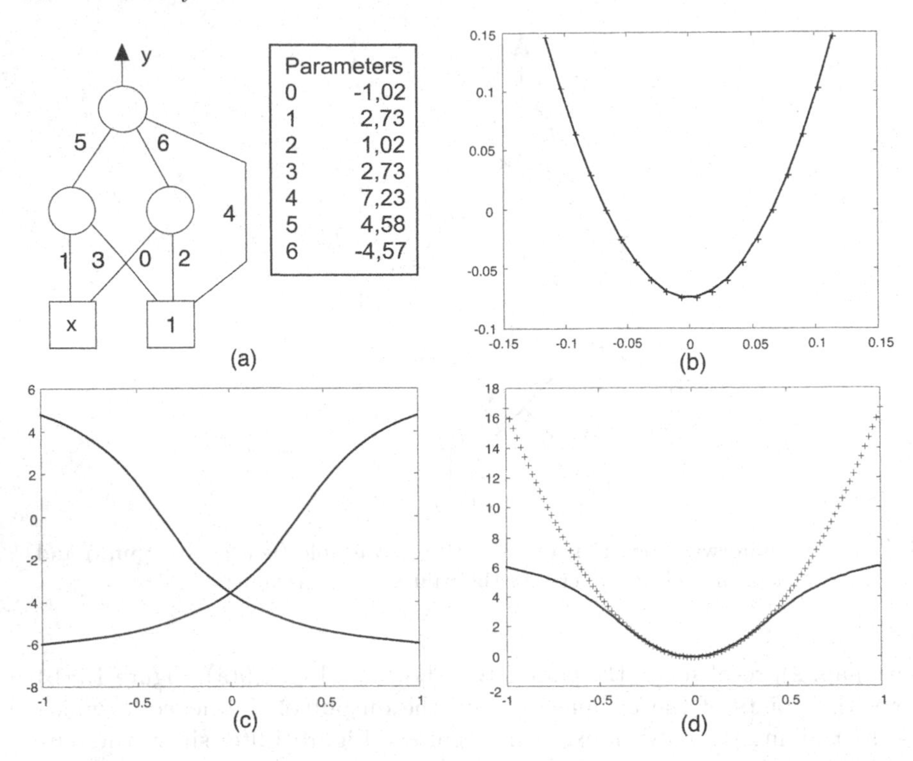

Fig. 1.9. Interpolation of a parabola by a neural network with two hidden neurons; (**a**) network; (**b**) training set (+) and network output (*line*) after training; (**c**) outputs of the two hidden neurons (sigmoid functions) after training; (**d**) test set (+) and network output (*line*) after training: as expected, the approximation is very inaccurate outside the domain of variation of the inputs during training

$k = 1$ to $N\}$ of a quantity of interest z_p related to a physical, chemical, financial, ..., process, are available. He assumes that there exists a relation between the vector of variables $\{\boldsymbol{x}\}$ and the quantity z_p, and he looks for a mathematical form of that relation, which is valid in the region of variable space where the measurements were performed, given that (1) the number of available measurements is finite, and (2) the measurements are corrupted by noise. Moreover, the variables that actually affect z_p are not necessarily measured. In other words, the engineer tries to build a *model* of the process of interest, from the available measurements only: such a model is called a *black-box* model. In neural network parlance, the observations from which the model is designed are called *examples*. We will consider below the "black-box" modeling of the hydraulic actuator of a robot arm: the set of variables $\{\boldsymbol{x}\}$ has a single element (the angle of the oil valve), and the quantity of interest $\{z_p\}$ is the oil pressure in the actuator. We will also describe an example of prediction of chemical properties of molecules: a relation between a molecular property (e.g., the boiling point) and "descriptors" of the molecules (e.g., the molecular mass, the number of atoms, the dipole moment, etc.); such a model allows predictions of the boiling points of molecules that were not synthesized before. Several similar cases will be described in this book.

Black-box models, as defined above, are in sharp contrast with knowledge-based models, which are made of mathematical equations derived from first principles of physics, chemistry, economics, etc. A knowledge-based model may have a limited number of adjustable parameters, which, in general, have a physical meaning. We will show below that neural networks can be building blocks of gray box or semi-physical models, which take into account both expert knowledge—as in a knowledge-based model—and data—as in a black-box model.

Since neural networks are not really used for function approximation, to what extent is the above-mentioned parsimonious approximation property relevant to neural network applications? In the present chapter, a cursory answer to that question will be provided. A very detailed answer will be provided in Chap. 2, in which a general design methodology will be presented, and in Chap. 3, which provides very useful techniques for the reduction of input dimension, and for the design, and the performance evaluation, of neural networks.

1.1.4.1 Static Modeling

For simplicity, we first consider a model with a single variable x. Assume that an infinite number of measurements of the quantity of interest can be performed for a given value x_0 of the variable x. Their mean value is the quantity of interest z_p, which is called the "expectation" of y_p for the value x_0 of the variable. The expectation value of y_p is a function of x, termed "regression function". Since we know from the previous section that any function can be approximated with arbitrary accuracy by a neural network, it may be expected that the black-box modeling problem, as stated above, can be solved by estimating the parameters of a neural network that approximates the (unknown) regression function.

The approximation will not be uniform, as defined and illustrated in the previous section. For reasons that will be explained in Chap. 2, the model will perform an approximation in the least squares sense: a parameterized function (e.g., a neural network) will be sought, for which the least squares cost function

$$J(\boldsymbol{w}) = \frac{1}{2}\sum_{k=1}^{N} \left[y_p(\boldsymbol{x}^k) - g(\boldsymbol{x}^k, \boldsymbol{w})\right]^2$$

is minimal. In the above relation, $\{\boldsymbol{x}^k, k = 1 \text{ to } N\}$ is a set of measured values of the input variables, and $\{Ly_p(\boldsymbol{x}^k), k = 1 \text{ to } N\}$ as set of corresponding measured values of the quantity to be modeled. Therefore, for a network that has a given architecture (i.e., a given number of inputs and of hidden neurons),

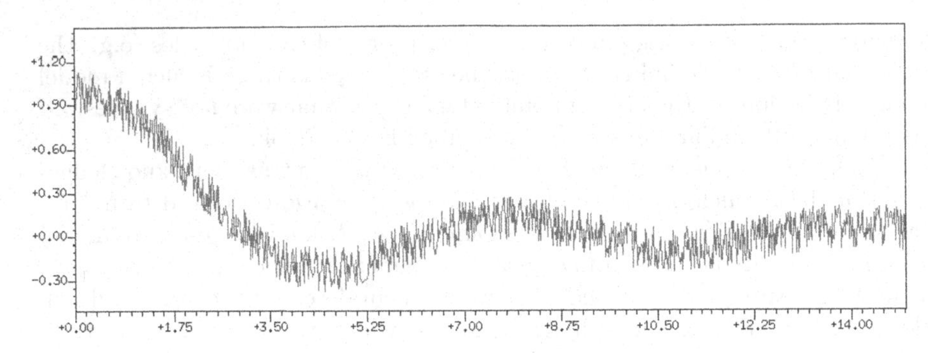

Fig. 1.10. A quantity to be modeled

training is a procedure whereby the least squares cost function is minimized, so as to find an appropriate weight vector $\boldsymbol{w}_0$.

That procedure suggests two questions, which are central in any neural network application, i.e.,

- for a given architecture, how can one find the neural network for which the least squares cost function is minimal?
- if such a neural network has been found, how can its prediction ability be assessed?

Chapter 2 of the present book will provide the reader with a methodology, based on first principles, which will answer the above questions.

These questions are not specific to neural networks: they are standard questions in the field of modeling, that have been asked for many years by all scientists (engineers, economists, biologists, and statisticians) who endeavor to extract relevant information from data [Seber 1989; Antoniadis 1992; Draper 1998]. Actually, the path from function approximation to parameter estimation of a regression function is the traditional path of any statistician in search of a model: therefore, we will take advantage of theoretical advances of statistics, especially in regression.

We will now summarize the steps that were just described.

- When a mathematical model of dependencies between variables is sought, one tries to find the regression function of the variable of interest, i.e., the function that would be obtained by averaging, at each point of variable space, the results of an infinite number of measurements; the regression function is forever unknown. Figure 1.10 shows a quantity $y_p(x)$ that one tries to model: the best approximation of the (unknown) regression function is sought.
- A finite number of measurements are performed, as shown on Fig. 1.11.
- A neural network provides an approximation of the regression function if its parameters are estimated in such a way that the sum of the squared

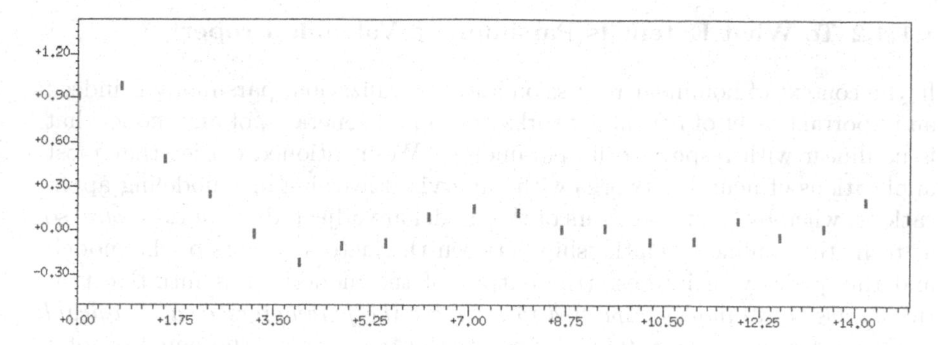

Fig. 1.11. A real-life situation: a finite number of measurements are available. Note that the measurements are equally spaced in the present example, but that is by no means necessary

differences between the values predicted by the network and the measured values is minimum, as shown on Fig. 1.12.

A neural network can thus predict, from examples, the values of a quantity that depends on several variables, for values of the variables that are not present in the database used for estimating the parameters of the model. In the case shown on Fig. 1.12, the neural network can predict values of the quantity of interest for points that lie between the measured points. That ability is termed "statistical inference" in the statistics literature, and is called "generalization" in the neural network literature. It should be absolutely clear that the generalization ability is necessarily limited: it cannot extend beyond the boundaries of the region of input space where training examples are present, as shown on Fig. 1.9. The estimation of the generalization ability is an important question that will be examined in detail in the present book.

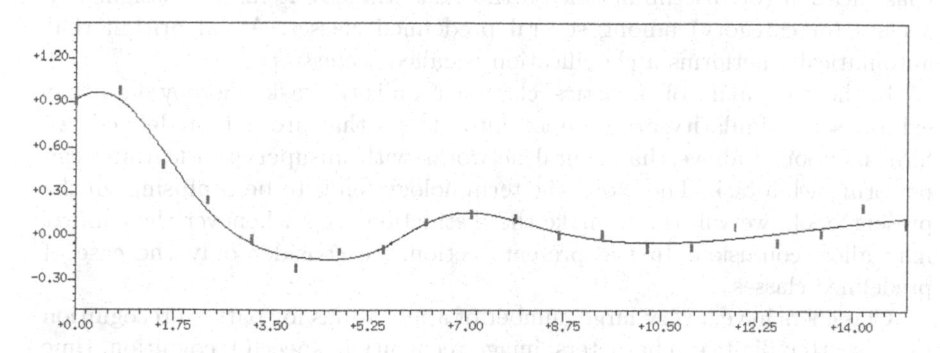

Fig. 1.12. An approximation of the regression function, performed by a neural network, from the experimental points of Fig. 1.11

1.1.4.2 To What Extent Is Parsimony a Valuable Property?

In the context of nonlinear regression and generalization, parsimony is indeed an important asset of neural networks and, more generally, of any model that is nonlinear with respect to its parameters. We mentioned earlier that most applications of neural networks with supervised learning are modeling applications, whereby the parameters of the model are adjusted, *from examples*, so as to fit the nonlinear relationship between the factors (inputs of the model) and the quantity of interest (the output of the model). It is intuitive that the *number of examples requested to estimate the parameters in a significant and robust way is larger than the number of parameters*: the equation of a straight line cannot be fitted from a single point, nor can the equation of a plane be fitted from two points. Therefore, models such as neural networks, which are parsimonious in terms of number of parameters, are also, to some extent, parsimonious in terms of number of examples; that is valuable since measurements can be costly (e.g., measurements performed on an industrial process) or time consuming (e.g., models of economy trained from indicators published monthly), or both.

Therefore, the actual advantage of neural networks over conventional nonlinear modeling techniques is their ability of providing models of equivalent accuracy from a smaller number of examples or, equivalently, of providing more accurate models from the same number of examples. In general, neural networks make the best use of the available data for models with more than 2 inputs.

Figure 1.42 illustrates the parsimony of neural networks in an industrial application: the prediction of a thermodynamic parameter of a glass.

1.1.4.3 Classification (Discrimination)

Classification (or discrimination) is the task whereby items are assigned to a class (or category) among several predefined classes. An algorithm that automatically performs a classification is called a *classifier*.

In the vocabulary of statistics, classification is the task whereby data that exhibit some similarity are grouped into classes that are not predefined; we have mentioned above that neural networks with unsupervised learning can perform such a task. Therefore, the terminology tends to be confusing. In the present book, we will try to make the distinction clear whenever the context may allow confusion. In the present section, we consider only the case of predefined classes.

Classifiers have a very large number of applications for pattern recognition (handwritten digits or characters, image recognition, speech recognition, time sequence recognition, etc.), and in many other areas as well (economy, finance, sociology, language processing, etc.). In general, a pattern may be any item that is described by a set of numerical *descriptors*: an image can be described by the set of the intensities of its pixels, a time sequence by the sequence of its values during a given time interval, a text by the frequency of occurrence of the significant words that it contains, etc. Typically, the questions whose answer a classifier is expected to contribute to are: is this unknown character a a, a b, a c, etc.? is this observed signal normal or anomalous? is this company a safe investment? is this text relevant to a given topic of interest? will there be a pollution alert to-morrow?

The classifier is not necessarily expected to give a full answer to such a question: it may make a contribution to the answer. Actually, it is often the case that the classifier is expected to be a decision aid only, the decision being made by the expert himself. In the first applications of neural networks to classification, the latter were expected to give a definite answer to the classification problem. Since significant advances have been made in the understanding of neural network operation, we know that they are able to provide a much richer information than just a binary decision as to the class of the pattern of interest: neural networks can provide an estimation of the probability of a pattern to belong to a class (also termed *posterior probability* of the class). This is extremely valuable in complex pattern recognition applications that implement several classifiers, each of which providing an estimate of the posterior probability of the class. The final decision is made by a "supervisor" system that assigns the class to the pattern in view of the probability estimates provided by the individual classifiers (committee machines).

Similarly, information filtering is an important problem in the area of data mining: find, in a large text data base, the texts that are relevant to a prescribed topic, and rank these texts by order of decreasing relevance, so that the user of the system can make a choice efficiently among the suggested documents. Again, the classifier does not provide a binary answer, but it estimates the posterior probability of the class "relevant." Feedforward neural networks are more and more frequently used for data mining applications.

Chapter 6 of the present book is fully devoted to feedforward neural networks and support vector machines for discrimination.

1.1.5 Feedforward Neural Networks with Unsupervised Training for Data Analysis and Visualization

Due to the development of powerful data processing and storage systems, very large amounts of information are available, whether in the form of numbers (intensive data processing of experimental results) or in the form of symbols (text corpuses). Therefore, the ability of retrieving information that is known to be present in the data, but that is difficult to extract, becomes crucial. Computer graphics facilitates greatly user-friendly presentation of the data, but the human operator is unable to visualize high-dimensionality data in an efficient way. Therefore, it is often desired to project high-dimensionality data onto a low-dimensionality space (typically dimension 2) in which proximity relations are preserved. Neural networks with unsupervised learning, especially self-organizing maps ("Kohonen maps"), are powerful data visualization techniques.

Chapter 7 of the present book is devoted to unsupervised learning, with emphasis on spectacular applications in satellite observation systems.

1.1.6 Recurrent Neural Networks for Black-Box Modeling, Gray-Box Modeling, and Control

In an earlier section, devoted to recurrent neural networks, we showed that any neural network can be cast into a canonical form, which is made of a feedforward neural network with external recurrent connections. Therefore, the properties of recurrent neural networks with supervised learning are strongly related to those of feedforward neural networks. The latter are used for static modeling from examples; similarly, recurrent neural networks are used for dynamic modeling from examples, i.e., for finding, from measured sequences of inputs and outputs, recurrent (discrete-time) equations that govern a process. A sizeable part of Chap. 2, and Chap. 4, are devoted to dynamic process modeling.

The design of a dynamic model may have several motivations.

- Use the model as a simulator in order to predict the evolution of a process that is described by a model whose equations are inaccurate.
- Use the model as a simulator of a process whose knowledge-based model is known and reliable, but cannot be solved accurately in real time because it contains many coupled differential or partial differential equations that cannot be solved numerically in real time with the desired accuracy: in such circumstances, one can generate a training set from the software code that solves the equations, and design a recurrent neural network that provides accurate solutions within a much shorter computation time; furthermore, it may be advantageous to take advantage of the differential equations of the knowledge-based model, as guidelines to the design of the architecture of the neural model: this is known as "gray-box" or "semi-physical" modeling, described in Sect. 1.1.6.1.
- Use the model as a one-step-ahead predictor, integrated into a control system.

1.1.6.1 Semiphysical Modeling

In the manufacturing industry, a knowledge-based model of a process of interest is often available, but is not fully satisfactory, and it cannot be improved through further analysis; this may be due to a variety of reasons:

- the model may be too inaccurate for the purpose that it should serve: for instance, if it is desired to perform fault detection by analyzing the difference between the state of the process that is predicted by the model

of normal operation, and the actual state of the process, the model of normal operation must be accurate and run in real time.
- The model may be accurate, but too complex for real-time operation (e.g., for an application in monitoring and control).

If measurements are available, in addition to the equations of the—unsatisfactory—knowledge-based model, it would be unadvisable to forsake altogether the accumulated knowledge on the process and to design a purely black-box model. Semi-physical modeling allows the model designer to have the best of both worlds: the designer can make use of the physical knowledge in order to choose the structure of the recurrent network, and make use of the data in order to estimate the parameters of the model. An industrial application of semi-physical modeling is described below, and the design methodology of a semi-physical model is explained in Chap. 2.

1.1.6.2 Process Control

The purpose of a control system is to convey a prescribed dynamics to the response of a process to a control signal or to a disturbance. In the case of a regulator system the process must stay in a prescribed state in spite of disturbances: the cruise control system of a car must keep the speed constant (equal to the setpoint speed) irrespective of the slope of the road, wind gusts, load variations, etc. A tracking system is designed to follow the variations of the setpoint, irrespective of disturbances: in a fermenting plant, the heating system must be controlled in order for the temperature to follow a prescribed temperature profile, irrespective of the temperature of ingredients that may be added during operation, of heat-producing chemical reactions that may take place, etc. In order to achieve such goals, a model of the process must be available; if necessary, the model must be nonlinear, hence be implemented as a recurrent neural network. Chapter 5 is devoted to nonlinear neural control.

1.1.7 Recurrent Neural Networks Without Training for Combinatorial Optimization

In the previous two sections, we emphasized the applications of recurrent neural networks that take advantage of their *forced dynamics*: the model designer is interested in the response of the system to control signals. By contrast, there is a special class of applications of recurrent neural networks that takes advantage of their spontaneous dynamics, i.e., of their dynamics with zero input.

Recurrent neural networks whose activation function is a step function (McCulloch-Pitts neurons), have a dynamics that features fixed points: if such a network is forced into an initial state, and is subsequently left to evolve under its spontaneous dynamics, it reaches a stable state after a finite transient sequence of states. This stable state depends on the initial state. The final state, i.e., the vector whose components are the (binary) states of the neurons of the network, can be considered as the binary code of a piece of information. Moreover, it can be shown that there exists a function, called the Liapunov function (or energy function), which always decreases during the spontaneous evolution of the state of the network; hence the stable states are the minima of the Liapunov function.

Now consider the inverse problem: in a combinatorial optimization problem, it is desired to find the minimum (or at least a good minimum) of a function (cost function) of binary variables. If there exists a recurrent neural network whose Liapunov function is identical to the cost function of the optimization problem, then the fixed points of the spontaneous dynamics of the recurrent neural network are solutions of the combinatorial optimization problem. If such a network can be constructed, then it will find a solution of the problem by evolving, under its spontaneous dynamics, from an arbitrary initial state.

Therefore, the resolution of a combinatorial optimization problem with a recurrent neural network requires

- finding a recurrent neural network whose energy function is identical to the cost function of the optimization problem,
- finding the parameters of that network,
- controlling the dynamics of the network so as to make sure that it will evolve to reach a good minimum of the cost function, for instance, by taking advantage of stochastic methods such as simulated annealing.

This powerful technique, together with some of its applications, will be described in Chap. 8 of the present book.

1.2 When and How to Use Neural Networks with Supervised Training

In the previous sections, we presented the theoretical arguments that support the use of neural networks in modeling applications. In the present section, we attack the practical questions raised by the design and training of a neural model. First, we will explain when neural networks can advantageously be used—and when they should not be used. In the subsequent section, we will emphasize *how* to use neural networks. An in-depth treatment of these important questions will be given in the next chapters.

1.2.1 When to Use Neural Networks?

We have shown earlier that the fundamental property of neural networks with supervised training is the parsimonious approximation property, i.e., their ability of approximating any sufficiently regular function with arbitrary accuracy.

Therefore, neural networks may be advantageous in any application that requires finding, in a machine learning framework, a nonlinear relation between numerical data.

Under what conditions is such an approach recommended?

- The first condition is necessary but not sufficient: since neural network design is essentially a problem in statistics, a set of examples, that sample the space of inputs appropriately, and that are in appropriate number, must be available.
- After gathering the data, one should make sure that a *nonlinear model* is necessary, since the design of a linear model is much simpler and faster than the design of a neural model. Therefore, if no prior knowledge on the quantity to be modeled is available, one should first try out a linear model; if it turns out that a linear model is too inaccurate, despite the fact that all relevant factors are present in the inputs, then the model designer may rightly resort to nonlinear models such as neural networks.
- If the appropriate examples are available, and if a nonlinear model is necessary, then one should decide whether the use of neural networks, instead of polynomials for instance, is advisable. Parsimony is the relevant choice criterion here: as mentioned above, the number of parameters of the first connection layer (between inputs and hidden neurons) increases linearly with the number of variables, whereas it increases exponentially for polynomial approximation (there exist, however, statistical tests that may, to some extent, limit the combinatorial explosion of parameters in polynomial modeling). Therefore, neural networks are advantageous when the number of variables is large, i.e., empirically, larger than or equal to 3.

To summarize, if appropriate data sets are available, neural networks can be used with advantage in all applications that require the estimation of the parameters of a regression function with three variables or more. If the number of variables is smaller, nonlinear models that are linear with respect to their parameters, such as polynomials, radial basis functions with fixed centers and standard deviation, wavelets with fixed translations and dilations, may be as accurate, and require a simpler implementation.

If the available data are not numerical (e.g., symbolic), they cannot be processed directly by a neural network. Some appropriate preprocessing is required in order to make data numerical (techniques evolved from the theory of fuzzy sets may be appropriate).

1.2.2 How to Design Neural Networks?

Neural networks are nonlinear parameterized functions, which can approximate any nonlinear function. Therefore, approaching a regression function from examples requires finding a neural network for which the sum, over all examples used for training, of the squared modeling errors (the least squares cost function) is minimum. As a consequence, the design of a neural network requires

- finding the relevant inputs, i.e., the factors that have a significant influence on the quantity to be modeled (i.e., an influence that is larger than the measurement noise),
- collecting the data that is necessary for training and testing the neural network,
- finding the appropriate complexity of the model, i.e., the appropriate number of hidden neurons,
- estimating the parameters for which the cost function is minimum, i.e., training the network,
- assessing the generalization ability of the neural network after training.

In view of the results, it may be necessary to iterate the whole procedure, or part of it.

These points will be considered in detail in the next sections.

1.2.2.1 Relevant Inputs

The selection of relevant inputs may have various requirements, depending on the application that is considered.

If the process to be modeled is an industrial problem that has been carefully engineered, the relevant factors and the causal relations between them are usually known. Consider, as an example, the industrial process of spot welding, which will be described in detail in a subsequent section: the metal sheets to be welded are melted together locally by passing a very large current (a few kiloamperes) during a few milliseconds, through two electrodes that are pressed onto the metal surfaces (Fig. 1.13). The quality of the joint is assessed from the diameter of the melted zone; it depends on the current intensity, on the duration of the current flow, on the stress applied to the electrodes while current flows and during cooling, on the surface state of the electrodes, on the nature of the metal sheets, and on a few additional factors. Thus, the desirable model inputs are essentially known from physics: however, it may be important to make a choice between these factors, so that only those factors that have a significant influence on the spot weld diameter, i.e., whose influence is larger than the uncertainty on the measurement of the diameter, are taken into account.

By contrast, if the process of interest is a complex natural process (e.g., in biology or ecology), or if it is an economic, financial or social process, the choice of the relevant inputs may be more difficult. An example of a complex natural process (the solubility of molecules in solvents), where the determination of the relevant factors is not trivial, will be described in a subsequent section. Similarly, great care must be exercised in the choice of relevant inputs for credit rating, an example that will also be described below.

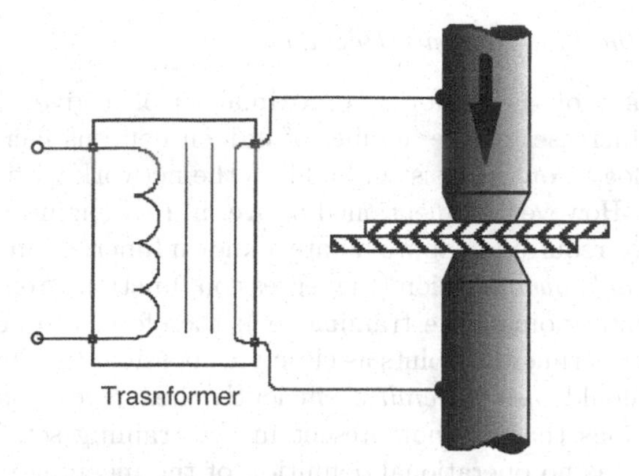

Fig. 1.13. A schematic representation of the spot welding process

The questions of input selection, and of model selection as well, are by no means specific to neural networks: they are of great importance for all modeling techniques, whether linear or nonlinear. It will be shown, in Chap. 2, that model selection techniques that were developed for linear models can be extended to nonlinear models such as neural networks.

1.2.2.2 Data Collection

Before training, observations must be collected in order to build the training set, as well as the validation and test sets, which will be defined below. Those observations must be numerous enough, and they must be typical of the situations that will be encountered by the network when in use. When the number of factors (model inputs) exceeds two or three, sampling the input domain in a regular and systematic way is generally not feasible because combinatorial explosion arises. Therefore, it is usually important to design the experiments as efficiently as possible: experimental design is an important part of model design. This is generally more difficult for nonlinear models than for linear ones. Some elements will be given in the "Experimental design" section of Chap. 2.

1.2.2.3 The Number of Hidden Neurons

The discrepancy between the neural approximation and the function to be approximated is inversely proportional to the number of hidden neurons [Barron 1993]; unfortunately, this result, as well as other theoretical results such as the Vapnik-Cervonenkis dimension (or VC-dimension) [Vapnik 1995] (described in Chap. 6) will only provide loose bounds or estimates of the number of hidden neurons. At present, no result allows the model designer to find the appropriate number of hidden neurons given the available data and the desired performance. Therefore, it is necessary to make use of a specific methodology. In the following, we will first define the problem of designing a nonlinear black-box static model, with emphasis on feedforward neural network design.

Overfitting and the Bias-Variance Dilemma

Since the accuracy of the uniform approximation of a given function by a neural network increases as the number of hidden neurons increases, a naïve design methodology would consist in building the network with as many neurons as possible. However, as mentioned above, in real engineering problems, the network is not required to approximate a known function uniformly, but to approximate an *unknown* function (the regression function) from a finite number of experimental points (the training set); therefore, the network should not only fit the experimental points as closely as possible (in the least squares sense), but it should also *generalize* efficiently, i.e., give a *satisfactory* response to situations that are not present in the training set. The difficulty here is that there is no operational definition of the meaning of *satisfactory*, since the regression function is unknown: the problem of generalization is an *ill-posed problem*. Therefore, the design problem is the following:

- if the neural network has too many parameters (it is said to be over-parameterized), it will be too "flexible," so that its output will fit very accurately all points of the training set (including the noise present in these points), but it will provide meaningless responses in situations that are not present in the training set. That is known as *overfitting*.
- by contrast, a neural network with too few parameters will not be complex enough to match the complexity of the (unknown) regression function, so that it will not be able to learn the training data.

This dilemma, known as the *bias-variance dilemma*, is the basic problem that the model designer is faced with.

Figure 1.14 shows the results obtained after training two different networks, with different numbers of hidden neurons (hence of parameters) with sigmoid activation functions, from the same training set: clearly, the most parsimonious model (i.e., the model with the smallest number of parameters) generalizes best. In practice, the number of parameters should be small with respect to the number of elements of the training set. The parsimony of neural networks with sigmoid activation functions is a valuable asset in the design of models that do not exhibit overfitting.

Figure 1.14 shows clearly which candidate neural network is most appropriate. When the model has several inputs, the result cannot be exhibited graphically in such a straightforward fashion: a quantitative performance index must be defined. The most popular way of estimating such an index is the following: in addition to the training set, one should build a *validation set*, made of observations that are distinct from those of the training set, from which a performance index is computed. The most frequently used criterion is the mean square error on the validation set (VMSE), defined as:

$$\mathrm{VMSE} = \sqrt{\frac{1}{N_V}\sum_{k=1}^{N_V}\left[y^k - g(\boldsymbol{x}^k, \boldsymbol{w})\right]^2}$$

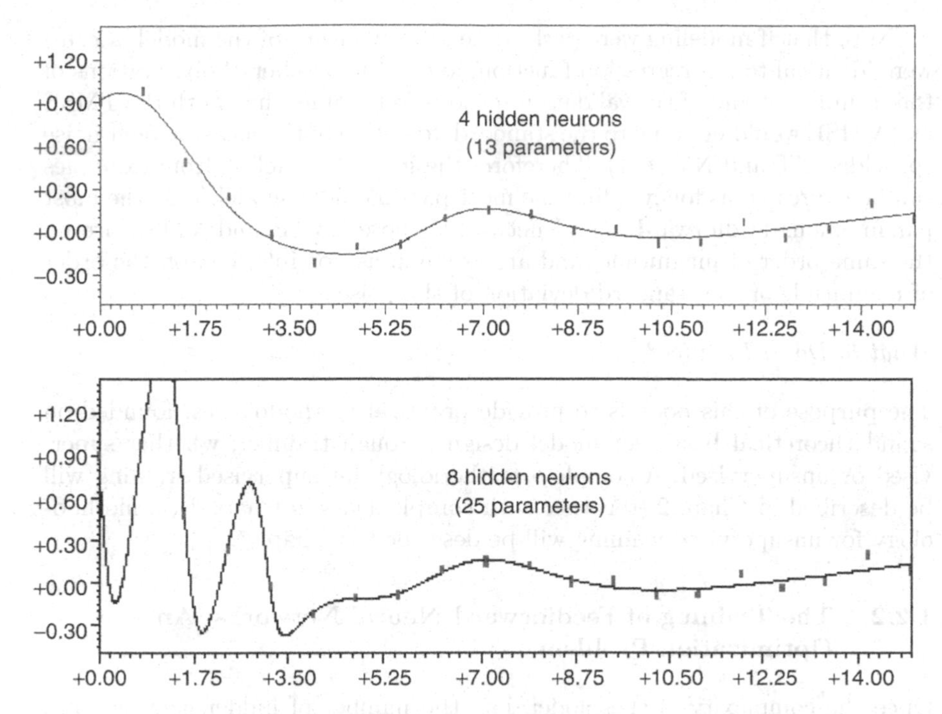

Fig. 1.14. The most parsimonious neural network has the best generalization abilities

where N_V is the number of observations present in the validation set, and where, for simplicity, y^k denote the measurements of the quantity to be modeled: $y^k = y_p(x^k)$. This relation is valid in the usual case of a model with a single output; if the model has several outputs, the VMSE is the sum of the mean square errors on each output.

This quantity should be compared with the mean square error on the training set (TMSE),

$$\text{TMSE} = \sqrt{\frac{1}{N_T} \sum_{k=1}^{N_T} [y^k - g(\boldsymbol{x}^k, \boldsymbol{w})]^2},$$

where N_T is the number of observations present in the training set.

Consider the example shown on Fig. 1.14, and assume that the observations of the validation set are the midpoints between the observations of the training set. Clearly, the TMSE of the second network is certainly smaller than the TMSE of the first network, whereas the VMSE of the second network is certainly larger than that of the first network. Therefore, if model selection were performed on the basis of the training mean square error, overparameterized networks would systematically be favored, thereby leading to models that exhibit overfitting.

Note that if modeling were perfect, i.e., if the output of the model $g(\boldsymbol{x}, \boldsymbol{w})$ were identical to the regression function, and if the number of observations of the training set and of the validation set were very large, then both the TMSE and VMSE would be equal to the standard deviation of the measurement noise (provided NT and NV $\gg 1$). Therefore, the goal of modeling from examples can be expressed as follows: find the most parsimonious model (e.g., the most parsimonious feedforward neural network) whose TMSE and VMSE are on the same order of magnitude, and are as small as possible, i.e., on the order of magnitude of the standard deviation of the noise.

What to Do in Practice?

The purpose of this book is to provide practical methodologies, founded on sound theoretical bases, for model design through training, whether supervised or unsupervised. A complete methodology for supervised training will be described in Chap. 2 (together with complements in Chap. 3), a methodology for unsupervised training will be described in Chap. 7.

1.2.2.4 The Training of Feedforward Neural Networks: An Optimization Problem

Once the complexity of the model, i.e., the number of hidden neurons of a feedforward neural network, is chosen, training can be performed: one has to estimate the parameters of the neural network that, given the number of parameters that are available to him, has a minimum mean square error on the training set. Therefore, training is a *numerical optimization problem.*

For simplicity, we consider a model with a single output $g(\boldsymbol{x}, \boldsymbol{w})$. The training set contains N examples. The least squares cost function was defined above as

$$J(\boldsymbol{w}) = \frac{1}{2}\sum_{k=1}^{N}\left[y_p(\boldsymbol{x}^k) - g(\boldsymbol{x}^k, \boldsymbol{w})\right]^2,$$

where $\boldsymbol{x}^k$ is the vector of the values of the variables for example k, $y_p(\boldsymbol{x}^k)$ is the corresponding measured value of the quantity to be modeled, $\boldsymbol{w}$ is the vector of the parameters (or weights) of the model, and $g(\boldsymbol{x}^k, \boldsymbol{w})$ is the output value of the model with parameters $\boldsymbol{w}$ for the vector of variables $\boldsymbol{x}^k$. Therefore, the cost function is a function of all adjustable parameters $\boldsymbol{w}$ of the model. Training consists in finding the parameter vector $\boldsymbol{w}$ for which $J(\boldsymbol{w})$ is minimum.

- For a model that is linear with respect to its parameters (e.g., radial basis functions with fixed centers and widths, polynomials, etc.), the cost function J is quadratic with respect to the parameters: the ordinary least squares methods can be used. They are simple and efficient. However, the resulting models are not parsimonious.

- For a model that is not linear with respect to its parameters (e.g., a feedforward neural network, or a RBF network with adjustable centers and widths), the optimization problem is multivariable nonlinear, which makes ordinary least squares inapplicable. The techniques that solve such problems are described in detail in Chap. 2; those are iterative techniques that make sequences of estimations of the parameters until a minimum is reached, or a satisfaction criterion is met.

In the latter case, the optimization techniques are gradient methods; they are based on the computation, at each iteration, of the gradient of the cost function with respect to the parameters of the model. The gradient thus computed is subsequently used for updating the values of the parameters found at the previous iteration. *Backpropagation* is a popular, computationally economical way of computing the gradient of the cost function (described in Chap. 2). Therefore, backpropagation is *not* a training algorithm: it is simply a technique for computing the gradient of the cost function, which is very frequently an ingredient of neural network training. It has been often stated that the invention of backpropagation made feedforward neural network training possible; that is definitely not correct: methods for computing the gradient of cost functions were used in signal processing long before the introduction of neural networks. Such methods can be used for feedforward neural network training [Marcos 1992].

Training algorithms have been tremendously improved during the past few years. At the beginning of the 1990's, publications would frequently mention tens or hundreds of thousands of iterations, requiring days of computing on powerful computers. At present, typical trainings require tens or hundreds of iterations. Figure 1.15 displays the training of a model with a single variable. Crosses are the elements of the training set. Parameters are initialized to "small" values (see Chap. 2 for the description of the initialization procedure), so that the output of the network is essentially zero. The result obtained after 13 iterations is "visually" satisfactory; quantitatively, the TMSE and VMSE (the points of the validation set are not shown) are of the same order of magnitude, which is of the order of the standard deviation of the noise, so that the model is appropriate.

1.2.2.5 Conclusion

In this section, we have explained how and why neural networks with supervised training should be used. To summarize, neural networks are useful whenever a nonlinear relation between numerical data is sought. Therefore, neural networks are statistical tools for nonlinear regression. An overview of the tasks implied in nonlinear model design was presented, together with conditions for successful applications. In Chap. 2, the reader will find all necessary details for neural network training, for input selection and for model selection, both for static models (feedforward neural networks) and for dynamic model (recurrent neural networks).

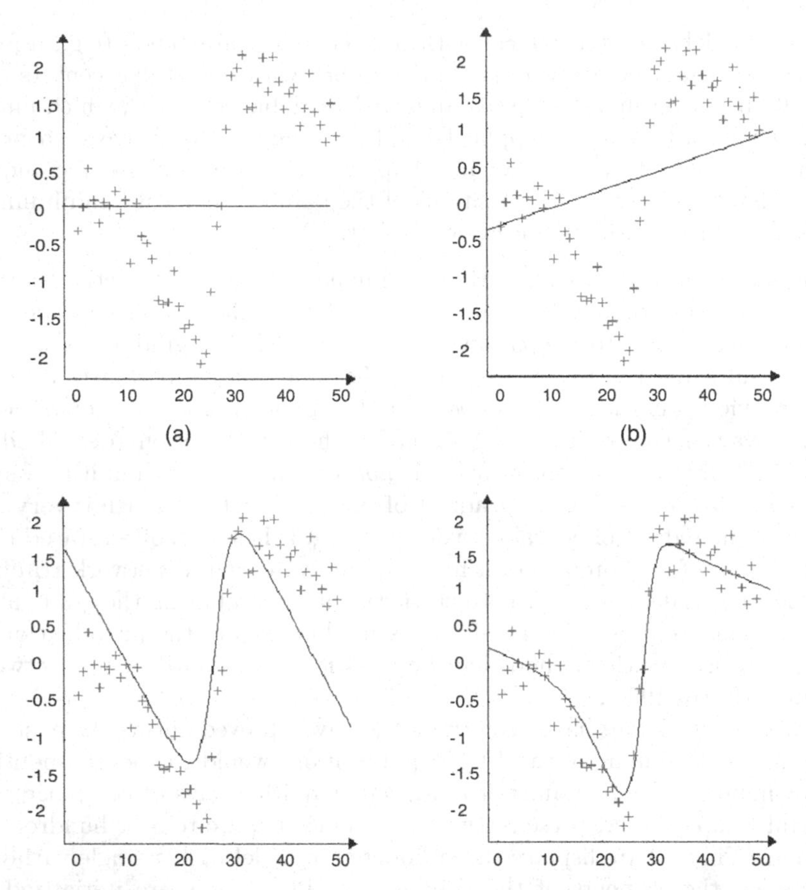

Fig. 1.15. Training of a feedforward neural network with one input variable and three hidden neurons. The line is the output of the model, the crosses are the elements of the training set. (**a**) initial state; (**b**) after one iteration; (**c**) after 6 iterations; (**d**) after 13 iterations (results obtained with the NeuroOne software package by NETRAL S.A.)

1.3 Feedforward Neural Networks and Discrimination (Classification)

In the early stages of the development of neural networks (in the years 1960's), the main incentive was the development of pattern recognition applications, as evidenced by the term perceptron that was used for the ancestor of present-day neural networks. Indeed, the first nontrivial industrial applications of neural networks, at the beginning of the 1980's, were related to pattern or signal recognition. Therefore, the present section is devoted to a general presentation of classification (or, equivalently, discrimination); it will be shown that many classification problems can be viewed as nonlinear regression problems, which

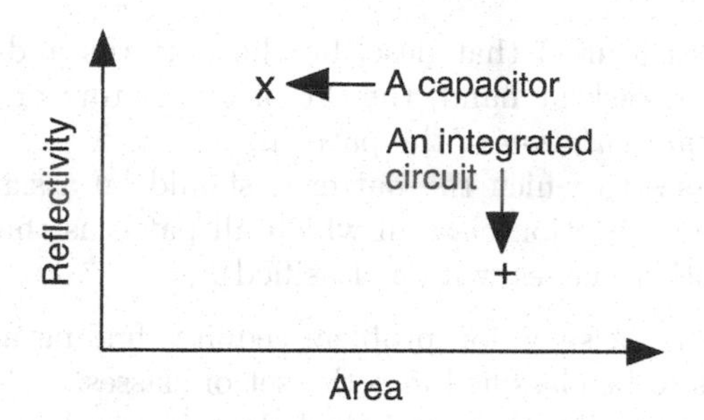

Fig. 1.16. Each sample is represented as a point in the area-reflectivity plane. Capacitors are shown as x's and integrated circuits as +'s

explains why feedforward neural networks are efficient classifiers. The purpose of the present section is to provide a general presentation of classification in its relation to nonlinear regression. Chapter 6 provides a much more detailed view of neural network classification and of techniques that evolved from neural networks.

1.3.1 What Is a Classification Problem?

A classifier is an algorithm that automatically assigns a class (or category) to a given pattern.

Before considering the specific case of "neural" classifiers, it is important to understand the basic characteristics of classification problems. Consider the following illustrative example: in an automatic sorting application, capacitors must be discriminated from integrated circuits, from a black-and-white picture provided by a video camera, so that a robotic arm can grab either a capacitor or an integrated circuit as requested. Roughly, capacitors appear in the picture as bright, small rectangular objects, whereas integrated circuits are large, dark objects. Therefore, the area A and the reflectivity R can be considered as relevant features for discriminating the objects, i.e., for assigning a given object either to the class "integrated circuit" or to the class "capacitor". Assume that samples of capacitors and of integrated circuits have been collected, and that their areas and reflectivities have been measured: then each sample can be represented by a point in a two-dimensional space, whose coordinates are its area and its reflectivity, as shown on Fig. 1.16.

1.3.2 When Is a Statistical Classifier such as a Neural Network Appropriate?

The above example shows that the ingredients of a classification problem are

- a set of N patterns;

- n variables (or features) that describe the patterns and are relevant to the classification task at hand, the set of descriptors of a given pattern building the *representation* of the pattern;
- a set of C classes to which the patterns should be assigned (one of the classes may be a rejection class in which all patterns that cannot be assigned to the other classes will be classified).

Therefore, solving a classification problem requires finding an application of the set of patterns to be classified into the set of classes.

It is important to realize that statistical classifiers such as neural networks are not appropriate for solving all classification problems: many alternative classification methods are available. The following (more or less academic) examples (from [Stoppiglia 1997]) illustrate the area of application of neural networks in classification. For each example, the following questions will be asked:

- Does prior knowledge suggest relevant features?
- Are those features measurable (or can they be computed from measurements)?
- What is the role of the rejection class?

Any vending machine can recognize the coins automatically, and reject fake or foreign coins. The answers to the above questions are

- Relevant features can easily be found: the coin diameter, its weight, its thickness, the alloy composition, etc.; there is a small number of such features, and new coins are actually designed in order to facilitate automatic discrimination.
- The features are measurable quantities.
- In feature space, the classes are small hyper-parallelepipeds defined by the manufacturing tolerances; the rejection class is the rest of feature space.

In such circumstances, a simple decision tree that operates with simple logical rules, derived from the analysis of the problem, can readily solve the classification problem. In such a case, statistical tools such as neural networks are not appropriate.

Vehicle comfort assessment can be viewed as a classification problem. In order to anticipate the reactions of customers to a new vehicle, car manufacturers resort to panels of customers, who are asked to express an opinion. Comfort is an ill-defined concept, which depends on many factors such as noise, seat design, etc. Assessing the comfort, for instance, by classifying it into three classes (very good, fair, poor), is a process that is difficult to formalize because it is based on feelings rather than on measurements.

- The relevant features are not necessarily known and clearly expressed by the customers; even if features could be defined, the assessments might be difficult to relate to the features; two customers, under the same conditions, could give very different assessments.

- The features are not measurable.
- There is no rejection class: all customers have an opinion on the comfort of a vehicle.

The fact that the features are not measurable precludes the use of a statistical method. In such a situation, a fuzzy classification method would be more appropriate.

Handwritten digit recognition, for instance zip code recognition, has been investigated in detail, and many applications are in routine operation. Consider the answers to the three questions that were asked in the previous two examples.

- In sharp contrast with the example of the vending machines, the huge diversity of handwriting styles makes the choice of features nontrivial, but feasible; in contrast to the vehicle comfort assessment problem, different persons who read the same digit will assign it to the same class (except if the digit is almost illegible).
- Features are numbers that can be extracted from the picture: in a typical low-level representation, the features would be the intensities of the pixels; in a high-level description, the features would be the location of horizontal, vertical or diagonal segments, the presence and location of loops, etc.
- The size of the rejection class can be defined, and in some cases, it is a performance criterion: for a given error rate, the rejection rate should be as low as possible. In mail processing, a rejected envelope requires a manual operation, which is less costly than sending a letter to the wrong address. Hence, the performance requirement is expressed as follows: for a given error rate (typically 1%) the rejection rate should be as low as possible. Clearly, it would be easy to design a classifier that never gives a wrong answer, by simply rejecting all patterns: by contrast, given the economic constraints of the problem of zip code reading, a "good" classifier makes a decision as often as possible, while making no more than 1% mistakes. If economic constraints were different, i.e., if a mistake was less costly than a human operation, a classifier should have the smallest possible error rate for a given maximum rejection rate (this is the case for large-scale automated medical diagnoses, where resorting to a medical doctor is more costly than delivering a wrong diagnostic).

In the latter example, statistical classification methods such as neural networks are perfectly appropriate, provided a suitable database is available. As in most nonacademic problems, the central question is that of data representation: a thoughtful representation design, together with data pre-processing techniques such as described in Chap. 3, is often as important as the classification algorithm itself.

1.3.3 Probabilistic Classification and Bayes Formula

Assume that, after analyzing a classification problem, a statistical classification approach has been deemed preferable to, for instance, a decision tree. Probabilistic classification methods are based on the idea that both features and classes may be modeled as random variables (readers unfamiliar with random variables will find more information at the beginning of Chap. 2). In that context, if a pattern is picked randomly from the patterns to be classified, the class to which it belongs is the realization of a discrete random variable. Similarly, the values of the features of a randomly chosen pattern can be viewed as realizations of random variable, which are usually continuous. For instance, in the example of discrimination between capacitors and integrated circuits (Fig. 1.16), the random variable "class" may be equal to 0 for a capacitor and to 1 for an integrated circuit, while the reflectivity R at the area A may be viewed as continuous random variables.

In that context, the classification problem can be simply stated as follows: given a pattern whose class is unknown, whose reflectivity is equal to r and whose area is equal to a (within measurement uncertainties), what is the probability that the random variable "class" be equal to 0 (i.e., that the pattern be a capacitor)? This probability is the *posterior probability* of class "capacitor" given the measured reflectivity and area, denoted by

$$\Pr(\text{class} = 0 \mid \{r, a\}).$$

Consider a set of capacitors and integrated circuits that have been labeled with the labels (0 or 1) of their classes, and whose feature values are also known. That information can be used for deriving two very important quantities,

- the *prior probability* of each class: a pattern picked randomly from the set of patterns has a probability $\Pr(C_i)$ of belonging to class C_i. It we assume that each pattern belongs to one of the classes, then one has $\sum_i \Pr(C_i) = 1$. That information is relevant to classification: assume that the prior probability of the class "capacitor" is known to be 0.9 (hence the probability of the class "integrated circuit" is 0.1); then a dumb classifier that would *always* choose the class "capacitor," irrespective of the pattern features, would exhibit an error rate on the order of 10%.
- the *conditional probability density of each feature*: if an integrated circuit is picked randomly, what is the probability for its area A to lie in an interval $[a - \delta a, a + \delta a]$? Clearly, that probability is proportional to δa. The *probability density of feature A conditioned to class* C_i, or *likelihood of* C_i *given feature* a is denoted as $p_A(a \mid C_i)$: the probability that feature A be in the interval $[a - \delta a, a + \delta a]$ given that it belongs to class C_i is equal to $p_A(a \mid C_i)\delta a$. Since the pattern whose feature A is measured belongs to class C_i, one has $\int p_A(a \mid C_i)\mathrm{d}a = 1$.

Figure 1.17 shows an estimate of the probability density $p_A(a \mid \text{Class} = \text{integrated circuit})$ as a function of a. Similarly, one could draw the conditional

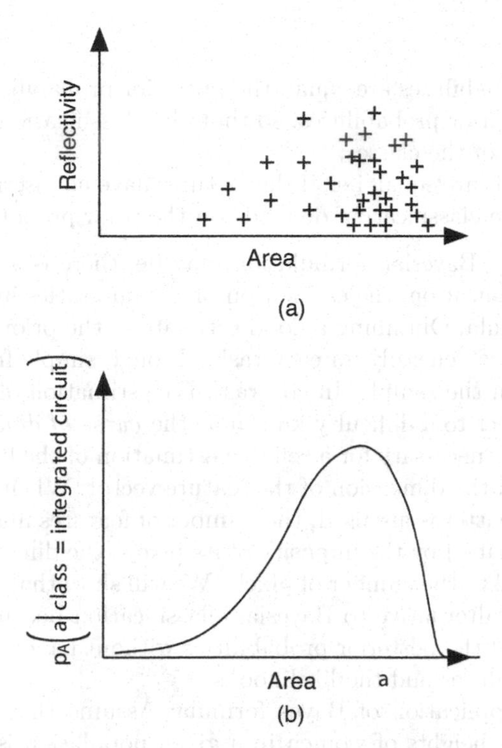

Fig. 1.17. (**a**) Representation of a sample of the class "integrated circuit" in the reflectivity-area plane. (**b**) Estimate of the conditional probability density of the area of the pattern if the latter is an integrated circuit

probability density of the reflectivity R, for the class integrated circuit, as a function of r.

Thus, given a sample of the population of patterns to be classified, estimates of the prior probabilities of the classes $\{\Pr(C_i)\}$, and of the conditional probability densities $p_X(\boldsymbol{x} \mid C_i)$ of their features, are available. Then, by Bayes formula, the solution of the classification problem, i.e., the posterior probability of a class given an unknown pattern, is given by

$$\Pr(C_i \mid \boldsymbol{x}) = \frac{p_X(\boldsymbol{x} \mid C_i)\Pr(C_i)}{\sum_j p_X(\boldsymbol{x} \mid C_j)\Pr(C_j)}.$$

Clearly, that estimate is relevant only if the features of the unknown pattern have the same conditional density probabilities as the patterns that were used to estimate the likelihoods.

Note that

- if the prior probabilities are equal, the posterior probabilities are independent from the prior probabilities, so that the classification relies solely on the likelihoods of the classes;
- if the likelihoods are equal, i.e., if the features have no discriminative power whatsoever, the classification depends on the prior probabilities only.

Elegant though the Bayesian formulation may be, there is a major difficulty in its practical application: the estimation of the quantities in the right-hand side of Bayes formula. Obtaining a good estimate of the prior probabilities of the classes $\Pr(C_i)$ is generally an easy task, through simple frequency counting of each class in the sample. In contrast, the estimation of the likelihoods $p_X(x \mid C_i)$ is subject to a difficulty known as the *curse of dimensionality*: the number of patterns necessary for a reliable estimation of the likelihoods grows exponentially with the dimension of the feature vector. When low-level representations of the patterns are used, the number of features may be very large: if a picture is described by the intensity of its pixels, the dimension of the feature vector is equal to the number of pixels. We will show that neural networks are an interesting alternative to Bayesian classification because they provide a direct estimate of the posterior probabilities without having to estimate the prior class probabilities and the likelihoods.

Consider an application of Bayes formula: Assume that the probability distribution of the heights of women in a given population is Gaussian with mean 1.65 m and standard deviation 0.16 m,

$$p_H(h \mid W) = \frac{1}{0.16\sqrt{2\pi}} \exp\left(-\frac{1}{2}\left(\frac{h-1.65}{0.16}\right)^2\right),$$

and that the probability distribution of the heights of men in that population is a Gaussian with mean 1.75 m and standard deviation 0.15 m:

$$p_H(h \mid M) = \frac{1}{0.15\sqrt{2\pi}} \exp\left(-\frac{1}{2}\left(\frac{h-1.75}{0.15}\right)^2\right).$$

The above probability densities are shown on Fig. 1.18. The Gaussians exhibit strong overlapping, which shows that the feature height is not very discriminant. In a real application, such curves would be a strong incentive for the designer to find one or more alternative features.

In addition, assume that there are as many men and women in the population. Given a person whose height is 1.60 m, what is the probability that it is a woman? The answer is provided by Bayes formula

$$\Pr(W \mid 1.60) = \frac{0,5 p_H(1.60 \mid W)}{0.5 p_H(1.60 \mid W) + 0.5 p_H(1.60 \mid M)} \approx 60\%.$$

Clearly, $\Pr(M \mid 1.60) = 40\%$.

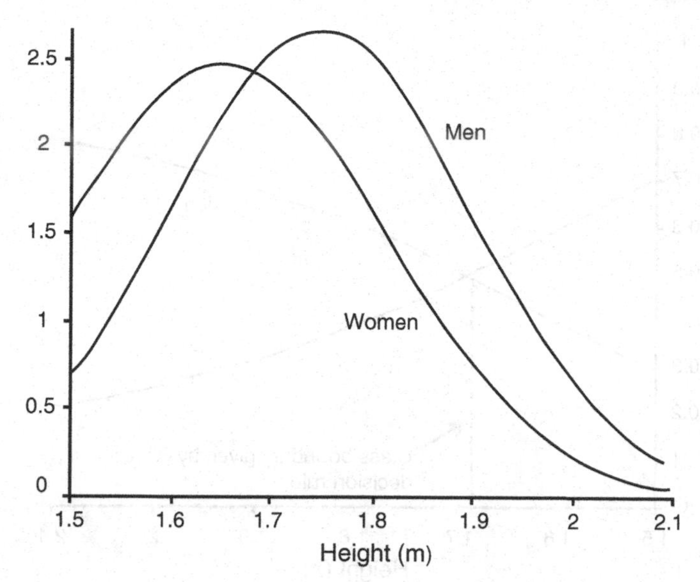

Fig. 1.18. Probability densities of heights of men and women for the population under investigation

In view of the above results, it is natural to assign the person to class W, which has the larger probability. This is an application of Bayes decision rule, which will be explained below. The boundary between the classes thus defined is shown on Fig. 1.19.

Because the prior probabilities are assumed to be equal, the discrimination relies solely on the likelihoods.

Now assume that the person is not a member of the general population, but is picked among the audience of a football match. Then the likelihoods of the classes, given the height, are the same as above, but the prior probabilities are different, since men are generally more numerous than women in the audience of football matches; assume that the proportions are: 30% of women and 70% of men. The posterior probabilities, as computed from Bayes formula, become $\Pr(W \mid 1.60) = 39\%$ and $\Pr(M \mid 1.60) = 61\%$. The results are very different from the previous ones: the observed person is assigned to the class man if Bayes rule is used as above; that important change results from the fact that the likelihoods are not very different because the feature height is not very discriminant, so that the classification relies heavily on the prior probabilities. That result is illustrated by Fig. 1.20.

That simple example shows how to use Bayes formula for estimating posterior probabilities, which are subsequently used for assigning each pattern to a class through Bayes decision rule.

It is important to realize that, in practice, and in contrast with the above examples, prior probabilities and probability densities are not known analytically, but are estimated from a finite set of observations O. Therefore,

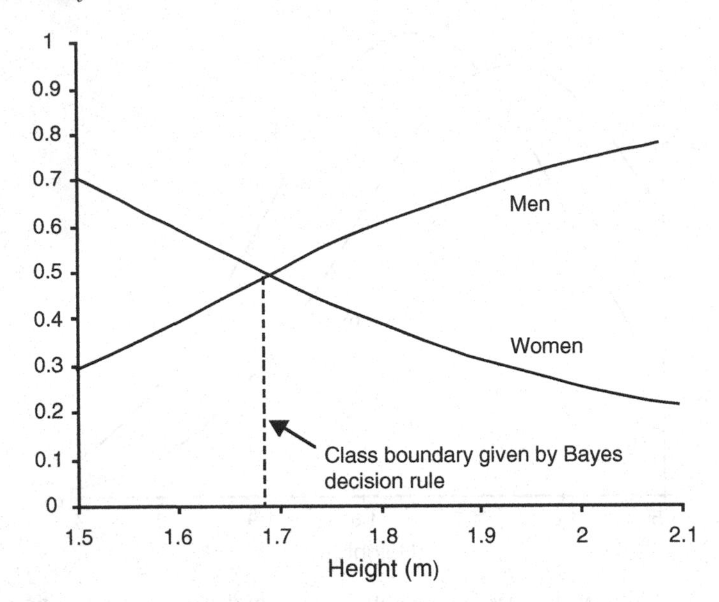

Fig. 1.19. Posterior probabilities of the classes man and woman, as a function of height, and boundary between the classes, when the person under investigation is drawn from the general population

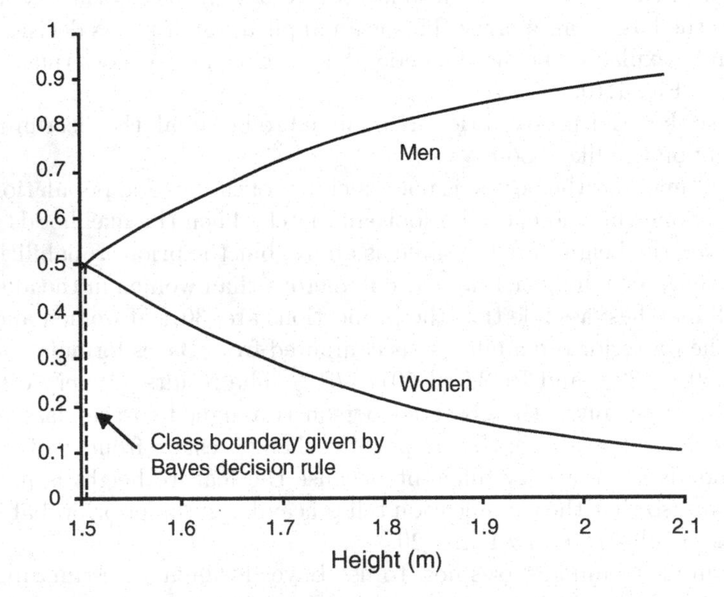

Fig. 1.20. Posterior probabilities of the classes man and woman, as a function of height, and boundary between the classes, when the person under investigation is drawn from the audience of a football match

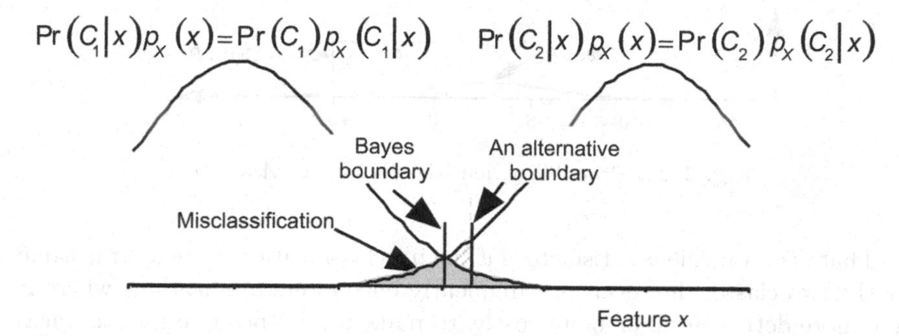

Fig. 1.21. A geometrical interpretation of Bayes decision rule; the *gray area* is the probability of misclassification when Bayes rule is used; the *striped area* is the increase of misclassification probability resulting from a different boundary choice

the likelihood should be denoted as $p_X(x \mid C_i, O)$, and the posterior probabilities should be denoted as $\Pr(C_i \mid x, O)$, since their estimates depend on the observation set O. For simplicity, we will not use these notations, but it should be remembered that the estimates of probabilities and of probability distributions are always conditioned to the observations from which they are estimated.

1.3.4 Bayes Decision Rule

When assigning a pattern to a class, the risk of making a classification error is minimum if the pattern is assigned to the class whose posterior probability is highest.

Consider a classification problem with two classes C_1 and C_2, and one feature. Clearly, the probability of misclassification is larger if the pattern lies close to the class boundary. However, during the normal operation of the classifier, it will handle patterns that are described by a large range of values of the feature, so that what one would really like to do is to find the boundary that minimizes the global error probability, i.e., the boundary that minimizes the quantity $\Pr(M) = \int_{-\infty}^{+\infty} \Pr(M \mid x) p_X(x) \mathrm{d}x$, where M denotes the event "misclassification". Since the probability density $p_X(x)$ is positive, the integral is minimal if $\Pr(M \mid x)$ is minimal for all x. $\Pr(M \mid x)$ is the posterior probability of C_1 if the decision is made of assigning the pattern to C_2, and the posterior probability of C_2 if the decision is made of assigning the pattern to C_1. Therefore, $\Pr(M \mid x)$ is minimized if the decision is to assign the pattern to the class with higher probability.

A geometrical interpretation of that argument is shown on Fig. 1.21: if Bayes rule is used, the misclassification probability is represented by the gray area. Any other boundary choice would increase that area.

The result can be easily extended to the multi-class case and the multi-feature case.

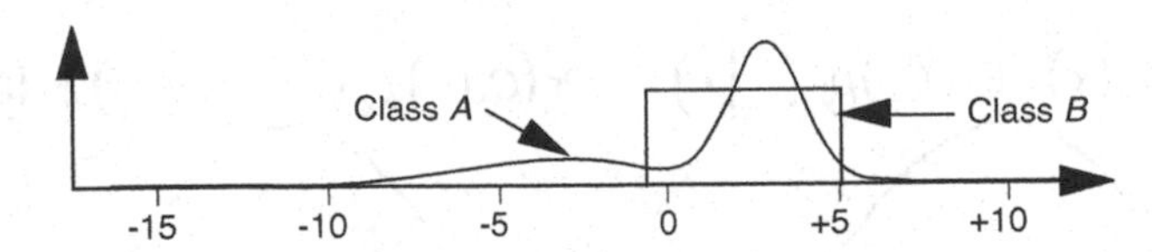

Fig. 1.22. Probability densities for classes A and B

That decision rule is satisfactory if the misclassification costs are the same for the two classes; however, one frequently encounters applications where it may more detrimental, or more costly, to make a false-positive misclassification (the pattern is considered to belong to class A whereas it actually belongs to class B) than a false-negative misclassification (the pattern is considered to belong to class B whereas it actually belongs to class A). In data mining applications for instance, a company that provides information filters may find it more suitable to market filters that reject documents whereas they are relevant to the chosen topic, than to market a filter that does not filter irrelevant documents (the user spots immediately documents that are irrelevant, whereas he may never find out that the filter missed a relevant text). In practice, such considerations are an important part of classifier design, whether in pattern recognition, data mining, credit scoring, etc.). Therefore, it is generally very desirable, in practical applications, to estimate posterior probabilities and subsequently make decisions: classifiers that determine class boundaries directly may lead to serious misconceptions.

The combination of Bayes formula and of Bayes decision rule is called the Bayes classifier, which has the best achievable performance if the prior probabilities and the likelihoods are known exactly. Since the latter condition is not frequently fulfilled in practice, Bayes classifier is essentially of theoretical interest. For instance, it may serve as a reference for assessing the quality of a classifier, by applying it to an academic problem where prior probabilities and likelihoods are known exactly.

As an illustrative example, consider a problem with two classes and one feature; the patterns of class A are generated from a mixture of two Gaussians; the patterns of class B are generated from a uniform distribution in a bounded interval (Fig. 1.22). Therefore, the posterior probabilities can be computed exactly (Fig. 1.23), and so are the boundaries between classes (Fig. 1.24). In

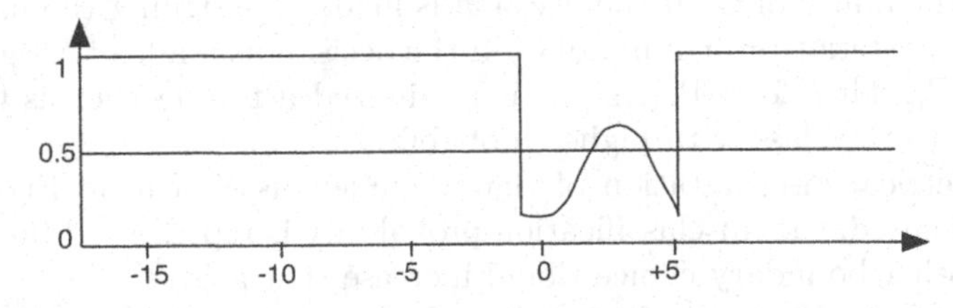

Fig. 1.23. Posterior probability of class A, from Bayes formula

Fig. 1.24. Classification achieved by Bayes classifier

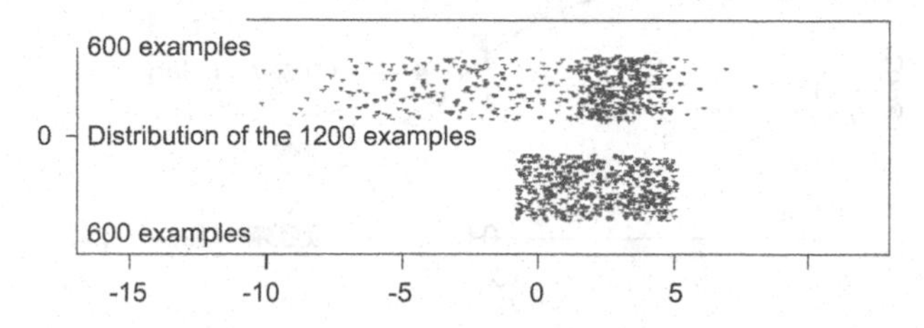

Fig. 1.25. Examples used for estimating the misclassification rate. *Top*: class A; *bottom*: class B

order to estimate the misclassification rate of the resulting Bayes classifier, a large number of realizations of examples of each class are generated, and the proportion of misclassified examples is computed. 600 examples of each class were generated (Fig. 1.25), and, by simple counting, the misclassification rate was estimated to be equal to 30.1%. Therefore, it can be claimed that no classifier, however carefully designed, can achieve a classification performance higher than 69.9%. The best classifiers are the classifiers that come closest to that theoretical limit.

1.3.5 Classification and Regression

The previous section was devoted to the probabilistic foundations of classification. We are going to show why neural networks, which are regression tools, are relevant to classification tasks.

1.3.5.1 Two-Class Problems

We first consider a problem with two classes C_1 and C_2, and an associated random variable Γ, which is a function of the vector of descriptors $\boldsymbol{x}$; that random variable is equal to 1 when the pattern belongs to class C_1, and 0 otherwise. We prove the following result: the regression function of the random variable Γ is the posterior probability of class C_1.

The regression function $y(\boldsymbol{x})$ of variable Γ is the expectation value of Γ given $\boldsymbol{x} : y(\boldsymbol{x}) = E(\Gamma \mid \boldsymbol{x})$. In addition, one has:

$$E(\Gamma \mid \boldsymbol{x}) = \Pr(\Gamma = 1 \mid \boldsymbol{x}) \times 1 + \Pr(\Gamma = 0) \times 0 = \Pr(\Gamma = 1 \mid \boldsymbol{x})$$

which proves the result.

Neural networks are powerful tools for estimating regression functions from examples. Therefore, neural networks are powerful tools for estimating posterior probabilities, as illustrated on Fig. 1.26 this is the rationale or performing

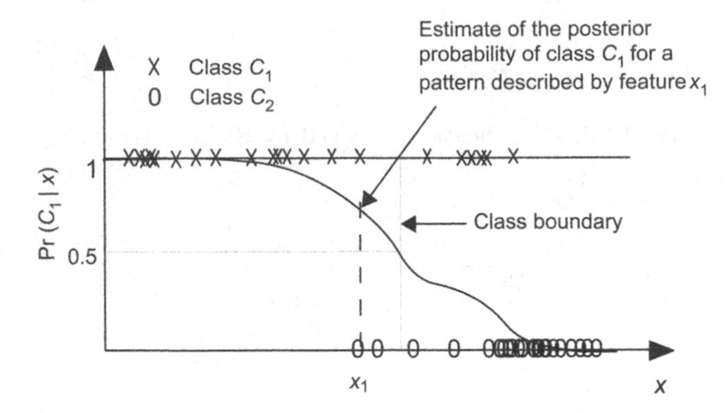

Fig. 1.26. Estimate of the posterior probability of class C_i, and boundary between classes from Bayes decision rule

classification by neural networks. A lucid and detailed description of that approach is given in C. Bishop's excellent book [Bishop 1995].

1.3.5.2 *C*-Class Problems

When the number of classes involved in a classification problem is larger than two, two strategies can be implemented, i.e.,

- find a global solution to the problem by simultaneously estimating the posterior probabilities of all classes;
- split the problem into two-class subproblems, design a set of pairwise classifiers that solve the subproblems, and combine the results of the pairwise classifier into a single posterior probability per class.

We will consider those strategies in the following subsections.

Global Strategy

That is the most popular approach, although it is not always the most efficient, especially for difficult classification tasks. For a C-class problem, a feedforward neural network with C outputs is designed (Fig. 1.27), so that the result is encoded in a 1-out-of-C code: the event "the pattern belongs to class C_i" is signaled by the fact that the output vector $\boldsymbol{g}$ has a single nonzero component, which is component number i. Similarly to the two-class case, it can be proved that the expectation value of the components of vector $\boldsymbol{g}$ are the posterior probabilities of the classes.

In neural network parlance, a *one-out-of-C* encoding is known as a *grandmother code*. That refers to a much-debated theory of data representation in nervous systems, whereby some neurons are specialized for the recognition of usual shapes, such as our grandmother's face.

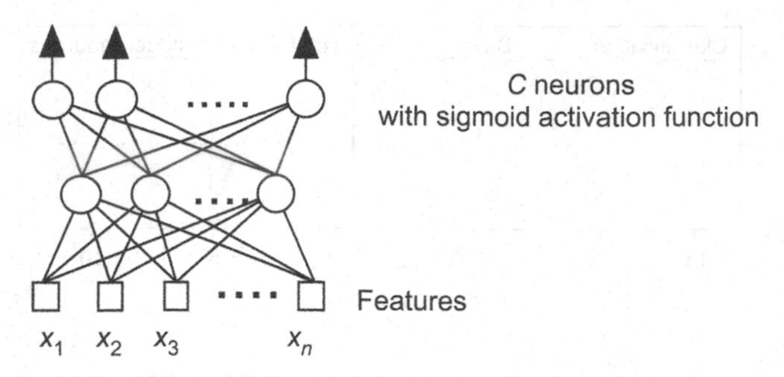

Fig. 1.27. A multilayer Perceptron with C outputs for classification. The activation functions of the output neurons are sigmoids

There are several important differences between a multilayer perceptron for classification and a multilayer perceptron for regression.

- The activation functions of the output neurons of neural networks for modeling is usually linear; by contrast, the output neurons of neural networks for classification have nonlinear activation functions such as sigmoids: since the outputs of the neural network are probabilities, they must lie between 0 and 1 (readily amenable to $[-1,+1]$); in Chap. 6, a theoretical justification for the use of the tanh function as an activation function of output neurons will be given,
- For classification, minimizing the cross-entropy cost function is more natural than minimizing the least squares cost function [Hopfield 1987; Baum 1988; Hampshire 1990]; the training algorithms that will be described in Chap. 2 can readily be applied to this cost function,

$$J = -\sum_k \sum_{i=1}^{C} \gamma_i^k \mathrm{Log}\left[\frac{g_i(\boldsymbol{x}^k)}{\gamma_i^k}\right] + (1-\gamma_i^k)\mathrm{Log}\left[\frac{1-g_i(\boldsymbol{x}^k)}{1-\gamma_i^k}\right].$$

where γ_i^k is the desired value (0 or 1) for output i when the classifier's input is example k, described by feature vector $\boldsymbol{x}^k$, and $g_i(\boldsymbol{x}^k)$ is the value of output i of the classifier. That function is minimum when all examples are correctly classified.

After training, it is safe to check that the sum of the outputs is equal to 1 for all examples. The Softmax technique [Bridle 1990] guarantees that the above condition is fulfilled automatically. Of course, that is not a problem for pairwise classifiers, which have a single output.

The question of overfitting, which we have encountered in nonlinear regression, is also valid for discrimination. If the classifier is overparameterized, it separates very accurately the patterns of the training set and has a poor generalization ability. Model selection techniques, such as those described in Chap. 2, must be used in order to select the best model. Essentially, one must

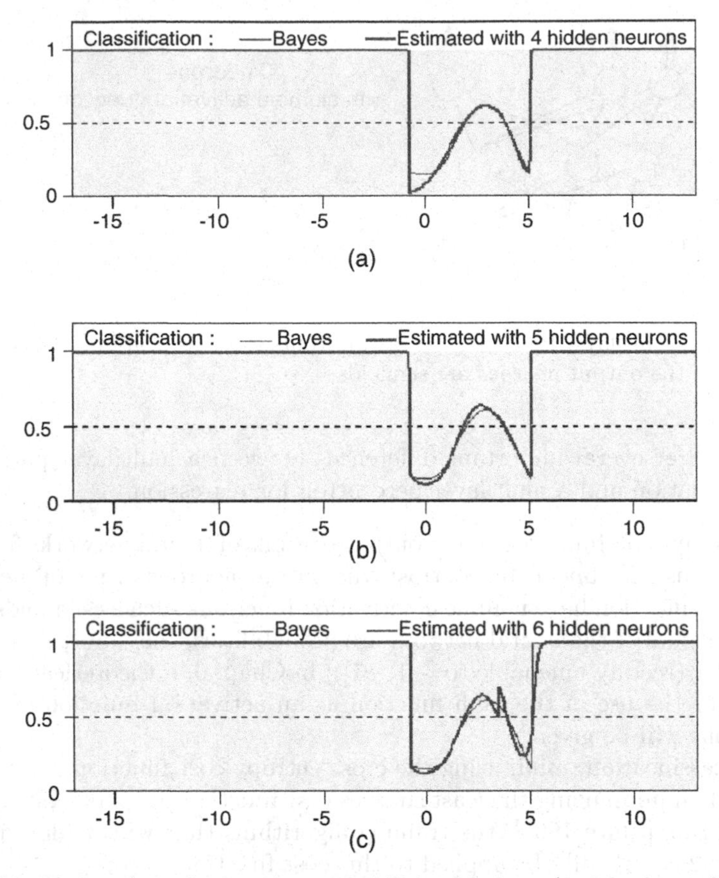

Fig. 1.28. Estimation of posterior probabilities of class A with three classifiers: (**a**) 4 hidden neurons (too low complexity), (**b**) 5 hidden neurons (performance very close to the best achievable correct classification rate, (**c**) 6 hidden neurons (strong overfitting)

find a classifier whose classification error rates are of the same order of magnitude, and as small as possible, on the training set and on an independent validation set. Figure 1.28 shows an example of overfitting in the estimation of the posterior probability of class A in the example shown on Fig. 1.23; clearly, the network with 4 hidden neurons is not complex enough for representing the posterior probability, whereas a neural network with 6 hidden neurons fits the fluctuations of the densities of points of the training set. The neural network with 5 hidden neurons has a misclassification rate of 30.3% (estimated on a test set of several million points), while the minimum achievable error rate, from Bayes classifier, is 30.1%. Therefore, neural networks are among the best classifiers.

Pairwise Classification

For difficult problems, it is often much safer to split a C-class classification problem into $C(C-1)/2$ pairwise classification problems, for the following reasons:

- When performing pairwise classification, the designer can take advantage of many theoretical results and algorithms, pertaining to linear class separation; they are fully developed in Chap. 6; we give a cursory introduction to that material in the next section, entitled linear separability.
- The resulting networks are much more compact, with fast training and simple analysis; since each network has a single output, its probabilistic interpretation is trivial.
- The features that are relevant for separating class A from class B are not necessarily identical to the features that are relevant for separating class A from class C; therefore, each classifier has only the inputs that are relevant to its own task, whereas a multilayer Perceptron for global separation must have all input features that are relevant for the discrimination of all classes; the feature selection techniques that are described in Chap. 2 can be used in a very straightforward fashion.

Once the $C(C-1)/2$ posterior probabilities are estimated, possibly with simple linear separators (neural networks with no hidden neuron), the posterior probability of class C_i for a feature vector $\boldsymbol{x}$ is computed as

$$\Pr(C_i \mid \boldsymbol{x}) = \frac{1}{\displaystyle\sum_{j=1,\, j\neq i}^{C} \frac{1}{\Pr_{ij}} - (C-2)},$$

where C is the number of classes and $\Pr_{ij}$ is the posterior probability of class i or class j, as estimated by the neural network that separates class C_i from class C_j.

Linear Separability

Two sets of patterns, described in an n-dimensional feature space, belonging to two different classes, are said to be "linearly separable" if they lie on different sides of a hyperplane in feature space.

If two sets of examples are linearly separable, a neural network made of a single neuron (also termed perceptron can separate them. Consider a neuron with a sigmoid activation function with n inputs; its output is given by $y = \mathrm{th}\left[\sum_{i=1}^{n} w_i x_i\right]$. The simple relation $P = (y+1)/2$ provides an interpretation of the output of the classifier as a posterior probability. From Bayes decision rule, the equation of the boundary between the classes is given by $P = 0.5$, or equivalently $y = 0$. Therefore, the separating surface is a hyperplane in

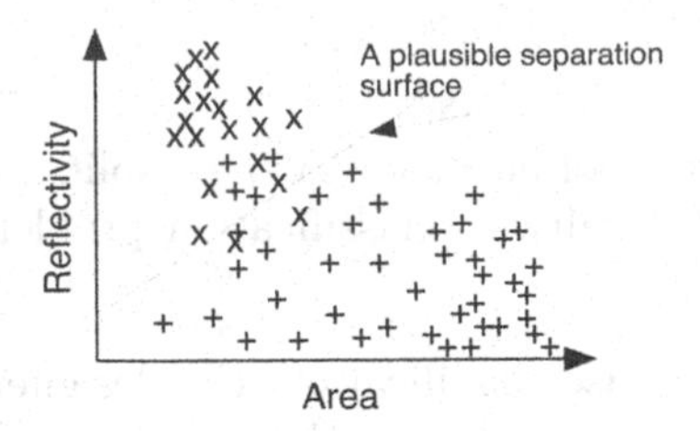

Fig. 1.29. Linear separation by a Perceptron (neural network with a single output, without hidden neurons: 10% misclassification rate

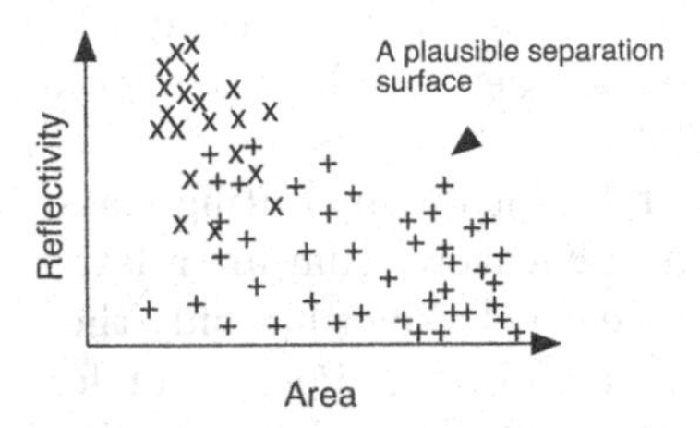

Fig. 1.30. Separation by a network with a small number of hidden neurons. Three examples per class are misclassified

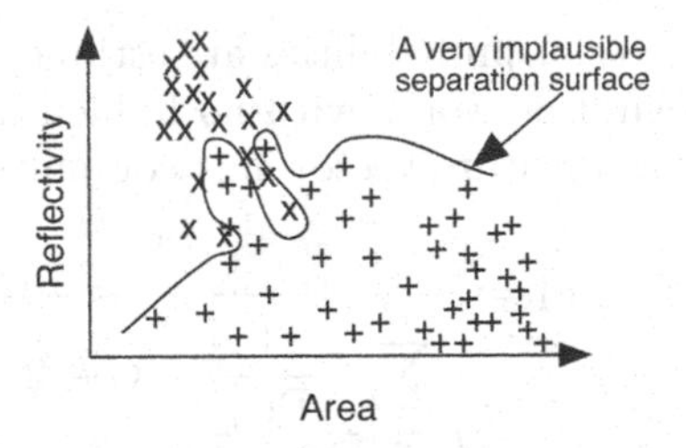

Fig. 1.31. Separation by an overparameterized neural network. All examples are correctly classified, but the generalization capacity is low

n-dimensional space,

$$v = \sum_{i=1}^{n} w_i x_i = 0.$$

Hence, $v > 0$ for all examples of one of the classes, and $v < 0$ for all examples of the other class. Figure 1.29 shows a separation surface that can be defined by a Perceptron, for the example of capacitors and integrated circuits.

Hidden neurons allow multilayer Perceptrons to define more complex separation surfaces, as shown on Fig. 1.30.

As usual, one can obtain zero misclassifications if enough hidden neurons are added, but that is detrimental to generalization because of overfitting. Fig. 1.31 illustrates a case of blatant overfitting.

When a multi-class problem is split into pairwise separation problems, linear separation between two classes is often complex enough; very frequently, in multi-class problems that are claimed to be difficult, the examples turn out to be pairwise linearly separable. In such cases, powerful, elegant algorithms give excellent solutions, as explained in Chap. 6: therefore, the first step in the design of a classifier consists, in checking whether the classes are pairwise linearly separable. Ho and Kashyap's algorithm [Ho 1965], which was discovered long before neural networks came into existence, provides an answer to that problem in finite time:

- If the examples are linearly separable, the algorithm finds a separating hyperplane in finite time.
- If the examples are not linearly separable, the algorithm signals it infinite time (see algorithmic complements at the end of this chapter).

The postal code database provided by the National Institute of Standards and Technology has served as a basis for many classifier designs. It turns out that, even with low-level representations such as a pixel representation, the classes of examples are pairwise linearly separable [Knerr 1992]! Similarly, a famous sonar signal data base has been investigated in great detail by many authors, and many complicated classifiers were designed to solve that two-class problem; actually, in less than ten minutes on a PC, the Ho and Kashyap algorithm, implemented as an uncompiled Matlab program, proves that the examples of the two classes are linearly separable. Therefore, a simple Perceptron can solve the problem, without resorting to any hidden layer. That application is investigated in more detail in Chap. 6.

1.3.5.3 Classifier Design Methodology

From the previous section, the following strategy for the design of a neural classifier can be derived (as discussed above, one should first ascertain that statistical classification is relevant to the problem at hand):

- Find an appropriate representation of the patterns to be classified, especially for pattern recognition (the techniques described in Chap. 3 can be especially useful in that respect); this is a crucial step, since a representation that has a high discriminative power is likely to make the classification problem trivial; this is illustrated in applications described below.
  If the number of examples available for training the classifier is not larger than the dimension of the feature vector, there is no point in pursuing the design any further, according to Cover's theorem [Cover 1965], which is explained in Chap. 6: before proceeding to the next steps, either a more "compact" representation must be found, or additional examples must be collected, or a very stringent regularization method such as weight decay (described in Chap. 2), must be implemented.
- For each pair of classes, select the relevant features with the feature selection methods described in Chap. 2; obviously, the discrimination of class A from class B may not require the same features as the discrimination of class A from class C.

- For each pair of classes, test the pairwise linear separability of the examples with the Ho and Kashyap algorithm.
- For all pairwise linearly separable classes, make use of linear separation methods (described in Chap. 6) and derive an estimate of the posterior probabilities.
- For nonlinearly separable classes, design small multilayer Perceptrons, or spherical perceptrons as described in Chap. 6, with probability estimations; use leave-one-out or cross-validation techniques for model selection (see Chap. 2).
- Estimate the global posterior probabilities of each class from the pairwise probabilities estimated at the previous steps, using the relation indicated in Sect. 1.3.5.2, subsection "pairwise classification" above.
- Determine the decision thresholds in order to define rejection classes.

That strategy is a variant of the STEPNET procedure [Knerr 1990, 1991], which allowed the design of several industrial applications.

In the planning of a classification project, the time required by the first and the last steps of the above strategy should definitely not be underestimated; for nontrivial applications, those are frequently the lengthiest and most painful steps.

The applicability of that strategy is limited by the fact that the number of pairwise classifiers grows as the square of the number of classes. However, each classifier is usually very simple, so that the procedure can be applied with up to a few tens of classes. For larger number of classes, hierarchical strategies must be resorted to.

1.4 Some Applications of Neural Networks to Various Areas of Engineering

1.4.1 Introduction

The present book is intended to assist the engineer or researcher in answering the following question: can neural networks solve my problem, and can they do it *more efficiently* (in terms of accuracy, computation time, etc.) than other techniques?

Contributions to a rational answer were provided at the beginning of the present chapter, where we explained the mathematical foundations and principles underlying the operation of neural networks. Although some elements may look somewhat technical, they are mandatory for getting an in-depth understanding of what one can and cannot do with neural networks. Since the software implementation of neural networks is straightforward with present-time tools, one might be tempted to apply neural networks without prior thinking, which may lead to disappointing results.

In addition to mathematical arguments, it may be helpful, in order to illustrate the use and limitations of neural networks, to describe some typical industrial applications. This is by no means intended to be an exhaustive presentation of neural network applications, which would require several books. The point here is to show some typical examples, and to stress the reason why neural networks made important, possibly decisive, contributions.

1.4.2 An Application in Pattern Recognition: The Automatic Reading of Zip Codes

Character recognition is definitely the application area where neural networks made their first significant contributions to engineering, proving to be reliable alternatives to classical pattern recognition methods. In the present section, some examples and results will be described, relying on the elements of theory and methodology provided in the previous section.

The automatic reading of postal codes is probably one of the most widely investigated problems in picture recognition. The automatic reading of printed envelopes and parcels is a relatively simple problem; by contrast, the huge variety of handwritings made the recognition of handwritten addresses a truly challenging problem. For each item handled by the postal service, a machine must either recognize the code, or resort to human inspection when it fails to identify the code. As indicated above, correcting a sorting mistake made by a machine is more costly than resorting to human inspection for reading and typing in the correct code; therefore, the most frequently used performance criterion for such machines is the following: given a maximum misclassification rate (say 1%), what fraction of the mail must be read by a human operator? At present, typical performances are 5% rejection rate for 1% misclassification.

The development of automatic zip code reading was primarily spurred by the industrial importance of the problem, but also by the fact that, as early as 1990, large-scale data bases were made available to the general public by the United States Postal Service (USPS), and later by the National Institute of Science and Technology (NIST). That policy allowed many laboratories, both in industry and in universities, to improve the state of the art, and to validate, in a statistically significant way, the methods and procedures that they developed; it had a general positive impact on the development of powerful classification methods.

Figure 1.32 displays some examples from the USPS database, which features 9,000 digits (which is not a very large number, considering the variety of handwriting styles). The difficulty of the problem is immediately apparent. Consider the postal code in the upper right corner of the picture; one reads 68544 effortlessly, but one notes

- that the digit 6 is split into two parts,
- that digits 8 and 5 are linked together,
- that digit 5 is split into two parts, the right part being linked to digit 4!

Thus, if one decides to base the recognition of the code on the recognition of its individual digits, the problem of segmentation must be solved first: how does

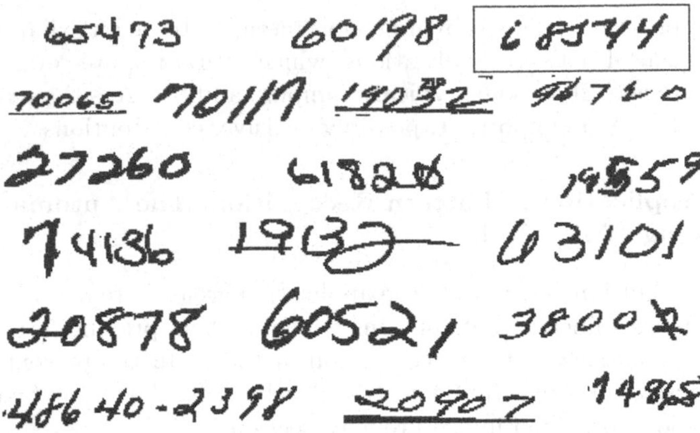

Fig. 1.32. Some excerpts from the NIST handwritten digit database

one decompose a zip code into five separate digits? Having solved that difficult problem, one must cope with the variety of styles, sizes, and orientations, of the isolated digits. To that end, an appropriate representation must be found; the design of an appropriate representation is completely problem-dependent, and requires new efforts for each new application. Clearly, one cannot use the same kind of representations for pictures such as handwritten or printed digits, satellite pictures, or X-ray medical images.

Despite the diversity of image processing techniques, some basic operations are found in essentially all applications, as well as in the human visual system: edge detection, contrast enhancement, etc. In the field of handwritten character recognition, normalization is mandatory, in order to apply the recognition algorithm to characters of similar sizes. We have already mentioned that the design of a real application requires a tradeoff between the complexity of the preprocessing that is necessary to yield the chosen representation, and the complexity of classification: a carefully designed preprocessing, which leads to very discriminant features, may allow the use of a very simple classifier, but the preprocessing must not be too demanding in terms of computation time. By contrast, a simple preprocessing, such as normalization alone, may be very fast but will not alleviate the task of the classifier. Thus, one must engineer the best tradeoff that allows meeting the requirements of the application. Two very different approaches to the same application will be described below.

The first approach was developed at the former AT&T Bell Labs. It consists of a neural network, known as LeNet [Le Cun 1991], which makes use of a pixel representation (after normalization). The first layers of the network perform local preprocessings that aim at automatically extracting features that are relevant to classification, while the final layers perform classification. The network is shown on Fig. 1.33.

The network is input with a 16×16 pixel intensity matrix. A first layer of hidden neurons is made of 12 sets of 64 neurons each, where each set of

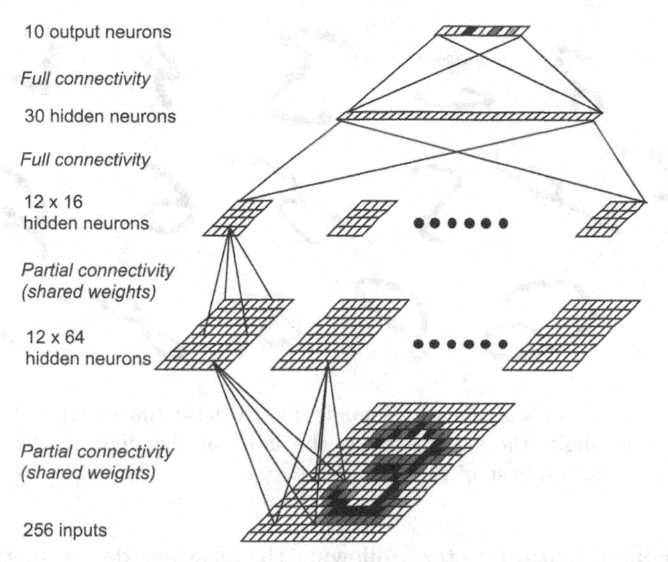

Fig. 1.33. LeNet, a neural network that performs feature extraction and classification

64 hidden neurons receives information from a "receptive field" of 5×5 pixels. Those sets of 64 neurons are called *feature maps*, for the inputs to a given map have the same weights (this is known as the "shared weights" technique, described in Chap. 2): thus, the same operator acts locally on a 25-pixel area of the picture, so that the outputs of a group of 64 neurons are the results of the application of the same operator to the receptive fields. The local operator technique is classical in picture processing, but the present approach is original in that these operators are not engineered, but are "discovered" through training by examples. The same technique is iterated by a second layer of operators that act on the results of the first layer. Thus, 12 maps of 16 hidden neurons are produced by 192 neurons that provide the representation of the digit. Classification is performed by a final layer of 30 hidden neurons, followed by 10 output neurons using a 1-out-of-N code: the number of outputs is equal to the number of classes, output neuron number i must be active if the input digit belongs to class i, and inactive otherwise.

Thus, the network performs, automatically and simultaneously, feature extraction and classification, whereas those operations are usually performed in a sequential fashion. The flexibility of the method has a price: given the size of the network, training is demanding, and, because of the large number of parameters, the network will be prone to overfitting.

In order to solve the same problem, a very different approach was implemented [Knerr 1992], which consists in performing a more elaborate preprocessing of the picture, in order to extract discriminant characteristics that lead to a relatively simple classifier. Preprocessing consists of edge extraction, followed by normalization, which produce 4 feature maps of 64 elements, hence

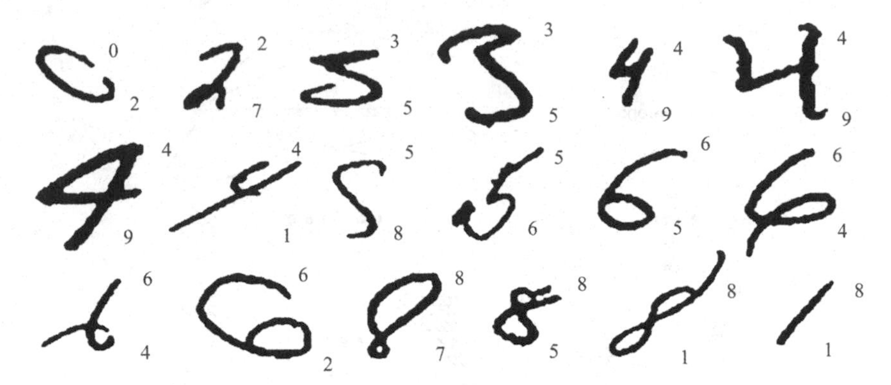

Fig. 1.34. The 18 misclassifications made by pairwise linear separation of the classes. For each digit, the superscript is the label of the digit in the base, and the subscript is the response of the classifier

a 256-component feature vector. Following the classifier design methodology described in the previous section, pairwise classification was performed by 45 different classifiers. Since the training sets were pairwise linearly separable, each classifier consisted of a single neuron; the classifiers were trained separately.

Figure 1.34 shows the 18 misclassifications made by the classifier on the 9,000 digits of the USPS database. Note that the bottom right digit is (correctly) recognized as a 1, whereas it was mistakenly labeled as an 8 in the database!

We have emphasized the impact of the choice of the representation on the efficiency of classification. This point can be nicely illustrated with this application. For the two representations that were mentioned above (representation by the intensities of 256 pixels, and representation by 4 feature maps of 64 elements each after edge detection), the euclidean distances between the centers of gravity of the classes were computed, and reported on Fig. 1.35. The inter-class distances are systematically larger, with the feature-based representation, than with the pixel-based representation. Thus, the feature-based representation increases the inter-class distances in input space, thereby making classification easier.

Table 1.1 illustrates the performance improvement resulting from the use of a more appropriate representation: after adjusting the decision thresholds so as to get, for both representations, a 1% misclassification rate, the rejection rate is much higher for the pixel representation than for the feature-map based representation. Note that since, in both cases, the dimension of input space is 256, the improvement does not result from the fact that the representation is more compact, but from the fact that it is more appropriate to the task. As usual, better engineering provides better results.

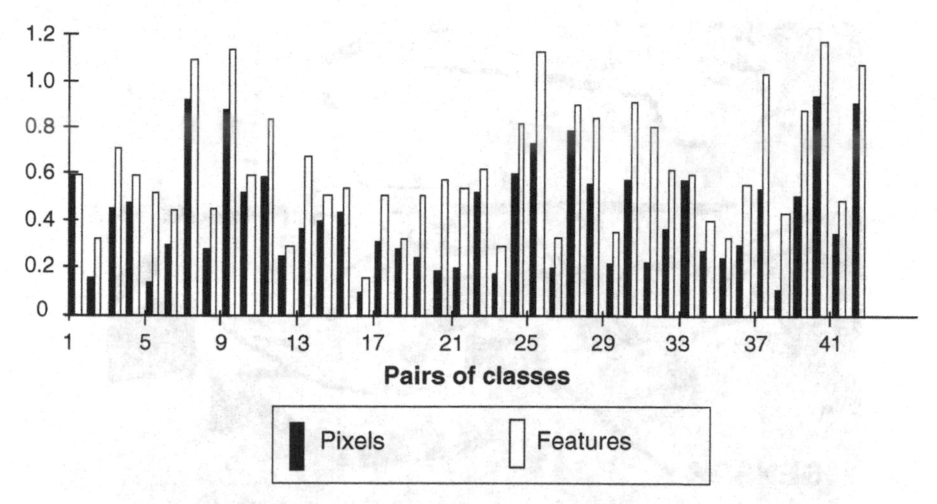

Fig. 1.35. Distances between classes for two representations: the feature-map based representation makes the classes more widely separated in input space, thereby making the classification task easier

1.4.3 An Application in Nondestructive Testing: Defect Detection by Eddy Currents

The example that was presented in the previous section used classification for picture recognition. Of course, patterns that can be recognized automatically vary widely in nature. The application that we describe in the present section pertains to nondestructive testing, where the patterns to be classified are signals. The objective is the automatic detection of defects in the rails of the Paris subway. It was developed by the National Research Institute on the Safety of Transportation Systems for RATP, the company that operates the Paris underground system [Oukhellou 1997].

Defect detection in metal parts by eddy currents is a standard nondestructive testing technique. An electromagnetic coil creates an alternating magnetic field, which generates eddy currents in the metal part to be tested. Those currents are detected by a second coil, and the presence of defects in the metal alters the amplitude and the phase of the resulting signal. Thus, that signal

Table 1.1.

	Correct classification rate (%)	Rejection rate (%)	Misclassification rate (%)
Pixel representation	70.9	28.1	1
Feature-map-based representation	90.3	8.7	1

Fig. 1.36. The eddy current generation and detection system

contains a "signature" of the defects. Since there are different categories of defects, which may be more or less detrimental to the operation of the system, classifying the defects is generally desirable. In the present case, it is also important to be able to discriminate between real defects and normal phenomena that are also detected by the eddy current technique, such as the presence of a weld joint or of a switch (the position of the latter is known, which makes discrimination easier).

In the present application, the system that generates and detects eddy currents is mounted below the carriage, a few tens of millimeters above the rail, as shown on Fig. 1.36.

As usual, the choice of the descriptors of the signal is crucial for the efficiency of discrimination. In the present case, a relatively small number of features, derived from Fourier components of the signal, are usually sufficient, provided they are chosen appropriately. Feature selection was performed by the "probe feature method," which is described in Chap. 2 [Oukhellou 1998].

1.4.4 An Application in Forecasting: The Estimation of the Probability of Election to the French Parliament

After the elections to the French parliament, all candidates must make an official statement of the amount of the expenses incurred during the campaign, and of the breakdown of those expenses. Making use of the data pertaining to the 1993 elections, it was possible to assess the probability of being elected as a function of the expenses and of their breakdown. This is a two-class classification problem, and neural networks provide an estimate of the probability

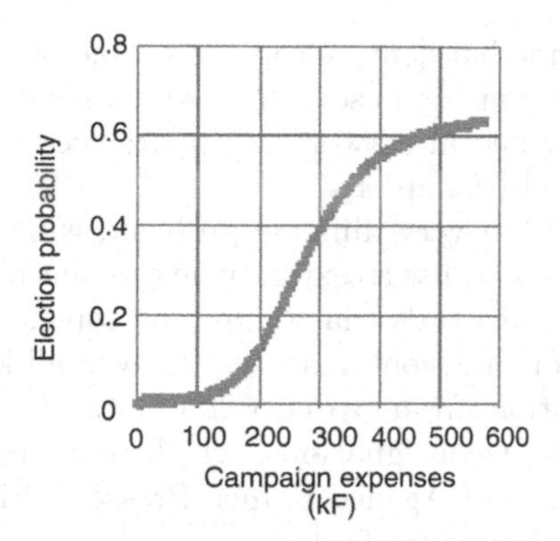

Fig. 1.37. Neural estimate of the election probability as a function of the total campaign expenses (data for the 1993 elections)

of being elected. Figure 1.37 shows the probability of election as a function of the total expense.

That application, although in the area of classification, is somewhat different from the previous two applications: in the latter, the classifier was intended to assign an existing pattern to a class, while, with high probability, the actual class of the unknown pattern would never be known with absolute certainty. In the present application, the situation is different, since the class of each pattern (candidate running for election) will be known unambiguously immediately after the election. This application falls in the class of forecasting by simulation: in order to optimize the probability of success, the candidate can estimate his success probability as a function of the strategy of expenses that he uses, and derive from those results the strategy that is most suitable to his situation.

In the next chapters, some sections will be devoted to forecasting by simulation: it will be shown that it is an area of excellence of neural networks.

1.4.5 An Application in Data Mining: Information Filtering

The rapid increase of the volume of available information, especially by electronic means, makes it mandatory to design and implement efficient information filtering tools, which allow the user to access relevant information only. Since such tools will address the needs of professionals, they must be reliable and user-friendly. The user can access relevant information either by being provided with full texts by the machine (text search), or by being provided text excerpts or answers to questions (information extraction).

Text categorization, also known as filtering, consists in finding, in a text corpus (e.g., of press releases, or of Web pages), the texts that are relevant to a predefined topic. The user can thus be provided with information that is important for his professional duties. In a machine-learning based system, the user does not express his topic of interest through a query, but by providing a set of relevant documents that define a topic or category. For a given topic, text categorization therefore consists in solving a two-class discrimination problem, which can be solved by neural networks, support vector machines (Chap. 6) or hidden Markov models (Chap. 4).

Text categorization is a very difficult problem, which goes much beyond text search by keywords, because a text may be relevant to a topic even though it contains none of the keywords that define the topic, or, conversely, a text may be irrelevant although it contains some or even all keywords.

The present application (from [Stricker 2000]) was developed by the French bank Caisse des dépôts et consignations, which provides an Intranet service for filtering press releases of Agence France Presse (AFP) in real time. The objective of the application is twofold:

- to develop an application that allows the user to create automatically an information filter on any topic of interest to him, under the condition that he provides examples of texts that are relevant to his topic of interest;
- to develop a machine-learning based tool that monitors the obsolescence of classical, rule-based information filters.

In the latter development, a neural-based filter is designed on the same topic as the rule-based filter. Since the neural network does not generate a binary response, but estimates a relevance probability, the largest discrepancies between the two filters can be analyzed and possibly be traced to vocabulary obsolescence: documents that are rated as relevant by the rule-based method, but whose relevance probability, estimated by the neural network, is very low, and documents that are rated as irrelevant by the rule-based filter and having an estimated relevance probability close to one as estimated by the neural filter [Wolinski 2000].

The former development consists in designing and implementing an automatic filter production system, whose major feature is the fact that it does not require any assistance from an expert, as opposed to rule-based filters. Therefore, a two-class discrimination system must be designed, from a database of texts that are labeled as relevant or irrelevant, that requires

- finding a representation of texts by real numbers, which should be as compact as possible,
- designing and implementing a classifier that uses that representation.

Thus, the problem of text representation, hence of input selection, is crucial for that application.

1.4.5.1 Input Selection

The most popular approach to text representation is the bag-of-words representation, whereby a text is represented by a vector, each component of which is a number that is a function of the presence or absence of the word in the text, or of its frequency in the text. Clearly, the main difficulty is the dimension of that vector, which is, in principle, equal to the number of words in the vocabulary. Nevertheless, all words are not equally discriminant: most frequent words (of, the, and) are not useful for discrimination, nor are very rare words. Therefore, the first step of the design of a filter is the determination of the vocabulary that is specific to the topic.

Word Encoding

The words are encoded in the following way: we denote by $FT(m,t)$ the frequency of occurrence of word m in text t, and by $FT(t)$ the average frequency of the terms in text t. Then the word m is described by [Singhal 1996]

$$x(m) = \frac{1 + \log(FT(m,t))}{1 + \log(FT(t))}.$$

Zipf's Law

Zipf's law [Zipf 1949] is helpful for finding discriminant words: given a corpus of T texts, we denote by $FC(m)$ the frequency of occurrence of word m in corpus T. A list of words, ranked in order of decreasing values of $FC(m)$, is built; we denote the rank of word m in that list by $r(m)$. Zipf's law states that $FC(m)r(m) = K$, where K is a corpus-dependent quantity. Hence, there is a very small number of very frequent words, and a large number of very rare words that occur once or twice in the corpus; between those extremes, there is a set of words in which discriminant words ought to be sought.

Extraction of the Specific Vocabulary

In order to extract the vocabulary that is specific to the topic, the ratio $R(m,t) = FT(m,t)/FC(m)$ is computed for each word m of each relevant text t. The words of the text are ranked in order of decreasing values of $R(m,t)$, the second half of the list is deleted, and a boolean vector $\boldsymbol{v}(t)$ is defined, such that $v_i(t) = 1$ if word i is present in the list, 0 otherwise. Finally, the vector $\boldsymbol{v} = \sum_t \boldsymbol{v}(t)$, is computed, where the summation is performed on all relevant documents: the specific vocabulary of the topic is the set of words that have a nonzero component in vector $\boldsymbol{v}$. Figure 1.38 shows that Zipf's law is obeyed on the corpus of Reuters releases, and that the words of the vocabulary specific to the topic *Falkland petroleum exploration* are indeed located in the middle of the distribution.

Final Selection

Within the specific vocabulary thus defined, which may be still large (one to several hundred words), a final selection is performed by the probe feature

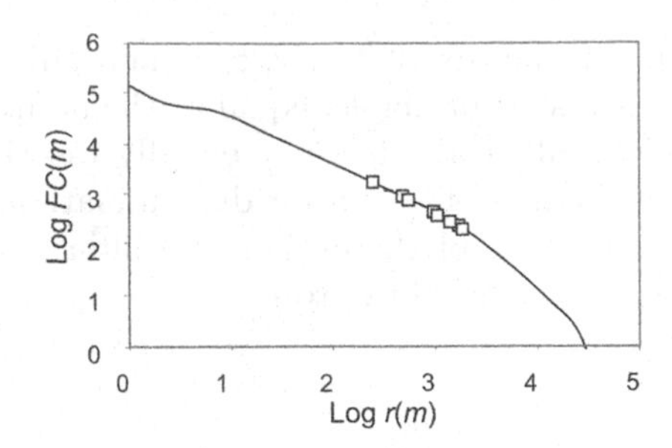

Fig. 1.38. An experimental verification of Zipf's law on the Reuters corpus, and location of the words of the vocabulary specific to the topic "Falkland petroleum exploration"

method, described in Chap. 2. After completion of that step, it turns out that, on the average over 500 different topics, the size of the specific vocabulary of a given topic is 25 words, which is a reasonable dimension for the input vector of a neural network. That representation, however, is not fully satisfactory yet. Since isolated words are ambiguous in such an application, the context must be taken into account.

1.4.5.2 Context Determination

In order to take into account the context in the representation of the texts, context words are sought in a window of 5 words on both sides of each word of the specific vocabulary.

- The words that are in the vicinity of the words of the specific vocabulary, in relevant texts, are defined as positive context words.
- The words that are in the vicinity of the words of the specific vocabulary, in irrelevant texts, are defined as negative context words.

In order to select the context words, the procedure that is used is identical to the selection procedure for the specific vocabulary. On the average over 500 topics, a topic is defined by 25 specific words, each of which having 3 context words.

1.4.5.3 Filter Design and Training

Filters Without Context

If the context is not taken into account, the inputs of the filter are the words of the specific vocabulary, encoded as indicated above. In accordance with the classifier design methodology described above, the structure of the classifier depends on the complexity of the discrimination problem. On the corpuses tested in the course of the development of the present application, the examples are linearly separable, so that networks made of a single neuron with sigmoid activation function solve the problem.

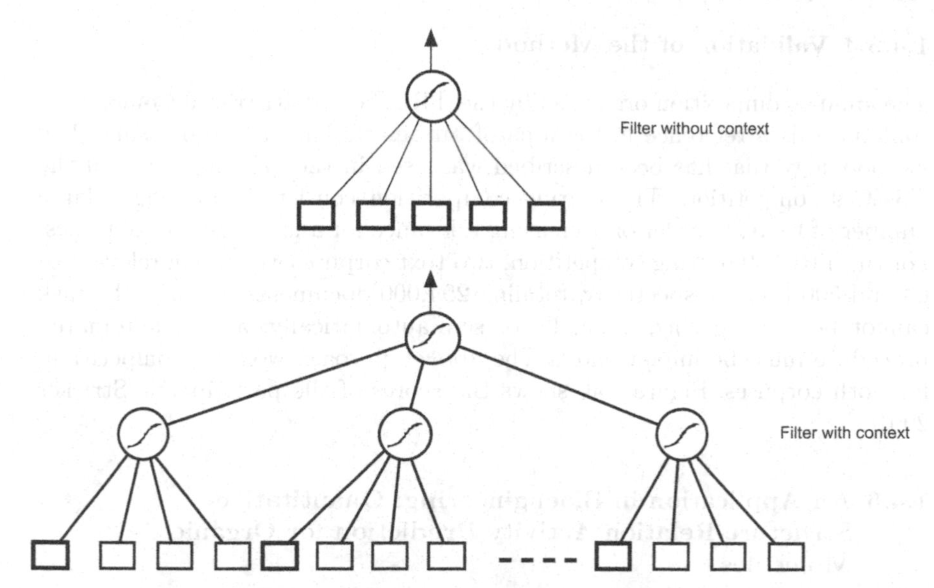

Fig. 1.39. A filter without context is a linear classifier whose inputs are the features that encode each word of the specific vocabulary (*rectangles in thick lines*); in a filter with context, the inputs are the features that encode the words of the specific vocabulary (*boxes in thick lines*), and, in addition, the features that encode the context words (*boxes in thin lines*)

Filters with Context

The context must have an influence of the feature that encodes each word of the specific vocabulary. Therefore, the filter represents each word of the specific vocabulary by a neuron with sigmoid activation function, whose inputs are

- the feature that encodes the word of interest,
- the features that encode the context words of that word.

The outputs of those neurons are separated linearly by a neuron with sigmoid activation function. Figure 1.39 shows a filter with context and a filter without context.

The introduction of context words as inputs increases the number of parameters of the classifier. Typically, for a topic with 25 words of specific vocabulary, and 3 context words per word of the specific vocabulary, the filter has 151 parameters. Since some topics are described by a small number of relevant texts, the use of a regularization method is mandatory. The weight decay method (described in Chap. 2) proved useful in the present application; its effect and implementation are explained in Chap. 2, in the section devoted to regularization techniques.

1.4.5.4 Validation of the Method

The annual competition organized by the TREC (Text REtrieval Conference) conference is a reference in the area of automatic language processing. The methodology that has been described was used in the "routing" task of the TREC-9 competition. The routing competition consists in ranking a large number of texts, in order of decreasing relevance for a large number of topics. For the TREC-9 routing competition, two text corpuses were used, relevant to 63 and 500 topics respectively, totaling 294,000 documents. Clearly, the task cannot be accomplished manually or semiautomatically: a fully automated procedure must be implemented. The above approach won the competition, for both corpuses. Figure 1.40 shows the scores of the participants [Stricker 2001].

1.4.6 An Application in Bioengineering: Quantitative Structure-Relation Activity Prediction for Organic Molecules

The investigation of quantitative structure-activity relations (QSAR) of molecules is a rapidly growing field thanks to progress in molecular simulation. The objective of QSAR is the prediction of chemical properties of molecules from structural data that can be computed *ab initio*, without actually synthesizing the molecule; thus, costly organic syntheses, leading to molecules that turn out not to have the desired property, can be avoided [Hansch 1995]. That approach is especially useful in the field of bio-engineering, for the prediction of pharmacological properties of molecules and for computer-aided drug discovery. It is also extremely valuable for solving conceptually similar problems, such as the prediction of properties of complex materials from their formulation, the prediction of thermodynamic properties of mixtures, etc.

Why are neural networks useful in that context? If there exists a deterministic relation between some features of the molecule and the property that must be predicted, then QSAR is amenable to a regression problem, i.e., to the determination of that unknown relation, from examples. If that relation is nonlinear, then neural networks can be advantageous, as argued above.

A prerequisite for such an approach is the availability of databases for training and testing the model. Because of the industrial importance of the problem, many databases of existing molecules for such properties as the boiling point, water solubility, or water-octanol partition coefficients (known as "LogP") are available. The latter property is important in pharmacology, because it gives a quantitative assessment of the ability of the molecule to cross biological barriers in order to be active; similarly, in the field of environment, the value of LogP of pesticides contributes to assessing their impact on environment.

Once the availability of appropriate databases is guaranteed, the relevant features that should be the inputs of the model must be determined. In the

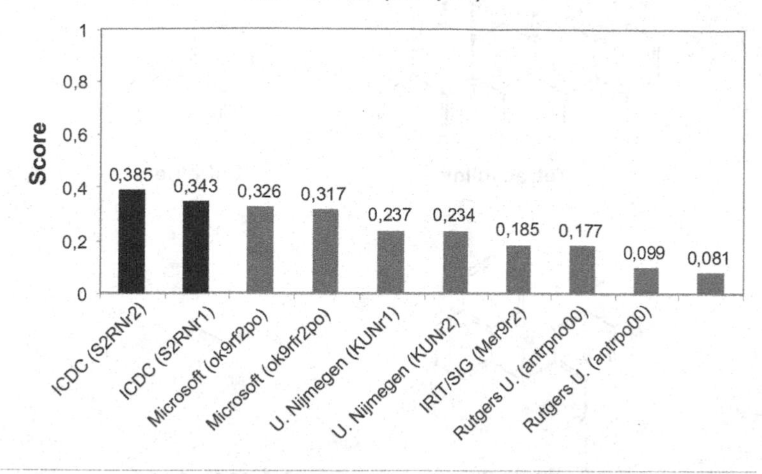

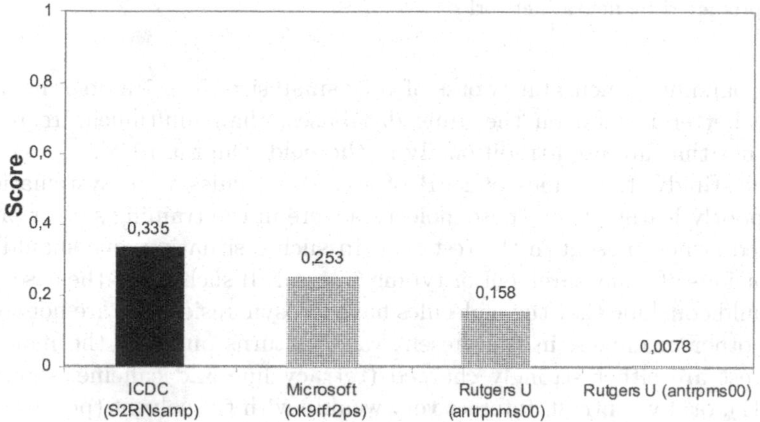

Fig. 1.40. Results of the TREC-90 routing task; *black*: results obtained by the above method; *grey*: results obtained by other methods

present case, the chemist's knowledge is of utmost importance. Classically, three categories of features are considered, i.e.,

- chemical features such as the molecular weight, the number of carbon atoms, etc.;
- geometrical features such as the volume of the molecule, its area, etc;
- electrical features such as the electric charges borne by each atom, dipole moments, etc.

For each property to be predicted, a set of candidate descriptors must be built, and selection techniques such as described in Chap. 2 must be applied. Because

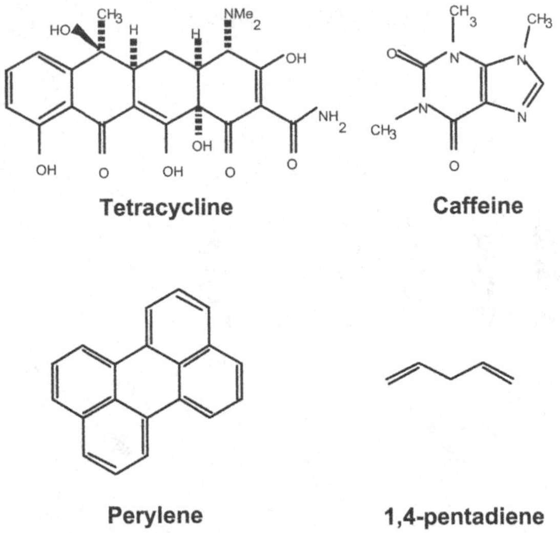

Tetracycline

Caffeine

Perylene

1,4-pentadiene

Fig. 1.41. Molecules that have chemical idiosyncrasies, whose properties may be poorly predicted by neural networks

of their parsimony, neural networks of very small size (5 to 7 hidden neurons) provide better results, on the same databases, than multilinear regression techniques that are used traditionally in the field [Duprat 1998].

Interestingly, the values of logP of some molecules were systematically either poorly learnt (when those molecules were in the training set) or poorly predicted (when present in the test set). In such a situation, one should first be suspicious of a measurement or typing-in error. If such is not the case, then one should conclude that the molecules have idiosyncrasies that are not shared by the other examples; in the present vase, it turns out that the molecules of interest are either strongly charged (tetracycline and caffeine, shown on Fig. 1.41), or, by contrast, interact very weakly with the solvent (perylene, 1-4 pentadiene, see Fig. 1.41). Thus, neural networks are able to detect anomalies; anomaly detection is actually one of the main areas of applications of neural networks.

1.4.7 An Application in Formulation: The Prediction of the Liquidus Temperatures of Industrial Glasses

In the same spirit as the previous application, albeit in a completely different field, thermodynamic parameters of materials can be predicted as a function of their formulation. Of specific interest is the prediction of the liquidus temperatures of oxide glasses. That temperature is the maximal temperature at which crystals are in thermodynamic equilibrium with the liquid; the prediction of the liquidus temperature is important for the glass industry, because the value of the viscosity at the liquidus temperature has a strong impact of the process of glass forming. Since the phase diagrams of glasses exhibit strong variations in the temperature domain of interest, many attempts at such predictions have been made (see for instance [Kim 1991]), and databases are available. Neural networks have been successfully used for the prediction of liquidus temperatures [Dreyfus 2003], especially (as expected) for glasses with more than three components.

Figure 1.42 illustrates, on the present industrial example, the parsimony of neural networks. It shows scatter plots, which are very convenient for assessing graphically the accuracy of the model: on the horizontal scale, the measured value of the quantity of interest is displayed, whereas the predicted values are displayed on the vertical scale. If prediction were perfect, all points should be aligned on the first bisector; actually, due to measurement inaccuracy and prediction errors, the points are more or less scattered; a good model should generate equivalent scatterings for the points of the training set and those of the validation or test set, and the vertical scattering should be on the order of the standard deviation of the noise. Clearly, such a tool is no substitute to the computation of the TMSE and VMSE as defined above, or of the leave-one-out score defined in Chap. 2, but it provides a quick means of comparing different models, for instance.

The model inputs are the contents of the glass in various oxides; the output is the estimated liquidus temperature. Figure 1.42(a) shows the results obtained on a silica glass (made of potassium oxide K_2O and aluminum oxide Al_2O_3, in addition to silicon oxide SiO_2, which is the main component), obtained with a network having 6 hidden neurons (25 parameters), and Fig. 1.42(b) shows the result obtained with a polynomial of degree 3, with a similar number of parameters (19). Clearly, with roughly the same number of parameters, the neural network performs much better. For comparison, Fig. 1.42(c) shows the scatter plot for a linear model.

1.4.8 An Application to the Modeling of an Industrial Process: The Modeling of Spot Welding

Spot welding is the most widely used welding process in the car industry: millions of such welds are made every day. The process is shown schematically on Fig. 1.13: two steel sheets are welded together by passing a very large current (tens of kiloamperes) between two electrodes pressed against the metal surfaces, typically for a hundred milliseconds. The heat thus produced melts a roughly cylindrical region of the metal sheets. After cooling, the diameter of the melted zone—typically 5 mm—characterizes the effectiveness of the process; a weld spot whose diameter is smaller than 4 mm is considered mechanically unreliable; therefore, the spot diameter is a crucial element in the safety of a vehicle. At present, no fast, nondestructive method exists for measuring the spot diameter, so that there is no way of assessing the quality of the weld immediately after welding. Therefore, a typical industrial strategy consists

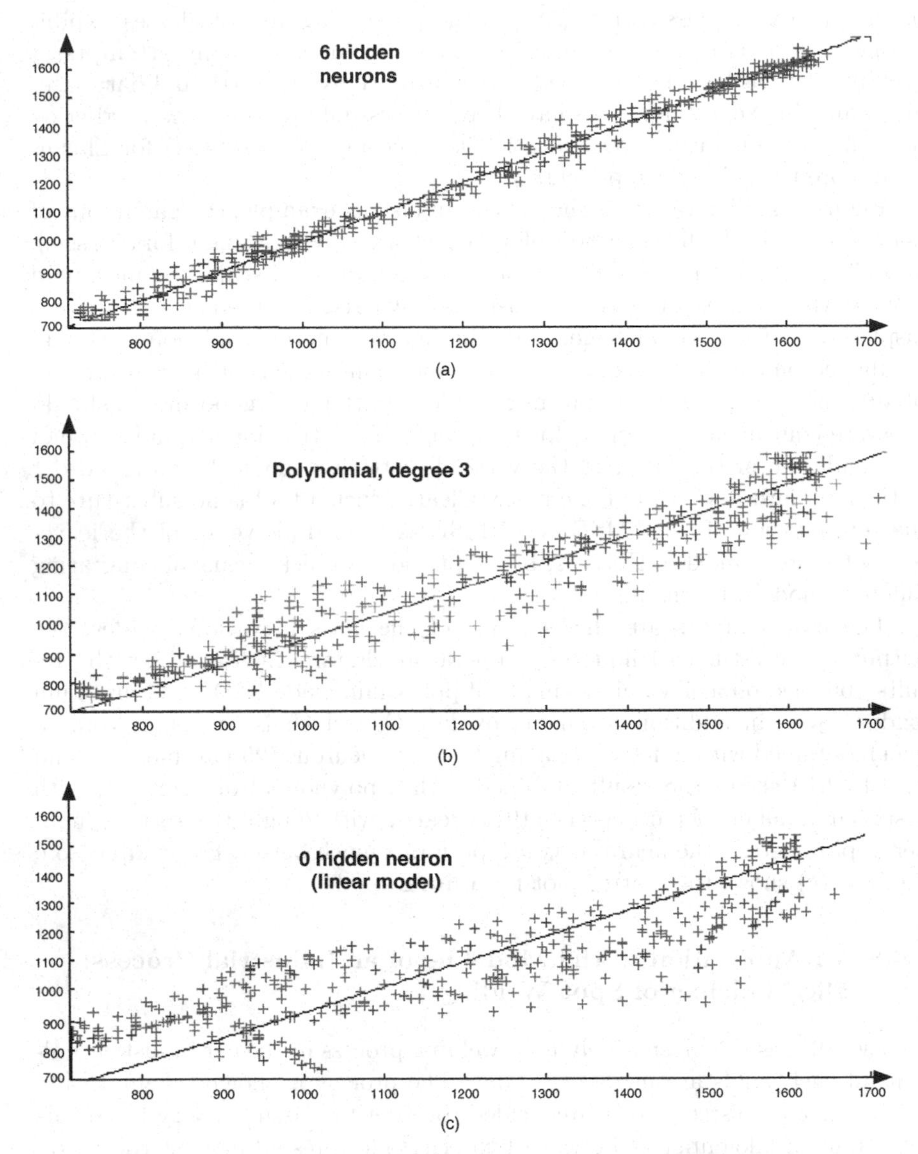

Fig. 1.42. Scatter plots for the prediction of the liquidus temperature of an oxide glass, as a function of its composition, for three different models

- in using an intensity that is much larger than actually necessary, which results in excessive heating, which in turn leads to the ejection of steel droplets from the welded zone (hence the sparks that can be observed during each welding by robots on car assembly chains);

- in making a much larger number of welds than necessary, just to be sure that a sufficient number of valid spots are produced.

Both the excessive current and the excessive number of spots result in a fast degradation of the electrodes, which must be changed or redressed frequently.

For all the above reasons, the modeling of the process, leading to a reliable on-line prediction of the weld diameter, is an important industrial challenge. Modeling the dynamics of the welding process from first principles is a very difficult task, for several reasons, including

- the computation time necessary for the integration of the partial differential equations of the knowledge-based model is many orders of magnitude larger than the duration of the process, which precludes real-time prediction of the spot diameter;
- many physical parameters appearing in the equations are not known reliably.

Those arguments lead to considering black-box modeling as an alternative. Since the process is nonlinear and has several input variables, neural networks are natural candidates for predicting the spot diameter from measurements performed during the process, immediately after weld formation, for on-line quality control [Monari 1999].

The main concerns for the modeling task are the choice of the model inputs, and the limited amount of examples available in the database, because gathering data is costly.

In [Monari 1999], the quantities that were candidates for input selection were mechanical and electrical signals that can be measured during the welding process. Input selection was performed by the techniques described in Chap. 2. The experts involved in the knowledge-based modeling of the process validated that set.

Because no simple nondestructive weld diameter measurement exists, the database is built by performing a number of welds in well-defined condition, and subsequently tearing them off; the melted zone, remaining on one of the two metal sheets, is measured. That is a lengthy and costly process, so that the initial training set was made of 250 examples only. Using experimental design through the confidence interval estimates described in Chap. 2, a training set extension strategy was defined in order to increase the database size. Half of the resulting data was used for training, and the other half for testing (the model selection method was virtual leave-one-out which, as explained in Chap. 2, does not require any validation set).

Figure 1.43 shows typical scatter plots, where each prediction is shown together with its confidence interval. The estimated generalization error, estimated from the virtual leave-one-out score defined in Chap. 2, was 0.27 mm, whereas the TMSE was 0.23 mm. Since those quantities are on the order of the measurement uncertainty, the results are satisfactory.

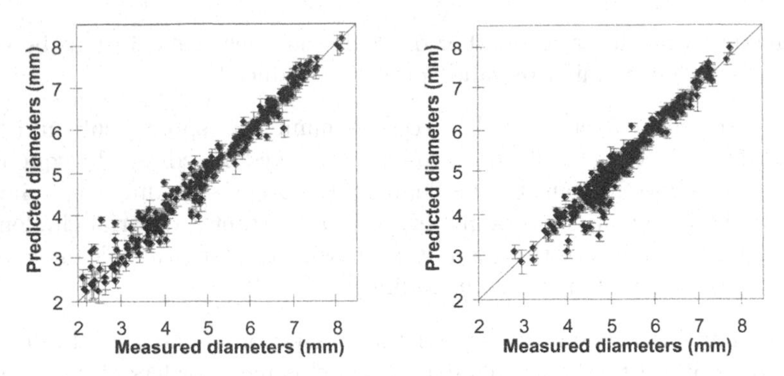

Fig. 1.43. Scatter plots for the prediction of the diameters of welding spots

1.4.9 An Application in Robotics: The Modeling of the Hydraulic Actuator of a Robot Arm

The previous applications involved feedforward neural networks only. We now turn to dynamic modeling, with recurrent neural networks.

We consider a hydraulic actuator that controls the position of a robot arm; therefore, the position of the arm depends on the hydraulic pressure in the actuator, which in turn depends of the angular position of a valve. A dynamic model of the relation between the hydraulic pressure and the opening of the valve was sought, in the framework of an informal competition between research groups involved in nonlinear modeling. A control sequence $\{u(k)\}$, i.e., a sequence of angles of the valve, and the corresponding sequence of the quantity to be modeled $\{y_p(k)\}$, i.e., the hydraulic pressure, are shown on Fig. 1.44. That sequence contains 1,024 samples, the first half of which, according to the rule of the game, was to be used as a training set and the second one as the validation set. Since no prior information was available from the physics of the process, a black-box model must be designed.

A cursory look at the data shows that a linear model of the process would certainly not be appropriate; the oscillations observed as responses to control variations that are almost steps suggests that the model is at least of second order. The training and validation sequences are approximately of the same type and same amplitude, but the amplitude of the control signal is larger in the validation set (around times 600 and 850) than in the training set. Thus, the conditions are not very satisfactory.

The example is analyzed in detail in Chap. 2. The best results [Oussar 1998] were obtained with a second-order state-space model, one state variable of which is the model output, of the form

$$y(k+1) = x_1(k+1) = \psi_1(x_1(k), x_2(k), u(k))$$
$$x_2(k+1) = \psi_2(x_1(k), x_2(k), u(k)),$$

with two hidden neurons. It is shown on Fig. 1.45.

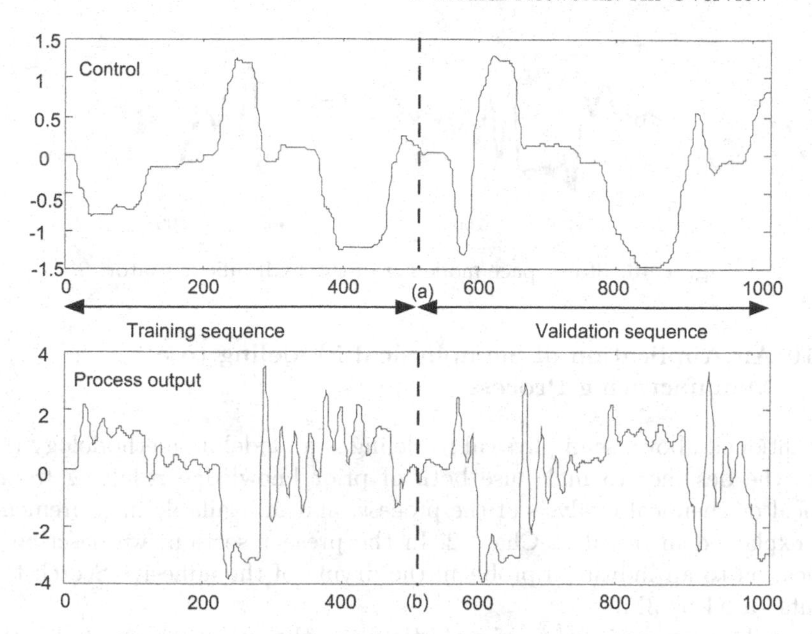

Fig. 1.44. Training and test sequences for a robot arm

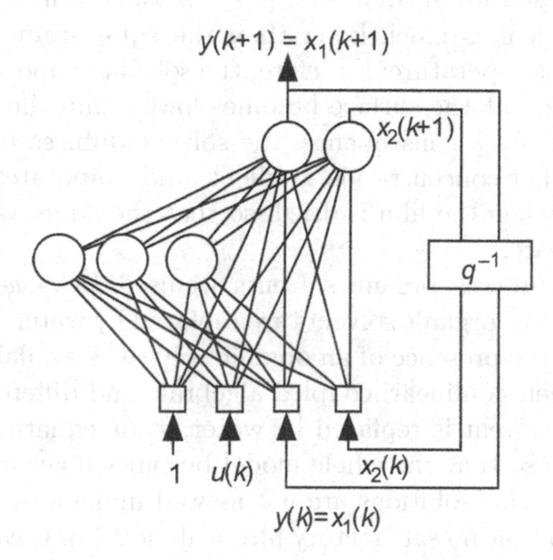

Fig. 1.45. State-space neural model of the hydraulic actuator. The output is one of the state variables

The mean square error obtained with that model is 0.07 on the training set and 0.12 on the validation set, which is very satisfactory given the available sequences. The modeling errors may be due to disturbances that are not measured, hence not present as inputs of the model. The results are shown on Fig. 1.46.

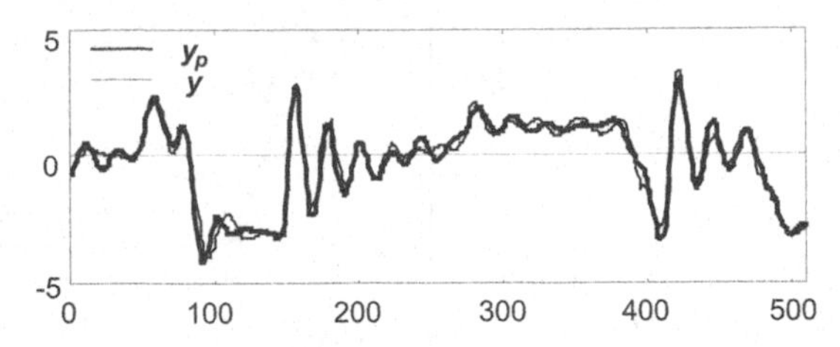

Fig. 1.46. State-space modeling of the hydraulic actuator

1.4.10 An Application of Semiphysical Modeling to a Manufacturing Process

As mentioned above, semi-physical modeling is a modeling methodology that allows the designer to make use both of prior knowledge resulting from a physical or chemical analysis of the process, and of available measurements. It is explained in detail in Chap. 2. In the present section, we describe its application to an industrial problem: the drying of the adhesive Scotch tape manufactured by 3M.

An adhesive tape is made of a plastic film—the substrate—coated with a liquid, which is passed in an oven, in a gas atmosphere in which the partial pressure of the solvent is much lower than the equilibrium pressure of the solvent at the oven temperature; therefore, the solvent evaporates, so that the solvent concentration at the surface becomes lower than the solvent concentration in the bulk. As a consequence, the solvent diffuses to the surface so as to compensate that concentration gradient, and evaporates at the surface. The process stops when the film is dried, so that the adhesive polymer alone stays on the substrate.

In a traditional process, organic solvents are used. However, for safety and environmental reasons, organic solvents are replaced by water. A very accurate model of drying in the presence of an organic solvent is available [Price 1997]; it is made of thirteen nonlinear, coupled algebraic and differential equations. When the organic solvent is replaced by water, some equations of the model are no longer valid, so that the whole model becomes inaccurate.

Polymers in aqueous solutions are not as well understood as polymers in organic solvents, so that no satisfactory physical model of the drying of water-based adhesives is available. However, sequences of measurements of sample weight as a function of time and oven temperature are available: the design of a semi-physical model is therefore possible and appropriate.

The equations of the model express the following phenomena:

- mass conservation in the bulk of the solvent: that equation is naturally still valid in the case of water-based adhesives;
- the diffusion of solvent towards the free surface (Flick's law); the validity of that equation is not arguable, but it involves a quantity (the diffusion

coefficient) whose variation as a function of concentration and temperature is given by the free-volume theory, whose validity can be disputed;
- mass conservation at the surface: any solvent molecule that reaches the surface, and evaporates, gives a contribution to the solvent partial pressure in the gas; that law remains valid;
- the boundary condition at the coating-substrate interface: since the substrate is impermeable to organic solvents and to water alike, that condition does not change;
- the value of the partial pressure of the solvent in the gas is the "driving force" of the whole process; it is given by an equation whose validity is not disputed.

Therefore, it turns out that the variation of the diffusion coefficient must be expressed by a black-box neural network, within the whole physical model. That has been done with the methodology that is described in detail in Chap. 2. Note that the equations of the model are not ordinary differential equations, but partial differential equations; that does not preclude the application of the method.

The reader interested in the details of the model and in the results will find them in [Oussar 2001]. Another industrial application of semi-physical modeling—the automatic detection of faults in an industrial distillation column—can be found in [Ploix 1997]. It is worth mentioning that applications or semi-physical modeling are in routine use in a major French manufacturing company for the design of new materials and products.

1.4.11 Two Applications in Environment Control: Ozone Pollution and Urban Hydrology

The two applications that are described in the present section are related to the prevision of nonlinear phenomena in environmental science.

1.4.11.1 Prevision of Ozone Pollution Peaks

Ozone concentration measurements are more and more widespread, and elaborate knowledge-based models of atmospheric pollution become available, so that the prediction of ozone peaks becomes feasible. The present section reports an investigation that was carried out at ESPCI within a work group to which measurements related to industrial area of Lyon (France) were made available. The objective was to assess the efficiency of machine learning techniques for designing black-box models for the prediction of ozone pollution peaks in that area.

The available data set was made of hourly measurements of a reliable ozone sensor between 1995 and 1998. Data pertaining to years 1995 to 1997 were used for training, and data of 1998 for validation. The task was to predict, 24 hours ahead of time, whether pollution would excess the legal alert threshold (180 $\mu g/m^3$ at the time of the investigation).

Two different approaches could be considered:

- classification: assign the next day to one of the classes "polluted" (threshold will be exceeded) or "non polluted," as a function of the data available at 2 pm GMT,
- prevision: predict the ozone concentration, 24 hours ahead.

Since the definition of the class "polluted" depends on the legal definition of the alert threshold, which may vary for administrative, political or economical reasons, it was deemed preferable by the ESPCI group to follow the second approach. A black-box model was designed, which performs the prediction of ozone concentration during the next 24 hours, from information available at 2 pm GMT.

The naive idea consisted in using a dynamic nonlinear model such as a recurrent neural network. However, it turned out that such an approach would not be appropriate, because the process is not time-invariant: the physico-chemical phenomena that determine ozone concentration depend on the time of the day. Therefore, a set of 24 cascaded neural networks was designed, each network specializing in the prediction of ozone concentration at a given hour of the day (Fig. 1.47): network $\#N$ predicts the concentration at 2 pm $+N$ GMT; for each network, the candidate inputs are:

- the predictions of the previous $N-1$ networks;
- the set of available data:
  1. the measurements of sensors of NO and NO_2 at 2 pm GMT,
  2. the temperature at 2 pm on day D,
  3. the maximal temperature measured on day D, and the maximal temperature predicted for day $D+1$ by the national weather forecast service,
  4. geopotentials on day D,
  5. the time series of ozone concentrations performed before 2 pm on day D.

For each network, input selection among the above candidate variables was performed with techniques described in Chap. 2. Thus, the inputs of a given network, specialized in a given time of the day, are the appropriate inputs for that time of the day only.

Clearly, that approach can be adapted to other data sets, and can integrate expert knowledge, in a semi-physical model, when it will be available.

The mean prediction error on the validation year (1998) is 23 $\mu g/m^3$. Figure 1.48 illustrates the difficulty of the problem: despite a very accurate prediction during 20 hours, that day appears as a "false negative" since the measurement exceeds (by a very small amount) the alert threshold. Presumably, when such tools will be in routine operation (which was not yet the case when the present book was written), more subtle alert procedures than the simple threshold strategy will be implemented.

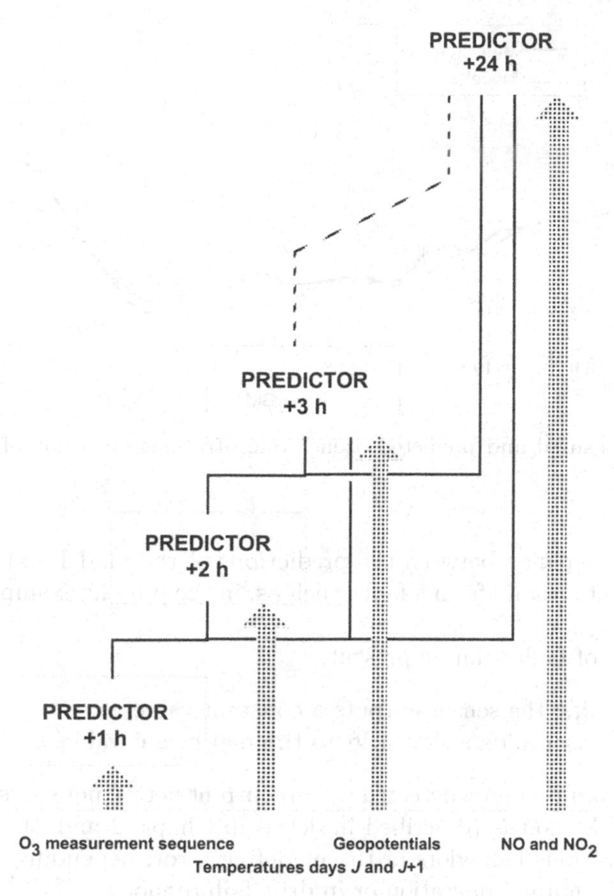

Fig. 1.47. The structure of a neural network for the prediction of ozone concentration, 24 hours ahead

1.4.11.2 Modeling the Rainfall-Water Height Relation in an Urban Catchment

The Direction de l'Eau et de l'Assainssement has developed a sophisticated system for measuring water heights in the sewers of an area of the suburbs of Paris, and has performed systematic measurements of rainfalls and of the corresponding water heights. The objective is to optimize the sewer network and to anticipate serious problems that are likely to arise in the case of severe rains. Hence the reliability of the water height sensors in the sewers is crucial for the reliability of the whole system: therefore, the automatic detection of faults in the water height sensors is mandatory [Roussel 2001].

Neural networks can be accurate models of nonlinear phenomena, which makes them useful tools for fault detection: if an accurate, real-time model of normal operation of a process is available, the observation of a statistically

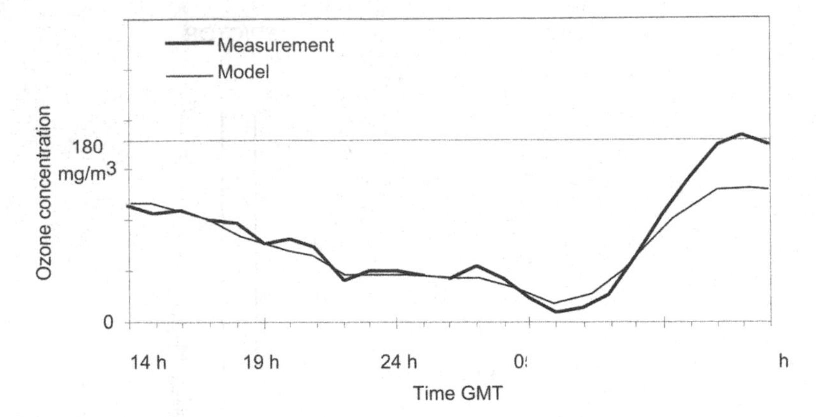

Fig. 1.48. Measured and predicted ozone concentrations on a day of 1998 (false negative)

significant discrepancy between the predictions of the model and the results of measurements results from a fault, such as, in the present example, a sensor failure.

Two kinds of faults can be present,

- stuck-at faults: the sensor outputs a constant value,
- drift: the sensor adds a slow drift to the real height value.

Both types of faults can be detected with recurrent neural networks, especially with NARMAX models (described in detail in Chaps. 2 and 4). Figure 1.49 displays the various behaviors of the modeling error, depending on whether the sensor is in normal operation or in drift failure mode.

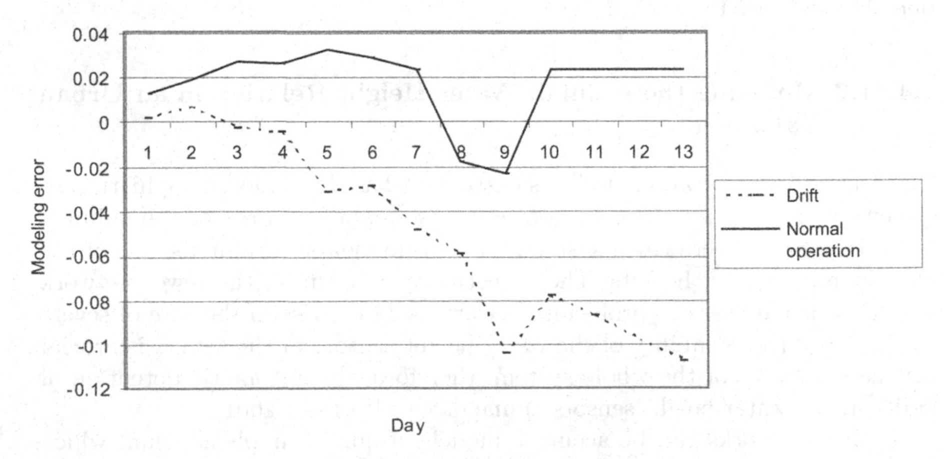

Fig. 1.49. Sensor fault detection in a sewer system

1.4.12 An Application in Mobile Robotics

Process control is the science that determines the control actions to which a process must be submitted in order to guarantee that it will operate in a prescribed fashion, in spite of unmeasured and unpredictable disturbances.

As an example of the role that neural networks can play in mobile robotics, we describe the automatic control of a 4WD Mercedes vehicle, which was equipped by the French company SAGEM with sensors and actuators that are necessary for making the vehicle autonomous. Controlling the vehicle consists in sending the appropriate signals to the actuators of the steering wheel, the throttle, and the brakes, in order to keep the vehicle on a prescribed trajectory with a prescribed velocity profile, in spite of disturbances such as wind gusts, sliding, slopes, etc.

Neural networks are good candidates as ingredients in nonlinear process control systems. They can implement any (sufficiently regular) nonlinear function. As a result, they can be useful in two different ways,

- as models of the process, since the design of a control system generally requires the availability of a model; neural networks are particularly useful in internal model control, as described in Chap. 5;
- as controllers, i.e., for computing the control signals (e.g., the angle by which the driving wheel must turn, and the velocity at which it should turn) from the setpoint (e.g., the desired heading of the vehicle).

The vehicle that was controlled was a 4 wheel-drive vehicle equipped with actuators (electric motor for actuating the driving wheel, hydraulic actuator for brakes, electric motor for actuating the throttle) and two categories of sensors,

- sensors that measure the state of the vehicle (proprioceptive sensors): odometers on the wheels, angular sensors for the driving wheel and for the throttle, pressure sensor in the brake system;
- sensors that measure the position of the vehicle with respect to the universe (exteroceptive sensors): an inertial platform.

The navigation and piloting system is made of the following elements:

- a planning module, which determines the desired trajectory of the vehicle, and its velocity profile, given the start and arrival points and the existing roads;
- a guiding module, which computes the heading and speed setpoint sequences;
- a piloting module, which computes the desired positions of the actuators;
- a control module of the actuators.

In that structure, neural networks are present at the level of the piloting module, where they compute the desired position sequences of the actuators as a function of the heading and speed setpoint sequences [Rivals 1994, 1995].

The application requested the design and implementation of two control systems that must fulfill two tasks,

- the control of the driving wheel, in order to keep the vehicle on the desired trajectory: a neural controller was designed, that performs a maximum lateral error of 40 cm, for curvatures up to 0.1 m^{-1}, and lateral slopes up to 30% in rough terrain; some elements of that controller used semi-physical modeling;
- the control of the throttle and the brake, in order to comply with the desired velocity profile.

All neural networks implemented within that application, whether models or controllers, are very parsimonious (less than ten hidden neurons). Their implementation on board did not require any special-purpose hardware: they were implemented as software on a standard microprocessor board that was also used for other purposes.

1.5 Conclusion

In the present chapter, we endeavored to explain why, and for what purposes, neural networks can be advantageously used. Some typical applications were presented (others are described in various chapters), so that model designers can get an intuition of what they can expect from that technique.

Before proceeding to more mathematical topics, it may be useful to emphasize the main points that should always be kept in mind when designing neural networks, i.e.,

- Neural networks are machine learning tools, that allow to fit very general nonlinear functions to sets of experimental data; just as for any statistical method, the availability of appropriate data is mandatory.
- Neural networks with supervised learning are parsimonious approximators, that can serve as static models (feedforward neural networks) or as dynamic models (recurrent neural networks).
- Neural networks with supervised learning can be high-quality classifiers, whose performances can reach those of the theoretical Bayes classifier; however, in the framework of classification for pattern recognition, the representation of the patterns to be recognized is often crucial for the performance of the whole system; in that context, neural networks with unsupervised learning may provide very valuable information for designing an efficient data representation.
- it is generally desirable, and often possible, to take advantage of all existing mathematical knowledge on the process to be modeled or patterns to be classified: neural networks are not necessarily black boxes.

The next chapters provide the mathematical background and the algorithmic information that are necessary for an efficient design of neural network models.

The foreword and the reading guide will help the reader to navigate through the chapters as a function of his own topics of interest.

1.6 Additional Material

1.6.1 Some Usual Neurons

Two categories of neurons can be distinguished, depending on the role of their parameters.

1.6.1.1 Neurons with Parameterized Inputs

The most popular neurons are neurons with parameterized inputs, in which one parameter is assigned to each input. The output of a neuron having n inputs $\{x_i\}$, $i = 0$ to $n-1$, is therefore given by an equation of the form $y = f\{x_i, w_i\}$, $i = 0$ to $n-1$, where $\{w_i\}$, $i = 0$ to $n-1$ are the parameters of the model.

In most cases, function f is the composition of two operations,

- the computation of the potential v of the neuron, which is the sum of the inputs of the neuron, weighted by the corresponding parameters,

$$v = \sum_{i=0}^{n-1} w_i x_i;$$

- the computation of a nonlinear function of the potential, termed activation function; that function is generally s-shaped, hence the generic name of sigmoid; preferably the activation function is symmetric with respect to the origin, such as the tanh function or the inverse tangent function, except if some prior knowledge on the problem prompts the implementation of different, more appropriate functions.

The set of inputs of the neuron generally includes a specific input, termed bias, the value of which is constant, equal to 1. It is usually assigned the index 0, so that the potential is of the form

$$v = w_0 + \sum_{j=1}^{n-1} w_j x_j.$$

Thus, the expression of the output of the neuron is: $y = f[w_0 + \sum_{j=1}^{n-1} w_j x_j]$.

Figure 1.50 shows the output of a neuron with three inputs ($x_0 = 1$, x_1, x_2) with parameters $w_0 = 0$, $w_1 = 1$, $w_2 = -1$.

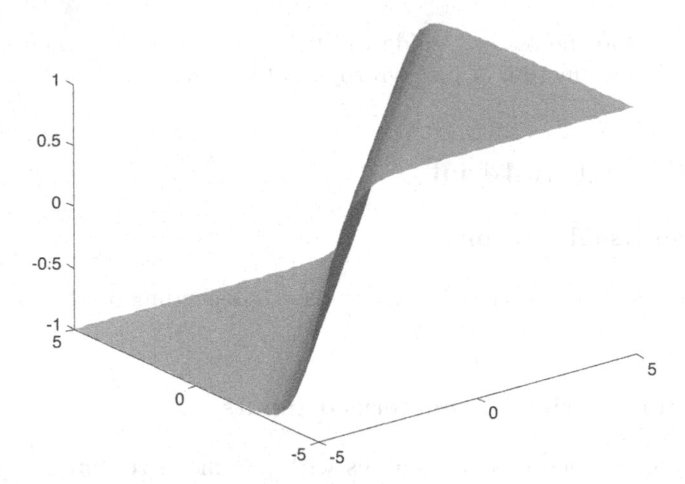

Fig. 1.50. Output of a neuron with 3 inputs $\{x_0 = 1, x_1, x_2\}$ with weights $\{w_0 = 0, w_1 = +1, w_2 = -1\}$, whose activation function is a tanh function: $y = \tanh(x_1 - x_2)$

Two variants of that type of neuron are

- high-order neural networks, whose potential is not an affine function of the inputs, but a polynomial function; they are the ancestors of the support vector machines (or SVM) used essentially for classification, described in Chap. 6;
- MacCulloch-Pitts neurons, or perceptrons, which are the ancestors of present-day neurons; Chap. 6 describes in detail their use for discrimination.

1.6.1.2 Neurons with Parameterized Nonlinearities

The parameters of those neurons are assigned to their nonlinearity: they are present in function f. Thus, the latter may be a "radial basis function" (RBF) or a wavelet.

Example

Gaussian radial basis function,

$$y = \exp\left[\sum_{i=1}^{n}(x_i - w_i)^2 \Big/ 2w_{n+1}^2\right].$$

The parameters $\{w_i, i = 1 \text{ to } n\}$ are the coordinates of the center of the Gaussian in input space; parameter w_{n+1} is its standard deviation. Figure 1.51 shows an isotropic Gaussian RBF, centered at the origin, with standard deviation $1/\sqrt{2}$.

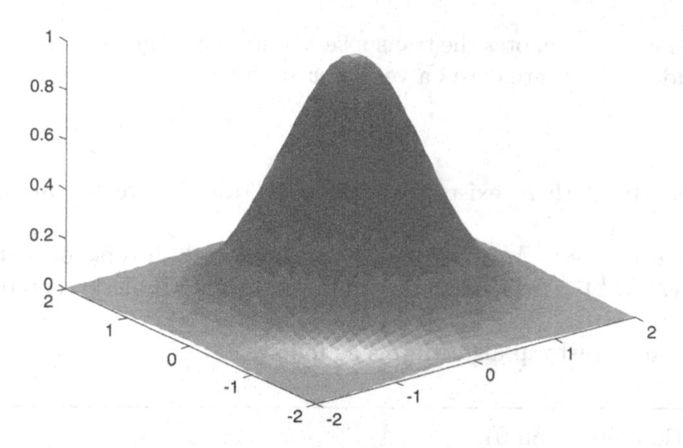

Fig. 1.51. Gaussian isotropic $RBF_y = \exp[-(x_1^2 + x_2^2)] : w_0 = w_1 = 0, w_3 = 1/\sqrt{2}$

The term radial basis function arises from approximation theory; they can be chosen so as to form a mathematical basis of functions. In regression, RBF's are generally not chosen so as to satisfy that requirement; however, following the current use, we will keep the term radial basis function.

1.6.2 The Ho and Kashyap Algorithm

The Ho and Kashyap algorithm finds, in a finite number of iterations, whether two given sets of observations are linearly separable in feature space. If they are, the algorithm provides a solution (among an infinity of possible solutions), which is not optimized (as opposed to algorithms that are explained in Chap. 6). Therefore, that algorithm is mainly used for finding out whether sets are linearly separable. If such is the case, it is advisable to use one of the optimized algorithms described in Chap. 6.

Consider two sets of examples, having n_A and n_B elements respectively, belonging to two classes A and B; if the examples are described by n features, each of them is described by an n-vector. We denote by $\boldsymbol{x}_k^A$ the vector that represents the k-th example of class A ($k = 1$ to n_A), and by $\boldsymbol{w}$ the vector of parameters of a linear separator; if such a separator exists, i.e., if the examples are linearly separable, then one has

$$\boldsymbol{x}_k^A \boldsymbol{w} > 0 \quad \text{for all } k,$$
$$\boldsymbol{x}_k^B \boldsymbol{w} < 0 \quad \text{for all } k.$$

We define matrix M whose rows are the vectors that represent the examples of A and the opposites of the vectors representing the examples of B, i.e.,

$$M = [\boldsymbol{x}_1^a, \boldsymbol{x}_2^a, \ldots, \boldsymbol{x}_{n_a}^a, \boldsymbol{x}_1^b, \boldsymbol{x}_2^b, \ldots, \boldsymbol{x}_{nb}^b]^{\mathrm{T}},$$

where superscript ^{T} denotes the transpose of a matrix. Then a linear separator exists if and only if there exists a vector $\boldsymbol{w}$ such that

$$M\boldsymbol{w} > 0,$$

or, equivalently, if there exists a vector $\boldsymbol{y} > 0$ and a vector $\boldsymbol{w}$ such that $M\boldsymbol{w} = \boldsymbol{y}$.

Then one has $\boldsymbol{w} = M^*\boldsymbol{y}$, where M^* is the pseudo-inverse of matrix M : $M^* = M^{\mathrm{T}}(MM^{\mathrm{T}})^{-1}$, which can be computed by the Choleski method [Press 1992].

The Ho and Kashyap algorithm is as follows:

Initialization (iteration 0):
$\boldsymbol{w}(0) = M^*\boldsymbol{y}(0)$ where $\boldsymbol{y}(0)$ is an arbitrary positive vector
Iteration i
$\alpha(i) = M^*\boldsymbol{w}(i) - \boldsymbol{y}(i)$
$\boldsymbol{y}(i+1) = \boldsymbol{y}(i) + \rho(\alpha(i) + |\alpha(i)|)$ where ρ is a positive scalar smaller than 1 $\boldsymbol{w}(i+1) = \boldsymbol{w}(i) + \rho M^*(\alpha(i) + |\alpha(i)|)$
If one of the components of $\boldsymbol{y}(i) < 0$ then the examples are not linearly separable.
If all components of $M\boldsymbol{w}(i) > 0$ then the examples are linearly separable and $\boldsymbol{w}(i)$ is a solution.

The algorithm converges after a finite number of iterations.

References

1. Antoniadis A., Berruyer J., Carmona R. [1992], *Régression non linéaire et applications*, Economica
2. Barron A. [1993], Universal approximation bounds for superposition of a sigmoidal function, *IEEE Transactions on Information Theory*, 39, pp 930–945
3. Baum E.B., Wilczek F. [1988], Supervised learning of probability distributions by neural networks, *Neural Information Processing Systems*, pp 52–61
4. Benveniste A., Juditsky A., Delyon B., Zhang Q., Glorennec P.-Y. [1994], Wavelets in identification, *10th IFAC Symposium on Identification*, Copenhague
5. Bishop C. [1995], *Neural networks for pattern recognition*, Oxford University Press
6. Bridle J.S. [1990], Probabilistic interpretation of feedforward classification network outputs, with relationship to statistical pattern recognition, *Neurocomputing: algorithms, architectures and applications*, pp 227–236 Springer
7. Broomhead D.S., Lowe D. [1988], Multivariable functional interpolation and adaptive networks, *Complex Systems*, 2, pp 321–355

8. Cover T.M. [1965], Geometrical and statistical properties of systems of linear inequalities with applications in pattern recognition, *IEEE Transactions on Electronic Computers*, 14, pp 326–334
9. Draper N.R., Smith H. [1998], *Applied regression analysis*, John Wiley & Sons
10. Duprat A., Huynh T., Dreyfus G. [1998], Towards a principled methodology for neural network design and performance evaluation in QSAR; application to the prediction of LogP, *Journal of Chemical Information and Computer Sciences*, 38, pp 586–594
11. Dreyfus C., Dreyfus G. [2003], A machine-learning approach to the estimation of the liquidus temperature of glass-forming oxide blends, *Journal of Non-Crystalline Solids*, 318, pp 63–78
12. Hampshire J.B., Pearlmutter B. [1990], Equivalence proofs for multilayer perceptron classifiers and the Bayesian discriminant function, *Proceedings of the 1990 connectionist models summer school*, pp 159–172, Morgan Kaufmann
13. Hansch C., Leo A. [1995], *Exploring QSAR, Fundamentals and applications in chemistry and biology*; American Chemical Society
14. Ho E., Kashyap R.L. [1965], An algorithm for linear inequalities and its applications, *IEEE Transactions on Electronic Computers*, 14, pp 683–688
15. Hopfield J.J. [1987], Learning algorithms and probability distributions in feed-forward and feedback neural networks, *Proceedings of the National Academy of Sciences*, 84, pp 8429–433
16. Hornik K., Stinchcombe M., White H. [1989], Multilayer feedforward networks are universal approximators, *Neural Networks*, 2, pp 359–366
17. Hornik K., Stinchcombe M., White H. [1990], Universal approximation of an unknown mapping and its derivatives using multilayer feedforward networks, *Neural Networks*, 3, pp 551–560
18. Hornik K. [1991], Approximation capabilities of multilayer feedforward networks, *Neural Networks*, 4, pp 251–257
19. Kim S.S., Sanders T.H. Jr. [1991], Thermodynamic modeling of phase diagrams in binary alkali silicate systems, *Journal of the American Ceramics Society*, 74, pp 1833–1840
20. Knerr S., Personnaz L., Dreyfus G. [1990], Single-layer learning revisited: a stepwise procedure for building and training a neural network, *Neurocomputing: algorithms, architectures and applications*, pp 41–50, Springer
21. Knerr S. [1991], *Un méthode nouvelle de création automatique de réseaux de neurones pour la classification de données: application à la reconnaissance de chiffres manuscrits*, Thèse de Doctorat de l'Université Pierre et Marie Curie, Paris
22. Knerr S., Personnaz L., Dreyfus G. [1992], Handwritten digit recognition by neural networks with Single-layer Training, *IEEE Transactions on Neural Networks*, 3, pp 962–968
23. LeCun Y., Boser B., Denker J.S., Henderson D., Howard R.E., Hubbard W., Jackel L.D. [1989], Backpropagation applied to handwritten zip code recognition, *Neural Computation*, 1, pp 541–551
24. Mallat S. [1989], A theory for multiresolution signal decomposition: the wavelet transform, *IEEE Transactions on Pattern Analysis and Machine Intelligence*, 11, pp 674–693
25. McCulloch W.S., Pitts W. [1943], A logical calculus of the ideas immanent in nervous activity, *Bulletin of Mathematical Biophysics*, 5, pp 115–133

26. Marcos S., Macchi O., Vignat C., Dreyfus G., Personnaz L., Roussel-Ragot P. [1992], A unified framework for gradient algorithms used for filter adaptation and neural network training, *International Journal of Circuit Theory and Applications*, 20, pp 159–200
27. Minsky M., Papert S. [1969] *Perceptrons*. MIT Press
28. Monari G. [1999], *Sélection de modèles non linéaires par leave-one-out; étude théorique et application des réseaux de neurones au procédé de soudage par points*, Thèse de Doctorat de l'Université Pierre et Marie Curie, Paris. Disponible sur le site *http://www.neurones.espci.fr.*
29. Moody J., Darken C.J. [1989], Fast learning in networks of locally-tuned processing units, *Neural Computation*, 1, pp 281–294
30. Nerrand O., Roussel-Ragot P., Personnaz L., Dreyfus G., Marcos S. [1993], Neural networks and nonlinear adaptive filtering: unifying concepts and new Algorithms, *Neural Computation*, 5, pp 165–197
31. Oukhellou L., Aknin P. [1997], Modified Fourier Descriptors: A new parametrization of eddy current signatures applied to the rail defect classification, *III International workshop on advances in signal processing for non destructive evaluation of materials*
32. Oukhellou L., Aknin P., Stoppiglia H., Dreyfus G. [1998], A new decision criterion for feature selection: application to the classification of non destructive testing signatures, *European Signal Processing Conference* (EUSIPCO'98)
33. Oussar Y. [1998], *Réseaux d'ondelettes et réseaux de neurones pour la modélisation statique et dynamique de processus*, Thèse de Doctorat de l'Université Pierre et Marie Curie, Paris. Disponible sur le site *http://www.neurones.espci.fr*
34. Oussar Y., Dreyfus G. [2000], Initialization by selection for wavelet network training, *Neurocomputing*, 34, pp 131–143
35. Oussar Y., Dreyfus G. [2001], How to be a gray box: dynamic semi-physical modeling, *Neural Networks*, vol. 14, pp 1161–1172
36. Ploix J.L., Dreyfus G. [1997], Early fault detection in a distillation column: an industrial application of knowledge-based neural modelling, *Neural Networks: Best Practice in Europe*, pp 21–31, World Scientific
37. Powell M.J.D. [1987], Radial basis functions for multivariable interpolation: a review, *Algorithms for approximation*, pp 143–167
38. Press W.H., Teukolsky S.A., Vetterling W.T., Flannery B.P. [1992], *Numerical recipes in C: the art of scientific computing*, Cambridge University Press
39. Price D., Knerr S., Personnaz L., Dreyfus G. [1994], Pairwise neural network classifiers with probabilistic outputs, *Neural Information Processing Systems*, 7, pp 1109–1116, Morgan Kaufmann
40. Price P.E., Wang S., Romdhane I.H. [1997], Extracting effective diffusion parameters from drying experiments. *AIChE Journal*, 43, pp 1925–1934
41. Rivals I., Canas D., Personnaz L., Dreyfus G. [1994], Modeling and control of mobile robots and intelligent vehicles by neural networks, *Proceedings of the IEEE Conference on Intelligent Vehicles*, pp 137–142
42. Rivals I. [1995], *Modélisation et commande de processus par réseaux de neurones: application au pilotage d'un véhicule autonome*, Thèse de Doctorat de l'Université Pierre et Marie Curie, Paris. Disponible sur le site *http://www.neurones.espci.fr.*
43. Roussel P., Moncet F., Barrieu B., Viola A. [2001], Modélisation d'un processus dynamique à l'aide de réseaux de neurones bouclés. Application à la

modélisation de la relation pluie-hauteur d'eau dans un réseau d'assainissement et à la détection de défaillances de capteurs, *Innovative technologies in urban drainage*, 1, 919–926, G.R.A.I.E

44. Seber G.A.F., Wild C.J. [1989], *Non-linear regression*, Wiley Series in Probability and Mathematical Statistics, John Wiley & Sons
45. Singhal A. [1996], Pivoted length normalization. *Proceedings of the 19th Annual International Conference on Research and Development in Information Retrieval (SIGIR'96)*, pp 21–29
46. Stoppiglia H. [1997], *Méthodes statistiques de sélection de modèles neuronaux; applications financières et bancaires*, Thèse de Doctorat de l'Université Pierre et Marie Curie, Paris. Disponible sur le site *http://www.neurones.espci.fr.*
47. Stricker M. [2000], *Réseaux de neurones pour le traitement automatique du langage: conception et réalisation de filtres d'informations*, Thèse de Doctorat de l'Université Pierre et Marie Curie, Paris. Disponible sur le site *http://www.neurones.espci.fr.*
48. Stricker M., Vichot F., Dreyfus G., Wolinski F. [2001], Training context-sensitive neural networks with few relevant examples for the TREC-9 routing, *Proceedings of the TREC-9 Conference.*
49. Vapnik V. [1995], *The nature of statistical learning theory*, Springer
50. Wolinski F., Vichot F., Stricker M. [2000], Using learning-based filters to detect rule-based filtering Obsolescence, *Conférence sur la Recherche d'Information Assistée par Ordinateur, RIAO'2000*, Paris
51. Zipf G. K. [1949], *Human Behavior and the Principle of Least Effort.* Addison-Wesley.

2

Modeling with Neural Networks: Principles and Model Design Methodology

G. Dreyfus

In the previous chapter, we showed that neural networks are nonlinear models, either static or dynamic, either "black-box" or "gray-box". The present chapter provides an in-depth treatment of the principles of modeling, together with a full model design methodology. For a new technology, the availability of a methodology is a proof of maturity, and it is a crucial asset for success in the development of applications.

2.1 What Is a Model?

A model is a representation of a part of the visible or observable world. In the present book, we consider only *mathematical* models, made of algebraic or differential equations that relate causes (called variables, factors, or model inputs) to effects (called quantities to be modeled, or quantities of interest, or model outputs); all these quantities are numbers. Symbolic or linguistic models, such as expert or fuzzy systems, will not be considered.

2.1.1 From Black-Box Models to Knowledge-Based Models

The black-box model is the most primitive form of a mathematical model: it is based only on observations; it may have some predictive value, but it does not provide any explanation. Thus, the Ptolemaic model of the universe was a black-box model: it did not provide any explanation of the motion of planets, but it did predict it, within the accuracy of experimental instruments that were available at that time.

By contrast, a knowledge-based model, or white-box model, results from an analysis of the physical, chemical, biological, etc., phenomena that generate the quantity to be modeled. Those phenomena are described by equations that depend on the theoretical knowledge that is available when the model is designed. Therefore, such a model has the abilities of predicting and of explaining. Scientific research strives to build knowledge-based models, whenever possible: the design of a knowledge-based model requires that a theory be available, whereas the design of a black-box model requires that measurements be available. Thus, Newton's theory of gravitation generated a knowledge-based model of the motion of celestial bodies.

Semiphysical or gray-box models stand between knowledge-based and black-box models: they embody both equations resulting from the application of a theory, and empirical results from a black-box model.

At present, most neural network applications are black-box models; therefore, the first part of the present chapter is devoted to black-box modeling. However, it will be shown that it can be very advantageous to use neural networks as semiphysical models.

2.1.2 Static vs. Dynamic Models

A static model is made of algebraic equations only (e.g., a feedforward neural network); by contrast, a dynamic model obeys differential (or partial differential) equations where time is the variable, and possibly algebraic equations as well. We will first consider the design of static models. The design principles for dynamic models (e.g., recurrent neural networks) will be explained next; Chaps. 4 and 5 deal in more detail with dynamic modeling and control.

2.1.3 How to Deal With Uncertainty? The Statistical Context of Modeling and Machine Learning

Before studying the design and implementation of a static black-box model, it may be useful, for the benefit of the reader who is not familiar with those techniques, to state the assumptions that underlie black-box modeling. Assume that the quantity of interest y_p is measurable, scalar[1], and that one knows, or suspects, that it depends, in some unknown, deterministic way, on one or several measurable quantities called factors that can be gathered into a vector $\boldsymbol{x}$ (which is a scalar if a single factor is involved in the modeling). In general, the measurable factors do not provide a complete description of the evolution of the quantity of interest: the latter is also subject to disturbances that are not measured (often not measurable). Two kinds of disturbances must be considered,

- deterministic disturbances: putting a cold dish into a temperature-regulated oven disturbs the temperature of the latter;
- noise: the noise inherent to the measurement of the quantity of interest y_p, for instance the noise of the sensor that measures the temperature of the oven, disturbs the measurement.

[1] The extension to the modeling of a vector does not involve any special difficulty.

Thus, if the same quantity is measured several times in conditions that are assumed to be identical, the results of the measurements are not identical. Black-box modeling aims at finding, from available measurements, a mathematical expression that provides an estimate of what the result of the measurement would be in the absence of disturbances, or, in other words, at finding a deterministic relation, if any, between the factors x and the quantity of interest y_p. Statistics provide the conceptual framework that is suitable for that task. Therefore, the chapter starts with the introduction of elementary vocabulary and concepts of statistics; some examples are developed in the additional material at the end of the chapter. The reader who has some familiarity with statistics may skip the next section.

2.2 Elementary Concepts and Vocabulary of Statistics

There are many classical textbooks in statistics (see or instance [Mood et al. 1974; Wonnacott et al. 1990]) to which the reader can refer for many more details and for the proofs of some results.

2.2.1 What is a Random Variable?

A random variable is a very convenient mathematical concept for dealing with quantities (such as results of measurements) whose values are uncertain: their values is considered as a realization of a random variable. The latter is fully defined by its probability density or distribution.

Denoting by $p_Y(y)$ the *probability distribution* function (pdf) of a random variable Y, the probability that the value of a realization of Y lie between y and $y + dy$ is equal to $p_Y(y)dy$.

Thus, modeling a measurable quantity y_p by a random variable $Y$is equivalent to assuming that the result of a measurement is the result of a random choice of a value y with a (generally unknown) probability distribution $p_Y(y)$. Modeling a quantity of interest by a random variable is definitely not equivalent to stating or assuming that the quantity of interest is not governed by a deterministic process: it is just a convenient mathematical trick for dealing with the fact that some factors that have an influence on the result of the measurement are not known, or are known but neither measured nor controlled (maybe because they are neither measurable nor controllable, such as wind gusts in the modeling of airplane flight).

Property. *The probability distribution function is the derivative of the cumulative distribution function:.* $p_Y(y) = (dF_Y(y))/(dy)$with $F_Y(y) =$ *Probability* $(Y \leq y)$

Because any realization y of the random variable Y lies between $-\infty$ and $+\infty$, one has: $\int_{-\infty}^{+\infty} p_Y(\mathrm{y})dy = 1$.

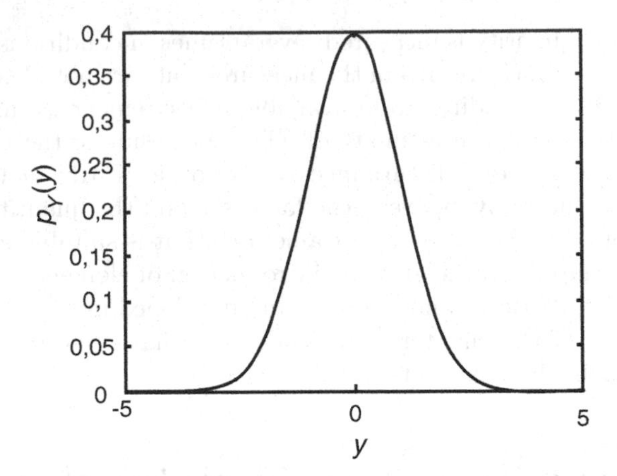

Fig. 2.1. Normal distribution

2.2.1.1 Examples of Probability Distributions

Uniform Distribution

A random variable Y has a uniform distribution if its density probability is $p_Y(y) = 1/(b-a)$ on a given interval $[a, b]$, and is zero elsewhere.

Gaussian Distribution

The Gaussian distribution $p_Y(y) = 1/(\sqrt{2\pi\sigma^2})\exp(-((y-\mu)^2)/(2\sigma^2))$ is very useful. μ is the mean of the Gaussian and σ (> 0) is its standard deviation. Figure 2.1 shows a *normal distribution*, with $\mu = 0$ and $\sigma = 1$.

Other Distributions

The Pearson (or χ^2) distribution, the Student distribution and the Fisher distribution are defined in the additional material at the end of the chapter.

2.2.1.2 Joint Distributions

Denoting by $p_{X,Y}(x, y)$ the joint density of two random variables, the probability that a realization of X lie between x and $x + \mathrm{d}x$ and that a realization of Y lie between y and $y + \mathrm{d}y$ is $p_{X,Y}(x, y)\mathrm{d}x\,\mathrm{d}y$.

Independent Random Variables

If two random variables X and Y are independent, one has:

$$p_{X,Y}(x, y) = p_X(x)p_Y(y).$$

2.2.2 Expectation Value of a Random Variable

The *expectation value of a random variable* Y is

$$E_Y = \int_{-\infty}^{+\infty} y p_Y(y) dy.$$

Therefore, the expectation value of a random variable is the first moment of its probability distribution .

2.2.2.1 Properties

- The expectation value of the sum of random variables is the sum of the expectation values of the random variables.
- If a variable Y is uniformly distributed in interval $[a, b]$, its expectation value is $(a+b)/2$.
- If a variable Y has a Gaussian distribution with mean μ, its expectation value is μ.

2.2.2.2 Example: Modeling the Result of a Measurement by a Random Variable

Assume that several measurements of the temperature of a fluid are performed, under conditions that are assumed to be identical, and that different results are obtained because of the intrinsic noise of the sensor and associated electronics, or because the conditions of the measurement are poorly controlled. Such a situation can be conveniently modeled by considering that the result T of the measurement is the sum of the true temperature T_0 (random variable with distribution $\delta(T_0)$) and of a random variable B with zero expectation value, $T = T_0 + B$. Then the expectation value of T is given by $E_T = T_0$ since the expectation value of B is equal to zero.

Clearly, the objective of performing a measurement of a quantity of interest, is to know its "true" value, i.e., within the above statistical framework, the expectation value of the quantity of interest. Therefore, the question that arises naturally is: how can one estimate that expectation value from the available measurements? To this end, the concept of *estimator* is useful.

2.2.3 Unbiased Estimator of a Parameter of a Distribution

An estimator is a random variable, which is a function of one or several measurable random variables.

An estimator H of a parameter of the distribution of an observable random variable G is said to be *unbiased* if its expectation value is equal to the parameter of interest. Then a realization of H is an unbiased estimate of the parameter of interest.

2.2.3.1 The Mean Is an Unbiased Estimator of the Expectation Value

Assume that N measurements of a quantity of interest G have been performed, under conditions that are assumed to be identical. The quantity of interest is modeled as a random variable whose expectation value γ is unknown. The result g_i of measurement i can be considered as a realization of the random variable G_i. If the experiment has been soundly designed, it can reasonably be assumed that the result of a given measurement is not affected by, and does not affect, other measurements: then the random variables G_i are mutually independent, and, since the measurements were performed in identical conditions, they have identical distributions, hence the same expectation γ.

Consider the random variable $M = (G_1 + G_2 + \cdots + G_N)/N$. Since the expectation of a sum of random variables is the sum of the expectations, one has $E_M = \gamma$: the expectation value of the random variable M ("mean") is equal to the expectation value of G, therefore the mean is an unbiased estimator of the expectation value. The quantity $m = (g_1 + g_2 + \cdots + g_N)/N$, which is a realization of the estimator of the expectation value of the random variable G, is an unbiased estimate of the latter.

Consider again the example of the estimation of the temperature of a fluid; we have shown

- that the expectation value of the variable T that models the temperature is equal to the "true" temperature T_0,
- that the mean is an unbiased estimator of the expectation value.

Therefore, if N temperature measurements are available, the mean of these measurements is an unbiased estimate of T_0.

However, the fact that the estimate is unbiased does not tell us anything about the accuracy of that result. If it is desirable, for instance, to know the temperature with 10% accuracy, does the estimate comply with that requirement? Clearly, the answer depends on the quality of the measurements, i.e., on the scattering of the measurements around the true value T_0. The concept of *variance* is useful in that context.

2.2.4 Variance of a Random Variable

The *variance of a random variable* Y with distribution $p_Y(y)$ is

$$\mathrm{var}_Y = \sigma^2 = \int_{-\infty}^{+\infty} [y - E_Y]^2 p(y) dy.$$

Hence, the variance is the centered second moment of the distribution.

2.2.4.1 Properties

- $\mathrm{var}_Y = E_{Y^2} - E_Y^2$.
- $\mathrm{var}_{aY} = a^2 \mathrm{var}_Y$.
- If a random variable is uniformly distributed on an interval $[a, b]$, its variance is $(b - a)^2/12$.
- If a random variable has a Gaussian distribution of standard deviation σ, its variance is σ^2.

2.2.4.2 Unbiased Estimator of the Variance of a Random Variable

In order to define the mean estimator M (unbiased estimator of the expectation value), we considered that N measurements of a quantity G were performed, and that the measurements were modeled as realizations of N independent identically distributed (i.i.d.) random variables G_i.

Unbiased Estimator of the Variance

The random variable

$$S^2 = \frac{1}{N-1} \sum_{i=1}^{N} (G_i - M)^2$$

is an unbiased estimator of the variance of G.

Therefore, if N measurement results g_i are available, the estimation of the variance requires

- first an estimation $m$of the mean, by relation $m = 1/(N) \sum_{i=1}^{N} g_i$,
- then an estimation of the variance by relation

$$s^2 = \frac{1}{N-1} \sum_{i=1}^{N} (g_i - m)^2.$$

Thus, the estimation of the variance provides a quantitative assessment of the scattering of the measurements around the mean. Since the mean itself is a random variable, it has a variance: the latter can be estimated by performing several sequences of measurements, under identical conditions, computing the mean of each sequence, then estimating the expectation value and the variance of the mean: this would provide an assessment of the scattering of the estimates of the temperature. However, this is indeed a heavy procedure, since it requires several sequences of measurements, in identical conditions.

2.2.5 Confidence Interval

The estimation of a confidence interval provides an elegant solution to the problem that has just been mentioned.

A *confidence interval*, with confidence threshold $1-\alpha$, for a random variable Y, is an interval that, with probability $1-\alpha$, contains the value of the expectation of Y.

Thus, instead of simply estimating the true value of the temperature by averaging the results of measurements performed presumably under identical conditions, one can estimate an interval within which the true value of the temperature is to be found, with probability $1-\alpha$. This is a much more useful and significant information: the smaller the confidence interval, the more confident one can be in the estimate of the quantity of interest.

The procedure for computing a confidence interval, and an example, are described in the additional material at the end of the chapter.

2.2.6 Hypothesis Testing

Hypothesis testing is a conventional statistical technique that aims at estimating whether a given hypothesis about a model is significantly in agreement, or in disagreement, with experimental data. In the field of modeling, the hypotheses that are tested are related to the model that is being designed.

A hypothesis, called "null hypothesis" H_0, and its complement H_1, are stated. A risk α of rejecting the null hypothesis H_0 although it is valid, is chosen. Then the design of a hypothesis test consists in

- finding a random variable, whose distribution is known if the null hypothesis is true, and a realization of which can be computed from the available experimental data;
- computing that realization.

If the probability of the latter lying in a given interval is too low given the distribution of the random variable, the null hypothesis has a low probability of being true, hence it is rejected. An example of hypothesis testing is provided in the additional material at the end of the chapter.

2.3 Static Black-Box Modeling

In the previous section, the basic elements of point estimation were explained: a measurable quantity was considered, and modeled as a random variable, its expectation value and variance were estimated, and a confidence interval was computed, from the available measurements performed under identical conditions. The process being the temperature of a fluid in an oven, it was assumed that all measurements were performed with a given heater intensity, a given external temperature, etc. Disturbances might be the intrinsic noise of the temperature measurement apparatus, variations of the external temperature, exo- or endo thermal reactions that may take place in the fluid. We did not try to model the relations between the measured temperature and the factors that may have an influence on the latter, since those factors were assumed to be constant.

The problem of modeling that is addressed in the present chapter is more complex. We want to find the mathematical relations between the quantity of interest and the factors that may have an influence on it. If such relations are available, then one can perform predictions about the evolution of the quantity of interest as a function of its factors: for instance, if a relation is found between the temperature of the oven and the intensity of the electrical current in the heating resistors, then one can predict the temperature that will be reached if a given intensity is flown into the resistors. One of the difficulties of modeling arises from the fact that all factors are not necessarily measured, and possibly are not measurable: therefore, the statistical framework is still appropriate, just as in the previous section.

2.3.1 Regression

Consider a measurable quantity y_p, which depends on a set of factors that are the components of a vector $\boldsymbol{x}$. As in the previous section, it is convenient to view the results of the measurements of y_p as realizations of a random variable Y, and to view the measured factors as realizations of a random vector[2] $\boldsymbol{X}$. Therefore, an estimate of the expectation value of the random variable $Y$for a given realization $\boldsymbol{x}$ of the random vector $\boldsymbol{X}$ is sought; it is denoted by $E_Y(\boldsymbol{x})$. That quantity is a function of $\boldsymbol{x}$, called regression function (or simply regression) of the random variable Y.

Since, as shown in the previous section, the expectation value is a quantity that can only be estimated, but cannot be known exactly, the regression function is also unknown and can only be estimated; some of its characteristics, such as the variance of Y for a given realization of $\boldsymbol{X}$, or a confidence interval on $Y$for a given realization of $\boldsymbol{X}$, can be estimated. Thus, the model that is sought is an estimation of the regression function; since neural networks with supervised training are nonlinear parsimonious approximators as shown in Chap. 1, they are good candidates as models of the quantity of interest if the regression function is nonlinear.

In order to estimate the regression function from measurements of the vector of factors (or input vector of the model, or variables of the model), one must first make an assumption as to the regression function: the simplest one is the linear (or affine) assumption: it is assumed that, in the domain of variation of the variables, a model that is linear or affine with respect to the latter can account satisfactorily for the behavior of the quantity of interest. If

[2] A random vector is a vector, the components of which are random variables; each component has its own probability distribution.

that assumption does not lead to a satisfactory model, then one must resort to a model that is nonlinear with respect to its variables, such as a polynomial, a neural network, etc.

Whatever the assumption on the mathematical form of the model, the problem of modeling, in the present context, is, by essence, the problem of estimating the parameters of the most satisfactory model, given the available data. How one can decide whether a model is satisfactory or not, is a major methodological problem that is considered in the present chapter.

2.3.2 Introduction to the Design Methodology

In all the following, a model whose vector of variables is $\boldsymbol{x}$ and whose vector of parameters is $\boldsymbol{w}$ will be denoted by $g(\boldsymbol{x}, \boldsymbol{w})$. If there exists a parameter vector $\boldsymbol{w}_p$ such that the model is identical to the regression function $g(\boldsymbol{x}, w_p) \equiv E_Y(\boldsymbol{x})$, then the family of functions $g(\boldsymbol{x}, \boldsymbol{w})$ contains the regression function, and the model $g(\boldsymbol{x}, \boldsymbol{w})$ is said to be "true". If such is not the case, then a model will be sought, that is as close as possible to the regression function $E_Y(\boldsymbol{x})$. *Training* is the algorithmic procedure whereby the parameters of such a model are sought, for a given family of functions (for instance, for the family of neural networks with three inputs and two hidden neurons).

Therefore, the design of a nonlinear black-box model requires the achievement of several tasks, including

- variable selection, i.e., the selection of the components of vector $\boldsymbol{x}$ in $g(\boldsymbol{x}, \boldsymbol{w})$; that task is carried out in two steps:
  1. the reduction of the dimension of the input vector;
  2. the selection of relevant variables, i.e., of the variables whose influence on the quantity to be modeled is larger than the influence of the disturbances;
- the estimation of the parameters $\boldsymbol{w}$ of the model $g(\boldsymbol{x}, \boldsymbol{w})$, i.e. the training of a model; this is also carried out in two steps:
  1. the choice of a family of functions within which the model is sought (for instance, the family of neural networks with three hidden neurons, the family of polynomials of degree 4, etc.);
  2. the training of one or several models within the chosen family;
- the selection of the best model and the estimation of its performances; if that best model is not satisfactory, another family of model is chosen (for instance, the number of hidden neurons is increased or decreased, the degree of the polynomial is increased or decreased, etc.), and the process is iterated to the second step of the previous task.

The latter step makes machine learning modeling different from conventional statistical modeling: in statistical modeling, the "best" model is the model whose parameters are estimated with the best accuracy. In machine learning, the best model is the model that generalizes best, the exact values of the parameters being of little or no interest.

The above three tasks are described in the next three sections, building up a complete design methodology that is essentially applicable to any nonlinear model, be it neural or otherwise.

2.4 Input Selection for a Static Black-Box Model

When a model is designed from measurements, the number of variables must be as small as possible, for each additional input generates additional parameters. In a neural model, each input gives rise to a number of parameters that is equal to the number of hidden neurons. Therefore, it is necessary

- to find an input representation that is as compact as possible,
- to select all relevant factors as inputs to the model, but only the relevant ones: the presence of input variables that are not relevant (i.e., whose contribution to the output is smaller than the contribution of disturbances) creates useless parameters and generates input variations that are not significant, hence will generate modeling errors.

Input selection has two different sides,

- reduction of the dimension of the representation space for the variables of the model,
- rejection of inputs that are not relevant.

2.4.1 Reduction of the Dimension of Representation Space

This first step of the input selection process considers only the inputs, irrespective of the quantity to be modeled; it aims at finding a data representation that is as compact as possible. Consider the example shown on Fig. 2.2: two data sets, corresponding to an input vector $\boldsymbol{x}$ of dimension 3, are displayed in that space; for the right-hand side data set, the points are essentially aligned, which means that the intrinsic data dimension is actually 1, instead of 3. Through an appropriate change of variables, after which all points are borne by a single axis, a one-dimensional representation of the data can be found. That change of variables can be obtained through *principal component analysis* (abbreviated as PCA, see for instance [Jollife 1986]). Similarly, for the second data set, each point can be described by its curvilinear abscissa on a curve: here again, the dimension of the representation can be reduced through an appropriate processing of the data, such as curvilinear component analysis or self-organizing maps [Kohonen 2001]. Those techniques are described in detail in Chaps. 3 and 7.

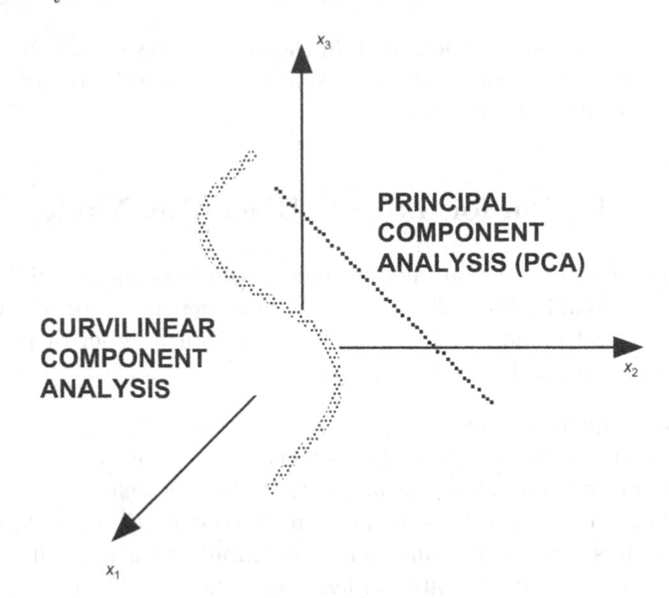

Fig. 2.2. Data dimensionality reduction

2.4.2 Choice of Relevant Variables

In the present chapter, the second task of input selection, i.e., the rejection of inputs whose influence on the output can be neglected, is described in more detail.

When modeling a physical or chemical process, the variables that have an influence on the quantity to be modeled are generally analyzed in detail, from first principles, by the experts; therefore, a systematic variable selection procedure is not necessary. By contrast, when modeling an economic, social, or financial process, or when modeling a very complex physical system, experts may give opinions about the relevant variables, but those are often more or less subjective, and need rigorous testing. Then the selection process starts with a large number of candidate variables, among which the factors that are really relevant should be selected. The results of the selection may disagree with current beliefs.

A large number of selection techniques were suggested (see for instance [McQuarrie et al. 1998], and, for a recent review, [Guyon et al. 2005]). The principles of the most popular technique are first described; then a technique that is intuitive and based on first principles is explained: the probe feature method.

2.4.2.1 Input Selection Strategies

The most natural strategy, for the choice of a set of inputs, consists in starting with an oversize set of candidate inputs (the model is said to be "complete"), in comparing the performance of the complete model with the performances of models whose inputs are subsets of the inputs of the complete model, and in choosing the best model with respect to an appropriate selection criterion. If q candidate variables are available, 2^q different combinations of inputs can be generated, hence at least 2^q models, whose performances should be compared: such an approach, whose complexity increases exponentially with the number of variables, is optimal but generally too demanding.

Two simpler, suboptimal strategies are used in practice:

- an elimination strategy (stepwise backward regression), whereby the less significant input is eliminated from the complete model: all submodels with $q-1$ inputs are compared, and the best of them (according to an appropriate criterion) is compared to the complete model. If the submodel is better than the complete model, that submodel is kept and the procedure is iterated; otherwise, the complete model is kept;
- a constructive strategy (stepwise forward regression), which starts with the simplest model, whose output is just the mean of the measured output values in the data set, hence is independent of the inputs: it is thus a model with zero variables; it is compared to the q models with 1 input; the best model is chosen, and the procedure is iterated until the addition of a new input no longer improves the quality of the model.

For both strategies, the maximum number of models is $1 + [q(q+1)/2]$: it grows as the square of the number of candidate variables, which is generally acceptable for practical purposes.

2.4.2.2 Comparison Criteria

The strategies described in the previous section rely on comparisons between models that have different numbers of inputs. Several comparison techniques may be used. We discuss two of them: hypothesis testing, and Akaike's information criterion.

Hypothesis Testing. Fisher's Test

The principle of hypothesis testing was discussed in a previous section. When comparing a submodel to the complete model in an elimination strategy, a model with q parameters is compared to a model with $q' < q$ parameters, which can be described as testing the null hypothesis "$q' - q$ parameters are equal to zero" to the alternative hypothesis. This can be done with Fisher's test, which is described in the additional material at the end of the chapter.

If the comparison to be performed is not between a complete model and a submodel, i.e., if the set of parameters of a model is not included in the set of parameters of the other, other tests may be used, such as the likelihood ratio test [Goodwin et al. 1977] and the LDRT test (logarithm determinant ratio test) [Leontaritis et al. 1987]. Those tests are asymptotically equivalent to Fisher's test [Söderström 1977].

Akaike's Information Criterion

In the above tests, the performance of the models is estimated through the mean square error on a set of examples. It may be desirable, for models that have similar performances, to take into account the complexity of the model, since the simplest models are generally preferable, as discussed in Chap. 1. Akaike's criterion [Akaike 1973, 1974; Norton 1986] is an example of such an approach. It consists in choosing the model for which the AIC (Akaike Information Criterion) is smallest,

$$\mathrm{AIC} = N \log(MSE) + 2(q+1),$$

where N is the number of examples, q is the number of variables of the model (linear with respect to the parameters), and where MSE is the mean square error on a data set. Thus, for a given performance as expressed by the mean square error, the most parsimonious models are favored.

A large number of variants of that criterion are discussed in [McQuarrie et al. 1998].

2.4.2.3 Variable Selection by the Probe Feature Method

The selection method that is described in the present section is intuitive, efficient, and based on simple principles [Stoppiglia et al. 2003]. It proceeds in two steps,

- ranking of the variables in order of decreasing relevance to the output,
- elimination of irrelevant variables.

We describe those two steps below.

Input Ranking through Gram-Schmidt Orthogonalization (Orthogonal Forward Regression)

In order to select the inputs of a neural model, it is convenient to perform input selection with a model that is linear with respect to its parameters (a polynomial model for instance), and to use the inputs thus selected as inputs of a neural network, because input selection is easier for a model that is linear with respect to its parameters.

Assume that p candidate variables (called primary variables) $x_i (i = 1$ to $p)$, are available, after discussions with the experts of the process to be modeled. If a nonlinear model is deemed necessary, one may consider, for instance, a polynomial model of degree 2; such a model is linear with respect to its parameters, its inputs being

- all combinations of 2 variables among the p candidate variables,
- the p candidate variables,
- a constant term.

Those inputs are the secondary variables ζ_i ($i = 1$ to $q = (p(p+1)/2) + p+1$); the set of secondary variables includes the primary variables. Then the model can be written as

$$g(\boldsymbol{\zeta}, \boldsymbol{w}) = \boldsymbol{\zeta}^{\mathrm{T}} \boldsymbol{w} = \sum_{i=1}^{q} w_i \zeta_i,$$

where $\boldsymbol{\zeta}^{\mathrm{T}}$ is the transpose of vector $\boldsymbol{\zeta}$ whose q components are the ζ_i (in the present chapter, the superscript T stands for the transposition of a vector of a matrix).

Assume that N measurements of each input are available, together with the corresponding measurements of the quantity to be modeled. We define the N-dimensional space (called *observation space*) in which each candidate variable is represented by a vector whose components are the N measured values of that input, and where, similarly, the process output is represented by the vector whose components are the measured values of the latter. We denote by $\boldsymbol{\xi}^i$ the vector whose components are the N values of the ith variable of the polynomial model, and by $\boldsymbol{y}_p$ the vector whose components are the N measured values of the quantity of interest. If the model is linear with respect to the parameters, the angle between the vector representing the ith variable and the vector representing the output decreases as the correlation between the ith variable and the output increases.

- If that angle is zero, i.e., if the output is proportional to variable i, the latter explains completely the output.
- If that angle is $\pi/2$, i.e., if the output is fully uncorrelated to variable i, the latter has no influence on the output.

Observation space is different from input space; the dimension of input space is equal to the number of variables of the model, whereas the dimension of observation space is equal to the number of measurements performed on the process prior to modeling.

In order to rank the inputs in order of decreasing relevance, it is not necessary to compute the angle θ_i between the vector that represents input i and the vector that represents the output $\boldsymbol{y}_p$: it is more convenient to compute the quantity $\cos^2 \theta_i = (((\xi^i)^{\mathrm{T}} \boldsymbol{y}_p)^2)/((\xi^i)^{\mathrm{T}} \xi^i)((\boldsymbol{y}_p)^{\mathrm{T}} \boldsymbol{y}_p)$.

In order to rank the inputs in order of decreasing relevance, the following orthogonalization procedure can be used [Chen 1989]:

- Choose the input that is most correlated to the output (with largest $\cos^2\theta$).
- Project the output vector and all other candidate inputs onto the null space of the selected input.
- Iterate in that subspace.

The procedure terminates when all candidate inputs are ranked, or when a maximal number of inputs are ranked (for models with many inputs, the full

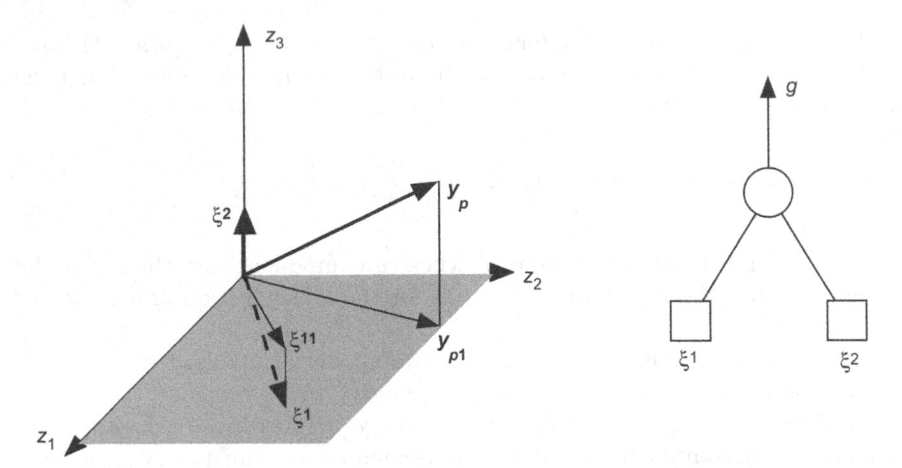

Fig. 2.3. Input orthogonalization by the Gram-Schmidt technique

ranking may be long or become numerically unstable for inputs that have very small correlations to the output).

The procedure is illustrated on Fig. 2.3, in a very simple case where three observations have been performed, for a model with two inputs (primary or secondary) $\boldsymbol{\xi}^1$ and $\boldsymbol{\xi}^2$: the three components of vector $\boldsymbol{\xi}^1$ are the three measured values of variable ζ_1 during the three observations.

Assume that vector $\boldsymbol{\xi}^2$ is the most correlated to vector $\boldsymbol{y}_p$. Therefore, $\boldsymbol{\xi}^2$ is selected, and $\boldsymbol{\xi}^1$ and $\boldsymbol{y}_p$ are orthogonalized with respect to $\boldsymbol{\xi}^2$, which yields vectors $\boldsymbol{\xi}^{11}$ and $\boldsymbol{y}_{p1}$. If additional candidate inputs were present, the procedure would be iterated in that new subspace until completion of the procedure. The orthogonalization can be advantageously performed with the modified Gram-Schmidt algorithm, as described for instance in [Björck 1967].

Input Selection in the Ranked List

Once the inputs (also called variables, or features) are ranked, selection must take place. This is important, since keeping irrelevant variables is likely to be detrimental to the performance of the model, and deleting relevant variables may be just as bad.

The principle of the procedure is simple: a random variable, called "probe feature" is appended to the list of candidate variables; that variable is ranked just as the others, and the candidate variables that are less relevant than the probe feature are discarded.

If the model were perfect, i.e., if an infinite number of measurements were available, that input would have no influence on the model, i.e., training would assign parameters equal to zero to that input. Since the amount of data is finite, such is not the case.

Of course, the rank of the random feature itself is a random variable. The decision thus taken must be considered in a statistical framework: there exists a nonzero *risk* of keeping an irrelevant variable, or of discarding a relevant variable. Therefore, the following procedure is used:

- Orthogonalize the output and the inputs with respect to the $m-1$ inputs selected during the previous $m-1$ iterations.
- In the subspace of dimension $q-m$, select the input that is most correlated to the projected output.
- Compute the probability that the rank of the probe feature be lower than or equal to the rank of the feature under examination, i.e., the probability that the probe feature be more relevant than the input under consideration. The computation of that quantity is explained in the additional material at the end of the chapter.
- If that probability is lower than the risk, chosen by the designer, that a variable be kept although it is less relevant than the probe feature, keep the feature under consideration and iterate the procedure; otherwise, discard the feature and terminate the procedure.

Example 1

In order to illustrate that input selection method, we consider a simulated process, described in [Lagarde 1983] and also investigated in [Stoppiglia 1998][3], [Stoppiglia et al. 2003]. Ten variables are candidate inputs, five of which only are relevant.

Figure. 2.4 shows the cumulative distribution function of the rank of the probe feature. It shows that if the five most relevant inputs are selected, the probability that the rank of the probe feature be smaller than or equal to 5 (i.e., that one of the 5 selected inputs be less relevant than the probe feature) is smaller than 10%. If 6 inputs are selected, the probability is larger than 10%. Therefore, if the designer is willing to accept a risk of 10%, then the first 5 inputs should be selected: that is exactly the number of relevant inputs. If the designer is willing to accept a higher risk of keeping an irrelevant input, 20% for instance, then the graph shows that the first 6 features should be kept. Thus, as in any statistical method, a tradeoff must be performed between the risk of designing an oversize model and the risk of designing too small a model.

Example 2

In a classification problem, synthetic data in which 2 variables only, out of 240 candidate variables, were relevant [Stoppiglia et al. 2003], and the other 238 variables were just random. The probe feature method was tested on 100 different such databases: it discovered at least 1 true variable in all cases, and discovered both true features in 74% of the cases. A hypothesis test showed that, when only one true variable is found, the classification performances of the model were not significantly different from the performances of models

[3] That thesis is available from URL http://www.neurones.espci.fr.

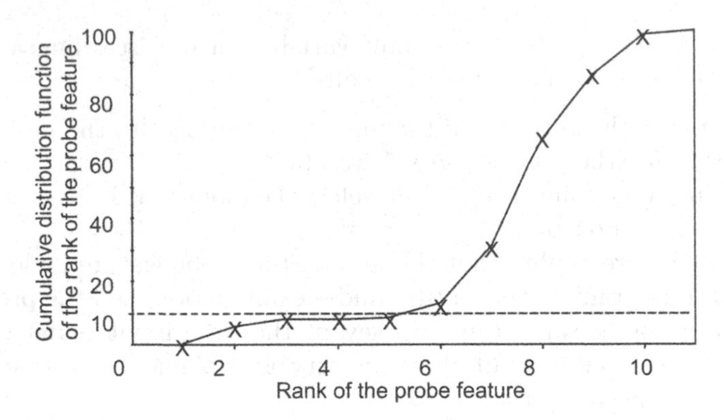

Fig. 2.4. Cumulative distribution function of the probe feature

having both true variables: the second variable happens to be, by chance, just as relevant as one of the true variables.

Example 3

In a classification problem, a data base was generated, containing 200 examples with 1,326 candidate variables, including 52 independent variables, among which 2 relevant variables were present. With a 1% risk, the probe feature method selected both relevant variables, and no other.

Once the inputs are selected, they can be used as inputs to a neural network.

That method is directly related to Fisher's test, which is discussed in the additional material at the end of the chapter.

2.4.2.4 Relation Between Fisher's Test and the Probe Feature Method

The interested reader will find in [Stoppiglia 1998; Stoppiglia et al. 2003] the proof of the following result: if the model under consideration at iteration k of the Gram-Schmidt orthogonalization procedure is complete, i.e., if it contains all relevant variables, and if it is *true,* i.e., if the regression function belongs to the family of functions within which the model is sought, then the selection procedure performed at iteration k is equivalent to a Fisher's test between the models examined at iterations k and $k-1$.

Therefore, the probe feature method has two advantages on Fisher's test: first, it gives a clear and intuitive interpretation to the selection criterion; second, it is applicable whether the complete model is available or not, and whether the model is true or not.

2.4.2.5 What to do in Practice?

Summary of the procedure for discarding irrelevant variables:

1. Choose the set of candidate inputs (primary and secondary variables).
2. Select the input that is most correlated to the output; in observation space, project all other inputs, and the output, onto the null subspace of the selected input.
3. In the null space of the $m - 1$ variables selected at previous iterations
   (a) Select the projected input vector that is most correlated to the projected output vector.
   (b) Compute the probability H_m for the probe feature to be more relevant than one of the m input selected previously, and compare it to the risk α chosen by the designer.
   (c) If H_m is smaller than the risk, project the projected output, and all remaining candidate inputs, onto the null space of the selected projected input and iterate to step 3.
   (d) If H_m is larger than the risk, proceed to step 4.
4. Use the selected variables as inputs of a neural network and train as indicated in the next sections.

2.4.3 Conclusion on Variable Selection

The first step in any model design procedure consists in reducing the dimension of the input space, by asking two questions.

- Is the intrinsic dimension of the input vector as small as possible, or is it possible to find a more compact input representation, while preserving the amount of relevant information?
- Are all candidate inputs relevant to the modeling of the quantity of interest?

The answer to the first question is provided by principle component analysis, or possibly by more complex operations such as curvilinear component analysis or self-organizing maps.

The answer to the second question is provided by statistical methods such as the probe feature method.

After performing input selection, the parameters of the model are estimated as discussed in the next section.

2.5 Estimation of the Parameters (Training) of a Static Model

We now turn to the problem of estimating the parameter of a model $g(\boldsymbol{x}, \boldsymbol{w})$: find the numerical values of the components of the parameter vector $\boldsymbol{w}$ that make the model satisfactory, with respect to a criterion that will be discussed below.

The basic principles of parameter estimation are the following:

- A set of N measurements $\{y_p^k\}$ ($k = 1$ to N) of the quantity to be measured is available, which corresponds to N values of the inputs $\{\boldsymbol{x}^k\} = \{[x_1^k, \ldots, x_q^k]\}$ ($k = 1$ to N). That set of observations is called training set.

- Because the training set is of finite size, the exact regression function cannot be derived; therefore, an approximation of the regression function is sought, within a family of functions that are deemed complex enough to account for the complexity of the data. The most reasonable approach consists in first trying to find an approximation of the regression function in the family of linear or affine functions (i.e., perform *linear regression*). In that case, the model is sought under the form $g(\boldsymbol{x}, \boldsymbol{w}) = \boldsymbol{x}^{\mathrm{T}}\boldsymbol{w} = \sum_{i=1}^{q} w_i x_i$; if the result of that model is not satisfactory, an approximation of the regression function must be sought in a more complex family of functions, either linear with respect to the parameters (polynomials, Gaussians with fixed centers and covariances, wavelets with fixed centers and dilations), or nonlinear with respect to the parameters (neural networks, Gaussians with adjustable centers and covariance matrices, etc.). If necessary, the complexity of the family of models is increased step by step, by increasing the degree of the polynomial, the number of Gaussians, the number of hidden neurons, etc.
- For a given family of functions, the values of the parameters $\boldsymbol{w}$ must be computed; this is done by minimizing a cost function that pictures the "distance" between the predictions of the model and the measured values. For each observation k of the training set, the residual is defined as $r_k = y_p^k - g(\boldsymbol{x}^k, \boldsymbol{w})$, where y_p^k is the kth measured value of the process output, and where $\boldsymbol{x}^k$ is the kth measured value of the input vector. The *least squares cost function*, as defined in Chap. 1, is the sum of the squared residuals of all observations of the training set: $J(\boldsymbol{w}) = \sum_{k=1}^{N}(y_p^k - g(\boldsymbol{x}^k, \boldsymbol{w}))^2 = r^{\mathrm{T}}r$, where $\boldsymbol{r}$ is the vector of residuals, of dimension N, whose components are the residuals r_k. If the modeling were perfect, the residual vector would be equal to zero, which is the absolute minimum of the cost function. However, since measurements have noise, it is not desirable to find a model that is so complex that the minimum of the cost function would be equal to zero: such a model would reproduce the noise, in addition to reproducing the deterministic behavior of the process, whereas the purpose of modeling is to find a model that captures the deterministic part of the process and filters out the noise. Since there is no point in finding a model whose predictions would be more accurate than the measurements from which it is designed, the model designer will not try to find a model with zero cost function, nor even the absolute minimum of the cost function in a given family of models: a model will be sought,

whose prediction error is on the order of the accuracy of the measurements. That crucial problem is discussed below, in the section devoted to model selection.

Empirical vs. Theoretical Cost Functions

The cost function $J(\boldsymbol{w})$ is sometimes called *empirical cost function*, as opposed to the *theoretical cost function* $\int (y_p(\boldsymbol{x}) - g(\boldsymbol{x}, \boldsymbol{w}))^2 p(\boldsymbol{x}) d\boldsymbol{x}$; the latter is the quantity that one would actually like to minimize, but it can obviously not be computed.

Global Minima and Local Minima

If the model is linear with respect to its parameters, the least squares cost function is quadratic with respect to them. If the model is not linear with respect to the parameters (e.g., a neural network), then the least squares cost function has several minima, one of which must be selected. This makes the model selection problem somewhat more complicated than in the case of models that are linear with respect to the parameters: that is the price to be paid for taking advantage of parsimony, which is an asset of models that are not linear with respect to their parameters.

The methods that can be used for minimizing the cost function fall into two categories:

- nonadaptive training, also called *batch training* or *off-line training*, whereby the cost function that is minimized takes into account all elements of the training set (as is the case for the least squares cost function defined above); such methods require that all elements of the training set be available when training starts;
- adaptive training, also called *on-line training*, whereby the parameters of the model are updated sequentially as a function of a *partial cost* related to each example k[4]: $J^k(\boldsymbol{w}) = (y_p^k - g(\boldsymbol{x}^k, \boldsymbol{w}))^2$. Such techniques are useful when new examples become available while training is already taking place.

Adaptive training can be performed even if all examples are available before training starts, whereas a nonadaptive technique cannot be used if all examples are not available. In practice, the following strategy is frequently used: the model is first trained nonadaptively, then it is updated by adaptive training during its operation, for instance to adapt the model to slow drifts of the parameters of the process (due to wear, ageing, etc.).

In the following, the training of models that are linear with respect to their parameters–the popular least squares method–will first be outlined. Then the training (nonadaptive and adaptive) of models that are nonlinear with respect

[4] The least squares cost function will also be called total cost, as opposed to the partial cost.

to their parameters, such as neural networks, will be discussed. Finally, regularization techniques, which aim at avoiding overfitting when training with a small number of examples, will be discussed.

2.5.1 Training Models that are Linear with Respect to Their Parameters: The Least Squares Method for Linear Regression

We assume that the measurements of the quantity to be modeled can be viewed as realizations of a random variable Y_p that is an affine function of variables which have been selected in an earlier step: $Y_p = \boldsymbol{\zeta}^{\mathrm{T}} \boldsymbol{w}_p + \boldsymbol{B}$, where $\boldsymbol{\zeta}$ is the vector of the variables of the model, of known dimension q,where $\boldsymbol{w}_p$ is the vector (non random but unknown) of the parameters of the model, and where $\mathbf{B}$ is a random vector whose expectation value is zero. Therefore, the regression function is linear with respect to the variables of the model

$$E(Y_p) = \boldsymbol{\zeta}^{\mathrm{T}} \boldsymbol{w}_p.$$

We want to design a model $g(\boldsymbol{\zeta}, \boldsymbol{w}) = \boldsymbol{\zeta}^{\mathrm{T}} \boldsymbol{w}$, given a set of Nmeasurements of the quantity of interest $\{y_p^k, k = 1 \text{ to } N\}$ that are a set of realizations of the random variable Y_p, and given a set of corresponding measurements of the inputs $\{\zeta^k, k = 1 \text{ to } N\}$.

2.5.1.1 Nonadaptive (Batch) Training of Models that are Linear with Respect to Their Parameters

Because there is a wealth of textbooks on the subject (see for instance [Seber 1977; Antoniadis et al. 1992; Draper et al. 1998], no proof will be given in the present section.

Minimizing the Least Squares Cost Function. The Normal Equations

The minimum of the following cost function is sought

$$J(\boldsymbol{w}) = \sum_{k=1}^{N} \left(y_P^k - g(\zeta^k, \boldsymbol{w}) \right)^2 ,$$

with $g(\boldsymbol{\zeta}, \boldsymbol{w}) = \boldsymbol{\zeta}^{\mathrm{T}} \boldsymbol{w}$. In such a model, the number of parameters q is equal to the number of inputs n.

The matrix of observations is the matrix $\boldsymbol{\Xi}$ whose column i ($i = 1$ to q) is the vector $\boldsymbol{\xi}^i$ whose components are the Nmeasurements of the ith input: therefore, it has N rows and q columns,

$$\Xi = \begin{bmatrix} \zeta_1^1 & \cdots & \zeta_n^1 \\ \cdots & \cdots & \cdots \\ \cdots & \cdots & \cdots \\ \cdots & \cdots & \cdots \\ \zeta_1^N & \cdots & \zeta_n^N \end{bmatrix} = \begin{bmatrix} (\zeta^1)^{\mathrm{T}} \\ \cdots \\ \cdots \\ \cdots \\ (\zeta^N)^{\mathrm{T}} \end{bmatrix} = \left[(\zeta^1) \ldots (\zeta^n) \right],$$

Therefore, the model can be written as $\boldsymbol{g} = \boldsymbol{\Xi w}$, and the least squares cost function becomes

$$J(\boldsymbol{w}) = \sum_{k=1}^{N} \left(y_P^k - g(\boldsymbol{\zeta}^k, \boldsymbol{w})\right)^2 = \| y_p - \boldsymbol{\Xi w} \|^2 = (y_p - \boldsymbol{\Xi w})^{\mathrm{T}}(y_p - \boldsymbol{\Xi w}).$$

In order to find the vector of parameters for which that function is minimum, one just has to write that the gradient of the cost function with respect to the parameters is equal to zero, and to solve the system of equations thus obtained. Since the cost function is quadratic with respect to the parameters, the gradient is linear with respect to the parameters. Therefore, the system of equations (called normal equations) is linear; its solution $\boldsymbol{w}_{\mathrm{LS}}$ is the least squares estimate of the parameters of the model,

$$\boldsymbol{\Xi}^{\mathrm{T}}\boldsymbol{\Xi}\boldsymbol{w}_{\mathrm{LS}} = \boldsymbol{\Xi}^{\mathrm{T}} y_p.$$

If the number of examples N is much larger than the number of inputs q, matrix $\boldsymbol{\Xi}$ is generally of rank q (i.e., q rows of $\boldsymbol{\Xi}$ are linearly independent). If Ξ has rank q, then it can be proved that $[\boldsymbol{\Xi}^{\mathrm{T}}\ \boldsymbol{\Xi}]$ also has rank q, hence is invertible. In that case, the unique *least squares solution* is readily obtained as

$$\boldsymbol{w}_{\mathrm{LS}} = (\boldsymbol{\Xi}^{\mathrm{T}}\boldsymbol{\Xi})^{-1}\boldsymbol{\Xi}^{\mathrm{T}} y_p.$$

By contrast if the number of experiments is too low ($N < q$), matrix Ξ may be of rank smaller than q, so that the problem has an infinite umber of solutions.

For an input vector $\boldsymbol{\zeta}$, the prediction of the model is given by $g(\boldsymbol{\zeta}, \boldsymbol{w}_{\mathrm{LS}}) = \boldsymbol{\zeta}^{\mathrm{T}}\boldsymbol{w}_{\mathrm{LS}}$. The vector of the predictions of the model related to the training examples is $\boldsymbol{g}(\boldsymbol{\zeta}, \boldsymbol{w}_{\mathrm{LS}}) = \boldsymbol{\Xi}\boldsymbol{w}_{\mathrm{LS}}$, and the vector of residuals (modeling errors on the training examples) is thus

$$r = y_p - \boldsymbol{\Xi}\boldsymbol{w}_{\mathrm{LS}}.$$

Example

The following is a very simple didactic example: a linear model must be deigned, with a single variable x (hence two inputs: the variable x and a constant input, equal to 1), from three observations. The three measured values of variable x are denoted by $\{x_1, x_2, x_3\}$, and the measured values of the quantity to be modeled by $\{y_p^1, y_p^2, y_p^3\}$. Thus, with the above notations, the input vector is $\boldsymbol{\zeta} = \binom{1}{x}$. The output vector is

$$y_p = \begin{pmatrix} y_p^1 \\ y_p^2 \\ y_p^3 \end{pmatrix}.$$

The vector of parameters is $\boldsymbol{w} = \binom{w_1}{w_2}$.

The model is of the form $g(\boldsymbol{x}, \boldsymbol{w}) = \boldsymbol{\zeta}^{\mathrm{T}}\boldsymbol{w} = w_1 + w_2 x$. The observation matrix is

$$\boldsymbol{\Xi} = \begin{pmatrix} 1 & x_1 \\ 1 & x_2 \\ 1 & x_3 \end{pmatrix}.$$

The least squares solution is given by relation

$$\begin{pmatrix} w_{mc1} \\ w_{mc2} \end{pmatrix} = \left[\begin{pmatrix} 1 & 1 & 1 \\ x_1 & x_2 & x_3 \end{pmatrix} \begin{pmatrix} 1 & x_1 \\ 1 & x_2 \\ 1 & x_3 \end{pmatrix} \right]^{-1} \begin{pmatrix} 1 & 1 & 1 \\ x_1 & x_2 & x_3 \end{pmatrix} \begin{pmatrix} y_p^1 \\ y_p^2 \\ y_p^3. \end{pmatrix}.$$

Clearly, the number of available observations is much too small for a reliable estimation of the two parameters of the model; this is just a didactic example, for which geometrical illustrations are feasible.

Geometrical Interpretation

The least squares method has a simple geometrical interpretation, which is sometimes useful for a better understanding of the results.

We have seen that the vector of the predictions of the model on the training set can be written as

$$g(\boldsymbol{\zeta}, \boldsymbol{w}_{mc}) = \boldsymbol{\Xi}\boldsymbol{w}_{mc} = \boldsymbol{\Xi}(\boldsymbol{\Xi}^{\mathrm{T}}\boldsymbol{\Xi})^{-1}\boldsymbol{\Xi}^{\mathrm{T}}y_p.$$

In observation space (whose dimension is equal to the number of observations available for training), matrix $\boldsymbol{\Xi}(\boldsymbol{\Xi}^{\mathrm{T}}\boldsymbol{\Xi})^{-1}\boldsymbol{\Xi}^{\mathrm{T}}$ is the orthogonal projection matrix onto the subspace spanned by the columns of matrix $\boldsymbol{\Xi}$ (called *solution subspace*): thus, the prediction of the model, for a training example, is the orthogonal projection of the process output onto the solution subspace, as shown on Fig. 2.5. Note that, among all vectors of solution subspace, the orthogonal projection of the process output vector is the closest vector to the process output vector itself: hence, the model obtained by the least squares solution provides the prediction vector that is closest to the actual output vector, given the available data.

As an illustration, consider the previous example of a model with one variable and three observations. The observation space is of dimension 3, and the subspace spanned by the columns of the observation matrix is of dimension $q = 2$. Figure 2.6 shows the three-dimensional observation space, and the two-dimensional solution subspace spanned by the vectors

$$\zeta^1 = \begin{pmatrix} 1 \\ 1 \\ 1 \end{pmatrix} \quad \text{and} \quad \zeta^2 = \begin{pmatrix} x_1 \\ x_2 \\ x_3 \end{pmatrix}.$$

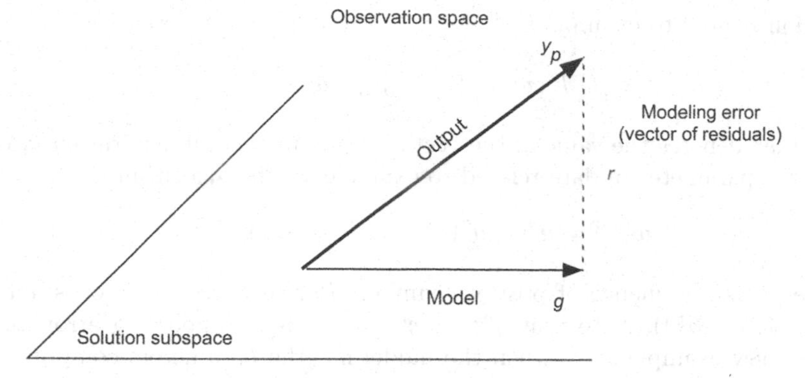

Fig. 2.5. Geometrical interpretation of the least squares method

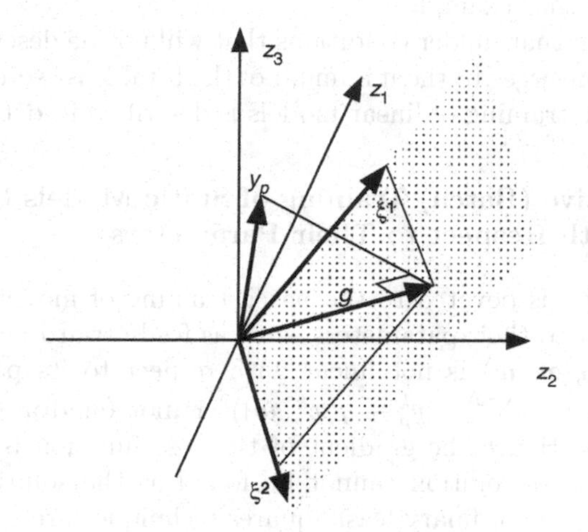

Fig. 2.6. Geometrical interpretation of the least squares method: a 3-dimensional example

2.5.1.2 Adaptive (On-Line) Training of Models that are Linear with Respect to Their Parameters: The Least Mean Squares Algorithm

In adaptive training, the parameters of the model are updated as a function of each example taken separately; this is especially useful for adaptive filtering or adaptive control, where the model must be adapted to the evolution of the process to be modeled. The *recursive least squares* algorithms find adaptively the least squares solution, for a model that is linear with respect to its parameters [Ljung 1987; Haykin 1994].

Among recursive least squares algorithms, the least mean squares (LMS) algorithm (widely used in linear adaptive filtering), also called Widrow-Hoff algorithm [Widrow 1960]) is also used for training neural networks adaptively. It updates the parameters as a function of the gradient of the partial cost function related to example k,

$$J^k(\boldsymbol{w}) = \left(y_p^k - g(\zeta^k, \boldsymbol{w})\right)^2,$$

where $\boldsymbol{w}^k$ denotes the value of the vector of parameters after iteration k, i.e., after the parameter update related to example k. The algorithm is

$$\boldsymbol{w}^{k+1} = \boldsymbol{w}^k + \mu^k(y_p^k - (\boldsymbol{x}^{k+1})^{\mathrm{T}}\boldsymbol{w}^k)\boldsymbol{x}^{k+1},$$

where μ^k is a sequence of positive numbers (for instance μ^k = constant or $\mu^k = 1/(\alpha + \beta k)$). Note that $y^{pk} - (\boldsymbol{x}^{k+1})^{\mathrm{T}}\boldsymbol{w}^k$ is the modeling error made on the new example x^{k+1} when the model has the parameters computed at iteration k. Hence, for each example, the weight update is proportional to the modeling error on that example.

It can be shown that, under conditions that will not be described here, the LMS algorithm converges to the minimum of the total least squares cost function. The adaptive training of linear models is described in detail in Chap. 4.

2.5.2 Nonadaptive (Batch) Training of Static Models that Are Not Linear with Respect to Their Parameters

The present section is devoted to the batch training of models that are not linear with respect to their parameters, such as feedforward neural networks. Since the model $g(\boldsymbol{x}, \boldsymbol{w})$ is not linear with respect to its parameters, the cost function $J(\boldsymbol{w}) = \sum_{k=1}^{N}(y_p^k - g(\boldsymbol{x}^k, \boldsymbol{w}))^2$ is not quadratic with respect to the parameters. Hence the gradient of the cost function is not linear, so that the least squares solution cannot be found as the solution of a linear system. Therefore, the ordinary least squares techniques are useless, and one has to resort to more elaborate minimization techniques, which update the parameters iteratively as a function of the gradient of the cost function with respect to the parameters.

Just as for linear models, training can be performed either adaptively or nonadaptively. Therefore, each training iteration (or *epoch*) requires two ingredients

- the computation of the gradient of the cost function,
- the updating of the parameters as a function of that gradient, in order to get closer to a minimum of the cost function.

Those two points are discussed in the following. As a preliminary, however, we consider the normalization of the inputs.

2.5.2.1 Input Normalization

Prior to training, the input variables must be normalized and centered: if the inputs have very different orders of magnitude, the smallest ones will not be taken into account during training. Therefore, for each input vector ζ_i, the mean μ_i and the standard deviation σ_i of its components must be computed, and the new variables $\zeta'_i = (\zeta_i - \mu_i)/\sigma_i$ (or any similar change) must be computed: hence the new variables are centered, and their standard deviation is on the order of 1.

It is also advisable especially for training dynamic models (recurrent neural networks), to center and normalize the outputs in a similar fashion.

2.5.2.2 Computation of the Gradient of the Cost Function

When the model is a feedforward neural network, the gradient of the cost function can be computed economically with an algorithm called backpropagation algorithm [Rumelhart et al. 1986; Werbos 1974] that has gained such popularity that it is sometimes considered as a training algorithm. Actually, backpropagation is not a training algorithm, but an ingredient in a training procedure. Furthermore, it will be shown that training can be performed without using backpropagation.

Phrases such as backpropagation neural network (or backprop net) are too often used as an equivalent to feedforward neural network. They are meaningless for two reasons. First, computing the gradient of the cost function without using backpropagation is perfectly feasible, and sometimes mandatory (see section "forward computation of the gradient of the cost function"); second, backpropagation is also useful for training recurrent networks. Thus, there is no relation whatsoever between the architecture of the network (feedforward or recurrent) and the computation of the gradient of the cost function by backpropagation.

Computation of the Gradient of the Cost Function by Backpropagation

We consider a feedforward neural network with hidden neurons and a single output neuron (the extension to neural networks with several output neurons is straightforward).Neuron i computes its output y_i, which is a nonlinear function of its potential v_i; v_i is the weighted sum of the inputs x_j, in which the value of input x_j is weighted by the parameter w_{ij},

$$y_i = f\left(\sum_{j=1}^{n_i} w_{ij} x_{ij}\right) = f(\nu_i).$$

The n_i inputs of neuron i may be either the outputs of other neurons, or inputs of the network. Therefore, in all the following, x_j will denote either the output y_j of neuron j or the input j of the network.

The cost function whose gradient must be computed is of the form

$$J(\boldsymbol{w}) = \sum_{k=1}^{N} \left(y_p^k - g(\boldsymbol{x}^k, \boldsymbol{w})\right)^2 = \sum_{k=1}^{N} J^k(\boldsymbol{w}).$$

In order to compute its gradient, one can compute the gradient of the partial cost function $J^k(\boldsymbol{w})$ related to observation k, and subsequently sum over all examples.

Backpropagation consists essentially in a repeated application of the rule of chained derivatives. First, one notices that the partial cost function depends of w_{ij} only through the value of the output of neuron i, which itself is a function of the potential of neuron i only; therefore, one has

$$\left(\frac{\partial J^k}{\partial w_{ij}}\right)_k = \left(\frac{\partial J^k}{\partial v_i}\right)_k \left(\frac{\partial v_i}{\partial w_{ij}}\right)_k = \delta_i^k x_j^k.$$

where

- $(\partial J^k)/(\partial vi)_k$ is the value of the gradient of the partial cost function with respect to the potential of neuron i when the inputs of the network are the variables of example k.
- $(\partial v_i)/(\partial w_{ij})_k$ is the value of the partial derivative of the potential of neuron i with respect to parameter w_{ij} when the inputs of the network are the variables of example k.
- x_j^k is the value of input j of neuron i when the inputs of the network are the variables of example k.

The computation of the last two quantities is straightforward. The only problem is the computation of δ_i^k on the right-hand side of the equation. These quantities can be advantageously computed recursively from the outputs to the inputs, as follows.

- For output neuron i,

$$\delta_i^k = \left(\frac{\partial J^k}{\partial v_i}\right)_k = \left(\frac{\partial}{\partial v_i}\left[(y_p^k - g(\boldsymbol{x}, \boldsymbol{w}))^2\right]\right)_k = -2g(\boldsymbol{x}^k, \boldsymbol{w})\left(\frac{\partial g(\boldsymbol{x}, \boldsymbol{w})}{\partial v_i}\right)_k.$$

  The output $g(\boldsymbol{x}, \boldsymbol{w})$ of the model is the output y_i of the output neuron; therefore the above relation can be written as $\delta_i^k = -2g(\boldsymbol{x}^k, \boldsymbol{w})f'(v_i^k)$ where $f'(v_i^k)$ is the derivative of the activation function of the output neuron when the network inputs are those of example k. Usually, for a feedforward neural network designed for modeling, the activation function of the output neuron is linear, so that the above relation reduces to $\delta_i^k = -2g(\boldsymbol{x}^k, \boldsymbol{w})$.
- For hidden neuron i, the cost function depends on the potential of neuron i only through the potentials of the neurons m that receive the value of the output of neuron i, i.e., of all neurons that are adjacent to neuron i in the graph of the connections of the network, and are located between that neuron and the output:

$$\delta_i^k \equiv \left(\frac{\partial J^k}{\partial v_i}\right)_k = \sum_m \left(\frac{\partial J^k}{\partial v_m}\right)_k \left(\frac{\partial v_m}{\partial v_i}\right)_k = \sum_m \delta_m^k \left(\frac{\partial v_m}{\partial v_i}\right)_k.$$

Furthermore, $v_m^k = \sum_i w_{mi} x_i^k = \sum_i w_{mi} f(v_i^k)$, therefore $(\partial v_m)/(\partial v_i)_k = w_{mi} f'(v_i^k)$.

Finally, one gets

$$\delta_i^k = \sum_m \delta_m^k w_{mi} f'(v_i^k) = f'(v_i^k) \sum_m \delta_m^k w_{mi}.$$

Thus, the quantities δ_i^k can be computed recursively from the outputs to the inputs of the network, hence the term *backpropagation.*

Once the gradients of the partial costs are computed, the gradient of the total cost function is obtained by a simple summation.

Summary of Backpropagation

For each example k, the backpropagation algorithm for computing the gradient of the cost function requires two steps,

- A propagation phase, where the inputs corresponding to example k are input to the network, and the potentials and outputs of all neurons are computed,
- A backpropagation phase, where all quantities δ_i^k are computed.

When those quantities are available, the gradients of the partial cost functions are computed as $(\partial J^k)/(\partial w_{ij})_k = \delta_i^k x_j^k$, and the gradient of the total cost function as $(\partial J)/(\partial w_{ij})_k = \sum_k (\partial J^k)/(\partial w_{ij})_k$.

The backpropagation algorithm can be interpreted graphically by defining the adjoint network of the network whose parameters must be estimated. This approach is sometimes useful; it is discussed in Chap. 4 for the modeling of dynamic systems.

Backpropagation was discussed here in the framework of the minimization of the least squares cost function. It can be adapted to the minimization of alternative cost functions, such as the cross entropy cost function, used for classification.

Forward Computation of the Gradient of the Cost Function

One of the most persistent myths in the field of neural networks is the following: the invention of backpropagation made the development of neural networks possible. Actually, it is definitely possible, albeit more computationally demanding, to compute the gradient of the cost function in the forward direction. That algorithm was extensively used for the estimation of the parameters of cascaded filters, long before backpropagation.

The forward algorithm proceeds as follows:

- For a neuron m, which receives the quantity x_j^k directly from input j of the network or from neuron j,

$$(\partial y_m/\partial w_{mj})_k = (\partial y_m/\partial v_m)_k(\partial v_m/\partial w_{mj})_k = f'(v_j^k)x_j^k,$$

where x_j^k is the value of input j of the network for example k,

- For a neuron m, which receives quantity x_j^k from input j of the network, or from neuron j, through other neurons of the network, located between input or neuron j and neuron m,

$$\left(\frac{\partial y_m}{\partial w_{ij}}\right)_k = \left(\frac{\partial y_m}{\partial v_m}\right)_k \left(\frac{\partial v_m}{\partial w_{ij}}\right)_k = f'(v_m^k)\sum_l \left(\frac{\partial v_m}{\partial y_l}\right)_k \left(\frac{\partial y_l}{\partial w_{ij}}\right)_k$$
$$= f'(v_m^k)\sum_I w_{ml}\left(\frac{\partial y_l}{\partial w_{ij}}\right)_k,$$

  where subscript l denotes all neurons that are adjacent to neuron m in the graph of connections, between neuron j (or input j) and neuron m.

By using those relations recursively, the derivatives of the output of each neuron with respect to the parameters can be computed, from the inputs to the outputs of the network.

Once those derivatives are computed, the gradient of the partial cost function can be derived as

$$\left(\frac{\partial J_k}{\partial w_{ij}}\right)_k = \left(\frac{\partial}{\partial w_{ij}}\left[(y_p^k - g(\boldsymbol{x},\boldsymbol{w}))^2\right]\right)_k = 2(y_p^k - g(\boldsymbol{x}^k,\boldsymbol{w}))\left(\frac{\partial g(\boldsymbol{x},\boldsymbol{w})}{\partial w_{ij}}\right)_k.$$

Furthermore, $g(\boldsymbol{x},\boldsymbol{w})$ is the output of a neuron of the network; therefore, the last derivative can be computed recursively by the same procedure. The gradient of the partial cost being computed for each example, the gradient of the total cost function is obtained by summation over all examples.

Comparison Between Forward Computation of the Gradient of the Cost Function and Backpropagation

The above discussion shows that backpropagation requires the evaluation of one gradient per neuron, whereas the forward computation requires the computation of one gradient per connection. Since the number of connections is roughly the square of the number of neurons, the number of gradient evaluations is larger for forward computation of the gradient than for backpropagation.

Therefore, backpropagation will be used for the evaluation of the gradient of the cost function in the training of feedforward neural networks. For recurrent neural networks, however, forward computation is sometimes mandatory, as shown in the section devoted to the training of recurrent neural networks.

Evaluation of the Gradient of the Cost Function under Constraint: The Shared Weight Technique

When training recurrent neural networks–as discussed in the section devoted to black-box dynamic modeling and in Chap. 4- and when training some feedforward neural networks for classification, a constraint must frequently be obeyed: some parameters of the model must have equal values at the end of training (this is known as the "shared weight" technique [Waibel et al. 1989]). Since the weights are updated, at each epoch of training, as a function of the gradient of the cost function, there is no reason why different weights, even if initialized at equal values at the beginning of training, should stay equal even after a single epoch. Therefore, a special procedure must be implemented.

We assume that, in a given network, v parameters must stay equal: $w_1 = w_2 = \cdots = w_v = w$.

The corresponding component of the gradient of the cost function can be written as

$$\frac{\partial J}{\partial w} = \frac{\partial J}{\partial w_1}\frac{\partial w_1}{\partial w} + \frac{\partial J}{\partial w_2}\frac{\partial w_2}{\partial w} + \cdots + \frac{\partial J}{\partial w_\nu}\frac{\partial w_\nu}{\partial w}.$$

Because

$$\frac{\partial w_1}{\partial w} = \frac{\partial w_2}{\partial w} = \cdots = \frac{\partial w_\nu}{\partial w} = 1, \text{ one has } \frac{\partial J}{\partial w} = \sum_{i=1}^{\nu} \frac{\partial J}{\partial w_i}.$$

Thus, when a network contains shared weights, backpropagation must be performed, at each epoch, in the conventional way, in order to compute the partial derivatives of the cost function with respect to those weights; then the sum of those partial derivatives must be computed, and that value must be assigned to the partial derivatives, before updating the parameters by one of the methods discussed in the next section.

2.5.2.3 Updating the Parameters as a Function of the Gradient of the Cost Function

In the previous section, the evaluation of the gradient of the cost function, at a given epoch of training, was discussed. The gradient is subsequently used in an iterative minimization algorithm. The present section examines some popular iterative schemes for the minimization of the cost function.

Simple Gradient Descent

The simple gradient descent consists in updating the weights by the following relation, at epoch i of training:

$$\boldsymbol{w}(i) = \boldsymbol{w}(i-1) - \mu_i \, \nabla J(\boldsymbol{w}(i-1)) \qquad \text{with } \mu_i > 0.$$

Thus, the descent direction, in parameter space, is opposite to the direction of the gradient. μ_i is called gradient step or learning rate.

This very simple, attractive method has several shortcomings:

- If the learning rate is too small, the cost function decreases very slowly; if the rate is too large, the cost may increase or oscillate; that situation is

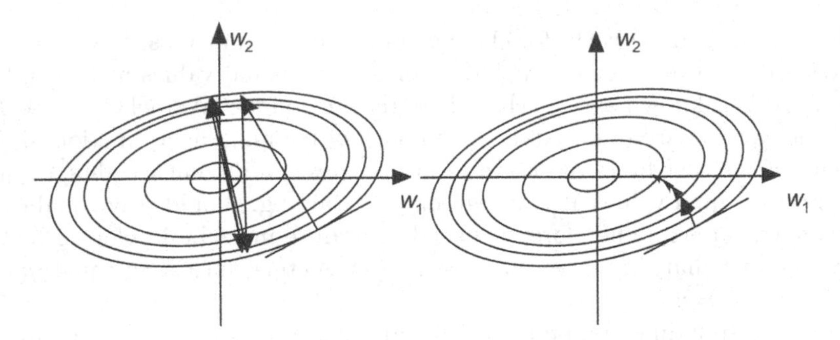

Fig. 2.7. Minimization of the cost function by simple gradient descent

illustrated on Fig. 2.7, which shows the iso-cost lines of the cost function (depending on two parameters w_1 and w_2), and the variation of vector $\boldsymbol{w}$ during the minimization.

- In the vicinity of a minimum of the cost function, the gradient becomes very small, so that the variation of the parameters becomes extremely slow; the situation is similar if the cost function has plateaus, so that, when training becomes very slow, there is no way to tell whether that is due to a plateau that may be very far from a minimum, or whether that is due to the presence of a real minimum.
- If the curvature of the surface is very nonisotropic, the direction of the gradient may be very different from the direction of the location of the minimum; such is the case if the cost surface has long narrow valleys as shown on Fig. 2.7.

In order to overcome the first drawback, a large number of heuristics were suggested, with varied success rates. Line search techniques (as discussed in the additional material at the end of the chapter) have solid foundations and are therefore recommended.

In order to overcome the other two difficulties, second-order gradient methods must be used. Instead of updating the parameters proportionally to the gradient of the cost function, one can make use of the information contained in the second derivatives of the cost function. Some of those methods also make use of a parameter μ whose optimal value can be found through line search techniques.

The most popular second-order techniques are described below.

Second-Order Gradient Methods

All second-order methods are derived from Newton's method, whose principle is discussed in the present section.

The Taylor expansion of a function $J(w)$ of a single variable w in the vicinity of a minimum w^* is given by

$$J(w) = J(w^*) + \frac{1}{2}(w - w^*)^2 \left(\frac{\mathrm{d}^2 J}{\mathrm{d}w^2}\right)_{w=w^*} + \mathrm{O}(w^3)$$

for the gradient of the cost function is zero at the minimum. Differentiating the above relation with respect to w gives an approximation of the gradient of the cost function in the neighborhood of a minimum,

$$\frac{\mathrm{d}J}{\mathrm{d}w} = (w - w^*) \left(\frac{\mathrm{d}^2 J}{\mathrm{d}w^2}\right)_{w=w^*}.$$

Therefore, if variable w is in the neighborhood of w^*, the minimum could be reached in a single iteration if the second derivative of the cost function at the minimum were known: $\boldsymbol{w}$ would simply be updated by an amount

$$\Delta w = -\frac{(\mathrm{d}J/\mathrm{d}w)}{(\mathrm{d}^2 J/\mathrm{d}w^2)_{w=w^*}}.$$

The same argument holds for a function of several variables, except for the fact that the second derivative becomes the Hessian matrix $H(\boldsymbol{w})$ of the cost function, whose general term is $(\partial^2 J)/(\partial w_i \partial w_j)$: in order to reach the minimum of the cost function in a single iteration, the weight vector should be updated (provided the Hessian matrix is invertible) by the amount

$$\Delta \boldsymbol{w} = -\left[H(\boldsymbol{w}^*)\right]^{-1} \nabla \cdot J(\boldsymbol{w}).$$

Thus, by contrast to simple gradient descent, the direction of motion, in parameter space, is not the direction of the gradient, but a linear transformation of the gradient.

Clearly, that relation is not applicable in practice, since vector $\boldsymbol{w}^*$ is not known. However, it suggests several iterative techniques that use an approximation of the Hessian matrix (or of its inverse). We discuss two of them in the additional material at the end of the present chapter: the Broyden-Fletcher-Goldfarb-Shanno algorithm (BFGS algorithm [Broyden 1970]) and the Levenberg-Marquardt algorithm ([Levenberg et al. 1944; Marquardt et al. 1963]). Obviously, those minimization methods are by no means specific to neural networks. Detailed descriptions are to be found in [Press et al. 1992], where the conjugate gradient method is also discussed.

What to Do in Practice?

First of all, one should by all means refrain from using simple gradient descent and its variants: their convergence times to a minimum (in number of iterations and in net computation time) are larger than those of second order methods by several orders of magnitude. Simple gradient should be used only in extreme cases for very large networks (several thousands of parameters) that may be useful in image processing with low-level picture representation, or for very large data bases (with millions of examples). In such cases, minimization is stopped before a minimum is reached, in order to prevent overfitting. That is a regularization method called early stopping), which will be discussed in the section devoted to training with regularization.

A heuristics called "momentum term" is often mentioned in the literature ([Plaut et al. 1986]); it consists in adding to the gradient term $-\mu_i \nabla J$, in simple gradient descent, a term that is proportional to the parameter update at the previous epoch $\lambda[\boldsymbol{w}(i-1) - \boldsymbol{w}(i-2)]$; that kind of low-pass filter may prevent oscillations and improve convergence speed if an appropriate value of λ is found.

The choice between BFGS and Levenberg-Marquardt is based on computation time and memory size. The BFGS method requires starting training with simple gradient descent in order to reach the vicinity of a minimum, then switching to BFGS to speed up the convergence; there is no principled method for finding the most appropriate number of iterations of simple descent before switching to BFGS: some trial-and error procedure is necessary. The Levenberg-Marquardt does not have that drawback, but it becomes demanding in memory size for large networks (about a hundred parameters), because of the necessary matrix inversions. Therefore, the Levenberg-Marquardt method will be preferred for "small" networks, and BFGS otherwise. If time is available, both should be tried.

Parameter Initialization

Since the above training methods are iterative, the parameters must be assigned initial values prior to training. The following arguments are guidelines for initialization:

- The parameters related to the bias inputs (constant inputs equal to 1) must be initialized to zero, in order to ascertain that the sigmoids of the hidden neurons are initialized around zero; then, if the inputs have been appropriately normalized and centered as recommended earlier, the values of the outputs of the hidden neurons will be normalized and centered too.
- Moreover, it should be ascertained that the values of the outputs of the hidden neurons are not too close to +1 or –1 (the sigmoids are said to be saturated). That is important because the gradient of the cost function, which is the driving force of minimization during training, depends on the derivatives of the activation functions of the hidden neurons with respect to the potential. If the outputs of the hidden neurons are initially near $+1$ or -1, the derivatives are very small, so that training starts very slowly, if at all.

If n is the number of inputs of the network, each hidden neuron receives $n-1$ variables x_i. The nonzero parameters should be small enough that the potential of the hidden neurons have a variance on the order of 1, in order to prevent the sigmoids from going into saturation. Assume that the inputs x_i can be viewed as realizations of random, identically distributed, centered and normalized variables X_i. The initial values of the parameters should be drawn from a centered distribution, whose covariance is unknown. The parameter related to the bias is equal to zero; the potential $\nu = \sum_{i=1}^{n} w_i x_i$ of each neuron is thus the sum of n-1 random variables that are the products of independent random variables, with zero mean, having the same distribution. It can be shown, from the elements of statistics provided at the beginning of the chapter, that one has

$$\mathrm{var}(V) = (n-1)\mathrm{var}(W_i)\mathrm{var}(X_i),$$

with $\mathrm{var}(X_i) = 1$ since the variables have been normalized prior to training.

Thus, if the desired variance of the potential is 1, the initial values of the parameters must be drawn from a centered distribution of variance $1/(n-1)$. For instance, it may be convenient to choose a uniform distribution between $-w_{\max}$ and $+w_{\max}$: $\mathrm{var}(W_i) = w_{\max}^2/3$, hence $w_{\max} = \sqrt{3/(n-1)}$.

The above discussion is valid for multilayer Perceptrons. For RBF or wavelet networks, the initialization problem is more critical, because those are localized functions; if they are initially located far from the domain of interest, or if their extension (standard deviation or dilation) is not appropriate, training will generally fail. The result of the teacher-student problem, described in the next section, depends critically on initialization for localized functions. The following strategy, described in detail [Oussar et al. 2002], should be implemented: a large library of RBFs or wavelets is created, and a selection method, analogous to the input selection methods described in a previous section, is applied. Training is subsequently applied to the wavelets or RBF's that were thus selected.

How to Test a Training Algorithm: The Teacher-Student Problem

The experience gained during years of teaching and research shows that it is very easy to design a faulty training algorithm, or to write a faulty training program, that nevertheless converges, sometimes very slowly, and produces a model that is not completely ridiculous. Algorithmic or software errors may pass unnoticed if care is not exercised. Therefore, it is important to test the validity of an algorithm or of a program that one has written or downloaded for free from the Web.

The following procedure, known as the teacher-student problem is convenient and simple to implement. A network is created (the teacher), whose parameters are random. That network is used for generating a training set, by using random inputs, and computing the corresponding outputs. That data set is used for training a second network (the student), which has the same number of inputs and of hidden neurons as the teacher network. If the training algorithm and the computer program are correct, the parameters of the teacher network should be retrieved by the student within roundoff errors: the mean square error is on the order of 10^{-30}, and each parameter of the student should be equal to a parameter of the teacher network, within roundoff errors. Otherwise, the training algorithm, or the program (or both) should be checked for errors.

The structure of the student network is identical to that of the teacher network within permutations of the hidden neurons. This is a consequence of the unicity theorem [Sontag 1993].

Two Test Problems

Problem 1: A network with 8 inputs, 6 hidden neurons and one output is generated by drawing weights uniformly in the interval [−20, +20]; a training set and a test set of 1,500 examples each are generated with random inputs from a uniform distribution in $[-1, +1]$; a network having the same structure is trained as follows: initialization of the parameters from a uniform distribution in $[-0.6, +0.6]$, computation of the gradient by backpropagation, minimization of the cost function by the Levenberg-Marquardt algorithm. The teacher network is retrieved exactly (TMSE and VMSE on the order of 10^{-31}) in 96% of trainings (for 48 trainings out of 50 trainings performed with different initializations).

Problem 2: A network with 10 inputs, 5 hidden neurons and an output is generated with weights drawn uniformly in $[-1, +1]$; a training set and a test set are generated with random inputs from a normal distribution; training is performed as in the previous example; the teacher network is retrieved in 96% of the trainings if the training set has 400 examples; it is retrieved in 100% of the trainings if the training set has 2,000 examples.

For the same problems, training *always* fails to retrieve the teacher network if simple gradient descent or stochastic gradient (see next section) are used, with or without momentum term.

Note that the teacher-student problem becomes difficult for some architectures because of a large number of local minima.

2.5.2.4 Summary

We summarize the procedure that must be used for training a feedforward neural network with a given number of inputs and hidden neurons:

- Initialize the parameters with the method described above.
- Compute the gradient of the cost function by backpropagation.
- Update the parameters iteratively with an appropriate minimization algorithm (simple gradient descent, BFGS, Levenberg-Marquardt, conjugate gradient, etc.).
- If a prescribed maximum number of epochs is reached, or if the variation of the module of the vector of parameters is smaller than a given threshold (the weights no longer change significantly), or if the module of the gradient is smaller than a given threshold (a minimum has been reached), terminate the procedure; otherwise, start a new epoch by iterating to the gradient evaluation.

2.5.3 Adaptive (On-Line) Training of Models that Are Nonlinear with Respect to Their Parameters

In the previous sections, we discussed methods that optimize the least squares cost function by using all the training data available at the beginning of training: the gradient of the total cost can be computed as the sum of the gradients of the partial costs.

In adaptive (on-line) training, parameters are updated by using the gradient of the partial cost for each example, so that training can start even before all training data is available. Such a procedure is often useful to update a model after an initial nonadaptive training. Those methods are discussed in detail in Chap. 4.

A variant of adaptive training algorithms consists in updating the parameters after reception of a block of data ("block training"): then the partial cost is not related to a single example but to a block of examples.

The most popular adaptive training technique is called stochastic gradient, whereby the parameter updates are proportional to the gradient of the partial cost,

$$\boldsymbol{w}^{k+1} = \boldsymbol{w}^k - \mu^k \nabla J^k(\boldsymbol{w}^k),$$

where $\boldsymbol{w}^k$ is the value of the vector of parameters after iteration k, i.e., after updating the parameters from example k. Note that the LMS algorithm, discussed in the framework of the training of linear models, is a particular case of stochastic gradient.

Some empirical results suggest that the stochastic gradient method avoids local minima more efficiently than simple gradient descent in batch learning.

An alternative technique, stemming from adaptive filtering, can be used for neural network training: the extended Kalman filter [Puskorius et al. 1994]. It is more efficient than stochastic gradient in terms of convergence speed, but the number of operations per iteration is higher. That approach is described in detail in Chap. 4.

2.5.4 Training with Regularization

As stated in Chap. 1, the objective of black-box modeling is the design of a model that is complex enough to learn the training data, but does not exhibit overfitting, i.e., does not adjust to noise. Two categories of strategies can be used.

- Passive techniques: several models, of different complexities, are trained as indicated in the previous section, and a selection between those models is performed after training, in order to discard models that exhibit overfitting; that is done by cross-validation or statistical tests as explained in the next section.

- Active techniques: training is performed in order to avoid designing models that exhibit overfitting, by limiting the magnitude of the parameters; regularization methods [Tikhonov et al. 1977; Poggio et al. 1985] are implemented, as discussed in the present section.

The latter techniques are of special importance when large networks need be designed; such is often the case in classification for visual pattern recognition, when a low-level representation is used (see the introduction to classification in Chap. 1). In such situations, overfitting cannot be avoided by limiting the number of parameters, since the number of inputs is a lower bound to the number of parameters: the only way of avoiding overfitting consists in limiting the amplitude of the parameters; it is even shown in [Bartlett et al. 1997] that, if a large network is designed, and if the training algorithms finds a small mean square error with parameters of small amplitudes, than the generalization performances depend on the norm of the vector of parameters, and is independent of the number of parameters.

There are essentially two families of regularization methods:

- Early stopping consists in stopping training before a minimum of the cost function is reached.
- Penalty methods consist in adding a penalization term in the cost function in order to favor regular models. The cost function has the form: $J' = J + \alpha\boldsymbol{\Omega}$, where J is, for instance, the least squares cost function, and $\boldsymbol{\Omega}$ is a function of the weights. The most popular penalty function is: $\boldsymbol{\Omega} = \sum_i \parallel w_i \parallel^2$. The method involving that penalty function is called weight decay.

Both techniques will be discussed below.

2.5.4.1 Early Stopping

Principle

As usual, training consists in minimizing iteratively a cost function, the least squares cost function for instance, whose value is computed on a training set. Regularization takes place through the stopping criterion: training is terminated before a minimum of the cost function is reached, so that the model does not fit the training data as well as it could, given the number of parameters that are available to him; thus overfitting is limited. The difficulty that arises is: when to stop training? The most popular method consists in monitoring the variation of the standard prediction error on a validation set, and in terminating training when the prediction error starts increasing.

Example

We discuss an academic example from [Stricker 2000]. It is a two-class classification problem; as explained in Chap. 1, the output of the classifier should

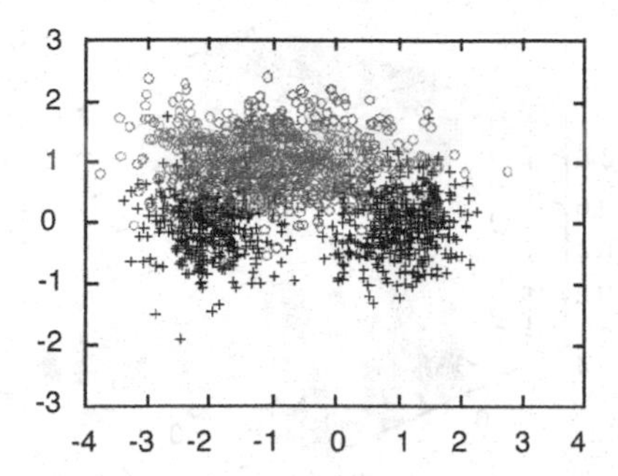

Fig. 2.8. The examples of class A (*circles*) are realizations of a random variable whose distribution is the product of two functions of x and y respectively; the distribution along x is the sum of two Gaussians with centers -2 and 0 respectively, and standard deviation 0.5, and the distribution along y is a Gaussian centered at 0, with standard deviation 0.5. The examples of class B (*crosses*) are drawn from a distribution that is the product of two Gaussian functions of x and y respectively; the distribution along x is centered at -1, with standard deviation 1, and the distribution along y is centered at 1, with standard deviation 0.5

be equal to 1 for all elements of one class (class A), and to 0 for all elements of the other class (class B). After training, the output is an estimate of the probability of the unknown pattern belonging to class A. In the present problem, feature space is of dimension 2, and the examples are drawn from overlapping distributions, as shown on Fig. 2.8.

A classifier must provide a graded response in the zone of overlapping between the classes, since the boundary between classes cannot be known with certainty given the limited amount of data. In the present academic example, the prior distributions are known, so that the posterior probability of the classes can be computed from Bayes formula (see Fig. 2.9),

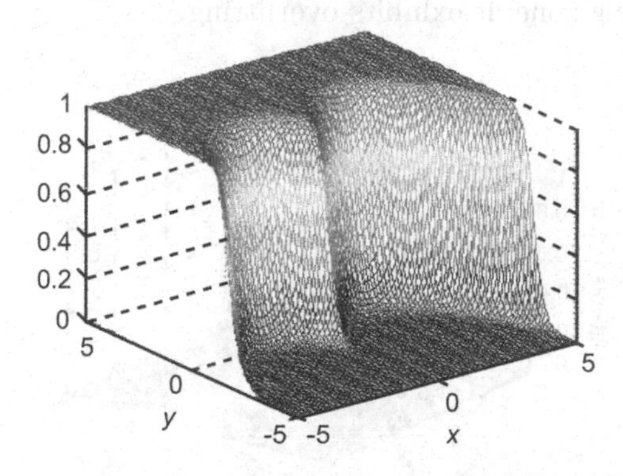

Fig. 2.9. Posterior probability computed by Bayes formula

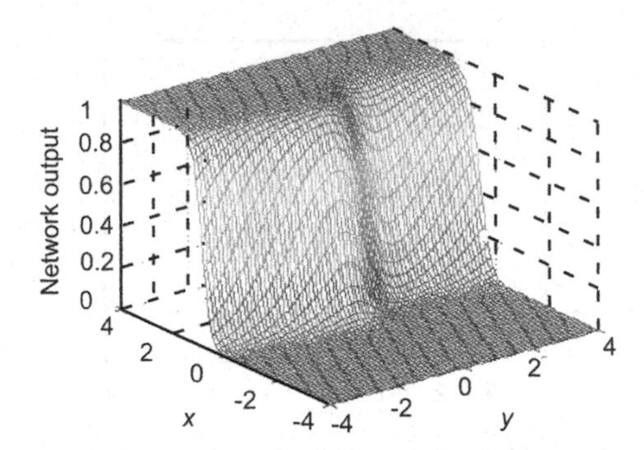

Fig. 2.10. Posterior probability computed by a neural network with 2 hidden neurons

$$\mathrm{P_R}(A \mid \boldsymbol{x}) = \frac{p_X(\boldsymbol{x} \mid A)\Pr(A)}{p_X(\boldsymbol{x} \mid A) + p_X(\boldsymbol{x} \mid B)},$$

where $\boldsymbol{x}$ is the vector $[x\ y]^{\mathrm{T}}$, $p_X(\boldsymbol{x} \mid A)$ is the distribution of the random vector X for the patterns of class A, and $\Pr(A)$ is the prior probability of class A. The estimation provided by the neural network from the examples shown on Fig. 2.8 should be as similar as possible to the surface shown on Fig. 2.9.

Training is performed with a set of 500 examples. A network with 2 hidden neurons provides the probability estimate shown on Fig. 2.10; the estimate provided by a neural network with 10 hidden neurons is shown on Fig. 2.11.

One observes that the result obtained with the network having 2 hidden neurons is very close to the theoretical probability surface computed from Bayes formula, whereas the surface provided with hidden neurons is almost binary: in the zone where classes overlap, a very small variation of one of the features generates a very sharp variation of the probability estimates. The 10-hidden neuron network is over-specialized on the examples that are located near the overlapping zone: it exhibits overfitting.

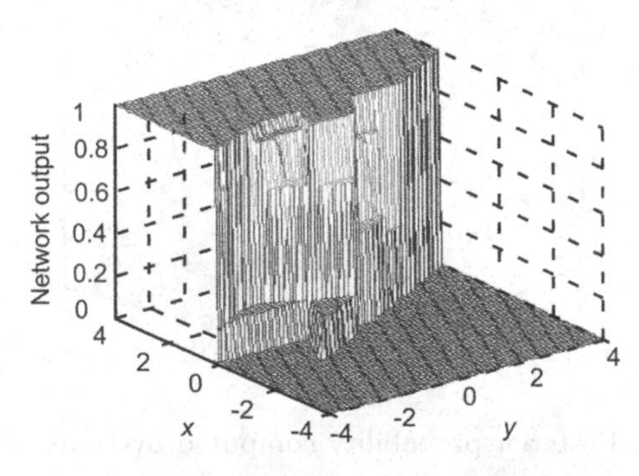

Fig. 2.11. Posterior probability computed by a neural network with 10 hidden neurons

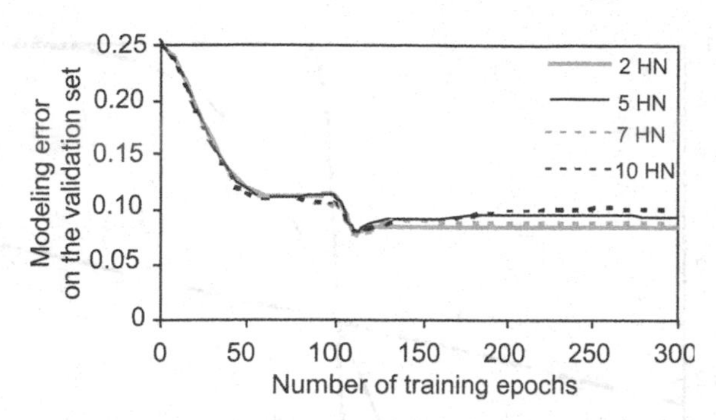

Fig. 2.12. Classification error on the validation set during training

The variation of the mean square error on a validation set of 300 examples, as a function of the number of epochs, is shown on Fig. 2.12, for various numbers of hidden neurons. Clearly, deciding when training should be terminated is difficult, because the error arises essentially from the examples that are close to the boundary zone, which corresponds to a relatively small number of points.

Therefore, that method is not very convenient, especially for classification. Therefore, regularization methods that involve penalizing large parameters are often preferred; it was proved [Sjöberg 1995] that early stopping is actually equivalent to the introduction of a penalty term in the cost function.

2.5.4.2 Regularization by Weight Decay

Large values of the parameters, for instance of the parameters of the inputs of hidden neurons, generate sharp variations of the sigmoids of the hidden neurons: that is illustrated on Fig. 2.13, which shows function $y = \tanh(wx)$, for three different values of w. The output of the network, which is a linear combination of the outputs of the hidden neurons, is therefore apt to exhibit sharp variations as well. Regular outputs therefore require that the sigmoids be in the vicinity of their linear zones, hence that the parameters not be too large. We consider again the classification example of the previous section: Fig. 2.14 shows the variation of the module of the vector of parameters, during training, for different architectures (2, 5, 7 and 10 hidden neurons). One observes that the norm of the vector of parameters increases sharply during training, except for the architecture with two hidden neurons: therefore, the sharp variations of the output surface after training the network with ten hidden neurons, as shown on Fig. 2.11, is not surprising.

Regularization by weight decay prevents the parameters from increasing excessively, by minimizing, during training, a cost function J' that is the sum of the least squares cost function J (or of any other cost function, such as cross entropy described in Chaps 1 and 6) and of a regularization term, proportional to the squared norm of the vector of parameters: $J^* = J + \frac{\alpha}{2} \sum_{i=1}^{q} w_i^2$,

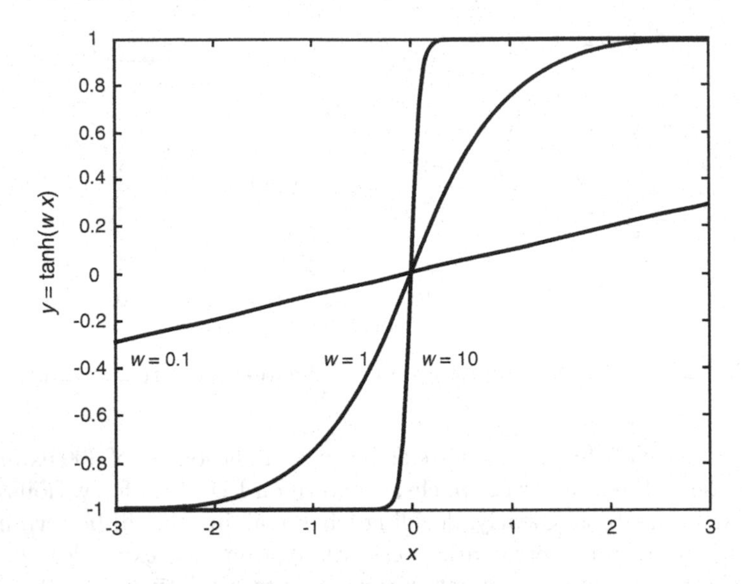

Fig. 2.13. Function $y = \tanh(wx)$ for 3 values of w

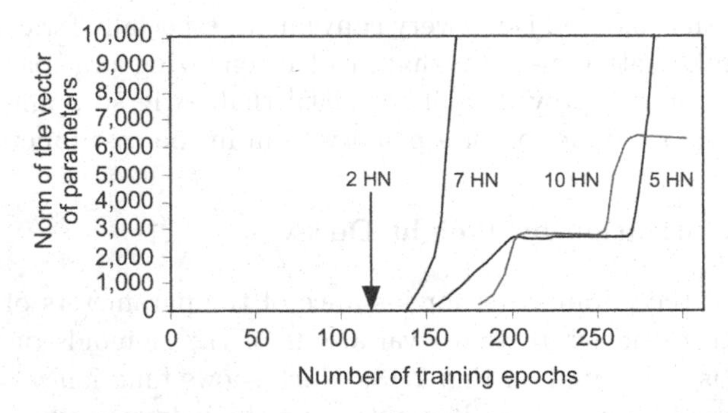

Fig. 2.14. Norm of the vector of parameters during training

where q is the number of parameters of the classifier, and α is a hyperparameter whose value must be found by performing a tradeoff: if α is too large, the minimization decreases the values of the parameters irrespective of the modeling error; by contrast, if α is too small, the regularization term has no impact on training, hence overfitting may occur.

The operation of the method is very simple: the gradient of J is computed by backpropagation, and the contribution of the regularization term is subsequently added,

$$\nabla J^* = \nabla J + \alpha \boldsymbol{w}.$$

Nevertheless, it should be noticed that the parameters of the network have different effects:

- The parameters of the connections between the variables of the model and the inputs of the hidden neurons control the slope of the sigmoids of hidden neurons.
- The parameters of the connections between the constant input (bias) and the inputs of the hidden neurons generate a horizontal shift of the sigmoids of hidden neurons.
- The parameters of the connections between hidden neurons and the inputs of the output neurons control the influence of each hidden neuron on the outputs.
- The parameters of the connections between the bias and the output neurons generate a vertical shift of the output of the network.

Therefore, it is natural to use different hyperparameters for those different types of parameters [McKay 1992]. Then the cost function becomes

$$J^* = J + \frac{\alpha_0}{2} \sum_{\omega \in W_0} w_i^2 + \frac{\alpha_1}{2} \sum_{\omega \in W_1} w_i^2 + \frac{\alpha_2}{2} \sum_{\omega \in W_2} w_i^2,$$

where W_0 is the set of parameters between the bias and the hidden neurons, where W_1 is the set of parameters between the inputs and the hidden neurons, and W_2 is the set of parameters of the inputs of the output neuron (including the bias of the output neuron). Therefore, the values of the three parameters α_1, α_2, α_3 must be found. A principled statistical method was proposed in [McKay 1992], but it relies on numerous assumptions and requires demanding computations. In practice, the values of the hyperparameters are not very critical; a heuristic approach, consisting in performing different trainings with different hyperparameters, is frequently sufficient.

We illustrate this discussion on an example of a real application, from [Stricker 2000].

Example

The application is a filtering task, as outlined in Chap. 1. In a corpus of texts (press releases of the Agence France Presse), the texts that are relevant to a given topic should be selected automatically. It is essentially a two-class problem: a press release is either relevant or irrelevant. A training set of 1,400 relevant press releases and 8,000 irrelevant ones is available. The performance measure is a quantity F that is a function of the precision of the classifier (the ratio of the number of documents that are really relevant to the number of documents that are considered as relevant by the classifier) and its recall (the ratio of the number of documents that are considered as relevant by the classifier to the number of relevant documents present in the database). The better the performance, the larger the value of F.

A linear classifier is used, i.e., a neural network with zero hidden neuron and an output neuron with sigmoid activation function. Since there are no hidden units, the number of parameters cannot be decreased without changing the data representation. Since it is not desired to change the latter (which

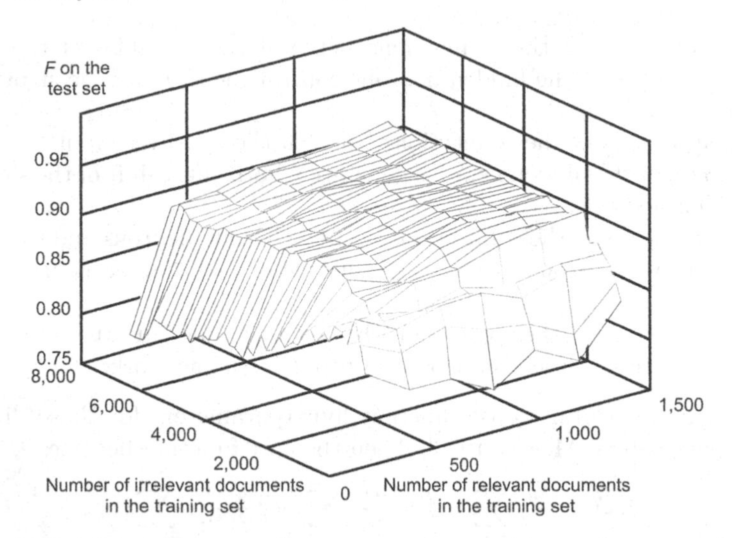

Fig. 2.15. Training without regularization: variation of the performance of a linear classifier as a function of the numbers of relevant and irrelevant documents in the training set

is discussed in detail in Chap. 1), regularization methods are mandatory to avoid overfitting. Figure 2.15 shows the variation of F on a test base, without regularization, as a function of the numbers of relevant and irrelevant documents present in the database. Clearly, the performance decreases, and the norm of the vector of parameters increases, when the number of examples of the training set decreases.

With the same training and test sets, training was performed with early stopping. The results (Fig. 2.17) show that the performance is improved for small numbers of examples in the training set, but it is decreased when numerous examples are available ($F < 0.9$), which is evidence that early stopping does not make the best of the available data. The norm of the vector of parameters (not shown) remains very small.

Weight decay was also implemented on the same example, with two hyperparameters: one for the bias ($\alpha_b = 0.001$) and one for the connections between the inputs and the output neuron ($\alpha_1 = 1$). The results are shown on Fig. 2.18; the performance is improved when the number of examples is small, and, by contrast with early stopping, it remains satisfactory for large numbers of examples. As in the previous case, the norm of the vector of parameters stays small.

Models whose outputs are not smooth enough can also be avoided, by penalizing large values of the derivatives of the output with respect to the inputs [Bishop 1993].

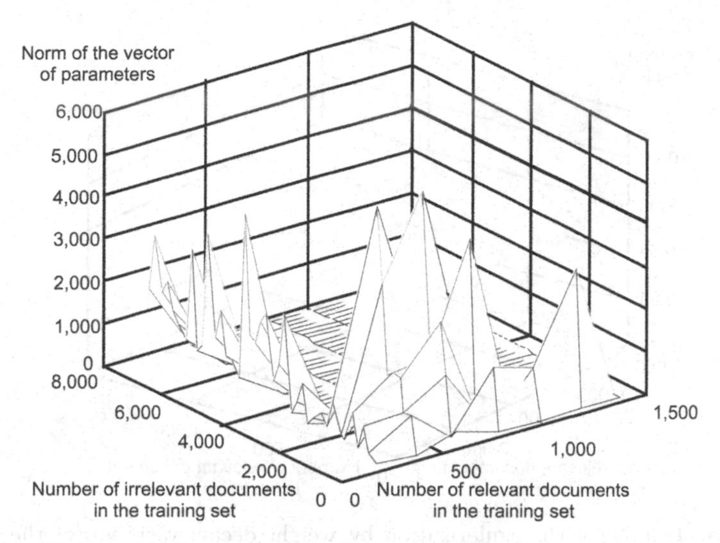

Fig. 2.16. Training without regularization: variation of the norm of the vector of parameters as a function of the numbers of relevant and irrelevant documents in the training set

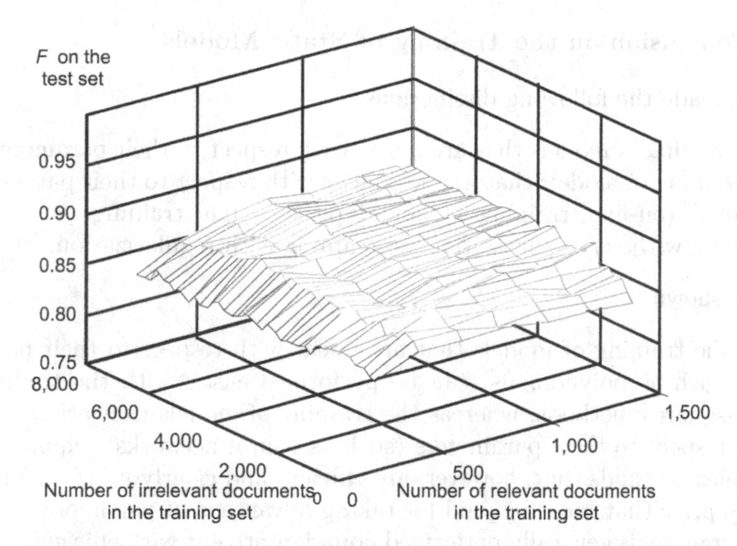

Fig. 2.17. Training with regularization by early stopping: variation of the performance of a linear classifier as a function of the numbers of relevant and irrelevant documents in the training set

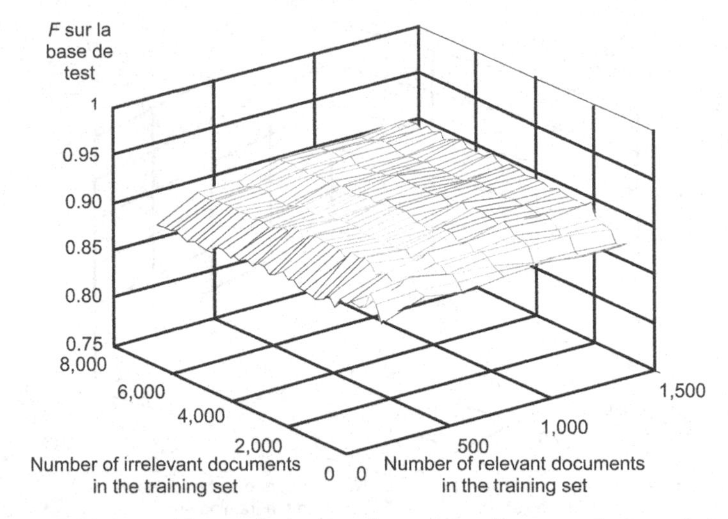

Fig. 2.18. Training with regularization by weight decay: variation of the performance as a function of the number of relevant and irrelevant documents in the training set

2.5.5 Conclusion on the Training of Static Models

We have made the following distinctions:

- The training of models that are linear with respect to their parameters vs. the training of models that are not linear with respect to their parameters.
- Adaptive (on-line) training vs. nonadaptive (batch) training.
- Training without regularization vs. training with regularization.

We have shown

- that the training of models that are linear with respect to their parameters (such as polynomials) can be performed easily with the traditional least-squares methods, whereas the training of models that are nonlinear with respect to their parameters (such as neural networks) requires more complex methods that, however, are efficient and clearly understood: that is the price that must be paid for taking advantage of parsimony;
- that training is generally performed nonadaptatively, with efficient second-order minimization algorithms; if necessary, the model can be updated by adaptive methods in order to take into account slow drifts of the characteristics of the process;
- that overfitting can be avoided by limiting the amplitude of the parameters of the model with a regularization method during training; that is especially necessary when the number of training examples is small.

The next section discusses the problem of overfitting in a more general framework: model selection.

2.6 Model Selection

After variable selection and training, model selection is the third important element of a model design methodology. We assume that several candidate models have been trained, one of which must be chosen. The model should be complex enough to find the deterministic relations between the quantity to be modeled and the factors that have a significant influence on it, yet not be overly complex in order to be free from overfitting. In other words, the selected model should embody the best tradeoff between learning capacity and generalization capacity: if the model learns too well, it fits the noise, hence generalizes poorly. That tradeoff has been formalized under the term bias-variance dilemma [Geman et al. 1992].

From a theoretical point of view, the model that is sought is the model for which the theoretical cost function $\int (y_p(\boldsymbol{x}) - g(\boldsymbol{x}, \boldsymbol{w}))^2 P_X(\boldsymbol{x}) d\boldsymbol{x}$ is minimal. That quantity may be split into two terms:

- the bias[5], which expresses the average, over all possible training sets (with all possible realizations of the random variables that model the noise) of the squared difference between the predictions of the model and the regression function;
- the variance, which expresses the sensitivity of the model to the training set (with its own realization of the noise).

Because the above theoretical cost function cannot be computed, the empirical least squares cost function is minimized during training, as discussed in the previous section.

Thus, a very complex model, with a large number of adjustable parameters, may have a very low bias, i.e., may have the ability of fitting the data whatever the noise present in the latter, but it is apt to have a very large variance, i.e., to depend strongly on the specific realization of the noise present in the training set. Conversely, a very simple model, with a small number of adjustable parameters, may be insensitive to the noise present in the training data, but turn out to be unable to approximate the regression function.

Figure 2.19 illustrates the behavior of two models $g_1(x)$ and $g_2(x)$, with the same complexity (linear models), which have too large a bias and too small a variance: the predictions of the two models, obtained with different training sets, are almost identical, but they are very different from the regression function. Conversely, Fig. 2.20 illustrates the behaviors of two models that have a low bias (they are close to the regression) but they have a large variance since their predictions depend on the training set.

The next two illustrations, and several elements of the present section, are excerpts from [Monari 1999].

[5] This should not be mistaken with the constant input of a model, unfortunately also called bias.

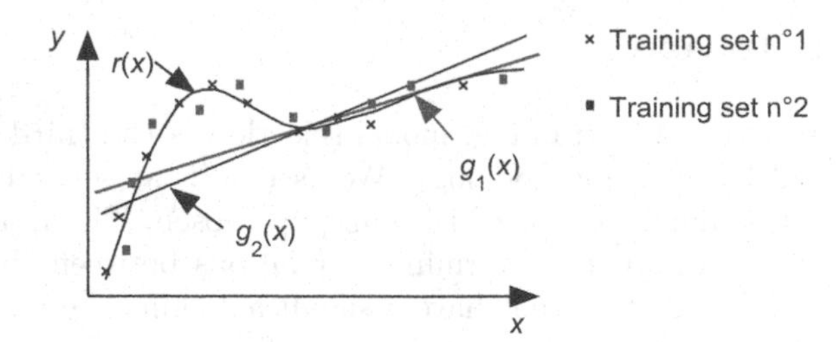

Fig. 2.19. Two models that have a large bias and a small variance

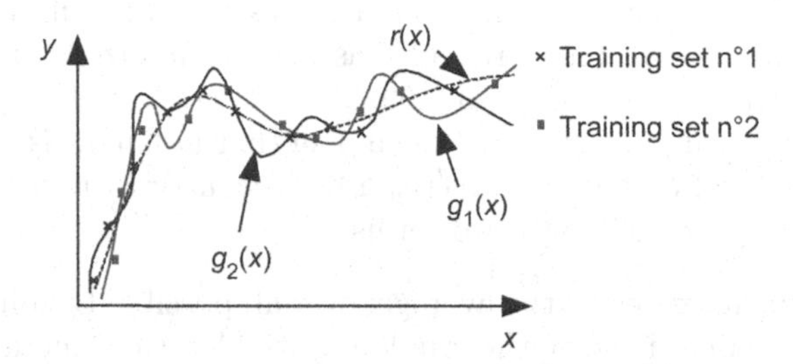

Fig. 2.20. Two models that have a small bias and a large variance

Unfortunately, bias and variance, just as the theoretical cost function, cannot be computed. Thus, the solution to the difficult problem of model selection is a tradeoff between two quantities that cannot be computed. The difficulty of the problem increases as the size of the training set decreases [Gallinari 1999].

The models, trained from the same training set, among which a choice is to be made, differ by two main characteristics:

- their complexity: the complexity of a model can be defined as the number of its elements (the number of monomials in a polynomial model, the number of hidden neurons in a neural network), hence the number of its adjustable parameters;
- the vector of parameters for a given complexity: for models that are non-linear with respect to the parameters, the cost function has several local minima; therefore, for a given complexity and a given training set, different trainings (with different initial values of the parameters) may provide different models corresponding to different minima of the cost function. Conversely, for models that are linear with respect to their parameters, the least squares cost function has a single minimum: for a given complexity and a given training set, there is a single vector of parameters for which the cost function is minimum.

Hence, for a model that is not linear with respect to its parameters, the model selection problem is actually twofold:

- Among models that have the same complexity, find the model that achieves the best bias-variance tradeoff.
- Among the best models that have different complexities, find the model that achieves the best bias-variance tradeoff.

All techniques that will be discussed below aim at (i) discarding models that are obviously prone to overfitting, and (ii) at estimating the generalization error (or theoretical cost function) in order to find the model that has the smallest generalization error. As a preliminary step, we show how to discard models that are prone to overfitting; subsequent sections will discuss two model selection techniques,

- a global method, which consists in estimating the generalization error: cross-validation;
- a local method whereby the influence of each example on the model is estimated: the local overfitting control via leverages (LOCL) method, which is based on the estimation of leverages and confidence intervals for the predictions of the model.

Finally, the above approaches will be combined into a complete model selection methodology for the selection of nonlinear models.

2.6.1 Preliminary Step: Discarding Overfitted Model by Computing the Rank of the Jacobian Matrix

2.6.1.1 Introduction

In the section devoted to the estimation of the parameters of a model that is linear with respect to its parameters, we have defined the matrix of observations Ξ; each column of that matrix has N elements, which are the values of a given variable for each example. Therefore, for a model with n variables, the matrix of observations is (N, n). For a model that is not linear with respect to its parameters, having a vector of q parameters $\boldsymbol{w}_{\mathrm{LS}}$, the equivalent of the observation matrix is the Jacobian matrix $Z(N, q)$; each column z_i of that matrix has N elements, which are the values of the partial derivatives of the output with respect to a given parameter: $z_i = (\partial g(\boldsymbol{x}, w)/\partial w_i)_{w=w_{\mathrm{IS}}}$. It can easily be checked that, for a model that is linear with respect to its parameters, the Jacobian matrix Z is identical to the observation matrix Ξ.

Thus, each column of the Jacobian matrix expresses the effect of the variation of a parameter on the output of the model. If the Jacobian matrix does not have full rank (i.e., if its rank is not equal to q), it can be concluded that the effect, on the model output, of two parameters (or more) are not independent. Therefore, there exist under-determined parameters in the model: the latter has too many parameters, hence its variance is certainly too large. Such a model should be discarded. Moreover, rank deficiency has an adverse effect on training [Saarinen et al. 1993] [Zhou et al. 1998].

2.6.1.2 Computation of the Jacobian Matrix

In the section devoted to the training of a model that is not linear with respect to its parameters, it was shown that the gradient of the cost function can easily be computed by backpropagation,

$$\left(\frac{\partial J}{\partial w_i}\right) = \left(\frac{\partial (y_p - g(\boldsymbol{x}, \boldsymbol{w}))^2}{\partial w_i}\right) = -2(y_p - g(\boldsymbol{x}, \boldsymbol{w}))\frac{\partial g(\boldsymbol{x}, \boldsymbol{w})}{\partial w_i}.$$

If the modeling error $y_p - g(\boldsymbol{x}, \boldsymbol{w})$ is equal to 1/2, then the gradient of the cost function is equal to the gradient of the output. Thus, the Jacobian matrix can easily be computed by backpropagating a modeling error equal to 1/2. The extra computation time incurred by the computation of the Jacobian matrix is marginal, since it is performed once per training, whereas backpropagation is performed at each training epoch.

2.6.1.3 Computation of the Rank of the Jacobian Matrix

The rank of the matrix can be computed by a variety of methods [Press et al. 1992]. They will not be described here. In the section devoted to the effect of withdrawing an example from the training set, we describe a technique that is convenient in the framework of model selection.

2.6.2 A Global Approach to Model Selection: Cross-Validation and Leave-One-Out

2.6.2.1 Introduction

As discussed in a previous section, model selection should be based on the comparison of the generalization errors of the candidate models, but the generalization error, just as the regression function, cannot be computed: therefore, it must be estimated.

The most natural idea consists in performing model selection on the basis of the mean square error on the training set (TMSE),

$$E_T = \sqrt{\frac{1}{N_T}\sum_{k=1}^{N_T}(r_k)^2},$$

where r_k is the modeling error on example k: $r_k = y_p^k - g(\boldsymbol{x}^k, \boldsymbol{w})$, and where the summation is performed over all N_T examples of the training set. That is a bad idea: as discussed previously, the modeling error on the training set can be made as small as desired by just adding hidden neurons, which is detrimental to generalization. Thus, the value of E_T is *not* a suitable selection criterion.

2.6.2.2 Cross-Validation

Cross-validation is a technique for estimating the generalization error of a model, from data that are not used for parameter estimation (training) [Stone 1974]. First, the set of available data is split into D disjoint subsets. Then, the following steps are performed, for a family of functions having the same complexity (e.g., neural networks with a given number of hidden neurons):

- iteration i, to be performed D times: build a training set with D-1 subsets of the available data; perform several trainings, with different initial values of the parameters; for each model, compute the mean square error (VMSE) on the validation set made of the N_V remaining examples,
$$E_V = \sqrt{\frac{1}{N_V}\sum_{k=1}^{N_V}(r_k)^2};$$
store in memory the smallest VMSE thus computed E_{Vi};
- compute the cross-validation score from the D quantities E_{Vi} at each of the D iterations
$$\sqrt{\frac{1}{D}\sum_{i=1}^{D}(E_{Vi})^2}.$$

That score is an estimate of the generalization error for the family of functions thus investigated.

For instance, if $D = 5$ is chosen (that is a typical value; the process is called 5-fold cross-validation), 5 different partitions of the database are constructed; for each partition, 80% of the data are on the training set and 20% in the validation set. As discussed above, the cross-validation score is the square root of the average of the VMSE's computed on each partition. That average must be performed because 20% of the database may not be a statistically significant sample of the distribution of all possible examples. In a heuristic fashion, the procedure may be simplified by performing a single partition of the database, choosing a validation set that is as close as possible to the distribution of the available examples. To that effect, one can estimate the Kullback-Leibler divergence ([Kullback et al. 1951; Kullback 1959] between two probability distributions p_1 et p_2,

$$D(p_1, p_2) = \int_{-\infty}^{+\infty} p_1(x)\ln\left(\frac{p_1(x)}{p_2(x)}\right).$$

Because the expression is not symmetrical, a more satisfactory distance is defined as

$$\Delta = \frac{1}{2}\left[D(p_1, p_2) + D(p_2, p_1)\right].$$

Several random partitions of the database are performed, and the partition for which the distance between the validation set and the training set is smallest is retained. That single partition can satisfactorily be used for estimating the generalization error. Since drawing randomly a large number of partitions and computing the Kullback-Leibler divergence is much faster than training a model, the computation time is divided roughly by a factor of 5 as compared to complete 5-fold cross-validation. Making the assumption that the distributions are two Gaussians $p_1(\mu_1, \sigma_1)$ and $p_2(\mu_2, \sigma_2)$, the Kullbak-Leibler distance can be written as

$$\Delta(p_1, p_2) = \frac{(\sigma_1^2 - \sigma_1^2) + (\mu_1 - \mu_2)^2}{4\sigma_1^2\sigma_1^2}(\sigma_1^2 + \sigma_1^2).$$

The proof of that relation is given in the additional material at the end of the chapter.

That heuristic procedure is very useful for fast prototyping of an initial model, which can be further refined by conventional cross-validation or by the virtual leave-one-out technique that is explained below.

2.6.2.3 Model Selection by Cross-Validation

Model design starts from simplest models (linear model), and gradually increases the complexity (for neural models, by increasing the number of hidden neurons).

One might also increase the number of hidden layers; for modeling problems, that can be considered in a second step of the design: if a satisfactory model has been found with one hidden layer, one can, time permitting, try to improve the performance by increasing the number of hidden layers, while decreasing the number of neurons per layer. That procedure sometimes leads to some improvement, usually a marginal one. Conversely, if no satisfactory model has been found with one hidden layer, increasing the number of layers will not do any good.

For each family of models, a cross-validation score is computed as explained above. When overfitting occurs, the cross-validation score increases when the complexity of the model increases. Therefore, the procedure is terminated when the score starts increasing. The model that has the smallest VMSE is selected.

2.6.2.4 Leave-One-Out

The estimation of the generalization error by leave-one-out is a special case of cross-validation, for which $D = N$: At iteration k, example k is withdrawn from the training set, trainings are performed (with different initial values of the parameters) with the $N - 1$ examples of the training set; for each model, the prediction error on the withdrawn example k is computed, and the smallest prediction error on the withdrawn example, denoted $r_k^{(-k)}$ is stored. The leave-one-out score

$$E_t = \sqrt{\frac{1}{N}\sum_{k=1}^{N}\left(r_k^{(-k)}\right)^2}$$

is computed. As in the case of cross-validation, models of increasing complexities are designed, until the leave-one-out score starts increasing with increasing complexity.

The main drawback of the leave-one-out technique is that it is computationally very demanding, but it can be shown that the leave-one-out score is an unbiased estimator of the generalization error [Vapnik 1995].

In the next section, we discuss a slightly different technique, whose computation time is roughly the computation time of leave-one-out divided by a factor N(the number of examples). It is based on the fact that the withdrawal of an example from the training set should not lead to a very different model, so that a model that is locally linear *in parameter space*, in the neighborhood of the minimum of the cost function, can be designed; therefore, powerful results from the theory of linear regression can be taken advantage of.

2.6.3 Local Least Squares: Effect of Withdrawing an Example from the Training Set, and Virtual Leave-One-Out

In the present section, we show that the effect of withdrawing an example from the training set on a nonlinear model can be predicted. Specifically, we prove that the modeling error made by the model on the withdrawn example can be accurately predicted without actually withdrawing the example (virtual leave-one-out), and that a confidence interval on the predictions of the model can be estimated. Finally, we show that the influence of an observation on the model can be summarized with a single parameter: the leverage of the observation.

2.6.3.1 Local Approximation of the Least Squares Method

Consider a model $\boldsymbol{g}(\boldsymbol{x}, \boldsymbol{w}^*)$. A first-order Taylor expansion of the model, in parameter space, in the neighborhood of $\boldsymbol{w}^*$, can be written as

$$\boldsymbol{g}(\boldsymbol{x}, \boldsymbol{w}) \cong \boldsymbol{g}(\boldsymbol{x}, \boldsymbol{w}^*) + Z(\boldsymbol{w} - \boldsymbol{w}^*),$$

where $\boldsymbol{g}$ is the vector of the N predictions of the model, and where Z is the Jacobian matrix of the model, as defined above. That model is linear with respect to its parameters, and matrix Z is equivalent to the matrix of observations.

In order to derive a local approximation, to first order in $\boldsymbol{w} - \boldsymbol{w}^*$, of the gradient of the least-squares cost function, a second-order approximation of the cost function, hence a second-order approximation of the model output, must be used ([Monari et al. 2000] the same result is derived in [Seber et al. 1989], albeit with an incorrect proof). The following approximation of the least-squares solution $\boldsymbol{w}_{\mathrm{LS}}$ is found:

$$\boldsymbol{w}_{\mathrm{LS}} \cong \boldsymbol{w}^* + (Z^{\mathrm{T}}Z)^{-1}Z^{\mathrm{T}}[\boldsymbol{y}_p - \boldsymbol{g}(\boldsymbol{x}, \boldsymbol{w}^*)].$$

That result is approximate for a nonlinear model, but is exact for a linear model: in the case of a linear model, Z is the matrix of observations Ξ, and $\boldsymbol{g}(\boldsymbol{x}, \boldsymbol{w}^*) = \Xi\boldsymbol{w}^*$. Then one gets $\boldsymbol{w}_{\mathrm{LS}} \cong \boldsymbol{w}^* + (\Xi^{\mathrm{T}}\Xi)^{-1}\Xi^{\mathrm{T}}y_p - (\Xi^{\mathrm{T}}\Xi)^{-1}\Xi^{\mathrm{T}}\Xi\boldsymbol{w}^*) = (\Xi^{\mathrm{T}}\Xi)^{-1}\Xi^{\mathrm{T}}y_p$, which is the exact result, as shown in the section devoted to the training of linear models.

2.6.3.2 The Effect of Withdrawing an Example on the Model

The results of the previous section are useful for estimating the effect, on the predictions of the model, of withdrawing an example from the training set. As defined in the section on leave-one-out, we use the superscript $(-k)$ for all quantities related to a model that was designed after withdrawing example k from the training set; the quantities that have no superscript are related to models whose training was performed with all available data.

The Effect of Withdrawing an Example on the Prediction: Virtual Leave-One-Out

Assuming that the withdrawal of example k has a small effect on the least-squares solution, the relation that was derived in the previous section can be used to compute the vector of the parameters of the model that is trained with the training set deprived of example k, as a function of the vector of the parameters of the model trained with the whole data set,

$$\boldsymbol{w}_{\mathrm{LS}}^{(-k)} \cong \boldsymbol{w}_{\mathrm{LS}} - (Z^{\mathrm{T}}Z)^{-1}\boldsymbol{z}^k \frac{r_k}{1 - h_{kk}},$$

where $\boldsymbol{z}^k$ is the vector whose components are the kth column of the Jacobian matrix Z, r_k is the predication error (or residual) on example k when the latter belongs to the training set

$$r_k = y_{pk} - f(\boldsymbol{x}^k, \boldsymbol{w}_{\mathrm{LS}}),$$

and where $h_{kk} = \boldsymbol{z}^{kT}(Z^{\mathrm{T}}Z)^{-1}\boldsymbol{z}^k$ is the leverage of example k [Lawrance 1995]. Geometrically, h_{kk} is the kth component of the projection, onto solution subspace, of the unit vector borne by axis k. Since these quantities are the diagonal elements of an orthogonal projection matrix, they obey the following relations:

$$\sum_{k=1}^{N} h_{kk} = q, \qquad 0 < h_{kk} < 1.$$

An efficient procedure for computing the leverages h_{kk} is discussed in the additional material at the end of the chapter.

In the section devoted to the rank of the Jacobian matrix, we have shown that it is useful to know whether that matrix has full rank. It can be checked as follows: the leverages are computed according to the procedure that is described in the additional material. That procedure can be performed accurately even if Z does not have full rank, and the above two relations are checked. If they are not obeyed, then matrix Z does not have full rank. Therefore, the model must be discarded.

A particularly useful result for the estimation of the generalization error is the following: the prediction error $r_k^{(-k)}$ on example k, when the latter is withdrawn from the training set, can be estimated in a straightforward fashion from the prediction error r_k on example k when the latter is in the training set

$$r_k^{(-k)} \cong \frac{r_k}{1 - h_{kk}}.$$

Here again, the result is exact in the case of a linear model (see for instance [Antoniadis et al. 1992]), and it is approximate for a nonlinear model.

A similar approach is discussed in [Hansen 1996] for models trained with regularization.

As an illustration, we describe an academic example: a set of 50 training examples is generated by adding a Gaussian noise, with zero mean and variance 10^{-2}, to the function $\sin x/x$. Figure 2.21 shows the training set and the output of a model with two hidden neurons. A conventional leave-one-out procedure, as described in a previous section, was carried out, providing the values of the quantities $r_k^{(-k)}$ (vertical axis of Fig. 2.22), and the previous relation was used, providing the values on the horizontal axis. All points are nicely aligned on the bisector, thereby showing that the approximation is quite accurate. Therefore, the virtual leave-one-out score E_p,

$$E_P = \sqrt{\frac{1}{N}\sum_{k=1}^{N}\left(\frac{r_k}{1 - h_{kk}}\right)^2},$$

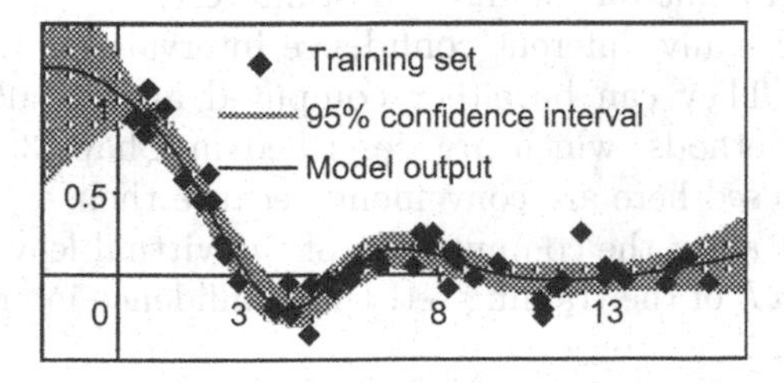

Fig. 2.21. Training set, output and confidence interval on the output, for a model with two hidden neurons

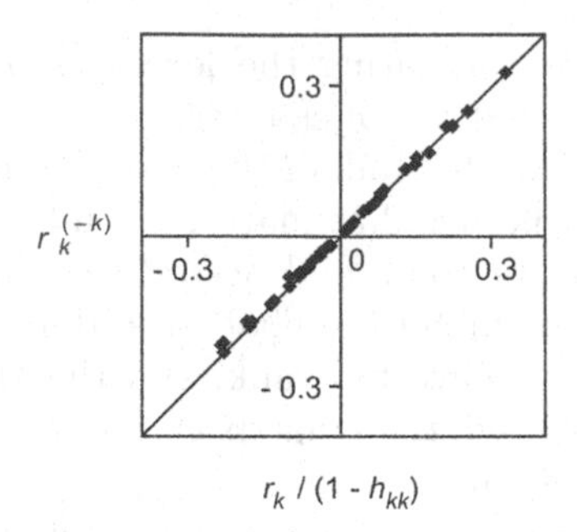

Fig. 2.22. Accuracy of residual estimation in virtual leave-one-out

can be used in a very computationally economic fashion in lieu of the leave-one-out score E_t defined above,

$$E_t = \sqrt{\frac{1}{N}\sum_{k=1}^{N}(r_k^{(-k)})^2},$$

insofar as E_p is a good estimate of the generalization error. It is an essential ingredient in the model selection procedure that will be described in the next section: it provides an estimate of the generalization error, with a computation time that is N times as small as conventional leave-one-out, since training is performed once instead of being performed Ntimes with $N-1$ examples.

Effect of Withdrawal of an Example on the Confidence Interval on the Prediction

In [Seber et al. 1989], an approximate confidence interval is derived for a nonlinear model, with confidence $1-\alpha$,

$$E(Y_p \mid x) \in g(\boldsymbol{x}, \boldsymbol{w}_{\mathrm{LS}}) \pm t_\alpha^{N-q} s \sqrt{\boldsymbol{z}^{\mathrm{T}}(Z^{\mathrm{T}}Z)^{-1}\boldsymbol{z}},$$

where t_α^{N-q} is the value of a Student variable with $N-q$ degrees of freedom and a confidence level $1-\alpha$, and s is an estimate of the variance of the prediction error of the model. Figure 2.22 shows the confidence interval computed from that relation, at all points of the interval of interest.

One can define many different confidence intervals for nonlinear models [Tibshirani 1996]. They can be either computed analytically, or estimated with resampling methods, which are described in Chap. 3. The confidence intervals that are used here are convenient because their expression involves the quantities that allow the computation of the virtual leave-one-out score.

For observation k of the training set, that confidence interval can be written as

$$\begin{aligned} E(Y_p \mid \boldsymbol{x}^k) &\in g(\boldsymbol{x}^k, \boldsymbol{w}_{\mathrm{LS}}) \pm t_\alpha^{N-q} s \sqrt{\boldsymbol{z}^{k\mathrm{T}}(Z^{\mathrm{T}}Z)^{-1}\boldsymbol{z}^k} \\ &= g(\boldsymbol{x}^k, \boldsymbol{w}_{\mathrm{LS}}) \pm t_\alpha^{N-q} s \sqrt{h_{kk}}. \end{aligned}$$

Thus, the confidence intervals on the prediction of the model involve the same quantities h_{kk} (leverages) as the prediction of the effect of the withdrawal of an example from the training set. That is not surprising since both groups of relations arise from a Taylor expansion of the output of the model.

The confidence interval on the prediction of an example that is withdrawn from the training set can also be estimated: given an input vector $\boldsymbol{x}^k$, the approximate confidence interval on the prediction of that example is given by

$$E^{(-k)}(Y_p \mid \boldsymbol{x}^k) \in g(\boldsymbol{x}^k, \boldsymbol{w}_{\mathrm{LS}}) \pm t_\alpha^{N-q-1} s^{(-k)} \sqrt{\frac{h_{kk}}{1-h_{kk}}}$$

[Seber et al. 1989]. In general, $s^{(-k)}$ can be approximated by s.

Interpretation of the Leverages

The leverages are the diagonal elements of an orthogonal projection matrix: they sum to the dimension of that matrix. In the present case, the orthogonal projection is onto the solution subspace, hence its dimension is equal to the number of parameters of the model: therefore, the sum of the leverages is equal to the number of the degrees of freedom of the model. That property can also be expressed as follows: the leverage of example k is the fraction of the degrees of freedom used for fitting example k [Monari et al. 2000, 2002].

Some specific cases are of interest:

- If all leverages are equal, they are equal to q/N: a fraction q/N of the parameters of the model is devoted to each example, and all examples have the same influence on the model: such a model should not exhibit overfitting since it is not "focused" on any example. That property can be used with advantage for model selection, as shown below.
- If a leverage is equal to zero, the model does not devote any degree of freedom to example k. That has a simple geometric interpretation: h_{kk} is the $k-$th component of the projection, onto solution subspace, of the unit vector borne by axis k in observation space; if that axis is orthogonal to the solution subspace, example k does not contribute to the model output, which lies in solution subspace (see Fig. 2.5); therefore, it has no influence on the parameters of the model. Whether that example is in the training set or has been withdrawn from it, the prediction of that example has the same error, as evidenced by relation: $r_k^{(-k)} = r_k/(1-h_{kk})$. The confidence interval on that prediction is zero; the prediction of the model is certainly equal to the expectation value of the quantity of interest.

The fact that the confidence interval is equal to zero does not mean that the prediction of the corresponding point is exact. It is not contradictory with the fact that the prediction error r_k is not zero: the prediction error is the difference between the *measured value* and the *predicted value*: it contains both the modeling error (difference between the *predicted* value and the unknown expectation value) and the noise (difference between the *measured* value and its unknown expectation value). If the model is perfect, the prediction error is due to the noise only. Therefore, one can obtain a leverage equal to zero if and only if the family of functions within which the model is sought contains the regression function.

If a leverage is very close to 1, the unit vector borne by axis k is very close to solution subspace; hence that example is almost perfectly learnt, and it has a large influence on the parameters of the model. The prediction error on that example is almost zero *when the example is in the training set* and it is very large *when the example is withdrawn from the training set.* Therefore, the model is overfitted to that example. The confidence interval on that example is very small when the example is in the training set, and very large when it is not in the training set.

The above interpretation of the leverages is central to the model selection methodology that is discussed in the next section.

2.6.4 Model Selection Methodology by Combination of the Local and Global Approaches

Assume that inputs have been selected as described in the sections devoted to input selection. We try to design the best model given the available data.

We discuss here a constructive procedure, whereby the complexity of the model is increased gradually until overfitting occurs. For didactic purposes, we split the procedure into two steps:

For a family of functions of given complexity, nonlinear with respect to its parameters (for instance, neural networks with a given number of hidden neurons), several trainings are performed with all available data, with different parameters initializations. Thus, several models are obtained; models whose Jacobian matrices do not have full rank are discarded. The next section, explains how to make a choice between the models that were not discarded because of the rank of their Jacobian matrices.

For a model that is linear with respect to its parameters, that step is very simple since the cost function has a single minimum: a single training is performed with all available data.

The previous step having been performed with families of models of increasing complexity, the best model is selected as explained in the section entitled "Selection of the best architecture".

2.6.4.1 Model Selection Within a Family of Models of Given Complexity: Global Criteria

For a given model complexity, several trainings are performed, and, at the end of each training, the rank of the Jacobian matrix of the model thus designed is computed. If that matrix does not have full rank, the model is discarded.

The fact that the global minimum of the cost function, for a family of models of given complexity, gives rise to a model whose Jacobian matrix does not have full rank does not mean that all models that have the same complexity must be discarded: a local minimum may give rise to a perfectly valid model whereas the global minimum gives rise to an overfitted model. That strategy is somewhat similar to early stopping: selecting a model that is not a global minimum of the cost function may be a form of regularization.

In order to perform a selection among the surviving models, the virtual leave-one-out technique is used. The leave-one-out score was defined above as

$$E_P = \sqrt{\frac{1}{N}\sum_{k=1}^{N}\left(\frac{r_k}{1-h_{kk}}\right)^2},$$

which is an unbiased estimate of the generalization error.

That score must be compared to the mean square error on the training set (TMSE),

$$E_T = \sqrt{\frac{1}{N_T}\sum_{k=1}^{N_T}(r_k)^2}.$$

It should be remembered that, in virtual leave-one-out, training is performed with all available data; hence the same quantity N is involved in E_p and $E_T$in the present case.

Generalization Error and TMSE

Since the leverages are positive and smaller than 1, E_p is larger than the TMSE; very overfitted models have numerous leverages on the order of 1, hence have a generalization error that is much larger than the TMSE.

The Case of Large Training Sets

If all leverages are equal to q/N, one has: $E_p = N/(N-q)E_T \cdot E_p$ and E_T are equal in the limit of very large training sets for a model without overfitting, which makes sense since the difference between TMSE and the generalization error stems from the fact that the number of elements in training set is finite: if an infinite amount of data were available, the regression would be known exactly.

As an illustration, consider a neural network with four hidden neurons, whose training was performed, with different initializations, with the Levenberg-Marquardt algorithm, with the training set shown on Fig. 2.21. Five hundred different trainings were performed. Figure 2.23 shows the results, with the following conventions:

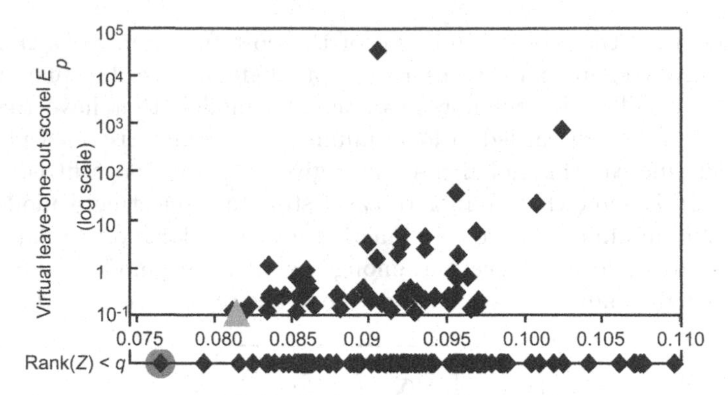

Fig. 2.23. Virtual leave-one-out scores for 500 different models

- For models whose Jacobian matrix does not have full rank, each model is shown as a point in a plane: the horizontal axis is the TMSE, and the vertical axis is the virtual leave-one-out score (estimation of the generalization error of the model); note that the vertical scale is logarithmic.
- For models whose Jacobian matrix does not have full rank, the corresponding points are shown below the graph, on a horizontal scale that shows the TMSE's of those models.

Note that

- The Jacobian matrix of the model with smallest TMSE does not have full rank: that model must be discarded.
- In the present example, 70% of the minima found do not have a Jacobian matrix with full rank.
- The estimate of the generalization error varies by several orders of magnitude, which requires a logarithmic scale for E_p. Models with very high virtual leave-one-out scores are very "specialized" on one or several points, with leverages very close to 1.

Figure 2.24 shows the outputs of the model that has the smallest value of E_T and of the model that has the smallest value of E_p (shown as a gray circle and a gray triangle respectively on Fig. 2.23). Note that the model with minimal E_T gives a prediction that is less smooth than the model with minimal E_p. Therefore, the latter is more satisfactory; however, it is the most satisfactory model among models that have four *hidden neurons.* In order to finalize the selection, that model must be compared to the best models found for different complexities

Figure 2.25 shows the virtual leave-one-out scores and the TMSE's of the best networks, found by the above procedure, for complexities increasing from 0 hidden neuron (linear model) to 5 hidden neurons. As additional information, the graph also displays the standard deviation of noise (which, in general, would be unknown in a real application). As expected, the TMSE

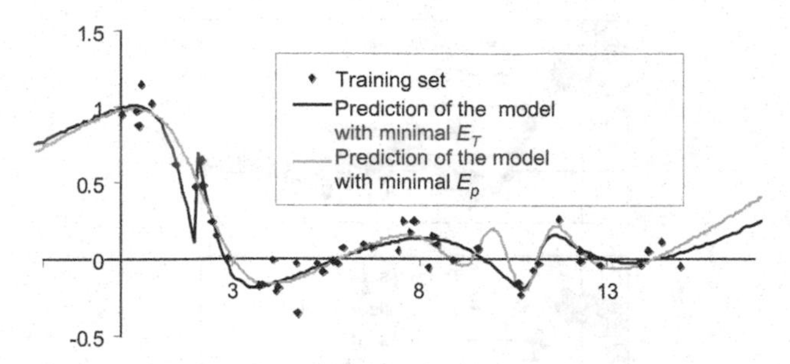

Fig. 2.24. Outputs of two models with 4 hidden neurons: the model that has the minimal TMSE, and the model that has the minimal virtual leave-one-out score

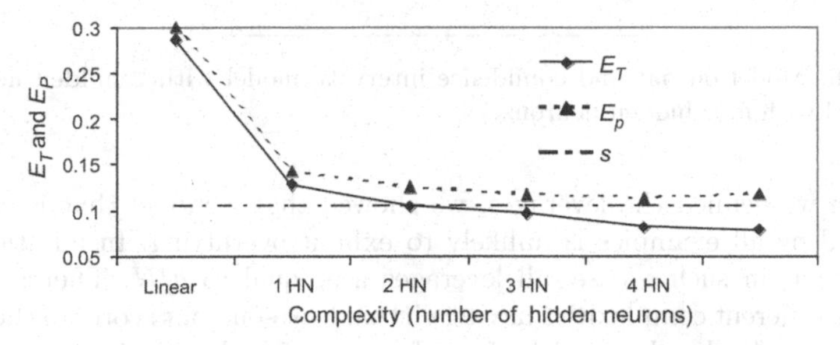

Fig. 2.25. Variation of the TMSE and of the virtual leave-one-out score as a function of the number of hidden neurons

decreases when the number of hidden neurons increases, whereas the virtual leave-one-out score seems to go through a minimum and subsequently to increase. However, the choice between 2, 3 and 4 hidden neurons is not perfectly clear, since the leave-one-out scores are not very different. The next section is devoted to the problem of the choice of the most appropriate architecture.

For more than three hidden neurons, the TMSE becomes smaller than the standard deviation of the noise; one can rightly conclude that models with more than three hidden neurons tend to be overfitted. However, that is not a practical selection criterion, since, in real applications, the standard deviation of the noise is generally unknown.

2.6.4.2 Selection of the Best Architecture: Local Criteria (LOCL Method)

In the previous section, a global criterion—the virtual leave-one-out score—was used for finding the model that is least prone to overfitting, among models having the same complexity. We have also shown that that criterion may not be sufficient for making a choice between models of different complexities. In such a case, it is advantageous to use the local overfitting control via leverages method (LOCL), based on the values of the leverages [Monari 1999; Monari et al. 2002].

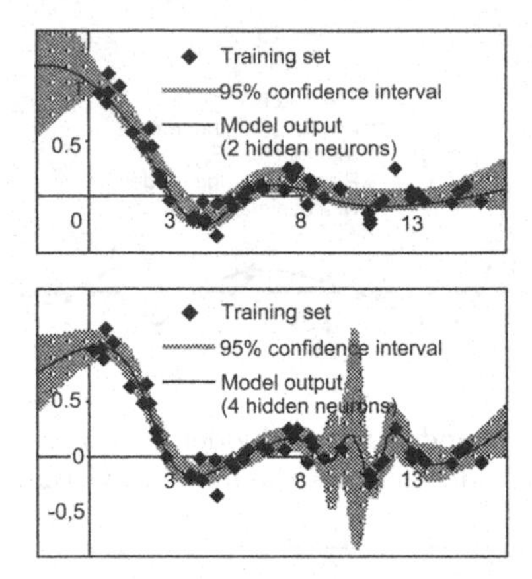

Fig. 2.26. Model output and confidence intervals; model with 2 hidden neurons, and model with four hidden neurons

When we defined the leverages, we showed that a model that is equally influenced by all examples is unlikely to exhibit overfitting. In addition, we showed that, in such a case, all leverages are equal to q/N. Therefore, for models of different complexities having virtual leave-one-out scores of the same order of magnitude, the model whose leverage distribution is most peaked around q/N will be favored, except in cases where it is known from prior knowledge that it is important that the model fit very accurately some specific observations.

Figure 2.26 shows, for the example that was discussed previously, the predictions of the best models selected with 2 and 4 hidden neurons. The same graphs display the 95% confidence intervals for the predictions of those models. For the two hidden neuron model, the confidence interval is roughly constant over the whole training domain, whereas, for the four-hidden neuron model, the confidence interval is large in [8, 12]; the output of that model oscillates, and it is not clear whether that oscillation is significant, or is just a consequence of fitting the model to the local realization of noise. The leverage distribution, shown on Fig. 2.27, reveals that the latter are more scattered for the model with 4 hidden neurons (gray), than for the model with two hidden neurons (black).

It is convenient to associate the quantity μ, defined as,

$$\mu = \frac{1}{N}\sum_{k=1}^{N}\sqrt{\frac{N}{q}h_{kk}},$$

to the leverage distribution. That quantity has the following properties:

- It is always smaller than 1.
- It is equal to 1 if and only if all leverages are equal to q/N.

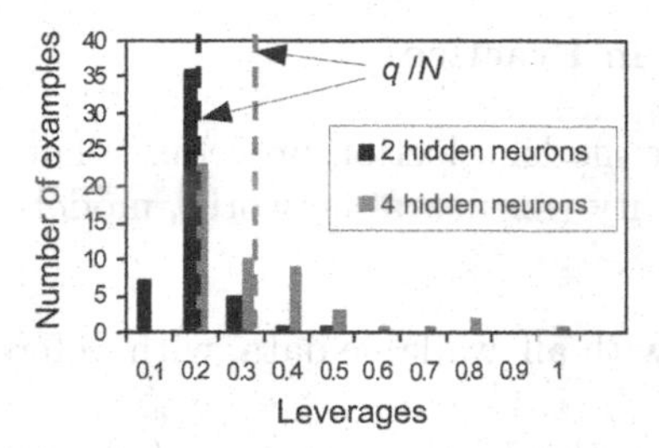

Fig. 2.27. Histogram of leverages for the best model with 2-hidden neurons and for the best model with 4 hidden neurons

Thus, μ is a normalized quantity that may characterize of the leverage distribution: the closer μ to 1, the more peaked the leverage distribution around q/N. Thus, among models of different complexities having virtual leave-one-out scores on the same order of magnitude, the model whose μ is closest to 1 will be favored.

We illustrate the usefulness of μ on the previous example. A test set of $N_G = 100$ examples was generated. The generalization error of the candidate models can be estimated by computing the mean square difference between the expectation value of the quantity to be modeled (which, in the present academic example, is known to be $\sin x/x$) and the prediction of the model,

$$E_G = \sqrt{\frac{1}{N_G}\sum_{k=1}^{N_G}\left(E_Y(\boldsymbol{x}^k) - g(\boldsymbol{x}^k, \boldsymbol{y})\right)^2}.$$

Figure 2.28 shows the quantities E_p, E_T, E_G and μ, as a function of model complexity. μ goes through a maximum that is very close to 1 for two hidden neurons; that is the architecture for which the generalization error E_G is minimum. Thus, μ is a suitable criterion for a choice between models whose virtual leave-one-out scores do not allow a safe discrimination.

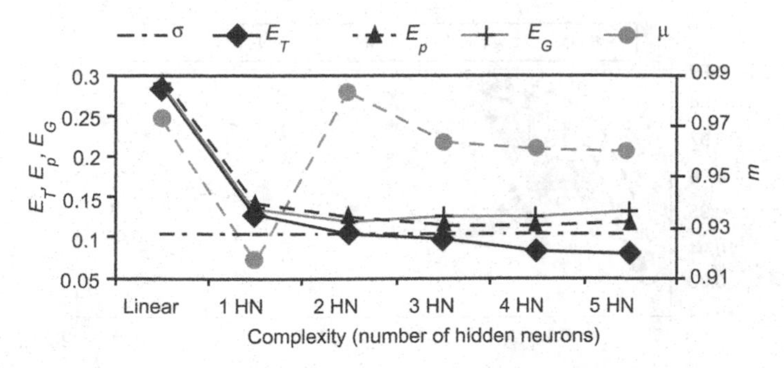

Fig. 2.28. TMSE, virtual leave-one-out score, generalization error and μ, as a function of the number of hidden neurons

2.6.4.3 What to Do in Practice?

We summarize here the model selection procedure that has been discussed.

For a given complexity (for neural networks, models with a given number of hidden neurons),

- Perform trainings, with all available data, with different parameter initializations.
- Compute the rank of the Jacobian matrix of the models thus generated, and discard the models whose Jacobian matrix does not have full rank.
- For each surviving model, compute its virtual leave-one-out score and its parameter μ.

For models of increasing complexity: when the leave-one-out scores become too large or the parameters μ too small, terminate the procedure and select the model. It is convenient to represent each candidate model in the $E_p - \mu$ plane, as shown, for the previous example, on Fig. 2.29. The model should be selected within the outlined area; the choice within that area depends on the designer's strategy:

- If the training set cannot be expanded, the model with the largest μ should be selected among the models that have the smallest E_p.
- If the training set can be expanded through further measurements, then one should select a slightly overfitted model, and perform further experiments in the areas where examples have large leverages (or large confidence intervals); in that case, select the model with the smallest virtual leave-one-out score E_p, even though it may not have the largest μ.

2.6.4.4 Experimental Planning

After designing a model along the guidelines described in the previous sections, it may be necessary to expand the database from which the model was

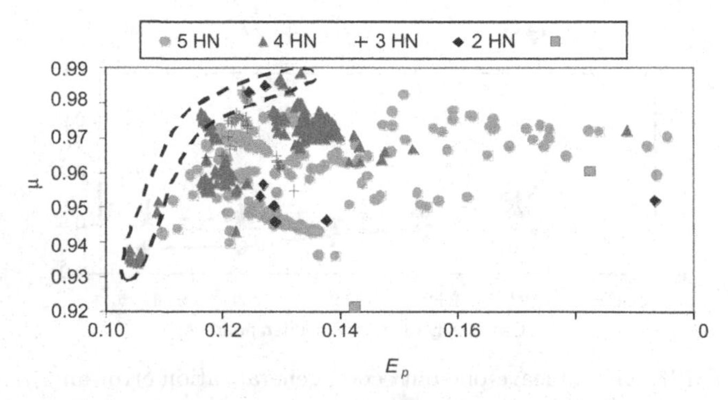

Fig. 2.29. Assessment of the quality of a model in the $E_p - \mu$. plane

designed. Then one should perform experimental planning, taking advantage of the results obtained during the design of the model, with emphasis on confidence intervals: the presence of large confidence intervals in an area of input space may be due to an inappropriate number of examples in that area. Therefore, measurements should be performed in the areas of input space where confidence intervals are too large.

2.6.4.5 Conclusion

The design of a good model requires a systematic, principled methodology. We have shown that such a methodology exists, which can be applied for designing essentially any nonlinear model, including, but not limited to, neural networks. Its principles are the following:

- Neural networks are parsimonious approximators, that can be advantageously used for models having more than two variables; for models with less than two variables, models that are linear with respect to their parameters, such as polynomials, give excellent results and are trained more easily.
- Whether the model is linear or nonlinear with respect to its parameters, the first step consists in an analysis of the input data, in order to find a data representation that is as compact as possible, and in a subsequent input selection in order to select only the candidate variables that are really relevant.
- A model architecture is subsequently chosen (number of monomials for a polynomial model, number of hidden neurons for a neural model, etc.), and the parameters of the model are estimated (training). Those tasks are performed from the simplest architecture (linear model), gradually increasing the complexity of the models.
- For each architecture the best model is selected, and the "best" models of the different structures are mutually compared, until the final choice is performed.

2.7 Dynamic Black-Box Modeling

The previous section discussed the design of static models, i.e., models that implement a static input-output mapping. Those models are very useful for modeling a process in a steady state, or for finding relations between time-independent data.

In the present section, we discuss dynamic models, whose inputs and outputs are related through differential equations, or, for discrete-time systems, by recurrent equations or difference equations. In the present chapter, we consider only discrete-time systems because the vast majority of real applications of neural networks involve computers or digital integrated circuits, which are sampled systems: the quantities of interest are measured at discrete times, which are integer multiples of a sampling period T.

For simplicity, the quantity T will be omitted in equations below: the value of a variable x at time kT, k positive integer, will be denoted as $x(k)$.

Chapter 4 offers a general view of nonlinear dynamic systems. In this chapter, the presentation will be restricted to a cursory introduction to continuous-state stochastic modeling, which derives directly from the previous discussions on static modeling. The elements of dynamic modeling that are presented here are sufficient for understanding the methodology of semiphysical modeling, which is very important for industrial applications.

2.7.1 State-Space Representation and Input-Output Representation

Dynamic modeling has several specific features, which are not relevant to static modeling.

The first specific feature is the existence of several *representations* for the dynamic model of a given process (see for instance [Kuo 1995] for an introduction to dynamic systems, and [Kuo 1992] for an introduction to discrete-time systems). In the following, the modeling of a single-output process is discussed; its extension to multiple-output systems is relatively straightforward. A model is said to be a state-space representation if its equations are in the form:

$$\begin{cases} \boldsymbol{x}(k) = \boldsymbol{f}(\boldsymbol{x}(k-1), \boldsymbol{u}(k-1), \boldsymbol{b}_1(k-1)) & \text{state equation} \\ y(k) = g(\boldsymbol{x}(k), \boldsymbol{b}_2(k)) & \text{observation equation or output equation,} \end{cases}$$

where vector $\boldsymbol{x}(k)$ is the state vector (whose components are the state variables), vector $\boldsymbol{u}(k)$ is the control input vector, $\boldsymbol{b}_1(k)$ and $\boldsymbol{b}_2(k)$ are the vectors of disturbances, and scalar $y(k)$ is the model output. $\boldsymbol{f}$ is a nonlinear vector function, and g is a nonlinear scalar function. The dimension of the state vector (i.e., the number of state variables) is called the "order" of the model. The state variables may be either measured or not measured.

For a single-input process with control input $u(k)$, the components of vector $\boldsymbol{u}(k)$ may be $u(k)$ and past values of the input control signal: $\boldsymbol{u}(k) = [u(k), u(k-1), \ldots, u(k-m)]^{\mathrm{T}}$.

The disturbances have an influence either on the output, or on the state variables, or on both. As opposed to control inputs, they are not measured. Therefore, they cannot be inputs of the model, although they do have an influence on the quantity to be measured. For instance, for an oven, the current intensity that flows in the heating resistor is a control input; the measurement noise of the thermocouple is a disturbance that can be modeled, if necessary, as a sequence of realizations of random variables.

The output may be one of the state variables (an example will be described in the section "What to do in practice?".

Thus, the designer of a state-space model seeks approximations of functions $\boldsymbol{f}$ and g, through training from sequences of inputs, of outputs and possibly of state variables if the latter are measured.

A model is in *input-output representation* if its equations are in the form

$$y(k) = h(y(k-1), \ldots, y(k-n), \boldsymbol{u}(k-1), \ldots, \boldsymbol{u}(k-m), \boldsymbol{b}(k-1), \ldots, \boldsymbol{b}(k-p)),$$

where h is a nonlinear function, n is the order of the model, m and p are two positive integer constants, $\boldsymbol{u}(k)$ is the vector of input control signals, $\boldsymbol{b}(k)$ is the vector of disturbances. Input-output representations are special forms of state-space representations, where the components of the state vector are $[y(k-1), y(k-2), \ldots y(k-n)]$.

In linear modeling, state-space representations and input-output representations are equivalent: one chooses the representation that is most convenient in view of the purpose that the model is intended to serve. By contrast, in nonlinear modeling, state-space representations are more general and more parsimonious than input-output models [Levin et al. 1993], as will be illustrated below on a real application; however, the design of a state-space model may be slightly more difficult than that of an input-output model, since two functions $\boldsymbol{f}$ and g must be approximated, while input-output models require the approximation of a single function h.

Once a choice has been made between state-space and input-output representation, an assumption must be made as to the influence of noise on the process. That is a basic fact that is often overlooked in the neural network literature, whereas it is common knowledge in linear dynamic modeling, as is shown in Chap. 4. In the present chapter, we show that the assumption on the noise has a deep influence on the *training* algorithm that must be used, on the structure of the model that must be implemented, and on its subsequent mode of operation. In the next section, the main assumptions on noise are discussed, and the resulting constraints on the training of the model, on its structure and on its operation are explained.

2.7.2 Assumptions on Noise and Their Consequences on the Structure, the Training and the Operation of the Model

In the present section, various assumptions on the influence of noise on the process are considered. We first discuss the assumptions and their consequences on the structure, training and operation of input-output models, then the consequences of the assumptions on state-space models.

2.7.2.1 Input-Output Representations

State Noise Assumption (Input-Output Representation)

We assume that the model can be appropriately described, in the desired validity domain, by a representation of the form

$$y_p(k) = \psi(y_p(k-1), \ldots, y_p(k-n), \boldsymbol{u}(k-1), \ldots, \boldsymbol{u}(k-m)) + b(k),$$

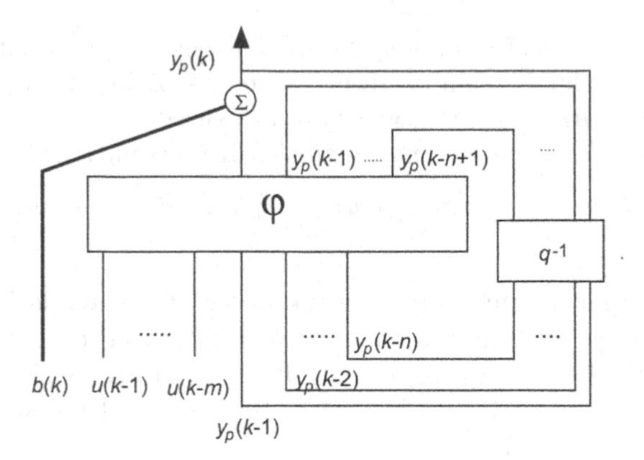

Fig. 2.30. Input-output representation, state noise assumption

where $y_p(k)$ is the measured process output. We assume that additive noise occurs at the process output (see Fig. 2.30), and that, at time k, noise influences the present output, and also the n past outputs. In nonlinear modeling, that assumption is known as NARX (Nonlinear Auto-regressive with eXogenous inputs) (see also Chap. 4) or *equation error* (see for instance [Ljung 1987; Goodwin et al. 1984]), or series-parallel [Narendra et al. 1989] in adaptive modeling.

Instead of the term assumption, the term postulated model is sometimes used in the statistics literature.

We assume that noise acts on the output, not only directly at time k, but also through the outputs at the n previous time steps; since the model that is sought should be such that the modeling error at time k is equal to noise at the same time step, it should take into account the *process outputs* at the n previous time steps. Consider the feedforward neural network shown on Fig. 2.31; it obeys the equation

$$g(k) = \varphi_{NN}(y_p(k-1), \ldots, y_p(k-n), \boldsymbol{u}(k-1), \ldots, \boldsymbol{u}(k-m), \boldsymbol{w}),$$

where $\boldsymbol{w}$ is a vector of parameters, and where function φ_{NN} is performed by the feedforward neural network. Assume that the neural network φ_{NN} has been trained, i.e., that a vector of parameters $\boldsymbol{w}$ has been found such that the network computes exactly function φ. Then relation $y_p(k) - g(k) = b(k)$ holds for all k. Thus, the model is such that the modeling error is equal to the noise of the process: it is the ideal model, since it captures all that is *deterministic* in the representation and does not model noise. Note that the inputs of the model are the control inputs and the measured *process* outputs: the ideal model (also called "predictor") is *not* trained as a recurrent neural network.

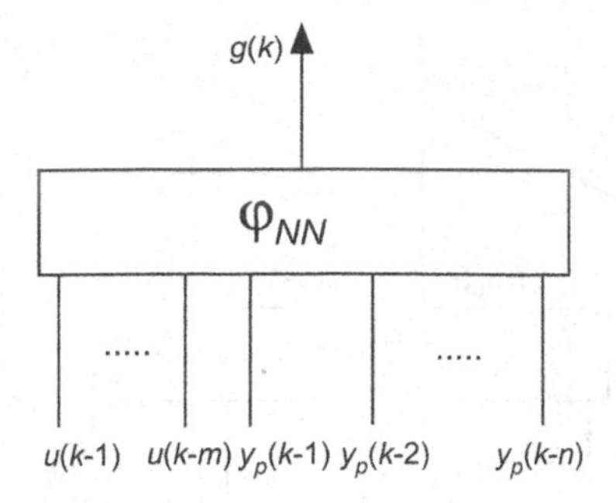

Fig. 2.31. The ideal model for an input-output representation with state noise assumption

Training of the Model: Directed (Teacher-Forced) Training

Since the ideal model is a feedforward neural network, it is trained with the techniques that were discussed in the section devoted to the training of static models. Training is called directed or teacher-forcing.

Operation of the Model

Since the inputs of the predictor are (in addition to control inputs) the measured outputs of the process, the output of the model can be computed only one step ahead of time; the predictor is said to be a "one-step ahead predictor". If the model is intended for use as a simulator, i.e., for predicting the process output on a time horizon that exceeds one sampling period, the inputs are necessarily the previous outputs of the predictor: the latter is no longer operated in optimal conditions.

Output Noise Assumption (Input-Output Representation)

Now we make a different assumption, namely, that the process can be appropriately described, in the desired validity domain, by a representation of the form

$$\begin{aligned} x_p(k) &= \varphi(x_p(k-1), \ldots, x_p(k-n), u(k-1), \ldots, u(k-m)) \\ y_p(k) &= x_p(k) + b(k). \end{aligned}$$

Therefore, the present assumption considers that the noise is additive on the output (Fig. 2.32). Thus, it appears outside the loop, hence it has an influence on the output at the same time step only. That assumption is known , in linear adaptive modeling, as "output error" or "parallel" [Narendra et al. 1989]. Since the output at time k is a function of the noise at the same time step only, the model that is sought should not involve the past process outputs. Therefore, we consider a recurrent neural network, shown on Fig. 2.33, which obeys the equation

$$g(k) = \varphi_{NN}(g(k-1), \ldots, g(k-n), u(k-1), \ldots, u(k-m), \boldsymbol{w}),$$

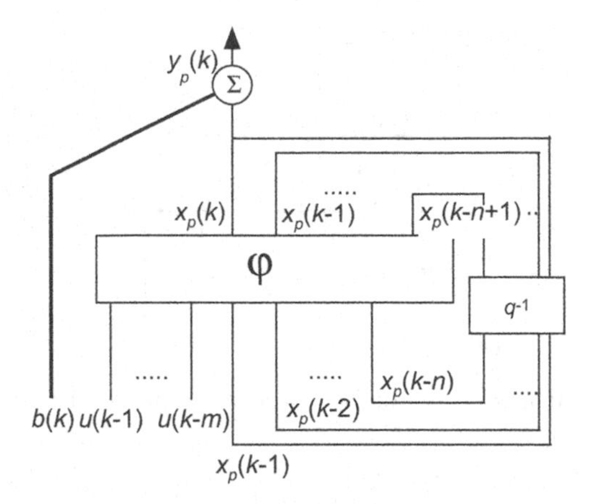

Fig. 2.32. Input-output representation, output noise assumption

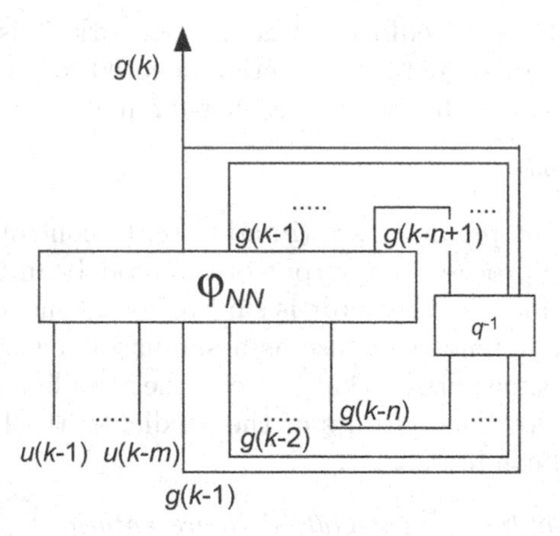

Fig. 2.33. The ideal model for an input-output representation with output noise assumption

where $\boldsymbol{w}$ is a vector of parameters, and where function φ_{NN} is computed by a feedforward neural network. Assume that the network has been trained so that φ_{NN} is exactly equal to φ. Moreover, assume that the prediction error is equal to the noise at the first n time steps: $y_p(k) - g(k) = b(k)$ for $k = 0$ to $n-1$. Then one has $y_p(k) - g(k) = b(k)$ for all k. Thus, the prediction error of the model is equal to the noise: the model is therefore ideal, since it accounts for all that is deterministic in the representation, and does not model noise.

If the initial condition is not obeyed, but nevertheless $\varphi_{RN} = \varphi$, and if the model is stable irrespective of the initial conditions, the modeling error vanishes as k increases.

Note that, in that case, the ideal model is *recurrent*.

Training of the Model: Semidirected Training

The training of a recurrent model can be cast into the framework of the training of a feedforward neural network, as will be shown below in the section devoted the training of recurrent neural networks ("semidirected training").

Operation of the Model

As opposed to the previous case, the model can be operated as a simulator in optimal conditions. Of course, it can also serve as a one-step-ahead predictor.

2.7.2.2 Illustration

Before carrying on with the main assumptions, we illustrate the importance of the proper choice of the training procedure depending on the influence of the noise on the process. This illustrative example is excerpted from [Nerrand 1992] and [Nerrand et al. 1994].

Modeling a Process with Output Noise

We consider a computer-simulated process that obeys the following equations:

$$x_p(k) = \left[1 - \frac{T}{a + b\, x_p(k-1)}\right] x_p(k-1) + \left[T\frac{c + d\, x_p(k-1)}{a + b\, x_p(k-1)}\right] u(k-1),$$

$$y_p(k) = x_p(k) + b(k),$$

with $a = -0.139$, $b = 1.2$, $c = 5.633$, $d = -0.326$, and sampling period $T = 0.1$ sec. $b(k)$ is a white noise with maximum amplitude 0.5. Thus it is a process with output noise. Figure 2.34 shows the response of the simulated process to a pseudo-random sequence of steps.

When modeling a real process, the influence of noise is generally not known. Therefore, several possible assumptions are made; trainings are performed according to each assumption, and the results are compared. We use that approach in the present academic example.

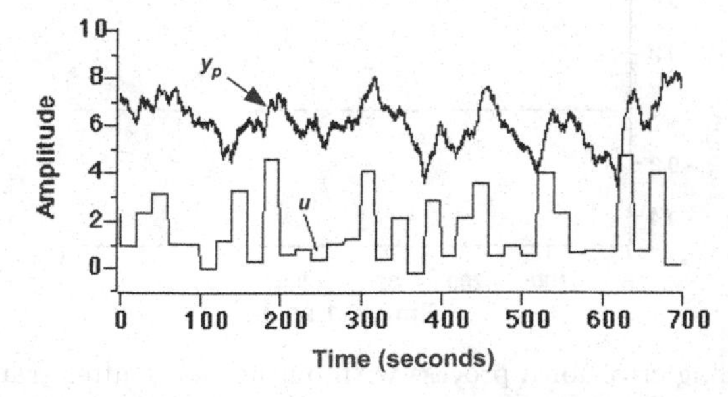

Fig. 2.34. Response of the simulated process to a pseudo-random step sequence

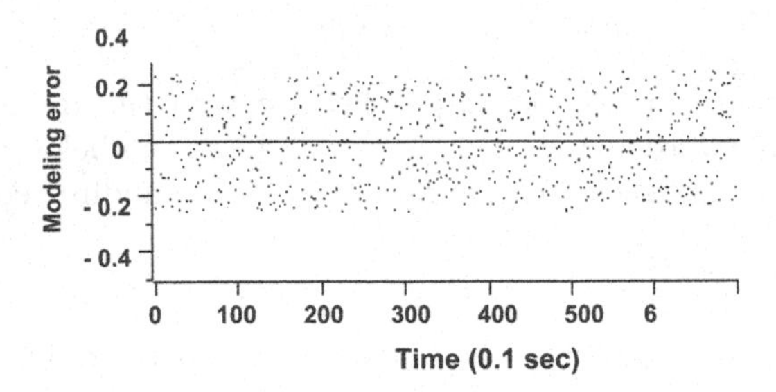

Fig. 2.35. Modeling error for a process with output noise, after training according to the output noise assumption

Output Noise Assumption

We first consider the (correct) assumption that noise can be modeled as output noise. Therefore, the ideal predictor is recurrent. Figure 2.35 shows the modeling error after training a recurrent neural network with 5 hidden neurons. The modeling error is white noise with amplitude 0.5: by making the right assumption and using the appropriate structure and training (recurrent neural network and semidirected training), the modeling error is equal to the noise, which is the best achievable result.

State Noise Assumption

Now, we consider the (wrong) assumption that the noise is state noise. According to that assumption, a feedforward neural network with 5 hidden neurons is trained. Figure 2.36 shows the resulting modeling error: its amplitude is larger than 0.5. As expected, the result is not as satisfactory as the result obtained with the output noise assumption, since we made the wrong assumption. It should be clearly understood that this is not a "technical" problem (too few or too many hidden neurons, inefficient optimization algorithm, inappropriate training set, etc.), but a basic problem: even with the best training algorithm,

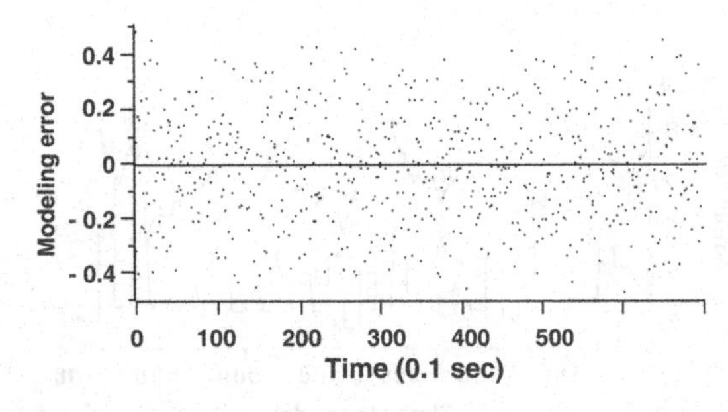

Fig. 2.36. Modeling error for a process with output noise, after training according to the state noise assumption

a perfectly appropriate architecture, and an arbitrarily large training set, one cannot obtain a modeling error equal to the noise if an inappropriate noise assumption is made.

Modeling a Process with State Noise

We consider a computer-simulated process, which obeys the following equation:

$$y_p(k) = \left[1 - \frac{T}{a + b\ y_p(k-1)}\right] y_p(k-1) + \left[T\frac{c + d\ y_p(k-1)}{a + b\ y_p(k-1)}\right] u(k-1) + b(k).$$

It is thus a process with state noise, whose deterministic part is the same as above: it will be modeled by a feedforward neural network with 5 hidden neurons, as above. Again we make the two noise assumptions (output noise and state noise).

Output Noise Assumption

We first make the (wrong) assumption that the noise is output noise. The ideal model would be a recurrent one. Figure 2.37 shows the modeling error after training a recurrent neural network with 5 hidden neurons. The modeling error is clearly not white noise: the modeling error contains deterministic information that the training of the model was unable to capture. Here again, the failure is not due to a technical problem (too few or too many neurons, inefficient training algorithm, inappropriate training data): it is due to the fact that the model has a wrong structure, following the wrong assumption that was made at the beginning.

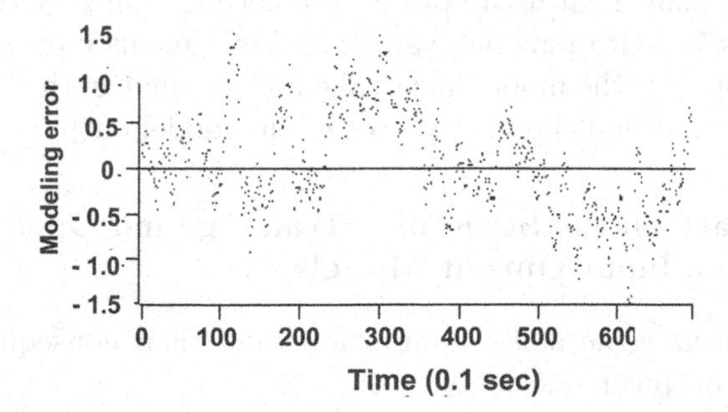

Fig. 2.37. Modeling error for a process with state noise after training according to the output noise assumption

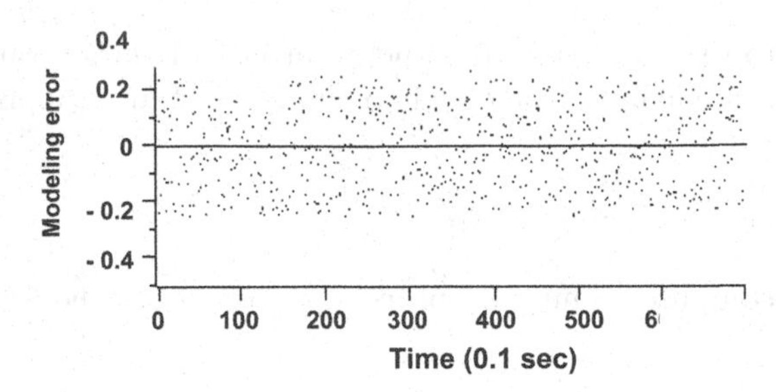

Fig. 2.38. Modeling error for a process with state noise after training according to the state noise assumption

State Noise Assumption

Finally, we make the (right) assumption that the noise is state noise. The ideal model is a feedforward neural network. Figure 2.38 shows that the modeling error is white noise with amplitude 0.5: the ideal predictor was thus obtained.

2.7.2.3 Output Noise and State Noise Assumption (Input-Output Representation)

Now we make the assumption that the noise has an influence both on the output and on the state; the process can be appropriately described by a model of the form

$$\begin{aligned}x_p(k) &= \varphi(x_p(k-1),\dots,x_p(k-n),u(k-1),\dots,u(k-m),b(k-1),\dots,\\&\qquad b(k-p))\\y_p(k) &= x_p(k)+b(k),\end{aligned}$$

as shown on Fig. 2.39. That assumption is sometimes called NARMAX (nonlinear autoregressive with moving average and exogenous inputs).

In the present case, the model must take into account both the past values of the process output and the past values of the model output.

2.7.2.4 Summary on the Structure, Training, and Operation of Dynamic Input-Output Models

Table 2.1 summarizes the noise assumptions and their consequences on the raining of input-output models.

2.7.2.5 State-Space Representations

We consider here the same assumptions as in the previous section, but we discuss their consequences on state-space models.

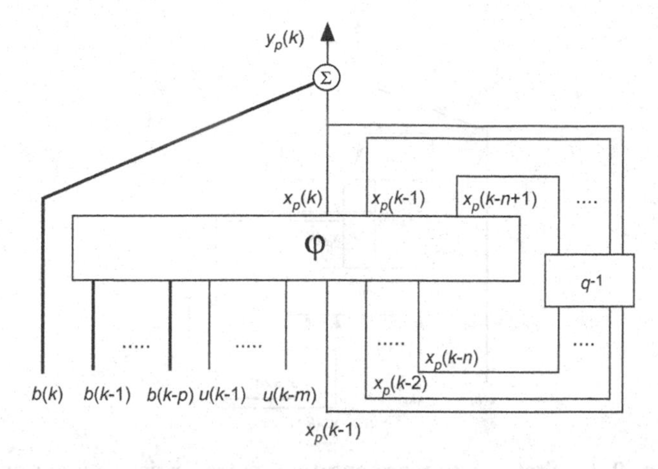

Fig. 2.39. NARMAX assumption

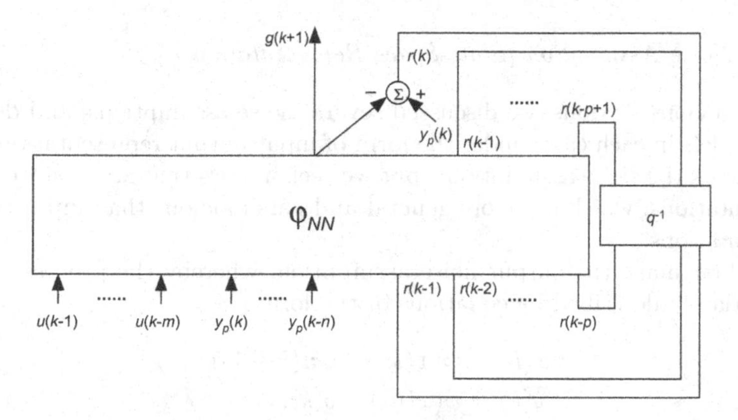

Fig. 2.40. NARMAX model

Table 2.1. Noise assumptions and their consequences on the training of input-output models

Assumption	Usual name in nonlinear	Equivalent in linear modeling	Training	Recommended operation
State noise		ARX	Directed	One-step-ahead predictor
Output noise		Output error	Semidirected	Simulator
State noise and output noise	NARMAX	ARMAX	Semidirected	One-step-ahead predictor

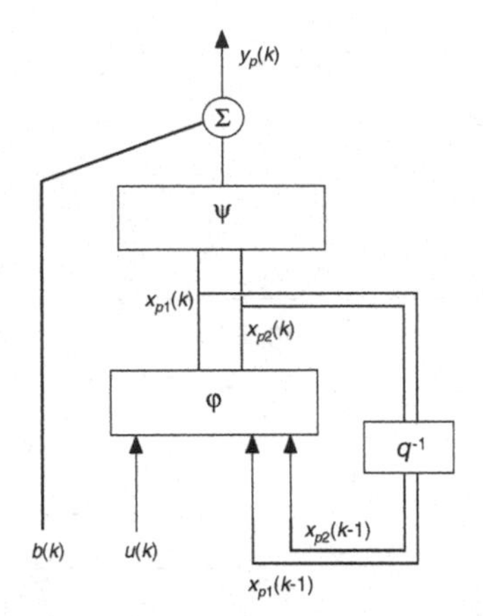

Fig. 2.41. State-space representation, output noise assumption

Output Noise Assumption (State-Space Representation)

In the previous sections, we discussed several noise assumptions, and derived ideal models in each case, under the form of input-output representations. We now discuss the same assumptions, but we seek models that are in state-space representations, which are more general and parsimonious than input-output representations.

We first make the output noise assumption, whereby the process can be appropriately described by equations of the form

$$\begin{aligned} \boldsymbol{x}(k) &= \boldsymbol{\varphi}(\boldsymbol{x}(k-1), \boldsymbol{u}(k-1)) \\ y(k) &= \psi(\boldsymbol{x}(k)) + b(k), \end{aligned}$$

as shown on Fig. 2.41 for a second-order model.

Because noise is present in the observation equation only, it has no influence on the dynamics of the model. From arguments similar to those we developed for input-output representations, the ideal model is recurrent, as shown on Fig. 2.42:

$$\begin{aligned} \boldsymbol{x}(k) &= \boldsymbol{\varphi}_{NN}(x(k-1)), \boldsymbol{u}(k-1)) \\ y(k) &= \boldsymbol{\psi}_{NN}(\boldsymbol{x}(k)), \end{aligned}$$

where $\boldsymbol{\varphi}_{NN}$ is exactly function $\boldsymbol{\varphi}$ et ψ_{NN} is exactly function ψ.

State Noise Assumption

We now assume that the process can be appropriately described by equations

$$\begin{aligned} \boldsymbol{x}(k) &= \varphi\left(\boldsymbol{x}\left(k-1\right), \boldsymbol{u}\left(k-1\right), \boldsymbol{b}\left(k-1\right)\right), \\ y(k) &= \psi\left(x(k)\right). \end{aligned}$$

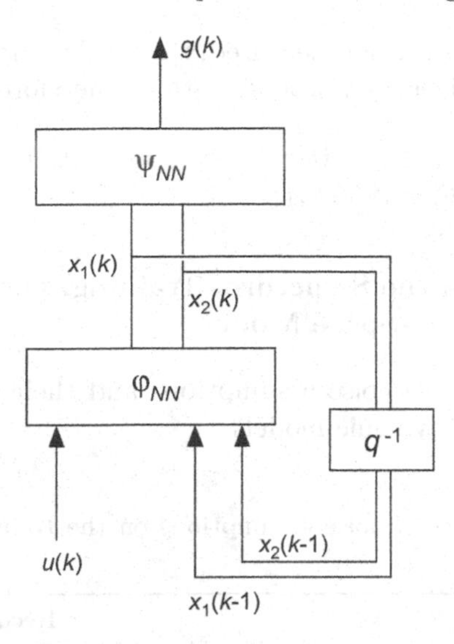

Fig. 2.42. Ideal model for a state-space representation with the "output noise" assumption

Then, from an argument that is similar to those developed for input-output models, the inputs of the ideal model should be, in addition to the control inputs $\boldsymbol{u}$, the state variables of the process. Two situations must be considered:

- Those variables are measured: then they can be considered as outputs, so that the problem is amenable to the design of an input-output model: the ideal model is a feedforward one, which can essentially be operated as a one-step-ahead predictor.
- Those variables are not measured: then the ideal model cannot be constructed; in such a case, one should either use an input-output representation (although not completely general), or design a feedback model (although non optimal).

Output Noise and State Noise (State-Space Representation)

Finally, we assume that the process can be appropriately described by the equations

$$\begin{aligned} \boldsymbol{x}(k) &= \varphi\left(\boldsymbol{x}\left(k-1\right), \boldsymbol{u}\left(k-1\right), \boldsymbol{b}_1\left(k-1\right)\right), \\ y(k) &= \psi\left(x(k)\right). \end{aligned}$$

Here again, two cases must be considered:

- If the state variables are measured, they can be regarded as outputs, so that the problem is amenable to the design of an input-output representation, as described previously.

- If the state variables are not measured, the ideal model should involve both the state and the measured process output; therefore, it is in the form

$$\begin{aligned} \boldsymbol{x}(k) &= \varphi\left(\boldsymbol{x}\left(k-1\right), \boldsymbol{u}\left(k-1\right), y_p\left(k-1\right)\right) \\ y(k) &= \psi\left(x(k)\right). \end{aligned}$$

2.7.2.6 Summary on the Structure, Training, and Operation of Dynamic State-Space Models

Table 2.2 summarizes the noise assumptions and their consequences on the training of state-space dynamic models.

Table 2.2. Consequences of noise assumptions on the training of dynamic state-space models

Assumption	Training	Recommended operation
State noise (measured state)	Directed	One-step-ahead predictor
State noise (state not measured)	Semidirected	Simulator (non optimal)
Output noise	Semidirected	Simulator
State noise and output noise	Semidirected	One-step-ahead predictor

2.7.3 Nonadaptive Training of Dynamic Models in Canonical Form

The previous sections have shown how to choose the structure of the dynamic model, as a function of the noise that is likely to be present in the process, so that one can hope to approach the ideal model, i.e., the model that accounts for the deterministic part of the process. We assume that appropriate sequence of inputs and outputs are available: we consider nonadaptive (batch) training.

In all the following, we assume that the model whose training must be performed is in canonical form, i.e., it is under the form

$$\begin{aligned} \boldsymbol{z}\left(k+1\right) &= \boldsymbol{\Phi}\left(z(k), \boldsymbol{u}(k)\right) \\ \boldsymbol{y}\left(k+1\right) &= \boldsymbol{\Psi}\left(z(k), \boldsymbol{u}(k)\right), \end{aligned}$$

where $\boldsymbol{z}(k)$ is the minimal set of ν variables, which allows the computation of the model at time $k+1$ knowing the state and the inputs of the model at time k, and where the vector functions $\boldsymbol{\Phi}$ and $\boldsymbol{\Psi}$ are feedforward neural networks.

ν is the order of the model. Therefore, that form is the minimal state-space representation; if the state vector is in the form

$$z(k) = [y(k), y(k-1), \ldots, y(k-\nu+1)]^{\mathrm{T}},$$

the canonical form is an input-output model: the output is the only quantity involved in the state vector.

Two cases must be considered:

- A black-box model is sought: then a model should be designed under the canonical form, since there is no reason to choose another form;
- A semiphysical model is sought, taking into account prior knowledge: the latter may lead to a model that is not in canonical form; then, prior to training, the predictor should be put in canonical form, which is always possible. The section entitled "Casting dynamic models into a canonical form" is devoted to that problem.

In the following sections, the model is always assumed to be in its canonical form.

We first discuss the training of feedforward models, then the training of recurrent models.

2.7.3.1 Nonadaptive (Batch) Training of Feedforward Input-Output Models: Directed (Teacher-Forced) Training

Under the state noise assumption, the ideal model is a feedforward (static) model whose inputs are the control inputs and the measured process outputs at the previous n time steps. The training is called directed by the process, or teacher-forced, since the measured process outputs are input to the model during, as shown on Fig. 2.43. Thus, the model is permanently "driven" by the process outputs. The training of that model is exactly similar to the training of a static model. The training set is a sequence of N input-output pairs $\{\boldsymbol{z}^k, \boldsymbol{y}^k\}$, where N is the length of the training sequence,

$$z^k = [u(k), u(k-1), \ldots, u(k-m+1), y_p(k), y_p(k-1), \ldots, y_p(k-n+1)]^{\mathrm{T}},$$
$$y^k = y_p(k+1).$$

The Dumb Predictor Pitfall

In directed training, the measured outputs of the process are input to the model, at each time step. Therefore, deceptively good results can readily be obtained, if the quality of the model is assessed by carelessly superimposing graphically the measured output and the predicted output. Actually, a "dumb predictor" made of a simple unit time delay, i.e., a predictor that states that the process output at time $k+1$ will be equal to the process output measured at time k, may provide excellent results, if the process output does not vary much during a sampling period, i.e., if the sampling frequency is high enough. Therefore, after training by teacher forcing, the results should always be compared to those of the dumb predictor. Disappointments are not infrequent.

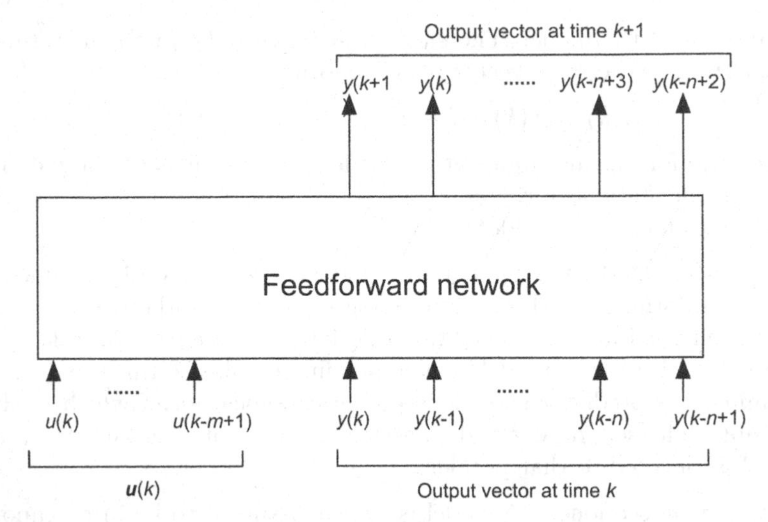

Fig. 2.43. Copy k of the feedforward neural network of the canonical form, for semidirected training

2.7.3.2 Nonadaptive (Batch) Training of Recurrent Input-Output Models: Semidirected Training

Under the output noise assumption, or the output noise and state noise assumption, the ideal model is a recurrent model, the inputs of which are

- the control inputs and the outputs of the model at the n previous time steps (under the assumption of output noise alone),
- the control inputs, the outputs of the model and the modeling errors on a suitable horizon p (under the NARMAX assumption).

Output Noise Assumption

Because the model is recurrent, its training, from a sequence of length N, requires unfolding the network into a large feedforward neural network, made of N identical copies of the feedforward part of the canonical form. The input of copy k (shown on Fig. 2.43) is

- the control input vector $\boldsymbol{u}(k) = [u(k), \ldots, u(k-m+1)]^{\mathrm{T}}$,
- the vector of outputs at time k and at the previous n time steps $[y(k), \ldots, y(k-n+1)]^{\mathrm{T}}$.

The output vector of copy k is the vector of the outputs at time $k+1$ and at the previous n time steps $[r(k+1), \ldots, r(k-n+2)]^{\mathrm{T}}$. Therefore, the network actually computes $r(k+1)$ only, the other components of the output vector being derived from the input vector by a unit delay. The output vector of copy k is part of the input vector of the next copy, corresponding to time $k+1$.

The designer must choose the input vector at time zero. If the process output is known during the first n time steps, those values are natural candidates for being the initial values. The process output is taken into account during the first n time steps only: that is why the present algorithm is called *semidirected*, as opposed to directed algorithms, whereby the process output is input to the model at each time step.

NARMAX Assumption

Because the predictor is recurrent, its training requires, as in the previous case, unfolding the recurrent network into a feedforward neural network, made of N identical copies of the feedforward part of the canonical form. All copies have the same vector of parameters. The input of copy k (shown on Fig. 2.44) is

- the control input vector $[u(k), \ldots, u(k-m+1)]^{\mathrm{T}}$,
- the vector $[y_p(k), \ldots, y_p(k-n+1)]^{\mathrm{T}}$,
- the vector of errors at time k and at the previous p time steps $[r(k), \ldots, r(k-p+1)]^{\mathrm{T}}$.

The output vector of copy k is the vector of errors at time $k+1$ and at the previous p time steps $[e(k+1), \ldots, e(k-p+2)]^{\mathrm{T}}$. Therefore, the network computes only $e(k+1)$, the other components of the error vector at time $k+1$ being derived from the errors at time k by a unit time delay. The vector of errors at time $k+1$ is part of the input vector of the next copy, corresponding to time $k+1$.

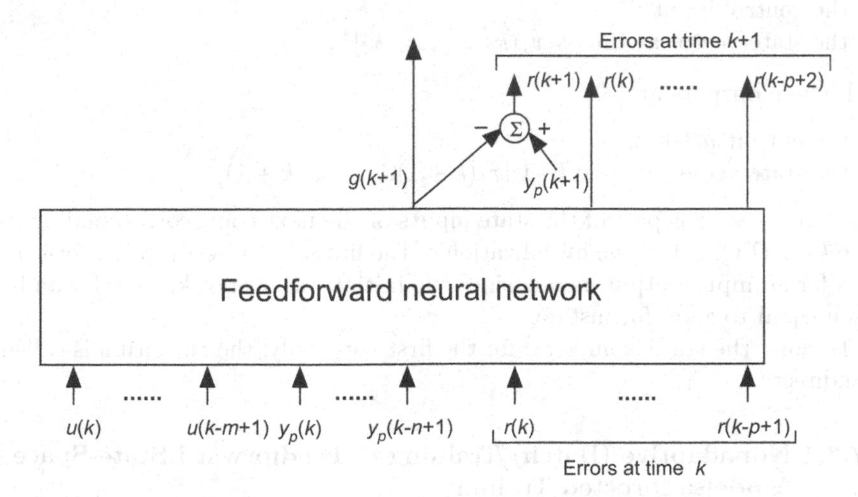

Fig. 2.44. Copy k of the feedforward neural network of the canonical form, for training a NARMAX model

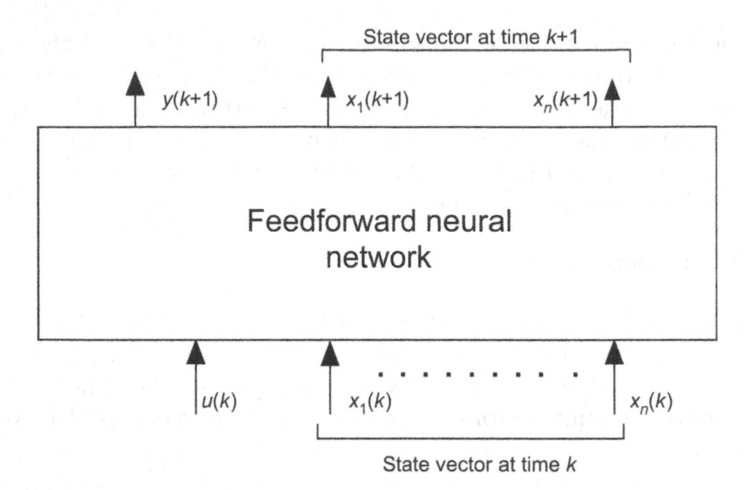

Fig. 2.45. Copy k of he feedforward neural network of the canonical form, for semidirected training of a state-space model

2.7.3.3 Nonadaptive (Batch) Training of Recurrent State-Space Models: Semidirected Training

Just as in the case of input-output models, training requires unfolding the model into a feedforward neural network made of N identical copies of the feedforward neural network of that canonical form, whose inputs are, for copy k,

- the control input $u(k)$,
- the state vector at time $k[x_1(k), \ldots, x_n(k)]^{\mathrm{T}}$,

and whose outputs are

- the output $y(k{+}1)$,
- the state vector at time $k{+}1$ $[x_1(k+1), \ldots, x_n(k+1)]^{\mathrm{T}}$.

The latter vector is part of the state inputs of the next copy, corresponding to time $k+1$ (Fig. 2.45) The initialization of the first copy is less straightforward than for an input-output model, since the initial state is not known. It can be taken equal to zero, for instance.

Because the state is imposed for the first copy only, the algorithm is called semidirected.

2.7.3.4 Nonadaptive (Batch) Training of Feedforward State-Space Models: Directed Training

Under the state noise assumption, and if state variables are measured, the ideal model is a feedforward model that predicts the state and the output, either with a single network, or with two different networks.

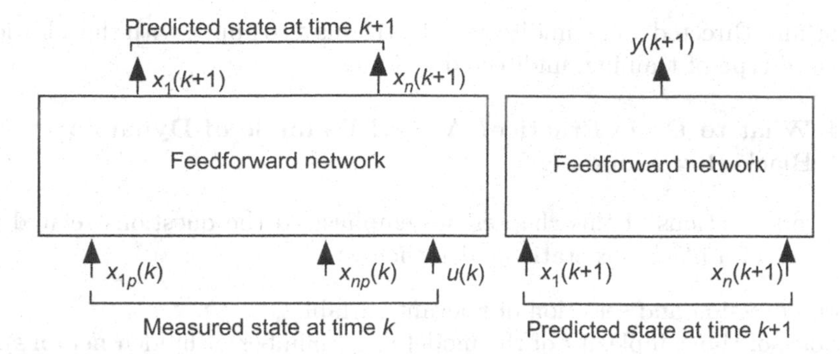

Fig. 2.46. Copy k for the training of a state-space model with two different networks for state and output prediction

Both the state predictor and the output predictor are feedforward. State prediction can be performed either by n different networks, which have identical inputs, but which predict different state variables, or by a single network that predicts all state variables,

- The state at time $k+1$ is computed from the measured state at time k and from the control inputs at time k.
- The output at time $k+1$ is computed from the state computed at time $k+1$.

Figure 2.46 shows the model if two different neural networks are used for computing the state variables and for computing the output.

Because the training of those networks is directed, it is performed as the training of a feedforward neural network.

The note related to the dumb predictor, in the section devoted to the directed training of input-output models, is also relevant to the training of state-space models.

Implementation of Directed and Semidirected Algorithms

All equations for the implementation of directed or semidirected algorithms can be found in Chap. 3, pages 64 to 69 (input-output models) and 72 to 81 (state-space models), of [Oussar 1998]. A very complete technical discussion can be found in that document.

2.7.3.5 Adaptive (On-Line) Training of Recurrent Neural Networks

Dynamic models, just as static ones, can be trained adaptively. Adaptive algorithms for dynamic models are described in Chap. 4, in the framework of stochastic approximation. The same principles as those described above for nonadaptive training apply (influence of noise on the choice of training algorithm). Directed and semidirected algorithms also apply, with the addition of a third type of training: undirected training.

2.7.4 What to Do in Practice? A Real Example of Dynamic Black-Box Modeling

In the first sections of this chapter, we emphasized the questions related to the design of a black-box static model, such as

- preprocessing and selection of relevant variables,
- choice of the complexity of the model (e.g., number of hidden neurons).

The design a dynamic models involves the following additional choices:

- choice of the representation (input-output or state),
- choice of the noise assumption (state noise, output noise, state and output noise),
- choice of the order of the model.

If no prior knowledge on the process is available, all combinations of assumptions and representations should be tested, and models of increasing order should be designed, until a satisfactory model is found. However, the following arguments should alleviate the designer's task:

- State-space models are more general and more parsimonious, but more difficult to train, than input-output models; therefore, it is recommended to first design input-output models, then, if the latter turn out to be unsatisfactory, try state-space models.
- Prior knowledge, however cursory, may give useful hints as to the influence of noise on the process.
- Similarly, a cursory analysis of the process response to typical inputs may provide valuable insights into the order of the model.

In order to illustrate the design methodology discussed above, the example of the black-box modeling of the hydraulic actuator of a robot arm is presented. Experimental data was gathered by the Linköping University (Sweden), and black-box modeling was performed by several groups (see for instance [Sjöberg 1995; Norgaard 2000]).

The control input is the opening of the liquid admission valve in the actuator, the output is the resulting hydraulic pressure. Sequences of input and output data are available for training (512 points) and testing (512 points). Figure 2.47(a) shows the available input data, and Fig. 2.47(b) shows the corresponding responses.

Because no validation set was provided, the performances reported here are the best performances obtained on the test set.

First, it appears clearly that he model must be nonlinear in order to account for the observations: input variations by a factor of 2 (for instance

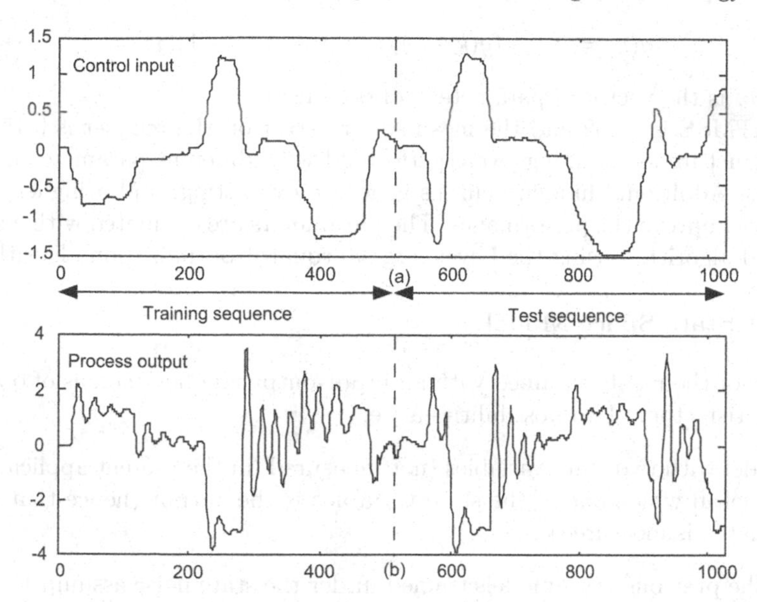

Fig. 2.47. Training and test sequences for the modeling of the hydraulic actuator of a robot arm

around times 10 and 380) do not elicit responses whose amplitudes have the same ratio.

No knowledge is available on the physics of the process, nor on any source of disturbances. Therefore, assumptions on state noise and on output noise must be tested.

Moreover, responses to stepwise inputs (for instance around time 220) suggest that the order of the model should be larger than 1.

Finally, since the application does not require adaptive training, we consider here batch training only.

2.7.4.1 Input-Output Model

Since input-output modeling is easier than state-space modeling, input-output models are designed first. Since no prior knowledge about noise and disturbances is available, the state noise assumption (directed training of a feedforward model, NARX model), the output noise assumption (semidirected training of a recurrent model), and the output noise and state noise assumption (training with both the model predictions and the process outputs present as model inputs), are tested.

The assumptions involving state noise lead to very poor results when the resulting models are operated as simulators, i.e., if they are asked to perform predictions more than one-step ahead; they are not shown here. Semidirected training of a recurrent model yields more satisfactory results. The best model is a second-order model with three hidden neurons with sigmoid activation function. Its equation is

$$g(k) = \varphi_{NNN}(g(k-1),\, g(k-2), u(k-1), \boldsymbol{w}),$$

where $\boldsymbol{w}$ is the vector of parameters, of dimension 19.

Its TMSE is 0.092 and the mean square error on the test set is 0.15. For each structure, 50 trainings were performed with different parameter initializations. Additional hidden neurons generate overfitting, and a higher order does not improve the performance. The parameters are estimated with a semi-directed algorithm using the Levenberg-Marquardt optimization algorithm.

2.7.4.2 State-Space Model

In view of the result obtained with an input-output model, models of order 2 seem satisfactory. Two possibilities arise,

- model with two state variables (not measured, in the present application),
- model in which one of the state variables is the output (hence that state variable is measured).

As in the previous case, models trained under the state noise assumption give poor results when operated as simulators.

Table 2.3 shows the best results obtained on the test set after semidirected training, for a network with three hidden neurons, optimized by the Levenberg-Marquardt algorithm.

Table 2.3. Results obtained after semidirected training of a network with three hidden neurons, with the Levenberg-Marquadt optimization algorithm

	Mean square error on the training set	Mean square error on the test set
Network with no measured state variable	0.091	0.18
Network whose output is one of the state variables	0.071	0.12

Therefore, the best model is a network whose output is one of the state variables. Its equation is

$$\begin{aligned} x_1(k) &= \varphi^1_{NN}(x_1(k-1), x_2(k-1), u(k-1)) \\ x_2(k) &= \varphi^2_{NN}(x_1(k-1), x_2(k-1), u(k-1)) \\ y(k) &= x_2(k). \end{aligned}$$

The network has 26 parameters, but is has better performances, on the test set, than an input-output network with 19 parameters. That is an experimental illustration of the parsimony of state-space models, which allow the use of a larger number of parameters without overfitting.

To the best of our knowledge, these are the best published results on the problem. The detail of the results, together with an application of wavelet networks to the same problem, can be found in [Oussar 1998].

2.7.5 Casting Dynamic Models into a Canonical Form

In the previous sections, we assumed that no prior knowledge of the process was available to the model designer, so that the form of the algebraic or differential equations that would be derived from a physical analysis was unknown. That is a typical black-box modeling situation.

In the next section, we show that any prior knowledge, available under the form of algebraic or differential equations, can be embodied into the structure of a neural network. The model thus designed is a "gray-box" or "semiphysical model". The design of such a model may lead to a complex recurrent network structure, which is neither an input-output representation, nor a state-space representation; since the training algorithms that are described in the previous section were applicable to state-space models or input-output models, how can one train networks that are neither? Should one design a special training algorithm for each specific architecture?

Similarly, Chap. 4 presents a set or network "models" (where the term model does not have its scientific meaning, but its commercial meaning—as in car or TV model), which are generally named from their author (Hopfield model, Jordan model, Elman model, etc.), whose structures are different from the architectures discussed above. Again, one may ask whether each specific architecture requires a specific training algorithm.

The answer to that question takes advantage of the following property.

Property. *Any recurrent neural network, however complex, can be cast into a minimal state-space form, called canonical form, to which the training algorithms discussed in the previous section can be applied.* The latter are therefore fully generic, since they can train any recurrent network, provided it has been cast into a canonical form.

Therefore, the present section shows how the canonical form of an arbitrary recurrent neural network, stemming from instance from a semiphysical modeling, can be derived. That task is performed in two steps,

- derivation of the order of the network,
- derivation of a state vector and of the corresponding canonical form.

A reminder: when designing a purely black-box model, without any prior knowledge, the model is sought directly in a canonical form.

2.7.5.1 Definition

The *canonical form* of a recurrent neural network is the *minimal* state-space representation

$$\begin{aligned} \boldsymbol{z}(k+1) &= \boldsymbol{\phi}(\boldsymbol{z}(k)), \boldsymbol{u}(k)) \\ \boldsymbol{y}(k+1) &= \boldsymbol{\psi}(\boldsymbol{z}(k), \boldsymbol{u}(k)), \end{aligned}$$

where $\boldsymbol{z}(k)$ is the minimal set, made of ν variables, that allows the derivation of the state of the model at time $k+1$, given the state of the model and its inputs at time k, and where functions $\boldsymbol{\phi}$ and $\boldsymbol{\psi}$ can be implemented as feedforward neural networks.

The order of the canonical form is ν. It is convenient, but not mandatory, to design the predictor as a single neural network, whose inputs are the control inputs and the state variables at time k, and whose outputs are the state variables at time $k+1$ (Fig. 2.48).

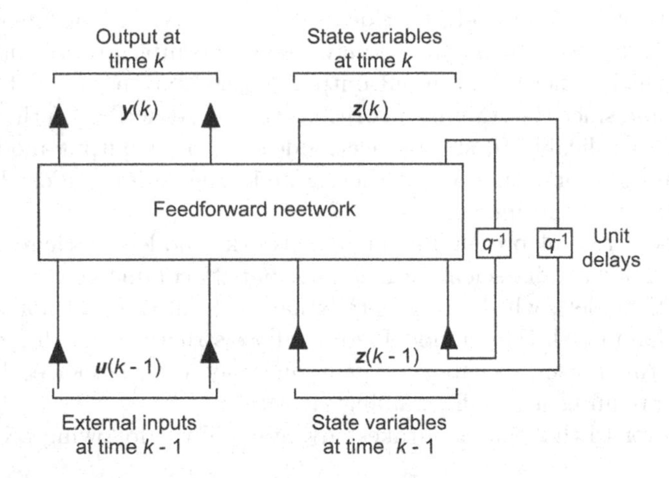

Fig. 2.48. Canonical form of a recurrent network

A general technique, which allows a fully automatic derivation of the canonical form of any recurrent network, is described in [Dreyfus et al. 1998]. An illustrative example is given below.

2.7.5.2 An Example of Derivation of a Canonical Form

The analysis of a process has led to the following model:

$$\begin{aligned} \ddot{x}_1 &= \phi_1(x_1, x_2, x_3, u) \\ x_2 &= \phi_2(x_1, x_3) \\ \ddot{x}_3 &= \phi_3(x_1, \dot{x}_2) \\ y &= x_3. \end{aligned}$$

Its discrete-time equivalent, derived with the explicit Euler discretization method, is given by

$$\begin{aligned} x_1(k+1) &= \psi_1[x_1(k), x_1(k-1), x_2(k-1), x_3(k-1), u(k-1)], \\ x_2(k+1) &= \psi_2[x_1(k+1), x_3(k+1)], \\ x_3(k+1) &= \psi_3[x_3(k), x_3(k-1), x_1(k-1), x_2(k), x_2(k-1)] \\ y(k+1) &= x_3(k+1). \end{aligned}$$

The explicit Euler discretization method consists in approximating the time derivative of a function $f(t)$ at time kT (where T is the sampling period, or integration step, and k is a positive integer) by

$$\{f[(k+1)T] - f(kT)\}/T$$

The question of the discretization of continuous-time differential equations is discussed in more detail in the section devoted to semiphysical modeling.

Clearly, the above equations are not in canonical form. For a clear analysis of the model, and for training it if the functions are parameterized, it is very desirable to know the minimum number of variables that are necessary for a complete description of the model, and to put it in canonical form. Note that a given recurrent neural network does not have a *unique* canonical form: generally, several different canonical forms can be derived; obviously, they all have the same number of state variables.

The network graph is useful for deriving the canonical form. Its nodes are the neurons, and its edges are the connections between neurons; each edge is assigned a length, which is the delay (possibly equal to zero), expressed as an integer multiple of the sampling time, and a direction, which is the direction of information flow in the edge. The length of a path in the graph is the sum of the lengths of the edges that belong to the path.

A cycle in a graph is a path that starts and ends at the same node, without going through the same node more than once, and complying with the directions of the edges. The length of a cycle is the sum of the lengths of its edges.

For a discrete-time neural network to be causal, its graph must have no cycle of length equal to zero. If a cycle had a length equal to zero, the value of the output of a neuron would be dependent on the value of the same output *at the same time step*.

Figure 2.49 shows a representation of the equations of the model as the graph of a recurrent neural network; nodes 1, 2 and 3 represent neurons whose activation functions are Ψ_1, Ψ_2 and Ψ_3, respectively. The figures in squares are the delays associated to each connection (number of sampling periods).

Vector $z(k) = [x_1(k), x_2(k-1), x_3(k), x_3(k-1)]^{\mathrm{T}}$ can be chosen as a state vector. The corresponding canonical form is shown on Fig. 2.49. It has a feedforward neural network with three hidden neurons (neuron 1, and neuron 2 which is duplicated in the canonical form with shared weights, and output neuron (neuron 3) which is also a state neuron. Since the order of the model is 4, there are 4 state outputs, which are connected back to the state inputs through unit delays, denoted by the conventional delay operator symbol q^{-1}.

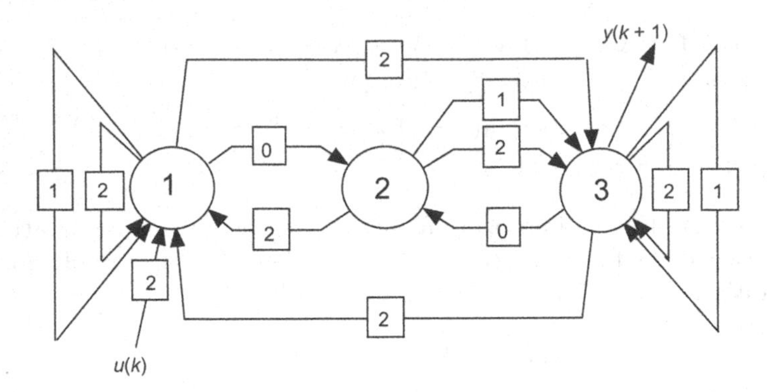

Fig. 2.49. Example of network graph

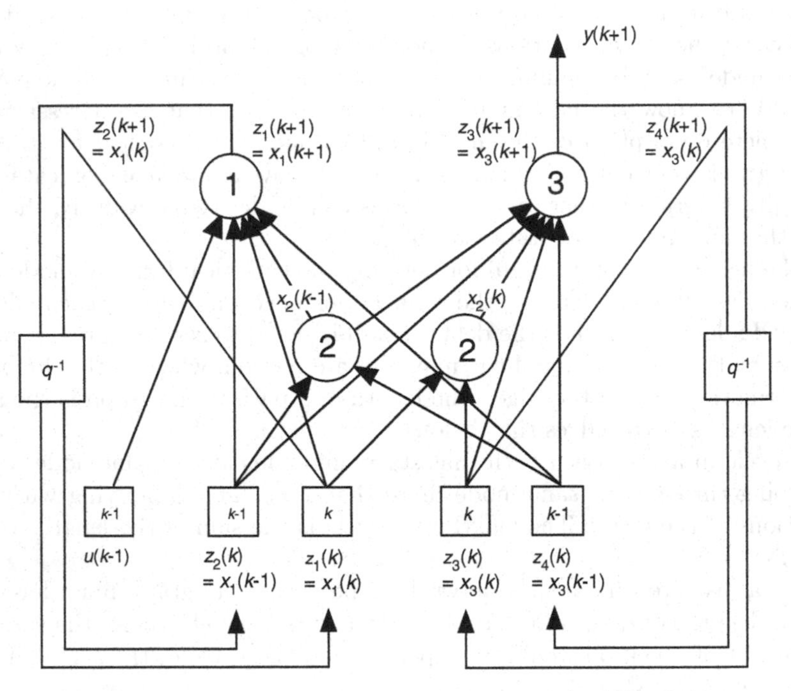

Fig. 2.50. The canonical form of the network of Fig. 2.49

The network shown on Fig. 2.50 is fully equivalent to the network shown on Fig. 2.49: it is simply re-written in a very convenient way, which makes the structure of the network more legible, and, even more importantly, that allows the training of the network with the conventional, generic training algorithms.

Algorithmic details for deriving the canonical form automatically can be found in [Dreyfus et al. 1998].

2.8 Dynamic Semiphysical (Gray Box) Modeling

In the previous sections of this chapter, the design methodology for black-box modeling was emphasized; it is the traditional view of neural networks, whereby models are designed from measurements only. Such an approach is very useful when no satisfactory knowledge-based model exists. However, it is frequently the case that a knowledge-based model does exist, but is not satisfactory, either because the computation time necessary for integrating the model with the requested accuracy is too long and precludes a real-time implementation of the model, or because the model is no accurate enough due to the present limitations of the knowledge of the phenomena that occur in the process. It is therefore desirable to take advantage of that knowledge, albeit not fully satisfactory, for the design of a more accurate, or faster, model making use of training from data: a semiphysical model is designed. Thus, one can design a model that combines the legibility of knowledge-based models with the flexibility and speed of black-box models.

In the following, we discuss a general design methodology for semiphysical models that takes advantage of the properties of neural networks. We emphasize the importance of the discretization of the continuous-time knowledge-based model, which has a strong impact on the *stability* of the resulting discrete-time model.

In the present section, we discuss the principles of the design of semiphysical models. An industrial application of that method is presented in Chap. 1.

2.8.1 Principles of Semiphysical Modeling

2.8.1.1 From Black-Box Modeling to Knowledge-Based Modeling

A knowledge-based model is a mathematical description of the phenomena that occur in a process, based on the equations of physics and chemistry (or biology, sociology, etc.); typically, the equations involved in the model may be transport equations, equations of thermodynamics, mass conservation equations, etc. They contain parameters that have a physical meaning (e.g., activation energies, diffusion coefficients, etc.), and they may also contain a small number of parameters that are determined through regression from measurements.

Conversely, a black-box model is a parameterized description of the process, all parameters of which are estimated from measurements performed on the process; it does not take into account any prior knowledge on the process (or a very limited one).

A semiphysical (or gray-box) model may be regarded as a tradeoff between a knowledge-based model and a black-box model. It may embody all the engineer's knowledge on the process (or a part thereof), and, in addition, it relies on parameterized functions, whose parameters are determined from measurements. This combination makes it possible to take into account all the phenomena that are not modeled with the required accuracy through prior knowledge. Since a larger amount of prior knowledge is used in the design of a semiphysical model than in the design of a black-box model, a smaller amount of experimental data is required to estimate its parameters reliably.

2.8.1.2 Design and Training of a Dynamic Semiphysical Model

Design Principle

The design of a semi-physical model requires the availability of a knowledge-based model, which is usually in the form of a set of coupled, possibly nonlinear, differential, partial differential, and algebraic, equations. We assume that model to be in standard state-space form,

$$\frac{\mathrm{d}\boldsymbol{x}}{\mathrm{d}t} = \boldsymbol{f}[\boldsymbol{x}(t), \boldsymbol{u}(t)]$$
$$\boldsymbol{y}(t) = \boldsymbol{g}[\boldsymbol{x}(t)],$$

where $\boldsymbol{x}$ is the vector of state variables, $\boldsymbol{y}$ is the vector of outputs, $\boldsymbol{u}$ is the vector of control inputs, and where $\boldsymbol{f}$ and $\boldsymbol{g}$ are known vector functions. That model may be unsatisfactory for various reasons: functions $\boldsymbol{f}$ and $\boldsymbol{g}$ (or some of their components) may be too inaccurate for the purpose that the model should serve, or they may involve parameters that are not estimated accurately, etc. In a black-box model, neural networks are used to approximate functions $\boldsymbol{f}$ and $\boldsymbol{g}$; they are trained from experimental data. In a semiphysical neural model, those functions that are not known accurately enough are implemented as neural models, whereas those functions, which are known reliably, are either kept under their analytic form, or implemented as a neural network with fixed parameters and nonlinearities.

In general, the design of a semiphysical neural model is performed in three steps:

- Step 1: construction of a discrete-time semiphysical model that is derived, by an appropriate discretization scheme (discussed below) from the knowledge-based model.
- Step 2: training of the semiphysical model, or of specific parts thereof, from results obtained by numerical integration of the knowledge-based model; that step is generally necessary in order to obtain appropriate initial values of the parameters, to be used in step 3.
- Step 3: training of the semiphysical neural model from experimental data.

That design strategy is exemplified in the next section.

An Illustrative Example

A knowledge-based model is described by the following equations:

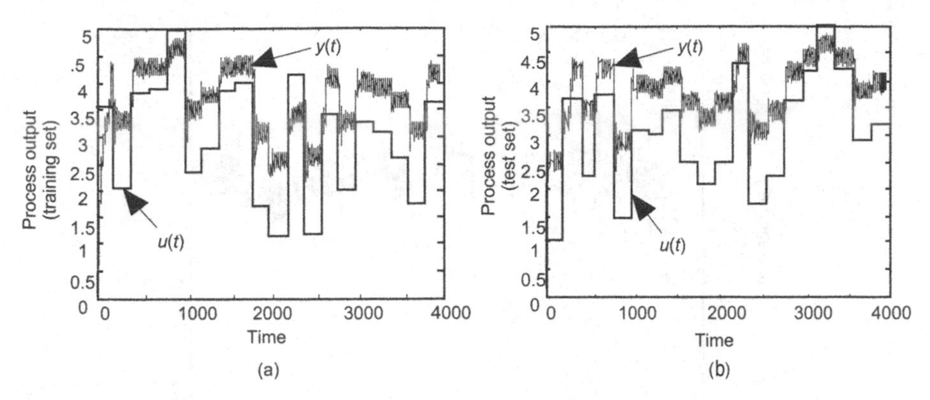

Fig. 2.51. process response to two input sequences: (**a**) training sequence, (**b**) test sequence

$$\frac{\mathrm{d}x_1(t)}{\mathrm{d}t} = (x_1(t) + 2x_2(t))^2 + u(t)$$
$$\frac{\mathrm{d}x_2(t)}{\mathrm{d}t} = 8.32x_1(t)$$
$$y(t) = x_2(t).$$

The state variables x_1 and x_2 are measurable. Figure 2.51 shows the process response to two sequences of input steps; throughout this section, the left-hand input and output sequences will be used as the training set, and the right-hand ones as the test set. The results obtained by numerical integration of the knowledge-based model are in poor agreement with experimental measurements of the output, as shown on Fig. 2.52. The mean square modeling error on the test set is equal to 0.17, which is much larger than the noise standard deviation of 0.01.

Experts of the process are reasonably confident that the first state equation is valid, but there are serious doubts about the second equation because

- The parameter 8.32 may be inaccurate.
- The linear dependence is controversial.
- It is even conjectured that the right-hand side of the second equation might depend on x_2.

Therefore, in order to build a more accurate model, it may be advantageous to use a semiphysical model. Actually, three different models, of increasing complexity, may be designed in order to meet the above three criticisms. We describe below the design of those models and the results thus obtained.

As mentioned above, the first step of the procedure consists in creating a discrete-time model from the knowledge-based model. Since data is gathered with a sampling period T, the latter is a natural candidate for being the discretization step of the equations. The simplest discretization method is Euler's explicit method, whereby the derivative $\mathrm{d}f(t)/\mathrm{d}t$ is approximated as

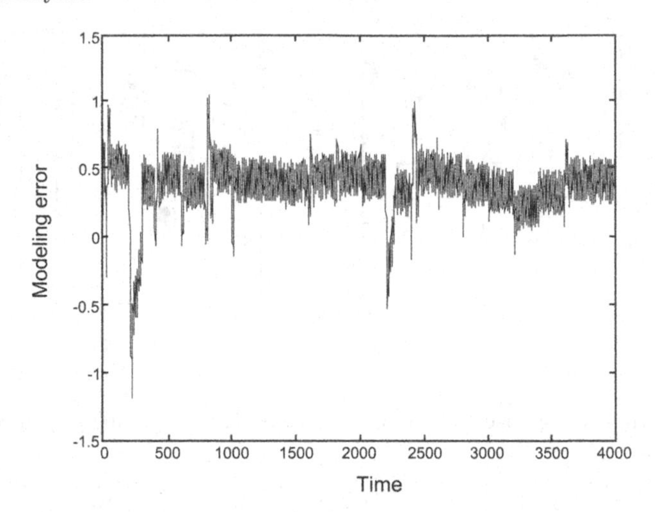

Fig. 2.52. Modeling error of the knowledge-based model

$[f[(k+1)T] - f(kT)]/T$ (where k is a positive integer). Thus the following discrete-time model is obtained:

$$x_1[(k+1)T] = x_1(kT) + T[-(x_1(kT) + 2x_2(kT))^2 + u(kT)]$$
$$x_2[(k+1)T] = x_2(kT) + T(8.32x_1(kT)).$$

Hence the simplest semiphysical model:

$$x_1[(k+1)T] = x_1(kT) + T[-(x_1(kT) + 2x_2(kT))^2 + u(kT)]$$
$$x_2[(k+1)T] = x_2(kT) + T(wx_1(kT)).$$

where w is a parameter that is estimated through appropriate training from experimental data. The equations are under the conventional form of a state-space model: it is therefore not necessary to cast the model into a canonical form; were that not the case, the model would have been cast into a canonical form as explained above. The model is shown on Fig. 2.53.

For simplicity, in all the following figures, the constant input (bias) will not be shown; furthermore, discrete time kT will be simply denoted by k. q^{-1} is the usual symbol for a unit time delay. On Fig. 2.53, neuron 1 performs a weighted sum s of x_1 and x_2, with the weights indicated on the figure, followed by the nonlinearity $-s^2$, and adds $u(k)$. Neuron 2 multiplies its input by the weight w. Neurons 3 and 4 just perform weighted sums. If w is taken equal to 8.32, then this network gives exactly the same results as the numerical integration of the discrete-time knowledge-based model by Euler's explicit discretization, with integration step T. If w is an adjustable weight, then its value can be computed by training the network from experimental data with any good training algorithm (evaluation of the gradient of the quadratic cost function by backpropagation through time, and gradient descent with the

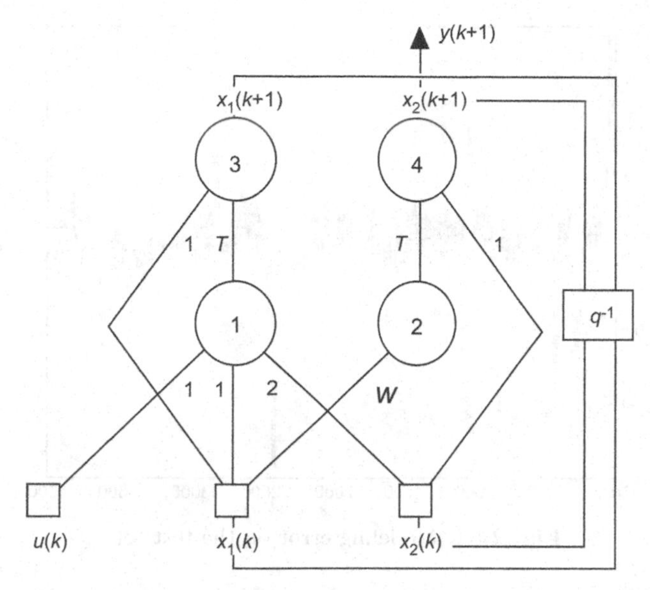

Fig. 2.53. Canonical form of the knowledge-based model discretized by the explicit Euler method

Levenberg-Marquardt or BFGS algorithm), using for instance a semidirected algorithm under the output noise assumption. For that training, it would be reasonable to initialize weight w to 8.32. Note that, in that very simple case, step 2 of the algorithm is bypassed.

Figure 2.54 shows the modeling error with that improved model. The mean square modeling error on the test sequence is 0.08 (instead of 0.17 for the knowledge-based model); since the noise variance is 0.01, further improvement may be expected from a more elaborate model.

Therefore, one considers the second level of criticism towards the knowledge-based model, i.e., the fact that the right-hand side of the state equation might be a nonlinear function of x_1. Therefore, neuron 2 is replaced by a feedforward neural network whose input is x_1, as shown on Fig. 2.53 with three hidden neurons (hence 6 parameters shown on the figure, and 4 parameters related to the bias, not shown).

The feedforward neural network made of the non-numbered neurons shown on Fig. 2.55 can be trained from data generated by the knowledge-based model (step 2 of the design procedure): although those values are known to be inaccurate, the weights resulting from that training are reasonable estimates, which are subsequently used for initializing the training of the neural network from experimental data (step 3 of the design procedure). Figure 2.56 shows the modeling error of that model, with two hidden neurons in the black-box part of the model (additional neurons generate overfitting). The mean square modeling error on the test sequence is 0.02, which is a sizeable improvement over the previous model.

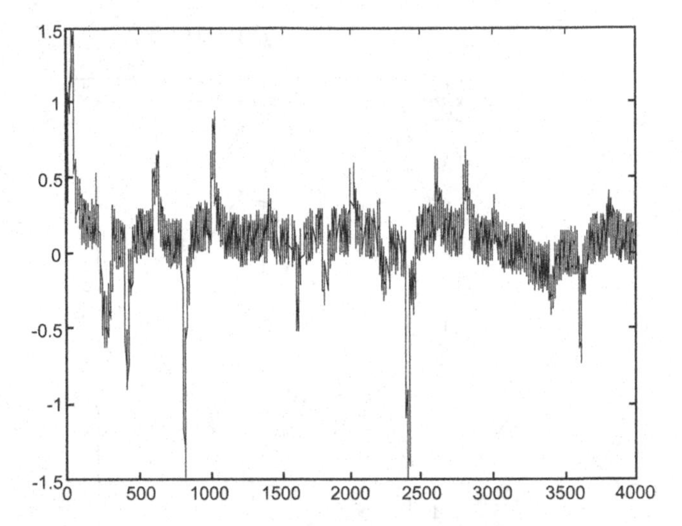

Fig. 2.54. Modeling error on the test set

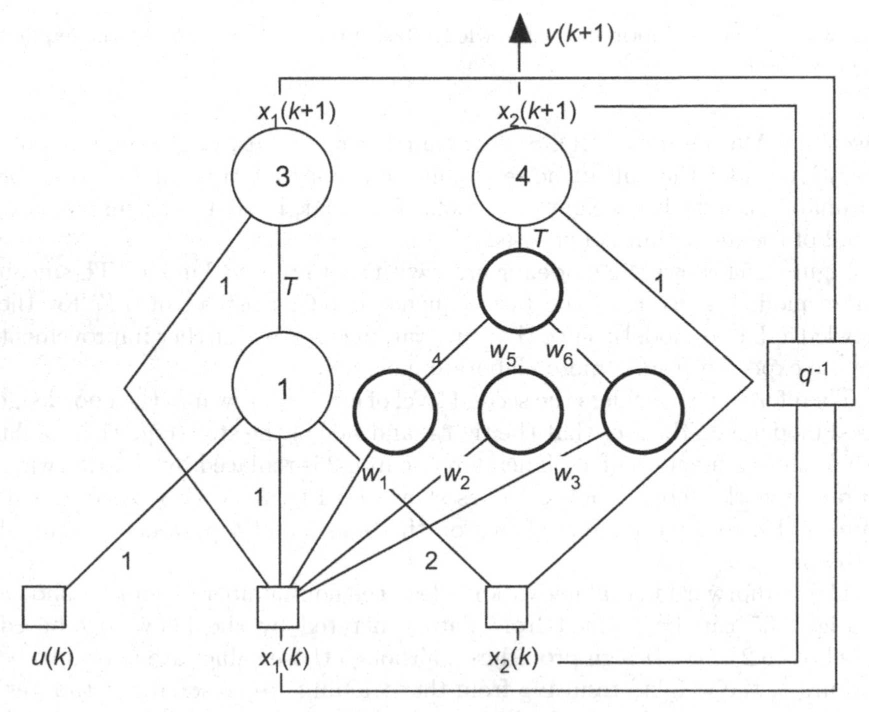

Fig. 2.55. Canonical form of a semiphysical model

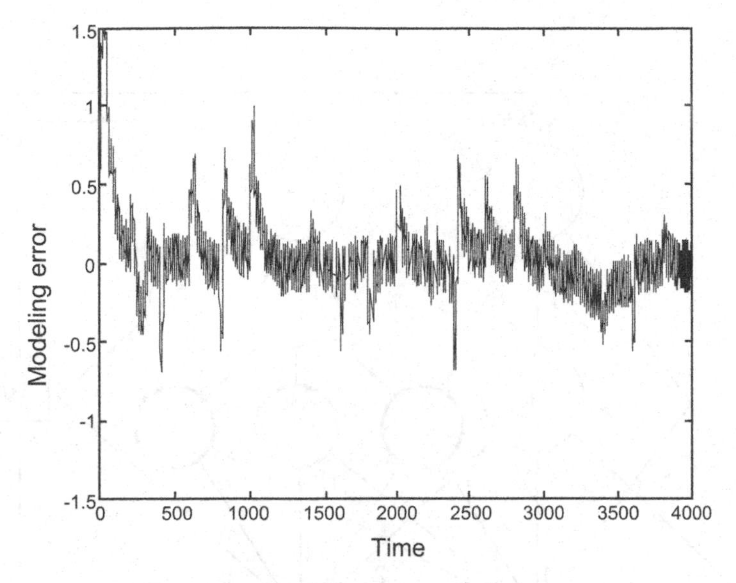

Fig. 2.56. Modeling error on the test set

Since the results are still unsatisfactory (the root mean square error on the test set is twice the standard deviation of noise), the conjecture that the right-hand side of the second state equation does not depend on x_1 only, but also depends on x_2, must be taken into account. Then a third knowledge-based neural model may be designed, where the right-hand side of the second state equation is implemented as a neural network whose inputs are x_1 *and* x_2. That is shown on Fig. 2.57 (with a feedforward network having three hidden neurons).

Steps 2 et 3 of the design are performed as for the previous model. The variance of the modeling error being on the order of the noise variance (see Fig. 2.58), the model can be considered satisfactory.

2.8.1.3 Discretization of a Knowledge-Based Model

The first step of the design of a semiphysical model is the discretization of the knowledge-based model, which is generally a continuous-time model, in order to find a discrete-time model whose structure is used for the design of the recurrent network. The choice of the discretization technique has important consequences regarding the stability of the model to be designed. The discretization of continuous-time differential equations is a basic chapter in any textbook of numerical analysis; we recall a few basic elements that are important for the design of a semiphysical model.

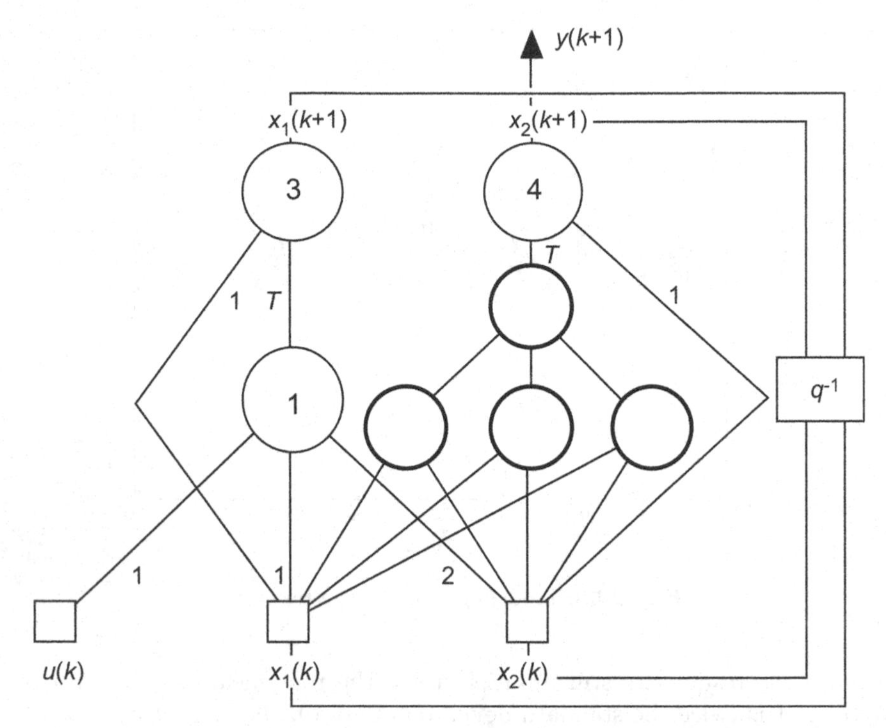

Fig. 2.57. Canonical form of a semiphysical model

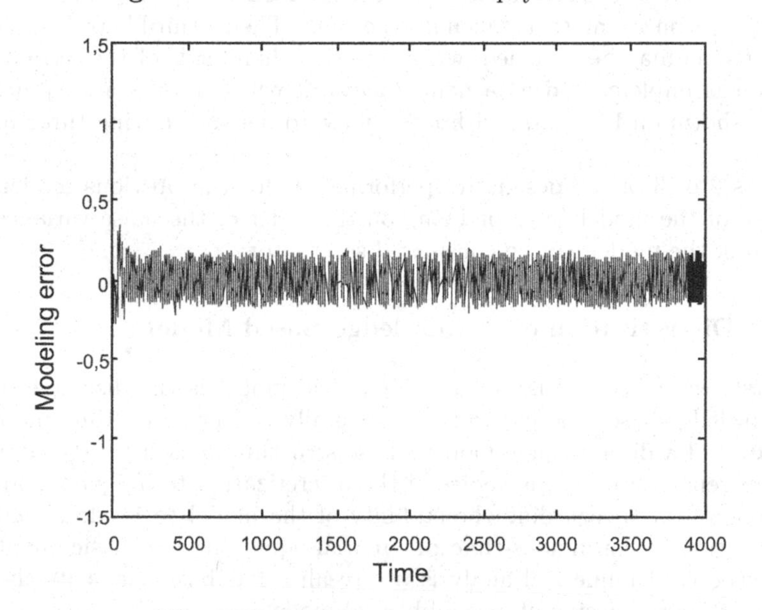

Fig. 2.58. Modeling error on the test set

Explicit (Forward) vs. Implicit (Backward) Discretization Schemes: Definitions

Consider a first-order differential equation,

$$\frac{\mathrm{d}x(t)}{\mathrm{d}t} = f[x(t)].$$

An explicit scheme discretizes it to

$$x[(k+1)T] = \varphi[x(kT), T],$$

- where T is the discretization (or integration) step, which is usually the sampling period of experimental data,
- where k is a positive integer,
- and where function φ depends on the discretization technique (examples are shown below).

An implicit scheme discretizes the same differential equation to

$$x[(k+1)T] = \psi[x[(k+1)T], x(kT), T].$$

The main difference is the fact that the quantity $x[(k+1)T]$ appears in the left-hand side only if an explicit scheme is used, whereas it appears on both sides if an implicit scheme is used. Therefore, if a one-step-ahead predictor for x is to be designed, the computation of $x[(k+1)T]$ from $x[kT]$ is trivial if an explicit scheme is used, whereas it requires solving a nonlinear equation if an implicit scheme is used.

More generally, consider a set of state-space equations written in vector form,

$$\frac{\mathrm{d}x}{\mathrm{d}t} = f[\boldsymbol{x}(t), \boldsymbol{u}(t)].$$

If an explicit discretization scheme is used, the discretized equations can be written under the general form,

$$K[\boldsymbol{x}(kT)]\boldsymbol{x}[(k+1)T] + \boldsymbol{\Psi}[\boldsymbol{x}(kT), \boldsymbol{u}(kT), T] = 0,$$

where K is a matrix and $\boldsymbol{\Psi}$ is a vector function, whereas, if an implicit discretization scheme is used, the discretized equation can be written under the general form,

$$K[\boldsymbol{x}[(k+1)T]]\boldsymbol{x}[(k+1)T] + \boldsymbol{\Psi}[\boldsymbol{x}[(k+1)T], \boldsymbol{x}(kT), \boldsymbol{u}[(k+1)T], T] = 0.$$

Again, the computation of $\boldsymbol{x}[(k+1)T]$ from $\boldsymbol{x}[kT]$ is trivial if an explicit scheme is used (provided matrix K is invertible):

$$\boldsymbol{x}[(k+1)T] = -K^{-1}[\boldsymbol{x}(kT)]\boldsymbol{\Psi}[\boldsymbol{x}(kT), \boldsymbol{u}(kT), T],$$

whereas it requires solving a system of nonlinear equations if an implicit scheme is used.

Examples

Consider again the first-order differential equation $dx/dt = f[\boldsymbol{x}(t), \boldsymbol{u}(t)]$.

Euler's explicit scheme consists in considering that function f is constant, equal to $f[x(kT)]$, between kT and $(k+1)T$, so that the integration of the differential equation between kT and $(k+1)T$ gives

$$x[(k+1)T] = x(kT) + Tf[x(kT)].$$

By contrast, Euler's implicit scheme consists in considering that function f is constant, equal to $f[x((k+1)T]$ between kT and $(k+1)T$, so that the integration of the differential equation between kT and $(k+1)T$ gives

$$x[(k+1)T] = x(kT) + Tf[x[(k+1)T]].$$

Similarly, Tustin's scheme consists in considering that function f varies linearly between kT and $(k+1)T$, so that the integration of the differential equation between kT and $(k+1)T$ gives

$$x[(k+1)T] = x(kT) + \frac{T}{2}\left[f\left[x[(k+1)T]\right] + f\left[x(kT)\right]\right].$$

Because the values of quantities at time $(k+1)T$ are present on both sides of the equation, the computation of $x[(k+1)T]$ requires solving a nonlinear equation.

Application

We consider again the knowledge-based model described by the equations

$$\begin{aligned}\frac{\mathrm{d}x_1(t)}{\mathrm{d}t} &= -(x_1(t) + 2x_2(t))^2 + u(t)\\ \frac{\mathrm{d}x_2(t)}{\mathrm{d}t} &= 8.32x_1(t)\\ y(t) &= x_2(t).\end{aligned}$$

Euler's explicit method discretizes it to

$$\begin{aligned}x_1[(k+1)T] &= x_1(kT) + T[-(x_1(kT) + 2x_2(kT))^2 + u(kT)]\\ x_2[(k+1)T] &= x_2(kT) + T(8.32x_1(kT)).\end{aligned}$$

Its discretization by Euler's implicit scheme discretizes it to

$$\begin{aligned}&[1 + Tx_1[(k+1)T] + 4Tx_2[(k+1)T]]x_1[(k+1)T] + 4Tx_2^2[(k+1)T]\\ &\quad = x_1(kT) + Tu[(k+1)T]x_2[(k+1)T] - T(8.32x_1[(k+1)T])\\ &\quad = x_2(kT).\end{aligned}$$

These equations are of the form

$$K[\boldsymbol{x}[(k+1)T]]\boldsymbol{x}[(k+1)T] + \boldsymbol{\Psi}[\boldsymbol{x}[(k+1)T], \boldsymbol{x}(kT), \boldsymbol{u}[(k+1)T], T] = 0,$$

with

$$K[\boldsymbol{x}[(k+1)]T] = \begin{pmatrix} [1 + Tx_1[(k+1)T] + 4Tx_2[(k+1)T]] & 4Tx_2[(k+1)T] \\ -Tw & 1 \end{pmatrix}$$

and

$$\boldsymbol{\Psi}[\boldsymbol{x}[(k+1)T], \boldsymbol{x}(kT), \boldsymbol{u}[(k+1)T], T] = \begin{pmatrix} x_1(kT) + Tu[(k+1)T] \\ x_2(kT) \end{pmatrix}.$$

Explicit vs. Implicit Discretization Scheme: Impact on Stability

The above examples show that an explicit discretization scheme makes the design of a semiphysical model simpler than an implicit scheme. The main incentive for using implicit scheme is the stability issue: implicit schemes may have better stability properties than implicit schemes. In order to illustrate that, we discuss the simple first-order linear differential equation

$$\frac{du(t)}{dt} = -\alpha u(t), \alpha > 0.$$

Euler's explicit method discretizes it to

$$\frac{u[(k+1)T] - u(kT)}{T} = -\alpha u(kT),$$

or equivalently

$$u[(k+1)T] = (1 - \alpha T)u(kT).$$

Thus, $u[(k+1)T]$ is computed from $u(0)$ recursively, and the recursion converges if and only if the magnitude of $(1 - \alpha T)$ is smaller than 1, or $T < 2/\alpha$. The computation time necessary for integrating that equation numerically is proportional to $1/\alpha$: if α is very large, numerical integration may become impossible since the integration step T must be very small.

Now we consider the discretization of the same equation by Euler's implicit method; one has

$$\frac{u[(k+1)T] - u(kT)}{T} = -\alpha u(kT),$$

or equivalently,

$$u[(k+1)T] = \frac{1}{(1+\alpha T)}u(kT).$$

Because the denominator on the right-hand side is larger than 1, the computation of $u[(k+1)T]$ converges irrespective of α.

However, the price to be paid is the fact that (in contrast to the previous very simple example), the computation of the quantities of interest at time $(k+1)T$ requires the resolution of a nonlinear equation. This has important consequences on the architecture of the corresponding neural model.

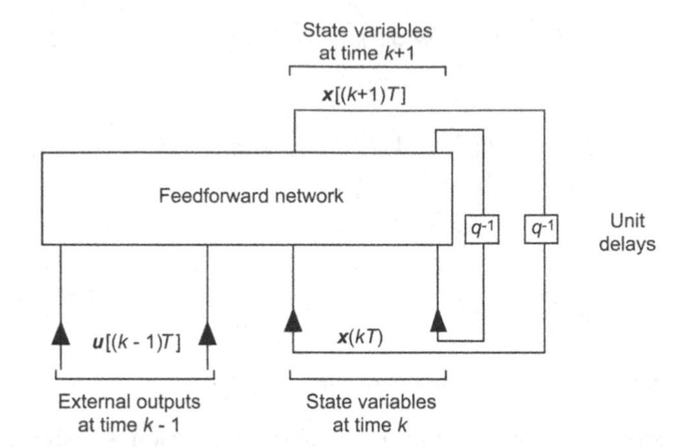

Fig. 2.59. Canonical form of the network resulting from the discretization by an explicit scheme

Explicit vs. Implicit Discretization Schemes: Impact on Neural Network Implementation and Training

The explicit discretization of a knowledge-based model provides equations that are readily put in neural network form, as shown in the above illustrative example: one has

$$\boldsymbol{x}[(k+1)T] = -K^{-1}[\boldsymbol{x}(kT)]\boldsymbol{\Psi}[\boldsymbol{x}(kT), \boldsymbol{u}(kT), T],$$

which is the canonical form of a recurrent neural network, as shown on Fig. 2.59, where the neural network is an approximation of function $-K^{-1}\boldsymbol{\Psi}$. The didactic example discussed above is an example of the design of a semiphysical model from a knowledge-based model discretized by an explicit scheme.

When an implicit discretization method must be used for stability reasons, the neural implementation of the resulting equations is less straightforward, but still feasible. A detailed description of that technique can be found in [Oussar et al. 2001].

2.9 Conclusion: What Tools?

This chapter gave a presentation of the basic concepts of modeling with neural networks. Elements of statistics were first provided, then a complete design methodology of nonlinear models, including but not limited to neural networks, was described. Static and dynamic models were discussed (the latter being considered in a deeper fashion in Chap. 4). Finally, the design methodology of semiphysical models was described.

For practical applications, the designer must understand the basic concepts in order to obtain reliable results, but he must also use appropriate tools (or build his own, which may be a lengthy process).

At present, available development tools fall into two categories:

- neural network toolboxes within general-purpose engineering software; typically, Matlab releases a toolbox that allows easy training of feedforward neural networks; the programming effort is very small for classical functions, but it may become important for the implementation of elements of methodology that are not specifically "neural" (leverage computation, input selection), or for recurrent neural networks;
- specific development tools that include a complete development methodology, requiring no programming effort; typical is the NeuroOne[6] package; such tools do not allow for the flexibility of personal programming, but they provide reliable results in a short time.

Some academic software packages are available freely on the Web. They are excellent for didactic purposes, but they may not stand up to the quality requirements of industrial applications.

Therefore, the model designer, whether in academy or industry, must choose his tools considering the time constraints, the development policy within the company, the size of the applications, etc. The best solution consists in having both types of tools available. Anyway, however powerful and user-friendly the programming tools, a good understanding of the basic concepts and methods, and the application of a principled methodology, are the keys to the development of successful applications.

2.10 Additional Material

This section is devoted to additional definitions, proofs, algorithms, which can be skipped on first reading.

2.10.1 Confidence Intervals: Design and Example

2.10.1.1 Design

In order to estimate a confidence interval for a random variable Y, one seeks a random variable Z, function of Y, whose distribution $p_Z(z)$ is known and independent of Y. Since the distribution $p_Z(z)$ is known and tabulated, the equation $\Pr(z_1 < z < z_2) = \int_{z_1}^{z_2} p_Z(z)\mathrm{d}z = 1-\alpha$ can be solved easily: one just has to compute the value of z_1 such that $\Pr(z < z_1) = \alpha/2$, and the value of z_2 such that $\Pr(z > z_2) = \alpha/2$. When z_1 and z_2 are found, function $Z(Y)$ is inverted in order to find the values of a et b such that $\Pr(a < y < b) = 1 - \alpha$.

[6] By NETRAL S.A.; several illustrations and applications described in Chaps. 1 and 2 were developed with that software.

2.10.1.2 Example

As an example, let us derive a confidence interval for the mean of N measurements: the latter quantity is an unbiased estimator of the expectation value. Assume that the N measurements are realizations of a random Gaussian variable G of mean μ and standard deviation σ. Using the distributions discussed in the next section, it can easily be shown that the random variable $(M-\mu)/(\sigma/\sqrt{N})$ has a normal distribution, and that the variable $\sum_{n=1}^{N}((G_1-M)^2)/(\sigma^2)$ has a Pearson distribution with $N-1$ degrees of freedom.

From the definition of the Pearson variable, one might conclude that the above variable has N (not $N-1$) degrees of freedom. One should note that the random variable M depends on the random variables G_i since one has $M=\sum_{n=1}^{N} G_i/N$: hence the variable has only $N-1$ degrees of freedom.

Those variables are independent. From a theorem stated below, the random variable

$$Z=\frac{M-\mu}{\sqrt{\sum_{i=1}^{N}(G_i-M)^2}}\sqrt{N(N-1)}$$

has a Student distribution with $N-1$ degrees of freedom. One can easily compute the value of z_1 and z_2 such that a realization of the random variable Z lie between those two values with probability 1- α, where α is a known quantity (e.g., $\alpha=0.05$ if a 95% confidence interval is sought). The quantity

$$z=\frac{m-\mu}{\sqrt{\sum_{i=1}^{N}(g_i-m)^2}}\sqrt{N(N-1)},$$

where m is the mean of the N measurements g_i, and where μ is the only unknown, is a realization of the random variable Z. Therefore, the only remaining task is the resolution of the two inequalities $z_1<z$ and $z<z_2$; they are linear in μ, hence the resolution is trivial. The two solutions $\mu_1=a$ and $\mu_2=b$ are the boundaries of the confidence interval

$$a=m-\sqrt{\frac{\sum_{i=1}^{N}(g_i-m)^2}{N(N-1)}}z_2 \quad\text{and}\quad b=m-\sqrt{\frac{\sum_{i=1}^{N}(g_i-m)^2}{N(N-1)}}z_1.$$

Because the Student distribution is symmetrical, z_1 and z_2 may be chosen symmetrically, e.g., $z_1=-z_0<0$ and $z_2=z_0>0$. The confidence interval is symmetrical around m:

$$a=m-\sqrt{\frac{\sum_{i=1}^{N}(g_i-m)^2}{N(N-1)}}z_0 \quad\text{and}\quad b=m+\sqrt{\frac{\sum_{i=1}^{N}(g_i-m)^2}{N(N-1)}}z_0,$$

where m, $\{g_i\}$ and N depend on the experiments, and z_0 depends on the chosen value of α only.

As expected, the width of the confidence interval depends both on the number of experiments N and on the noise through the scattering of the data around the mean value, as expressed by the summation under the square root. The larger the number of experiments, the smaller the confidence interval, hence the more reliable the estimation of the expectation value μ by the mean m. Conversely, the larger the variability of the results, the larger the confidence interval, hence the less reliable the estimation of μ by m.

2.10.2 Hypothesis Testing: An Example

N measurements $\{g_i\}$ have been performed, which can be modeled as independent realizations of a random Gaussian variable of mean μ and standard deviation σ. One would like to know, with a risk α of getting a wrong answer, whether the mean of the distribution has a given value μ_0. Thus, the null hypothesis H_0 is: $\mu = \mu_0$, and the alternative hypothesis H_1 is $\mu \neq \mu_0$. If the null hypothesis is true, then variable

$$Z = \frac{M - \mu_0}{\sqrt{\sum_{i=1}^{N}(G_i - M)^2}} \sqrt{N(N-1)}$$

is a Student variable with $N - 1$ degrees of freedom.

A realization of that random variable can be computed,

$$Z = \frac{m - \mu_0}{\sqrt{\sum_{i=1}^{N}(G_i - m)^2}} \sqrt{N(N-1)},$$

where m is the mean of the measurements. The values of z_1 and z_2 such that $\Pr(z < z_1) = \alpha/2$ and $\Pr(z > z_2) = \alpha/2$ can easily be computed. Then the null hypothesis can be rejected if z is outside the interval $[z_1, z_2]$.

In that particular case, the hypothesis test consists in checking whether the assumed value of the mean μ_0 is within the confidence interval computed in the previous section, and rejecting the null hypothesis if it is outside the confidence interval.

2.10.3 Pearson, Student and Fisher Distributions

2.10.3.1 χ^2(Pearson) Distribution

If a random variable S is the sum of the squares of N random independent Gaussian variables, then it has a χ^2 (or Pearson) distribution with N degrees of freedom. It can be shown that $E(S) = N$ and that $\text{var}(S) = 2\,N$.

2.10.3.2 Student Distribution

If Y_1 is a normal variable, and if Y_2 is a random variable, which is independent from Y_1 and which has a χ^2 (Pearson) distribution with N degrees of freedom, the random variable $Z = (Y_1)/(\sqrt{Y_2/N})$ has a Student distribution with N degrees of freedom.

2.10.3.3 Fisher Distribution

If Y_1 is a Pearson variable with N_1 degrees of freedom, and if Y_2 is a Pearson variable with N_2 degrees of freedom, the random variable $Z = (Y_1/N_1)/(Y_2/N_2)$ has a Fisher distribution with N_1 and N_2 degrees of freedom.

2.10.4 Input Selection: Fisher's Test; Computation of the Cumulative Distribution Function of the Rank of the Probe Feature

2.10.4.1 Fisher's Test

We first describe the use of Fisher's test for model selection.

We assume that the measurements of the quantity of interest can be modeled as the realizations of a random variable such that $Y_p = \boldsymbol{\zeta}^{\mathrm{T}} \boldsymbol{w}_p + \boldsymbol{\Omega}$, where $\boldsymbol{\zeta}$ is the vector of the variables of the model (of unknown dimension), where $\boldsymbol{w}_p$ is the vector (nonrandom but unknown) of the parameters of the model, and where $\boldsymbol{\Omega}$ is an unknown random Gaussian vector, with zero mean. Thus, one has

$$E(Y_p) = \boldsymbol{\zeta}^{\mathrm{T}} \boldsymbol{w}_p.$$

We want to find a model g, from a set of N measurements $\{y_p^k,\ k = 1 \text{ to } N\}$; $\boldsymbol{y}_p$ is the N-dimension vector whose components are the y_p^k. The model depends on the training set: therefore, it is also a realization of a random variable G.

Assume that a set of Q variables, which contains certainly the measurable variables that are relevant to the modeling of the quantity of interest, has been found. A model that contains all relevant variables is called a complete model. Then a model is sought, of the form

$$G_Q = \boldsymbol{\zeta}_Q^{\mathrm{T}} \boldsymbol{W}^Q,$$

where $\boldsymbol{\zeta}_Q$ is the input vector of the model (of dimension $Q+1$ since, in addition to the relevant variables, a component equal to 1 is present in the input vector), and where $\boldsymbol{W}$ is a random vector, which depends on the realization of the vector Y_p that is used for the design of the model. That model is said to be true: there exists certainly a realization w_p of the random vector $\boldsymbol{W}$ such that $g_Q = E(Y_p)$.

In the present chapter, the vector of the parameters of the model was always found by minimizing the least squares cost function (except when using weight decay) $J(\boldsymbol{w}) = \sum_{k=1}^{N} (y_p^k - g_Q(\boldsymbol{\zeta}^k, \boldsymbol{w}))^2 = \|\boldsymbol{y}_p - \boldsymbol{g}_Q(\boldsymbol{\zeta}, \boldsymbol{w})\|^2$, where $\boldsymbol{w}$ is a realization of the vector of parameters W, $\boldsymbol{\zeta}^k$ is the vector of the $Q+1$ inputs for example k, and where $\boldsymbol{g}_Q(\boldsymbol{\zeta}, w)$ is the vector of the realizations of G_Q for the N measurements.

We denote by $\boldsymbol{w}_{\mathrm{LS}}^{Q}$ the vector of parameters for which the least squares cost function J is minimum. The resulting model is thus of the form $g_Q = \boldsymbol{\zeta}^{\mathrm{T}}\boldsymbol{w}_{\mathrm{LS}}^{Q}$, and one can define vector $\boldsymbol{g}_Q = \boldsymbol{\Xi}\boldsymbol{w}_{\mathrm{LS}}^{Q}$, where

- $\boldsymbol{g}_Q$ is a vector whose N components are the predictions of the model for the N measurements.
- $\boldsymbol{\Xi}$ is the observation matrix: column i ($i = 1$ to $Q+1$) is the vector $\boldsymbol{\Xi}^i$, of which the components are the N measurements of the ith input: matrix $\boldsymbol{\Xi}$ has N rows and $Q+1$ columns,

$$\boldsymbol{\Xi} = \begin{bmatrix} \zeta_1^1 & \cdots & \zeta_n^1 \\ \cdots & \cdots & \cdots \\ \cdots & \cdots & \cdots \\ \cdots & \cdots & \cdots \\ \zeta_1^N & \cdots & \zeta_n^N \end{bmatrix} = \begin{bmatrix} (\zeta^1)^{\mathrm{T}} \\ \cdots \\ \cdots \\ \cdots \\ (\zeta^N)^{\mathrm{T}} \end{bmatrix} = \left[\zeta^1 \cdots \zeta^n\right].$$

The input selection problem is the following: are all Q candidate variables relevant? If a variable is irrelevant, the corresponding parameter in the complete model should be equal to zero. A submodel is a model that is obtained by setting to zero one or several parameters of the complete model. Thus, in order, to solve the problem, the complete model must be compared to all its submodels. We consider a submodel whose last q components (numbered from $Q - q + 2$ to $Q + 1$) are equal to zero: $\boldsymbol{g}_{Q-q} = \boldsymbol{\Xi}\boldsymbol{w}_{mc}^{Q-q}$, where $\boldsymbol{w}_{mc}^{Q-q}$ is the vector of parameters that is obtained by minimizing the least squares cost function $J = \|\boldsymbol{y}_p - \boldsymbol{g}_{Q-q}(\boldsymbol{\zeta}, \boldsymbol{w})\|^2$ under the constraints that the last q components of the vector of the parameters be equal to zero. We want to test the null hypothesis H_0: the last q parameters of the random vector $\boldsymbol{W}$ are equal to zero. If that hypothesis is true, then the random variable

$$Z = \frac{N-Q-1}{q}\frac{\|\boldsymbol{Y}_p - \boldsymbol{G}_{Q-q}\|^2 - \|\boldsymbol{Y}_p - \boldsymbol{G}_Q\|^2}{\|\boldsymbol{Y}_p - \boldsymbol{G}_Q\|^2} = \frac{N-Q-1}{q}\frac{\|\boldsymbol{G}_Q - \boldsymbol{G}_{Q-q}\|^2}{\|\boldsymbol{Y}_p - \boldsymbol{G}_Q\|^2}$$

is a Fisher variable with q and $N - Q$-1 degrees of freedom.

Proof. The quantity $\|\boldsymbol{Y}_p - \boldsymbol{G}_Q\|^2$ is the sum of the squares of the components of vector $\boldsymbol{Y}_p - \boldsymbol{G}_Q$, which is orthogonal to the subspace spanned by the $Q+1$ columns of the observation matrix $\boldsymbol{\Xi}$. Thus, it is the sum of $N - (Q + 1)$ squared independent Gaussian variables: it has a Pearson distribution with $N-Q$-1 degrees of freedom. Similarly, vector $\boldsymbol{G}_Q - \boldsymbol{G}_{Q-q}$ is in a q-dimensional space, hence the square of its norm is the sum of q squared independent Gaussian variables: therefore, $\|\boldsymbol{G}_Q - \boldsymbol{G}_{Q-q}\|^2$ is a Pearson variable with q degrees of freedom. The ratio Z of those Pearson variables is a Fisher variable, as mentioned above.

Assume that a very large number of measurements is available; if the null hypothesis is valid, the numerator of Z is very small since the minimization of the cost function gives values equal to zero to the "useless" parameters of the complete model, hence $\boldsymbol{g}_Q$ and $\boldsymbol{g}_{Q-q}$ are very close; if the null hypothesis is not valid, the two models cannot be very similar, even if the amount of data is large, since the submodel does not have the appropriate complexity for accounting for the data. That explains why the realization of Z must be small if the null hypothesis is valid.

Thus, Fisher's test consists in choosing a risk α, and computing, from the Fisher distribution, the value z_α such that $\Pr(z < z_\alpha) = \alpha$. Then the quantity

$$z = \frac{N-Q-1}{q} \frac{\left\|\boldsymbol{y}_p - \boldsymbol{g}_{Q-q}\left(\boldsymbol{w}_{mc}^{Q-q}\right)\right\|^2 - \left\|\boldsymbol{y}_p - \boldsymbol{g}_Q\left(\boldsymbol{w}_{mc}^{Q}\right)\right\|^2}{\left\|\boldsymbol{y}_p - \boldsymbol{g}_Q\left(\boldsymbol{w}_{mc}^{Q}\right)\right\|^2}$$

(realization of Z with the available data) is computed, and the null hypothesis is accepted if and only if $z < z_\alpha$.

2.10.4.2 Computation of the Cumulative Distribution Function of the Rank of the Probe Feature

In the present section, we discuss the computation of the probability for the probe feature to have a higher rank (rank 1 being the highest rank), at a given step of the selection procedure, than one of the features selected during the previous steps. The complete computation can be found in [Stoppiglia 1998].

We denote by H_{k-1} the probability for the probe vector to be ranked higher than one of the $k-1$ features selected at previous steps. The probability for the probe feature to have a lower rank than the first $k-1$ features is therefore $1 - H_{k-1}$. The probability for the probe feature to be ranked higher than the $k-1$ first features but lower than the kth feature is thus $P_{N-k}(\cos^2(\theta_k))[1 - H_{k-1}]$, where $P_{N-k}(\cos^2(\theta_k))$ is the probability for the angle of the projection of the feature k under consideration, onto the null subspace of the previously selected features, and the projection of the process output on the same subspace, to be smaller than θ_k. Therefore, the probability H_k for the probe feature to be more significant than one of the k selected features is given by: $H_k = H_{k-1} + P_{N-k}(\cos^2\theta_k)(1 - H_{k-1})$. Thus, H_k can be computed recursively, with $H_0 = 0$. That requires the computation of $P_{N-k}(\cos^2\theta_k)$, which is given by the following relations:

$P_n(\cos^2\theta) = 1 - fr_n(\cos^2\theta)$ (n positive integer), with

- for n even: $fr_n(x) = 2/\pi[\sin^{-1}\sqrt{x} + \sqrt{x(1-x)}P^{(n/2)-2}(x)]$, where, for $n \geq 6, P^{(n/2)-2}(x) = 1 + \sum_{k=1}^{(n/2)-2}[2^k(k!)/((2k+1)!!)(1-x)^k]$; for $n = 4: P^0 = 1$; for $n = 2: P^{-1} = 0$;

- for n odd: $fr_n(x) = \sqrt{x}P^{(n-3)/2}(x)$, with, for $n \geq 6$: $P(n-3)/2(x) = 1 + \sum_{k=1}^{(n-3)/2}[1/(2^k)((2k-1)!!)/(k!)(1-x)^k]$; for $n = 3$: $P^0(x) = 1$; for $n = 1$: $P^{-1}(x) = 0$.

2.10.5 Optimization Methods: Levenberg-Marquardt and BFGS

That presentation is from [Oussar 1998]. The algorithms are also described in [Press et al. 1992].

2.10.5.1 The BFGS Algorithm

The BFGS algorithm consists in updating the parameters, at iteration i, by: $\boldsymbol{w}(i) = \boldsymbol{w}(i-1) - \mu_i M_i \nabla_J(\boldsymbol{w}(i-1))$ where μ_i is positive, and where M_i is an approximation, computed iteratively, of the inverse of the Hessian matrix; the latter is computed, at each iteration, by

$$M_i = M_{i-1} + \left[1 + \frac{\gamma_{i-1}^{\mathrm{T}} M_{i-1}\gamma_{i-1}}{\delta_{i-1}^{\mathrm{T}}\gamma_{i-1}}\right]\frac{\delta_{i-1}^{\mathrm{T}}\delta_{i-1}}{\delta_{i-1}^{\mathrm{T}}\gamma_{i-1}} - \frac{\delta^{\gamma}_{i-1}{}^{\mathrm{T}}_{i-1}M_{i-1} + M_{i-1}\gamma^{\delta}_{i-1}{}^{\mathrm{T}}_{i-1}}{\delta_{i-1}^{\mathrm{T}}\gamma_{i-1}},$$

where $\gamma_{i-1} = \nabla J(\boldsymbol{w}(i)) - \nabla J(\boldsymbol{w}(i-1))$ and $\delta_{i-1} = \boldsymbol{w}(i) - \boldsymbol{w}(i-1)$. The initial value M_0 is the identity matrix. If, at some iteration, the matrix is not found to be definite positive, it is re-initialized to the identity matrix.

That approximation is valid only in the neighborhood of a minimum of the cost function. Therefore, it is recommended to use simple gradient descent (or stochastic gradient descent) at the beginning of training, in order to get close to a minimum, then switch to BFGS when the minimum is close enough.

2.10.5.2 The Levenberg-Marquardt Algorithm

The Levenberg-Marquardt algorithm consists in updating the parameters, at iteration i, by

$$\boldsymbol{w}(i) = \boldsymbol{w}(i-1) - [H(\boldsymbol{w}(i-1)) + \mu_i I]^{-1}\nabla(\boldsymbol{w}(i-1)),$$

where μ_i is positive. For small values of the step μ_i, the Levenberg-Marquardt algorithm is close to the Newton method, whereas, for large values of μ_i, the Levenberg-Marquardt algorithm is equivalent to simple gradient descent with step $1/\mu_i$.

The application of the algorithm requires the inversion of matrix $[H(\boldsymbol{w}(i-1)) + \mu_i I]$. The exact expression of the Hessian matrix of the total cost function $J(\boldsymbol{w})$ is

$$H(\boldsymbol{w}(i)) = \sum_{k=1}^{N}\left(\frac{\partial e^k}{\partial \boldsymbol{w}(i)}\right)\left(\frac{\partial e^k}{\partial \boldsymbol{w}(i)}\right)^T + \sum_{k=1}^{N}\frac{\partial^2 e^k}{\partial \boldsymbol{w}(i)\left(\partial \boldsymbol{w}(i)\right)^T}e^k,$$

with $e^k = y_p^k - y^k$.

The above relations are valid for a single-output model. The extension to multiple outputs is straightforward.

Because the second term of the above relation is proportional to the error e^k, it can be neglected in a first approximation, which yields

$$\tilde{H}\left(\boldsymbol{w}(i)\right)=\sum_{k=1}^{N}\left(\frac{\partial e^k}{\partial \boldsymbol{w}(i)}\right)\left(\frac{\partial e^k}{\partial \boldsymbol{w}(i)}\right)^T=\sum_{k=1}^{N}\left(\frac{\partial y^k}{\partial \boldsymbol{w}(i)}\right)\left(\frac{\partial y^k}{\partial \boldsymbol{w}(i)}\right)^T.$$

In the case of a model that is linear with respect to its parameters, y is a linear function of $\boldsymbol{w}$, so that the second term of the expression of is equal to zero.

Several techniques can be useful for inverting matrix $[\tilde{H}+\mu_i I]$.

- Indirect inversion.

The inverse of a matrix can be computed recursively by the following inversion lemma. If A, B, C and D denote four matrices, one has:

$$(A+BCD)^{-1}=A^{-1}-A^{-1}B\left(C^{-1}+DA^{-1}B\right)^{-1}DA^{-1}.$$

Moreover, with the notation $\boldsymbol{\zeta}^k=(\partial y^k)/(\partial \boldsymbol{w}(i))$, matrix $\tilde{H}$can be constructed recursively by defining "partial" matrices $\tilde{H}^k$, of dimension (k, k), by $\tilde{H}^k=\tilde{H}^{k-1}+\boldsymbol{z}^K(\boldsymbol{z}^k)^{\mathrm{T}}$ with $k=1,\ldots,N$. One has $\tilde{H}=\tilde{H}^N$ as desired.

If the inversion lemma is applied to the previous relation with $A=\tilde{H}$, $B=\boldsymbol{\zeta}^k$, $C=I$, and $D=\left(\boldsymbol{\zeta}^k\right)^{\mathrm{T}}$, one gets:

$$\left(\tilde{H}^k\right)^{-1}=\left(\tilde{H}^{k-1}\right)^{-1}-\frac{\left(\tilde{H}^{k-1}\right)^{-1}\zeta^k\left(\zeta^k\right)^T\left(\tilde{H}^{k-1}\right)^{-1}}{1+\left(\zeta^k\right)^T\left(\tilde{H}^{k-1}\right)^{-1}\zeta^k}$$

At the first step $(k:=1)$, one takes $\tilde{H}^0=\mu_i I$, which gives, at step N:

$$\left(\tilde{H}^N\right)^{-1}=\left[(\tilde{H})+\mu_i I\right]^{-1}.$$

- Direct inversion.

Many inversion methods exist. Since the algorithm is iterative, and since the line search procedure (described below) often requires several matrix inversions, an efficient inversion method is mandatory. Since the approximation of the Hessian matrix $\mu_i\ I$ is symmetric definite positive, it is advantageous to use Cholesky's method [Press et al. 1992].

As for simple gradient descent and for BFGS, μ_i must be adjusted at each iteration. A line search method can be used as discussed in the next section.

Note that the expression of the Hessian of the cost function is specific to the least squares cost function; in contrast to the BFGS method, the Levenberg-Marquardt algorithm cannot be used for minimizing arbitrary cost functions, in particular the cross-entropy cost function, often used for classification.

2.10.6 Line Search Methods for the Training Rate

At iteration i of an optimization method, an update direction must be computed; in BFGS for instance, the direction $\boldsymbol{d}^i = -M_i \nabla J(\boldsymbol{w}(i-1))$ is computed by evaluating the gradient of the cost function by backpropagation, and by computing matrix M_i as indicated in the previous section; in simple gradient, the update direction is $\boldsymbol{d}^i = -\nabla J(\boldsymbol{w}(i-1))$. The magnitude of the parameter update along that direction depends on the value of μ_i: one seeks the value of μ_i for which the cost function will be minimum after updating the parameter vector along that direction, for which $J(\boldsymbol{w})$ is minimum if $w = w(i-1) + \mu_i \boldsymbol{d}^i$. Insofar as μ_i is the only unknown, the problem is a line search problem. That search must be performed at each iteration of the training procedure: therefore, it must be fast; since the value of μ_i is not critical for second-order methods, a simple technique may be used. Nash's method produces satisfactory results; it seeks a step such that the updated value of the cost function is smaller than a given bound.

The technique seeks a step that complies with the descent condition,

$$J\left(\boldsymbol{w}(i-1) + \mu_i \boldsymbol{d}^i\right) \leq J(\boldsymbol{w}(i-1)) + m_1 \mu_i \left(\boldsymbol{d}^i\right)^{\mathrm{T}} \nabla J(\boldsymbol{w}(i-1)),$$

where m_1 is a factor, much smaller than 1 (typically $m_1 = 10^{-3}$). The research proceeds iteratively as follows: μ_i is given an arbitrary positive value. The upper boundary condition is checked. If it is obeyed, the parameter update is accepted. Otherwise, the step is multiplied by a quantity smaller than 1 (typically 0.2), and the condition is tested again. The procedure is iterated until satisfaction. If the step value becomes too small, e.g., on the order of 10^{-16}, without the condition being obeyed, or if the number of such search iterations exceeds a limit, the search is abandoned and the procedure is terminated.

An even simpler strategy, often used in conjunction with the Levenberg-Marquardt technique [Bishop 1995], is the following: we denote by $r > 1$ (generally equal to 10) a scale factor. At the beginning of the algorithm, μ_0 is initialized to a large value ([Bishop 1995] suggests 0.1). At iteration $i$of the algorithm:

1. Compute $J(\boldsymbol{w}(i))$ with μ_i computed at the previous step.
2. If $J(\boldsymbol{w}(i)) < J(\boldsymbol{w}(i-1))$, then accept the update and multiply μ_i by r.
3. Otherwise, retrieve $\boldsymbol{w}(i-1)$ and multiply μ_i by r. Iterate until a value of μ_i producing a decrease of J is found.

The above procedure requires a small number of matrix inversions at each iteration of the algorithm. However, the choice of the initial step has an influence on the rate of convergence of the algorithm. That drawback can be circumvented by a method that requires a larger number of matrix inversions:

1. Initialize μ_0 top an arbitrary value.
2. Compute $J(\boldsymbol{w}(i))$ with μ_i found at the previous step.

3. If $J(\boldsymbol{w}(i)) < J(\boldsymbol{w}(i-1))$, then retrieve $\boldsymbol{w}(i-1)$, divide μ_i by r and go to step 2.
4. Otherwise multiply μ_i by r. Iterate until a suitable value of μ_i is found.

2.10.7 Kullback-Leibler Divergence Between two Gaussians

The expression of the Kullback-Leiibbler divergence between two Gaussians with mean and standard deviation (μ_1, σ_1) and (μ_2, σ_2) respectively is derived.

The following relations are useful:

$$\frac{1}{\sigma\sqrt{2\pi}}\int_{-\infty}^{+\infty} \exp\left(\frac{(x-\mu)^2}{2\alpha^2}\right) dx = 1$$
$$\frac{1}{\sigma\sqrt{2\pi}}\int_{-\infty}^{+\infty} x \exp\left(\frac{(x-\mu)^2}{2\alpha^2}\right) dx = \mu$$
$$\frac{1}{\sigma\sqrt{2\pi}}\int_{-\infty}^{+\infty} (x-\mu)\exp\left(\frac{(x-\mu)^2}{2\alpha^2}\right) dx = \sigma^2.$$

The Kullback-Leibler divergence is defined as

$$D(p_1, p_2) = \int_{-\infty}^{+\infty} p_1(x) \ln\left(\frac{p_1(x)}{p_2(x)}\right) dx.$$

Because that definition is not symmetrical with respect to the two distributions, the following quantity is preferred:

$$\begin{aligned}
\Delta &= [D(p_1,p_2) + D(p_2,p_1)]/2 \\
D(p_1,p_2) &= \frac{1}{\sigma_1\sqrt{2\pi}}\int_{-\infty}^{+\infty} \exp\left(\frac{(x-\mu_1)^2}{2\sigma_1^2}\right) \\
&\qquad \times \left[\ln\frac{\sigma_2}{\sigma_1} - \frac{(x-\mu_1)^2}{2\sigma_1^2} + \frac{(x-\mu_2)^2}{2\sigma_2^2}\right] dx \\
&= \frac{1}{\sigma_1\sqrt{2\pi}}\left[\int_{-\infty}^{+\infty} \exp\left(\frac{(x-\mu_1)^2}{2\sigma_1^2}\right) \ln\frac{\sigma_2}{\sigma_1} dx \right. \\
&\qquad - \int_{-\infty}^{+\infty} \exp\left(\frac{(x-\mu_1)^2}{2\sigma_1^2}\right) \frac{(x-\mu_1)^2}{2\sigma_1^2} dx \\
&\qquad \left. + \int_{-\infty}^{+\infty} \exp\left(\frac{(x-\mu_1)^2}{2\sigma_1^2}\right) \frac{(x-\mu_2)^2}{2\sigma_2^2} dx \right]
\end{aligned}$$

The first two terms are equal to $\ln(\sigma_2/\sigma_1) - (1/2)$.

For the third term, one writes

$$\begin{aligned}
(x-\mu_2)^2 &= (x-\mu_1+\mu_1-\mu_2)^2 \\
&= (x-\mu_1)^2 + (\mu_1-\mu_2)^2 + 2(x-\mu_1)(\mu_1-\mu_2).
\end{aligned}$$

Hence,

$$\frac{1}{\sigma_1\sqrt{2\pi}}\int_{-\infty}^{+\infty}\exp\left(\frac{(x-\mu_1)^2}{2\sigma_1^2}\right)\frac{(x-\mu_1)^2}{2\sigma_2^2}dx=\frac{\sigma_1^2}{2\sigma_2^2}$$

$$\frac{1}{\sigma_1\sqrt{2\pi}}\int_{-\infty}^{+\infty}\exp\left(\frac{(x-\mu_1)^2}{2\sigma_1^2}\right)\frac{(\mu_1-\mu_2)^2}{2\sigma_2^2}dx=\frac{(\mu_1-\mu_2)^2}{2\sigma_2^2}$$

$$\frac{1}{\sigma_1\sqrt{2\pi}}\int_{-\infty}^{+\infty}\exp\left(\frac{(x-\mu_1)^2}{2\sigma_1^2}\right)\frac{(x-\mu_1)(\mu_1-\mu_2)}{2\sigma_2^2}dx=0.$$

Finally, one gets

$$D(p_1,p_2)=\ln\left(\frac{\sigma_2}{\sigma_1}\right)-\frac{1}{2}+\frac{\sigma_1^2}{2\sigma_2^2}+\frac{(\mu_1-\mu_2)^2}{2\sigma_2^2}.$$

Then Δ can be computed as

$$\begin{aligned}\Delta&=-\frac{1}{2}+\frac{\sigma_1^2}{4\sigma_2^2}+\frac{\sigma_2^2}{4\sigma_1^2}+\frac{(\mu_1-\mu_2)^2}{4\sigma_2^2}+\frac{(\mu_1-\mu_2)^2}{4\sigma_1^2}\\&=\frac{-2\sigma_1^2\sigma_2^2+\sigma_1^4+\sigma_2^4+(\mu_1-\mu_2)^2(\sigma_1^2+\sigma_2^2)}{4\sigma_1^2\sigma_2^2}\\&=\frac{(\sigma_1^2-\sigma_2^2)+(\mu_1-\mu_2)^2(\sigma_1^2+\sigma_2^2)}{4\sigma_1^2\sigma_2^2}.\end{aligned}$$

2.10.8 Computation of the Leverages

Many discussions of the computation of leverages can be found in the literature. The present one is from [Monari 1999].

Z is an $(N,\ q)$ matrix, with $N \geq q: Z=[z^1,\cdots,z^N]^{\mathrm{T}}$. The leverages to be computed are the diagonal terms of the orthogonal projection matrix $H=Z(Z^TZ)^{-1}Z$,

$$h_{kk}=\boldsymbol{z}^{k\mathrm{T}}(Z^{\mathrm{T}}Z)^{-1}\boldsymbol{z}^k.$$

As diagonal elements of an orthogonal projection matrix, the terms

$$\{h_{kk}\}_{k=1,\ldots,N}$$

are defined only if Z has full rank, i.e., if $Z^{\mathrm{T}}Z$ is invertible. If it is, the following relations are valid:

$$\forall\, k\in[1,\ldots,N],\quad 0\leq h_{kk}\leq 1,\quad \mathrm{trace}(H)=\sum_{k=1}^{N}h_{kk}=\mathrm{rank}(Z).$$

A first leverage computation technique consists in computing matrix $Z^{\mathrm{T}}Z$, inverting it with a classical method (Cholesky, LU decomposition, etc.), then in left and right multiplying by $\boldsymbol{z}^k$ and $\boldsymbol{z}^{kT}$. That method is satisfactory only if $Z^{\mathrm{T}}Z$ is well conditioned. Otherwise, that computation will give values of the leverages that are larger than 1, or negative.

A better solution consists in decomposing matrix Z as

$$Z = U\,W\,V^{\mathrm{T}},$$

where:

- U is an (N, q) matrix such that $U^{\mathrm{T}}U = I$.
- W is a diagonal (q, q) matrix, whose diagonal terms called singular values of Z, are positive or zero, and ranked in order of decreasing values.
- V is a (q, q) matrix, such that $V^{\mathrm{T}}V = VV^{\mathrm{T}} = I$.

That decomposition, known as singular value decomposition or SVD decomposition, is accurate and robust, even if Z is ill-conditioned, or has rank smaller than q (see [Press et al. 1992], and Chap. 3).

Thus, one has

$$Z^{\mathrm{T}}Z = VWU^{\mathrm{T}}UWV^{\mathrm{T}} = VW^2V^{\mathrm{T}},$$

then

$$(Z^{\mathrm{T}}Z)^{-1} = VW^{-2}V^{\mathrm{T}}.$$

That decomposition allows the direct computation of matrix $(Z^{\mathrm{T}}Z)^{-1}$, the elements of which can be written as

$$\left(Z^{\mathrm{T}}Z\right)^{-1}_{lj} = \sum_{k=1}^{q} \frac{V_{lk}V_{jk}}{W_{kk}^2}.$$

After some algebra, one gets

$$h_{kk} = \boldsymbol{z}^{k\mathrm{T}} \left(Z^{\mathrm{T}}Z\right)^{-1} \boldsymbol{z}^{k} = \sum_{l=1}^{q}\sum_{j=1}^{q} Z_{kl}Z_{kj} \left(Z^{\mathrm{T}}Z\right)^{-1}_{lj},$$

and, finally

$$h_{kk} = \sum_{i=1}^{q} \left(\frac{1}{W_{ii}} \sum_{j=1}^{q} Z_{kj}V_{ji} \right)^2.$$

Thus, the leverages can be computed without resorting to the computation of $(Z^{\mathrm{T}}Z)^{-1}$, which is important in the case of ill-conditioned matrices. Since the singular values are ranked in order of decreasing values, it is advantageous to compute the leverages by varying i from q to 1, not from 1 to q.

References

1. Akaike H. [1973], Information theory and an extension of the maximum likelihood princiaple, 2nd *International Symposium on Information Theory,* pp 267–281, Akademia Kiado
2. Akaike H. [1974], A new look at the statistical model identification, *IEEE Transactions on Automatic Control,* 19, pp 716–723
3. Antoniadis A., Berruyer J., Carmona R. [1992], *Régression non linéaire et applications,* Economica
4. Bartlett P.L. [1997], For valid generalization, the size of the weights is more important than the size of the network, *Neural Information Processing Systems,* 9, Morgan Kaufmann
5. Bishop C. [1993], Curvature-driven smoothing: a learning algorithm for feedforward networks, *IEEE Transactions on Neural Networks,* 4, pp 882–884
6. Bishop C. [1995], *Neural Networks for Pattern Recognition,* Oxford University Press.
7. Björck A. [1967], Solving linear least squares problems by Gram-Schmidt orthogonalization. *BIT,* 7, pp 1–27
8. Broyden C.G. [1970], The convergence of a class of double-rank minimization algorithms 2: the new algorithm, *Journal of the Institute of Mathematics and its Applications,* 6, pp 222–231
9. Chen S., Billings S.A., Luo W., Orthogonal least squares methods and their application to nonlinear system identification, *International Journal of Control,* 50, pp 1873–1896
10. Draper N.R., Smith H. [1998], *Applied Regression Analysis,* Wiley
11. Dreyfus G., Idan Y. [1998], The canonical form of discrete-time nonlinear models, *Neural Computation,* 10, pp 133–164
12. Gallinari P., Cibas T. [1999], Practical complexity control in multilayer perceptrons, *Signal Processing,* 74, pp 29–46
13. Geman S., Benenstock E., Doursat R. [1992], Neural networks and the bias/variance dilemma, *Neural Computation* 4, pp 1–58
14. Goodwin G.C., Payne R.L. [1977], *Dynamic System Identification: Experiment Design and Data Analysis,* Mathematics in Science and Engineering, Academic Press
15. Goodwin G.C., Sin K.S. [1984], *Adaptive Filtering Prediction and Control,* Prentice-Hall, New Jersey
16. Guyon I., Gunn S., Nikravesh M., Zadeh L., eds. [2005], Feature extraction, foundations and applications, Springer
17. Hansen L.K., Larsen J. [1996], Linear unlearning for cross-validation, *Advances in Computational Mathematics,* 5, pp 269–280
18. Haykin S. [1994], *Neural Networks: a comprehensive approach,* MacMillan
19. Jollife I.T. [1986], *Principal Component Analysis,* Springer
20. Kohonen T. [2001] *Self-Organizing Maps,* Springer
21. Kullback S., Leibler R. A. [1951], On information and sufficiency, *Annals of mathematical Statistics,* 22, pp 79–86
22. Kullback S. [1959], *Information Theory and Statistics,* Dover Publications
23. Kuo B. C. [1992], *Digital Control Systems,* Saunders College Publishing
24. Kuo B. C. [1995], *Automatic Control Systems,* Prentice Hall
25. Lagarde de J. [1983], *Initiation à l'analyse des données,* Dunod, Paris

26. Lawrance A.J. [1995], Deletion, influence and masking in regression, *Journal of the Royal Statistical Society,* B 57, pp 181–189
27. Leontaritis I.J., Billings S.A. [1987], *Model selection and validation methods for nonlinear systems*, International Journal of Control, 45, pp 311–341
28. Levenberg K. [1944], A method for the solution of certain nonlinear problems in least squares, *Quarterly Journal of Applied Mathematics*, 2, pp 164–168
29. Levin A., Narendra K.S. [1993], Control of nonlinear dynamical systems using neural networks: controllability and stabilization, *IEEE Transaction on Neural Networks,* 4, pp 1011–1020
30. Ljung L. [1987], *System Identification; Theory for the User,* Prentice Hall
31. McKay D.J.C. [1992], A practical Bayesian framework for backpropagation networks, *Neural Computation*, 4, pp 448–472
32. McQuarrie A.D.R, Tsai C., *Regression and Time Series Model Selection*, World Scientific
33. Marquardt D.W. [1963], An algorithm for least-squares estimation of nonlinear parameters, *Journal of the Society of Industrial and Applied Mathematics*, 11, pp 431–441
34. Monari G. [1999], *Sélection de modèles non-lineaires par leave-one-out; étude théorique et application des réseaux de neurones au procédé de soudage par points*, Thèse de Doctorat de l'Université Pierre et Marie Curie, Paris. *Available from* http://www.neurones.espci.fr
35. Monari G., Dreyfus G. [2000], Withdrawing an example from the training set: an analytic estimation of its effect on a nonlinear parameterised model, *Neurocomputing*, 35, pp 195–201
36. Monari G., Dreyfus G. [2002], Local Overfitting Control via Leverages, Neural Computation
37. Mood A.M., Graybill F.A., Boes D.C. [1974], *Introduction to the Theory of Statistics*, McGraw-Hill
38. Narendra K.S, Annaswamy A.M. [1989], *Stable Adaptive Systems*, Prentice-Hall
39. Nerrand O. [1992], *Réseaux de neurones pour le filtrage adaptatif, l'identification et la commande de processus*, thèse de doctorat de l'Université Pierre et Marie-Curie
40. Nerrand O., Urbani D., Roussel-Ragot P., Personnaz L., Dreyfus G. [1994], Training recurrent neural networks: why and how? An Illustration in Process Modeling, *IEEE Transactions on Neural Networks* 5, pp 178–184
41. Norgaard J.P., Ravn O., Poulsen N.K., Hansen L.K. [2000], *Neural Networks for Modelling and Control of Dynamic Systems*, Springer
42. Norton J.P. [1986], *An introduction to Identification*, Academic Press
43. Oussar Y. [1998], *Réseaux d'ondelettes et réseaux de neurones pour la modélisation statique et dynamique de processus*, Thèse de Doctorat de l'Université Pierre et Marie Curie, Paris. *Available from* http://www.neurones.espci.fr
44. Oussar y., Dreyfus G. [2002], *Initialization by selection for wavelet network training, Neurocomputing*, 34, pp 131–143
45. Oussar Y., Dreyfus G. [2001], How to be a gray box: dynamic semiphysical modeling, *Neural Networks*, vol. 14
46. Plaut D., Nowlan S., Hinton G.E. [1986], *Experiments on learning by back propagation*, Technical Report, Carnegie-Mellon University
47. Poggio T., Torre V., Koch C. [1985], Computational vision and regularization theory, *Nature, 317*, pp 314–319

48. Press W.H., Teukolsky S.A., Vetterling W.T., Flannery B.P. [1992], *Numerical Recipes in C: The Art of Scientific Computing*, Cambridge University Press.
49. Puskorius G.V., Feldkamp L.A. [1994], Neurocontrol of nonlinear dynamical systems with Kalman Filter trained recurrent networks, *IEEE Trans. on Neural Networks*, 5, pp 279–297
50. Rumelhart D.E., Hinton G.E., Williams R.J. [1986], Learning internal representations by error backpropagation, *Parallel Distributed Processing: Explorations in the Microstructure of Cognition*, pp 318–362, MIT Press
51. Saarinen S., Bramley R., Cybenko G. [1993], Ill-conditioning in neural network training problems, *SIAM J. Sci. Stat. Comp.*, 14, pp 693–714
52. Seber G.A.F., Wilde C.J. [1989], *Nonlinear Regression*, Wiley
53. Seber G.A.F. [1977], *Linear Regression Analysis*, Wiley
54. Sjö berg J., Zhang Q., Ljung L., Benveniste A., Delyon B., [1995], Nonlinear black–box modeling in system identification: a unified overview, *Automatica*, 31, pp 1691–1724
55. Soderstrom T. [1977], On model structure testing in system identification, *International Journal of Control*, 26, pp 1–18
56. Sontag E.D. [1993], Neural networks for control, *Essays on control: perspectives in the theory and its applications*, pp 339–380, Birkhäuser
57. Stone M. [1974], Cross-validatory choice and assessment of statistical predictions, *Journal of the Royal Statistical Society*, B 36, pp 111–147
58. Stoppiglia H. [1998], *Méthodes statistiques de sélection de modèles neuronaux; applications financières et bancaires*, thèse de doctorat de l'Université Pierre et Marie-Curie. *Available from* http://www.neurones.espci.fr
59. Stoppiglia H., Dreyfus G., Dubois R., Oussar Y. [2003], Ranking a Random Feature for Variable and Feature Selection, *Journal of Machine Learning Research*, pp 1399–1414
60. Stricker M. [2000], *Réseaux de neurones pour le traitement automatique du langage: conception et réalisation de filtres d'informations*, thèse de l'Université Pierre et Marie-Curie. *Available from* http://www.neurones.espci.fr
61. Tibshirani R.J. [1996], A comparison of some error estimates for neural models, *Neural Computation*, 8, pp 152–163
62. Tikhonov A.N., Arsenin V.Y. [1977], *Solutions of Ill-Posed Problems*, Winston
63. Vapnik V.N. [1995], *The Nature of Statistical Learning Theory*, Springer
64. Waibel , Hanazawa T., Hinton G., Shikano K., and Lang K. [1989], Phoneme recognition using time-delay neural networks, *IEEE Transactions on Acoustics, Speech, and Signal Processing*, 37, pp 328–339
65. Werbos P.J. [1974], *Beyond regression: new tools for prediction and analysis in the behavioural sciences*, Ph. D. thesis, Harvard University
66. Widrow B., Hoff M.E. [1960], Adaptive switching circuits, *IRE Wescon Convention Records,* 4, pp 96–104
67. Wonnacott T.H., Wonnacott R.J. [1990], *Statistique économie-gestion-sciences-médecine*, Economica, 4e édition, 1990
68. Zhou G., Si J. [1998], A systematic and effective supervised learning mechanism based on Jacobian rank deficiency, *Neural Computation*, 10, pp 1031–1045

3

Modeling Methodology: Dimension Reduction and Resampling Methods

J.-M. Martinez

3.1 Introduction

This chapter provides additional elements of methodology for neural network design. It provides answers to methodological questions raised by neural network modeling. As explained in the previous chapter, there is more to the design of a neural model than choosing the number of hidden neurons and implementing a training algorithm:

- Before using a neural network or any other statistical model, it may be necessary to construct new input variables to decrease their number whilst losing as little information as possible concerning their distribution.
- After estimating the parameters of the model (training if the model is a neural network), the user should assess the risk of using the model thus designed; that risk is linked to the generalization error, which cannot be computed, hence must be estimated. In the previous chapter, we discussed a method for estimating the generalization error by computation of the virtual leave-one-out score. In this chapter, we describe another recent statistical technique, based on resampling, which is used to estimate the statistical characteristics of the generalization error.

Therefore, the aspects of the methodology described in this chapter are related to

- the preprocessing to be performed on the data,
- the techniques for reducing the number of inputs, based on principal component analysis and curvilinear component analysis,
- the estimation of the generalization error using statistical resampling techniques, with emphasis on the bootstrap.

The reduction in size is not only intended to decrease the number of variables describing each example: it also attempts to design more compact data representations, thus making their analysis easier. In the context of linear modeling, the conventional method of input dimension reduction is termed principal component analysis (PCA): the latter consists in projections, and is limited to linear varieties. To process nonlinear representations, we will describe an alternative method, termed curvilinear component analysis (CCA), which may be considered as a "nonlinear" extension of PCA. It is similar to the "Kohonen maps" (Chap. 7), but it is more flexible than them, since the structure of the projection space is not imposed.

Resampling methods aim at performing estimations estimates when the probability distributions of the variables to be analyzed are not known. In the problems raised by regression, particularly regression by neural networks, they allow estimations of the generalization error, and they lead to efficient and robust assessments of the variability of the network with respect to the data; that is the key element of the bias-variance dilemma (described in Chap. 2), which arises in the generation of any statistical model. Those advanced techniques are computer-intensive, but the increased speed of computers makes them more and more popular. A new method will be described, combining the *bootstrap* and early stopping (described in the previous chapter) to automate and monitor the training of neural networks.

3.2 Preprocessing

3.2.1 Preprocessing of Inputs

In the previous chapter, we mentioned that the values of model variables are generally expressed in different units and have different orders of magnitude. It is therefore necessary to pre-process those values so that they have the same influence on the design of the model. Therefore, variables must be centered and reduced or at least normalized. The preprocessing described in the section "Input normalization" of Chap. 2 transforms the input components into variables with zero average and unit standard deviation.

Standardize or Reduce

For distributions with uniform and centered inputs, the ratio between standardization and reduction is only $\sqrt{3}$ for the standard deviation. The standard deviation of a uniform distribution over an interval l is $l/2\sqrt{3}$ and standardization over the same interval divides the variable by $l/2$.

Boolean Variables

The values 0 and 1 of Boolean variables should be transformed into -1 and $+1$ respectively; variables resulting from fuzzy encoding should be subject to similar processing.

Figure 3.1 shows the effect of preprocessing. It corresponds to a shift of the centre of gravity of the scatter diagram followed by standardization of the dispersion of values on each axis without altering the distribution of points.

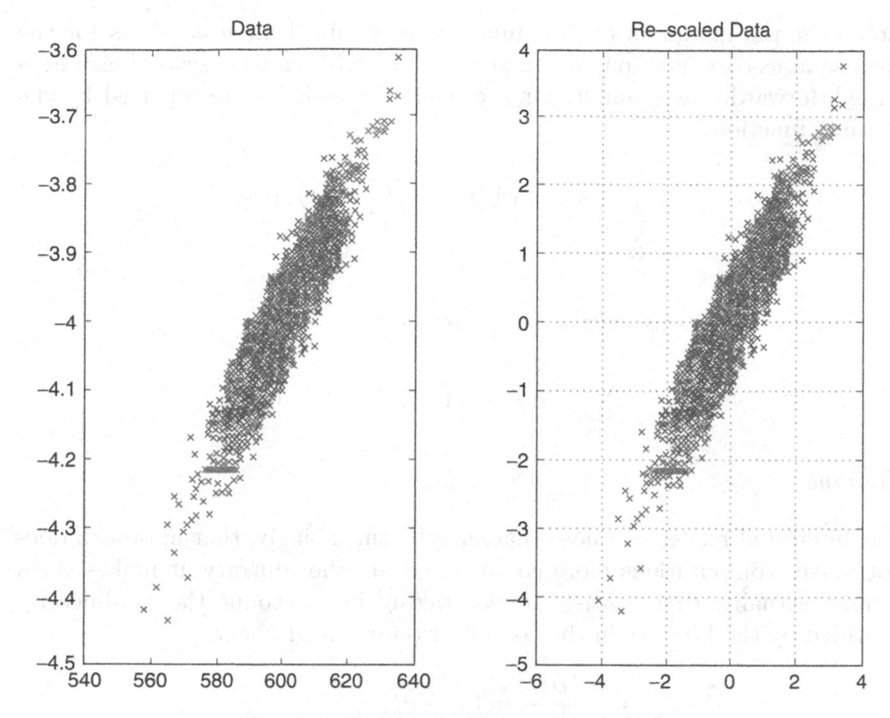

Fig. 3.1. Data centering and reduction

That simple preprocessing, applied to all components, is often used to detect anomalies in the database. A standard deviation that is too low may mean that the corresponding variable has too small variability to actually have an influence on model. Variables with zero standard deviation should of course be ignored, since they do not provide any information in the design of the model. For a more extensive diagnosis of such "anomalies", the advice of the process expert must be sought.

3.2.2 Preprocessing Outputs for Supervised Classification

Preprocessing of outputs is link to output encoding. For supervised classification (described in detail in Chap. 6), the encoding of outputs is associated with posterior probabilities, so that the problem of preprocessing is irrelevant: the encoding of posterior probability leads to representing each class by an output neuron with a logistic activation function. The associated cost is cross-entropy rather than the least-squares cost. For two-class discrimination, where y and y^* are the network output and the desired class code respectively, cross-entropy is defined by

$$J = y^* \ln y + (1 - y^*) \ln (1 - y) .$$

Note that the minimum of that function is obtained for $y = y^*$, as for the least squares cost function. The extension to problems with several classes is straightforward. For example, for n classes, the logistics are replaced by the softmax function,

$$y_i = \frac{e^{z_i}}{\sum_{j=1}^{n} e^{z_j}} \qquad \text{with } z_i = \sum_k w_{ik} x_k + w_{i0}.$$

For each example, cross-entropy is expressed by

$$E = \sum_{i=1}^{n} y_i^* \ln y_i + (1 - y_i^*) \ln(1 - y_i).$$

Training

The interested reader will note that, maybe surprisingly, that approach does not makes computations more complicated: on the contrary, it makes them simple: actually, that consists in not taking into account the nonlinearity provided by the logistics in the computation of the gradient,

$$\frac{\partial E}{\partial w_{ik}} = (y_i - y_i^*) x_k.$$

That is equivalent to Widrow-Hoff's training rule described in Chap. 2.

3.2.3 Preprocessing Outputs for Regression

In regression problems, the outputs represent conditional averages. The residues around the average value are assumed to follow a normal centered law. In order to optimize the design of the model, outputs are therefore centered and reduced; the averages and variances of the outputs are estimated on the basis of examples.

The average quadratic error EQM_r, computed in the reduced output space, corresponds to the average quadratic error EQM computed from raw data, divided by the estimated variance.

$$\mathrm{EQM}_r = \frac{1}{N} \sum_{k=1}^{N} (\tilde{y}_k - \tilde{y}_k^*)^2 \Rightarrow \mathrm{EQM} = \mathrm{EQM}_r \times \sigma_y^2.$$

Reduced Error and Coefficient of Nondetermination

The relation between the average quadratic error computed from the centered reduced variables is the "residual variance divided by total variance" used in linear regression to express the percentage of the variance not taken into account by the model. In that case, the one's complement of the average quadratic error, known as the coefficient of nondetermination, defines the contribution of the model: the least expensive and least powerful model is the model that predicts the output as the average value of the measured output, irrespective of the input. For that model, the average quadratic error EQM_r is 1.

3.3 Input Dimension Reduction

The design of the model $g(\boldsymbol{x}, \boldsymbol{w})$ may require a reduction in dimension of the input vector $\boldsymbol{x}$. That is particularly important when the number variables is too large to be handled conveniently; or when it is assumed that they are not mutually independent. In the latter case, their reduction simplifies the design of the model. The latter is therefore more robust with respect to the variability of the data, and is less sensitive to overfitting due to over-parameterization (see Chap. 2).

In order to explore the structure of multidimensional data, the analysis is based on the observation of the distribution of variables in the input space. When the number of factors is too high for visual analysis or digital processing, it must be decreased. In linear statistics, PCA (Principal Component Analysis) is used for reducing the number of factors. The method is based on a linear combination of factors by projection. It provides a more synthetic representation of the data.

In this section, we will review the principles of PCA; we will then discuss CCA (Curvilinear Component Analysis), which may be viewed as a nonlinear extension of PCA, well suited to representations of more complex data structures. A parallel will be drawn with self-organizing Kohonen maps, which are also used for nonlinear data analysis.

3.4 Principal Component Analysis

Principal component analysis is one of the oldest statistical analysis techniques. It was developed to study samples of individuals described by several factors. The method is therefore suited to the analysis of multidimensional data: in general, the separate study of each factor is not sufficient, since it does not allow for the detection of possible dependencies between factors.

3.4.1 Principle of PCA

To reduce the number of factors (components), PCA constructs sub-spaces of input space (also termed representation space), whose dimensions are therefore smaller than the number of factors, in which the distribution of observations (points) is as similar as possible to their distribution in representation space. The similarity criterion is the total inertia of the scatter diagram.

Therefore, PCA is a linear projection method that maximizes the inertia of the scatter diagram.

Before describing the theoretical developments, let us review, as a simple illustration, the example of the distribution of a scatter diagram in $\mathbb{R}^2$ shown in Fig. 3.1. The first main axis found by PCA is the axis with respect to which the inertia of the scatter diagram is maximal. The second axis, orthogonal to the previous one, is the axis with respect to which the inertia of the scatter diagram, in the null space of the first axis. The other axes are defined similarly.

PCA and Gram-Schmidt Orthogonalization

This procedure may be reminiscent of the Gram-Schmidt orthogonalization described in the previous chapter for the selection of inputs. That analogy, however, is deceptive. PCA is a procedure that is carried out in *representation space*, in which each observation is represented by a point, whose co-ordinates are the values of the factors that correspond to that observation. By contrast, Gram-Schmidt orthogonalization for the selection of inputs is carried out in *the observation space*, where each factor is represented by a vector, the components of which are observations of this factor in the database. The dimension of representation space is the number of factors of the model, whilst the dimension of observation space is the number of observations in the database.

Figure 3.2 shows the 2 main axes defined by the 1st and 2nd bisector respectively (the orthogonality of the axes is distorted by the scale of the graph). The main components will be represented by projections of points on the main axes. Linear transformation by PCA therefore consists in changing the variables, defined by the main axes, on the centered data.

We will show that the "mechanical" concept of total inertia of the scatter diagram is equivalent to the "statistical" concept of variance. The inertia of points is computed with respect to the centre of gravity of the scatter diagram. We denote by g the centre of gravity and by I_n the inertia of the scatter diagram defined in $\mathbb{R}^n$, we have

$$g_j = \frac{1}{N}\sum_{i=1}^{N} x_{ij} \Rightarrow I_n = \sum_{j=1}^{n}\sum_{i=1}^{N} \left(x_{ij} - g_j\right)^2 .$$

Inertia I_n is therefore equal to the trace of the variance-covariance matrix of the data X defined by

$$V = \left(X - Ig\right)^{\mathrm{T}} \left(X - Ig\right),$$

where I denotes the identity matrix.

Since inertia is shift-invariant, the data may be centered by $X' = X - Ig$, so that one has the following simple relation between the inertia and the variance-covariance matrix on the new centered data X':

$$I_n = \mathrm{Trace}\left(X'^{\mathrm{T}} X'\right).$$

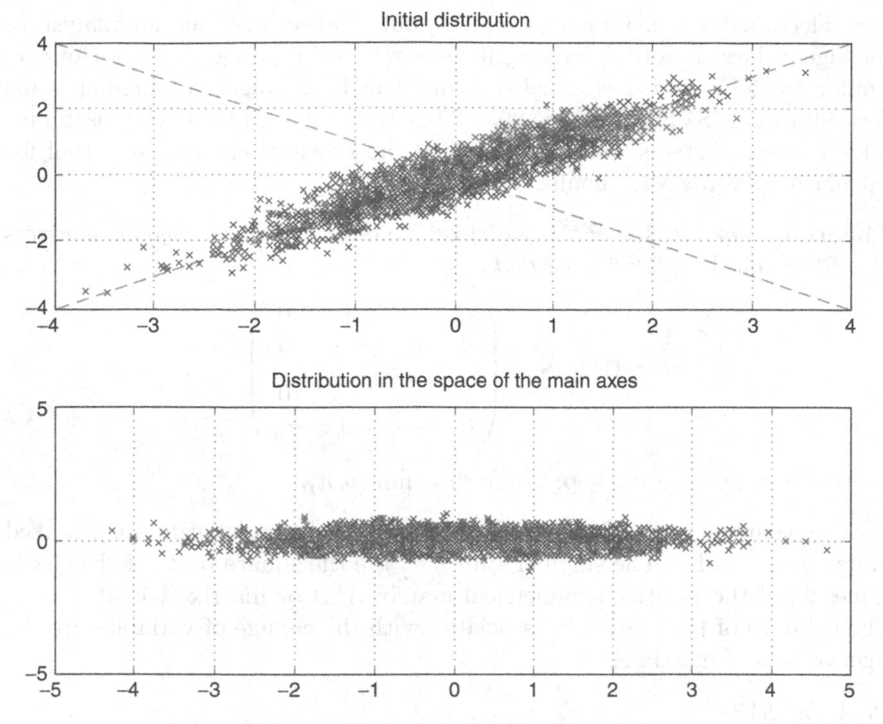

Fig. 3.2. Change of variables by PCA

Inertia on Centered and Reduced Data

For centered and reduced data, one has $\mathrm{Tr}\left(X^{\mathrm{T}}X\right) = n$.

Consider a sub-space of dimension $q < n$, and denote by $V_{n\times q}$ the matrix associated with the projector on $\mathbb{R}^q$; the scatter diagram projected on $\mathbb{R}^q$ is represented by matrix XV, the inertia of which is

$$I_q = \mathrm{Tr}\left(V^{\mathrm{T}}X^{\mathrm{T}}XV\right).$$

PCA defines the linear projection that maximizes I_q, the value of the inertia of the points computed in $\mathbb{R}^q$. That problem is solved by finding the first axis with respect to which the inertia is maximum, then a second axis, orthogonal to the previous one, to carry on with the maximization of the inertia, and so on up to the pth axis. The axes obtained are borne by the eigenvectors of matrix $X^{\mathrm{T}}X$, ranked in order of decreasing eigenvalues λ_i. The eigenvalues λ_j, $j = 1, \ldots, n$ are positive or zero, since matrix $X^{\mathrm{T}}X$ is positive symmetrical. The transformation to be performed on of the centered data to obtain the main components is

$$x \in \mathbb{R}^n \to V_{n\times q}^{\mathrm{T}}x \in \mathbb{R}^{q<n}.$$

The selection of the main components (q out of p) results from an analysis of the eigenvalues. Before describing it, it is worthwhile reviewing a technique, similar to PCA, which is extensively used in linear algebra: singular value decomposition (SVD) [Cichoki 1993]. That technique, which is very useful for solving linear systems, was mentioned in the previous chapter as a tool for calculating leverages for nonlinear models.

Theorem. *For all $A \in \mathbb{R}^{n\times p}$ matrices, there exist two orthogonal matrices $U \in \mathbb{R}^{n\times p}$ and $V \in \mathbb{R}^{p\times p}$ such that*

$$U^{\mathrm{T}}AV = S = \begin{pmatrix} \sigma_1 & 0 & \cdots & 0 \\ 0 & \sigma_2 & \cdots & 0 \\ \vdots & \ddots & \ddots & 0 \\ 0 & \cdots & 0 & \sigma_m \end{pmatrix},$$

with $\sigma_1 \geq \sigma_2 \geq \cdots \geq \sigma_m \geq 0$, where $m = \min(p, n)$.

The elements of the diagonal matrix S are the singular values σ_j, ranked in decreasing order. The singular values σ_j are the square roots of the eigenvalues λ_j of the positive symmetrical matrix $A^{\mathrm{T}}A$ or matrix AA^{T} if $m < n$. The columns of the matrix V associated with the change of variables are the eigenvectors of matrix $A^{\mathrm{T}}A$.

PCA and SVD

Therefore, PCA and SVD are equivalent when operated on centered data.

Unlike diagonalization techniques for square matrices, singular value decomposition applies to all types of matrices. The index for the 1st singular value equal to 0 is the rank of the matrix; its condition number is the ratio of the largest to the smallest singular value σ_1/σ_p.

From the orthogonality of matrices U and V, one has

$$U^{\mathrm{T}}AV = S \Rightarrow A = USV^{\mathrm{T}}.$$

In a modeling application, if A is the matrix of centered observations (defined in the previous chapter), matrix $US = AV$ describes the same observations in an "orthogonal" representation: the new inputs obtained after transformation are not subject to linear correlation. The same technique is used for "cleaning" signals [Davaud 1991]. In order to reduce the new inputs, matrix U is retained as a new base of examples: the linear transformation thus becomes $S^{-1}V^{\mathrm{T}}x$ instead of $V^{\mathrm{T}}x$.

Singular value decomposition of the matrix of centered data X is used to express the inertia with respect to the singular values σ_j or the eigenvalues λ_j of matrix $X^{\mathrm{T}}X$,

$$I_p = \mathrm{Tr}(X^{\mathrm{T}}X) \Rightarrow I_p = \sum_{j=1}^{p} \lambda_j \Rightarrow I_p = \sum_{j=1}^{p} \sigma_j^2.$$

Inertia and Matrix Norm

That result is familiar in linear algebra since the inertia of the scatter diagram corresponds to the Frobenius matrix norm, which is expressed as a function of singular values,

$$\|X\|_F = \sqrt{\sum_{i,j} x_{ij}^2} = \sqrt{\sum_j \sigma_j^2}.$$

The projection matrix $P_{p\times q}$ associated with the first q axes is therefore represented by the first q vectors of matrix $V_{p\times p}$. The contribution to the inertia of each main axis is given by the ratio of σ_j^2 to the sum $\sigma_1^2+\sigma_2^2+\cdots+\sigma_p^2$. The contribution of the first q axes is:

$$I_q = \sum_{j=1}^{q} \sigma_j^2 \Rightarrow I_q = I_p \frac{\sum_{j=1}^{q\leq p} \sigma_j^2}{\sum_{j=1}^{p} \sigma_j^2}.$$

The quality of the dimension reduction depends on the value of q. There is no general rule for determining the best value. A few rules used to determine the number q of components [Saporta 1990] may be mentioned:

- The part of the explained inertia to contribute at least a fixed percentage of the inertia,
- Kaiser's rule,which retains eigenvalues larger than the average of eigenvalues (for reduced centered data, that consists in retaining the eigenvalues that are larger greater than 1, since the sum of the eigenvalues is equal to n),
- The "scree test" which, from the curve of I_q as a function of $q = 1, 2, \ldots, n$, selects the value of q that corresponds to the 1st break in the gradient, as shown in the example given in Fig. 3.3 with a break in the gradient from the 4th eigenvalue.

Before applying PCA systematically, it must be remembered that the so-called principal component is defined from the criterion concerning the inertia of the scatter diagram. For certain problems, the principal component is by far not the most informative aspect. For example, in a set of human faces of several different races, the recognition of race is based more fully on the secondary components; the first component is more representative of the average characteristics of the faces.

3.5 Curvilinear Component Analysis

For more complex distributions, dimensionality reduction may require nonlinear processing. Curvilinear component analysis was proposed by [Demartines

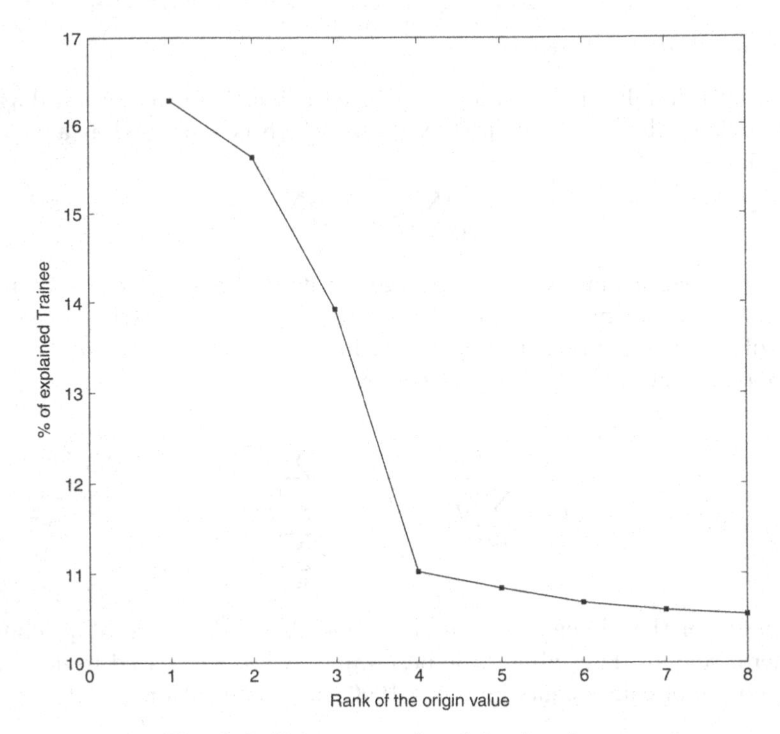

Fig. 3.3. Percentage of explained variance

1995] for analyzing and reducing the dimensions of nonlinear distributions. It may be viewed as a nonlinear extension of principal component analysis. CCA uses a more local criterion than PCA, which allows it to keep the local topology of the distribution of input points. An analysis of this method, together with examples of applications may be found in [Hérault 1993; Vigneron 1997].

Figure 3.4 shows CCA applied to dimension reduction of nonlinear data structures: on the left, the left-hand part shows a set of points defined in $\mathbb{R}^3$, and the right-hand part shows a representation in $\mathbb{R}^2$. The dimension reduction may therefore be seen as a "nonlinear" projection that retains the proximity of points and therefore the local topology of the distribution.

In closed structures, such as a sphere or a cylinder, dimensionality reduction will inevitably result in some local distortion, as shown in Fig. 3.5, which shows an example of the projection of a sphere on the plane. The main principle of CCA is the gradual control of local distortion, during training.

Since the main goal of CCA is a dimensionality reduction that preserves the local topology, it is ideally suited for the representation of nonlinear varieties. A variety in $\mathbb{R}^p$ may be defined roughly as a set of points, the local dimensions of which are smaller than p. The envelope of a sphere defined in $\mathbb{R}^3$ is an example: the dimension of the variety is 2. More strictly, a variety of dimension q in $\mathbb{R}^q$ is a sub-set of $\mathbb{R}^n$ obtained by applying a function defined by $\mathbb{R}^q$ in $\mathbb{R}^q$.

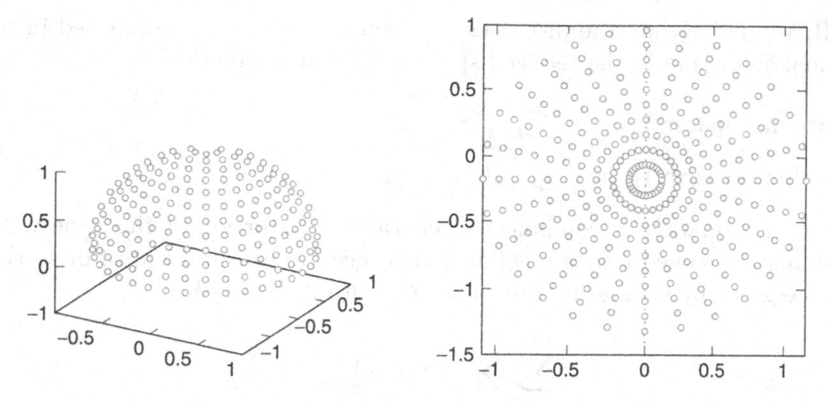

Fig. 3.4. CCA Projection of a hemisphere

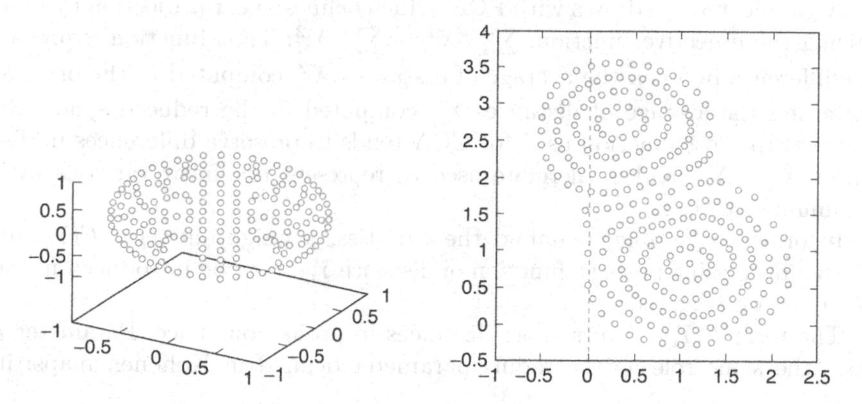

Fig. 3.5. CCA projection of a sphere

The rank of the differential of the application determines the local dimensions of the variety.

In relation to PCA, that method is therefore used to represent data structures that are distributed in a nonlinear manner. It is similar to methods based on Kohonen's self-organizing maps, but its principle is different. There are no constraints on the points in the projection space. In theory, no neighborhood is defined between the points in the projection space. That gives great flexibility to the method.

3.5.1 Formal Presentation of Curvilinear Component Analysis

The co-ordinates of the p points are defined

- by $x_i \in \mathbb{R}^n$, $i = \{1, \dots, p\}$ in the original space;
- by $y_i \in \mathbb{R}^{n'<n}$, $i = \{1, \dots, p\}$ in the reduced space.

If X_{ij} and Y_{ij} are the distances between points i and j, computed in the original space and in the reduced space respectively, one has:

- original space: $X_{ij} = \sqrt{\sum_{k=1}^{n} (x_{ik} - x_{jk})^2}$,
- reduced space: $Y_{ij} = \sqrt{\sum_{k=1}^{n'} (y_{ik} - y_{jk})^2}$.

The transformation of components generates a distortion of the variety. By retaining the same metrics (euclidean distance), a measurement of distortion may be given by comparing distances X_{ij} with distances Y_{ij}:

$$\sum_{i=1}^{p} \sum_{j=i+1}^{p} (X_{ij} - Y_{ij})^2 .$$

A parallel may be drawn with PCA, which defines linear projection by minimizing the objective function: $\sum_{i,j} X_{ij}^2 - \sum_{i,j} Y_{ij}^2$. That function expresses the difference between the average of distances X_{ij}^2 computed in the original space and the average of distances Y_{ij}^2 computed in the reduced space. By contrast, the cost function used for CCA tends to preserve differences in distances $X_{ij} - Y_{ij}$, and is therefore used to represent nonlinear varieties with minimum distortion.

In order to be able to unfold the varieties, a weighting term $F(Y_{ij}, \rho)$, which a decreasing positive function of distance Y_{ij}, may be introduced in the cost function (Fig. 3.6).

The term $F(Y_{ij})$ favors short distances in projection space. Parameter ρ plays the same role as the radius parameter defined in Kohonen maps: in

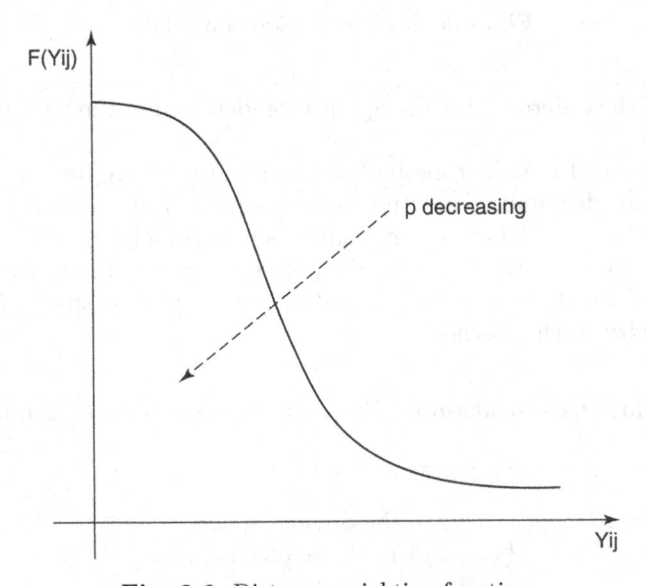

Fig. 3.6. Distance weighting function

output space, distances greater than ρ will no longer be taken into account. The decrease in parameter ρ during training allows the opening, and possibly the breaking, of certain nonlinear varieties. The projection of a sphere R^3 in R^2 (Fig. 3.4) shows an example of a variety for which the projection requires a breaking. The function is therefore used to open certain varieties by retaining the local topology as far as possible.

Therefore, the objective function minimized by CCA takes the following form:

$$E = \sum_{i=1}^{p} \sum_{j=i+1}^{p} (X_{ij} - Y_{ij})^2 F(Y_{ij}, \rho).$$

3.5.2 Curvilinear Component Analysis Algorithm

The algorithm consists in minimizing the above cost function with respect to the coordinates of each point in the database in reduced space. As for learning, we may use any of the optimization algorithms given in Chap. 2. Training can be performed by any minimization algorithm, as described in Chap. 2. For illustration, we describe the minimization of the cost function by stochastic gradient.

Thus, we compute the partial derivatives of the cost function with respect to each parameter; we denote by y_{ik} the k^{-ith} coordinate of point i,

$$\begin{aligned}\frac{\partial E}{\partial y_{ik}} &= \sum_{j \neq i} \frac{\partial E}{\partial Y_{ij}} \frac{\partial Y_{ij}}{\partial y_{ik}} \\ &= -\sum_{j \neq i} \frac{X_{ij} - Y_{ij}}{Y_{ij}} [2F(Y_{ij}) - (X_{ij} - Y_{ij})F'(Y_{ij})](y_{ik} - y_{jk}).\end{aligned}$$

Parameters are updated as follows, where μ is the gradient step:

$$\Delta y_i = \mu \sum_{j \neq i} \frac{X_{ij} - Y_{ij}}{Y_{ij}} [2F(Y_{ij}) - (X_{ij} - Y_{ij})F'(Y_{ij})] (y_i - y_j).$$

A condition should be provided to guarantee the convergence of the minimization. The term $\beta_{ij} = 2F(Y_{ij}) - (X_{ij} - Y_{ij}) F'(Y_{ij})$ must be positive. If Y_{ij} is too large with respect to X_{ij}, point j should be brought closer to point i. The functions $F(Y_{ij})$ should be selected in order to guarantee $\beta_{ij} > 0$. That condition is difficult to satisfy: for instance, for $F(Y_{ij}) = \exp(-Y_{ij}/\rho)$, the stability requires $\rho > (Y_{ij} - X_{ij})/2$. That condition cannot always be fulfilled because ρ decreases during training. The following simplification of the training rule guarantees, almost everywhere, that $\beta_{ij} = 2 > 0$:

$$\Delta y_i = \begin{cases} \mu \sum_{j \neq i} \frac{X_{ij} - Y_{ij}}{Y_{ij}} (y_i - y_j) & \text{if } Y_{ij} > \rho; \\ 0 & \text{otherwise.} \end{cases}$$

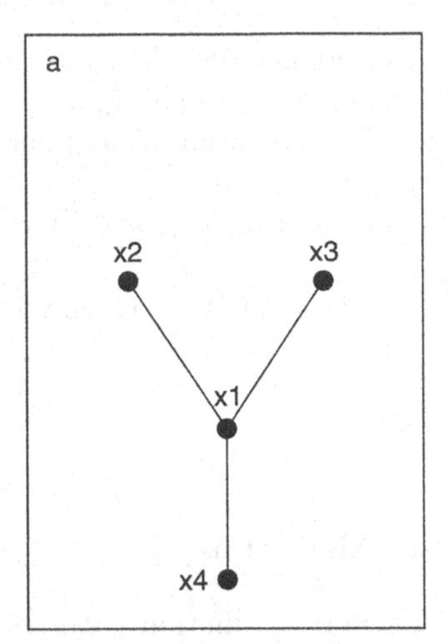

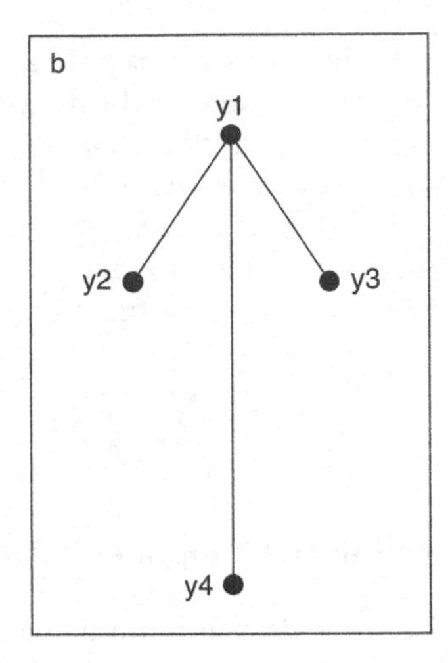

Fig. 3.7. Example of a dead end

The contribution of the $n-1$ points j on point i produces an average effect. In certain situations, this may lead to dead ends. Figure 3.7 shows an example of such a situation.

In input space (a) point $i = 1$ at the center of gravity of the other three points. In output space (b), the initial conditions located it outside the three points. With the exact rule, point 3 in output space will be blocked by points 2 and 3. Therefore, point 1 cannot reach the optimum position.

To overcome such problems, a simple empirical rule can be used. Instead of adapting point i to the other points, the new rule consists in adapting all other points to point i,

$$\Delta y_j = \begin{cases} \mu \sum_{j \neq i} \frac{X_{ij} - Y_{ij}}{Y_{ij}} (y_j - y_i) & \text{if } Y_{ij} > \rho; \\ 0 & \text{otherwise.} \end{cases}$$

To a certain extent, this stochastic version of the gradient is used to overcome problems of local minima, whilst guaranteeing the average minimization of the cost function.

3.5.3 Implementation of Curvilinear Component Analysis

The implementation of the method requires the selection of

- preprocessing of data x_{ij},
- the initial values of components y_{ij},
- a law for the decrease of parameter ρ.

Given the metrics used to compute distances, and for the same reasons as for PCA, the adapted preprocessing operations consists in a reduction of each component in order to standardize their importance in the computation of the distances. Although that is not mandatory, data may also be centered in order to obtain graphic representations around the origin.

As for Kohonen maps, the components y_{ij} of units in the output space, are initialized to random values. To standardize their distribution, each component may be uniformly distributed in $[-1, 1]$. Given the computations of the euclidean distances X_{ij} and Y_{ij} evaluated respectively in spaces of different dimensions, p and q, the comparison of distances is biased. To overcome that problem, especially for high dimension reduction rates, the recommended rule consists in assessing average distances while taking into account the dimensions of the spaces:

$$X_{ij} = \sqrt{\frac{\sum_{k=1}^{p} (x_{ik} - x_{jk})^2}{p}}, \qquad Y_{ij} = \sqrt{\frac{\sum_{k=1}^{p} (y_{ik} - y_{jk})^2}{q}}.$$

The selection of parameter ρ has a large impact on the quality of the projection. During the first iterations, all points $\boldsymbol{y}_i$ in output space should contribute to the cost function. The rule consists in initializing parameter ρ to the maximum of distances Y_{ij},

$$\rho(0) = \max_{ij} Y_{ij}.$$

The final value of ρ should correspond to the smallest value required on Y_{ij}, i.e., to the smallest of values X_{ij},

$$\rho(t_{\max}) = \min_{ij} X_{ij}.$$

Parameter decreases according to a law that depends on the number of iterations t from the initial value $\rho(0)$ to the final value $\rho(t_{\max})$,

$$\rho(t) = \rho(0) \left(\frac{\rho(t_{\max})}{\rho(0)} \right)^{t/t_{\max}}.$$

3.5.4 Quality of the Projection

One of the important aspects of curvilinear analysis is the criterion used to assess the quality of the result. That criterion is based on the comparison of the values X_{ij} and Y_{ij} that correspond to the distances between points, computed in the original space and in the reduced space respectively. The distances are represented in a plane $dx - dy$ by a point of coordinates $dx = Y_{ij}$ and $dy = X_{ij}$. The points close to the line $dx = dy$ correspond to neighboring distances. The distortion due to the dimensionality reduction is therefore proportional to the average distance from the points to the straight line $dx = dy$. Figure 3.8 shows the average distribution of the distances for the example of the hemisphere and for that of the sphere.

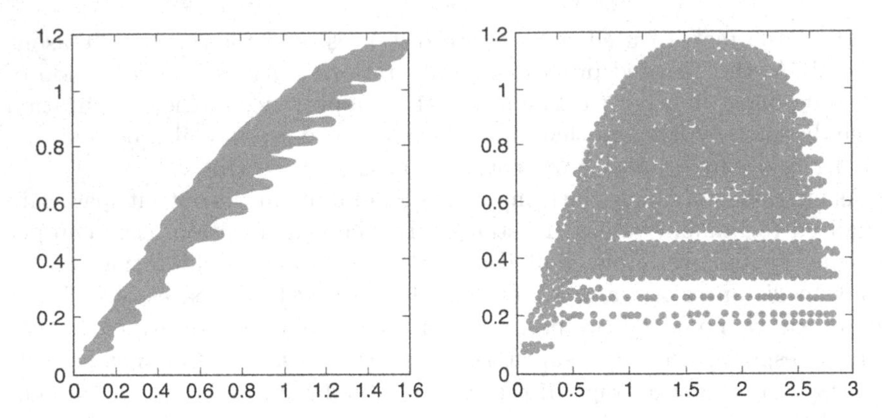

Fig. 3.8. Distribution of distances in the plane $(dy - dx)$ for the hemisphere and sphere

For the nonlinear varieties illustrated by those examples, the projection must exclude certain points. That is true for the map of the earth obtained using Mercator's projection. The "western" projection separates the coastlines of the Bering straits.

In the plane $dy - dx$, the scatter diagram takes the form of a bell: points close to each other in the original space (dx small) will be further apart (dy large) in projection space. The bell shape appears clearly for the projection of the sphere, where an opening separates the points on the large diameter (Fig. 3.5). Checking the projection consists in checking that the bell shape retains the local topology as far as possible: if two points are close in the reduced space, they must be close in the original space.

3.5.5 Difficulties of Curvilinear Component Analysis

Before describing an application, we outline the shortcomings of CCA. The first problem is that of computation time. The distances between points should be computed. If the number of points is too great, CCA cannot be applied directly to the data. A preliminary sampling step is necessary to reduce the number of examples.

The second problem is related to the use of CCA in on-line mode. Unlike PCA, reduced components cannot be computed directly. They are obtained by iterations of gradient descent. Let us describe the CCA procedure. Let x_0 be a new input; we want to find the corresponding component y_0. The algorithm consists in initializing y_0 by the center of gravity of 3 or 4 points y_k that correspond to the points x_k closest to x_0. The projection y_0 is calculated using the same algorithm:

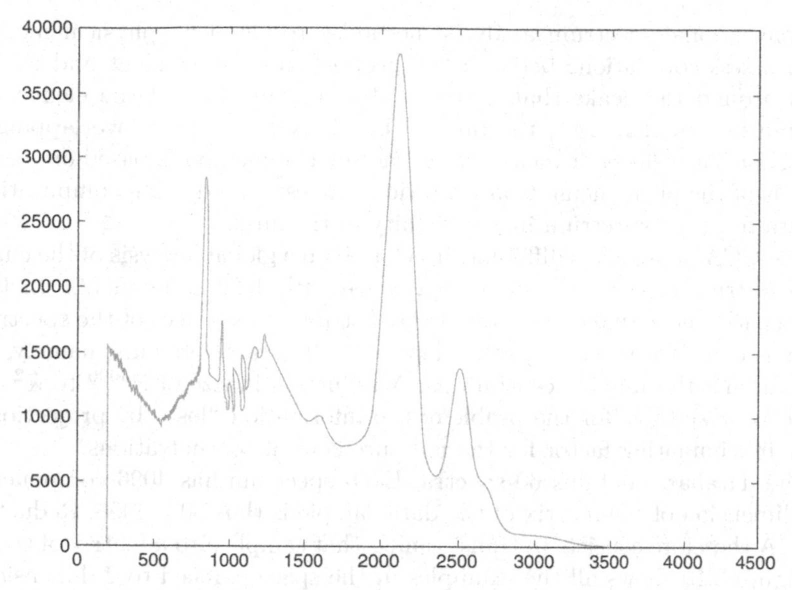

Fig. 3.9. Example of a spectrum

$$\Delta y_0 = \begin{cases} \mu \sum_{j \neq i} \frac{X_{i0} - Y_{i0}}{Y_{i0}} (y_0 - y_j) & \text{if } Y_{i0} > \rho; \\ 0 & \text{otherwise.} \end{cases}$$

That initialization method appears to be very efficient; convergence is obtained in a few iterations (less than ten) [Pilato 1998].

3.5.6 Applied to Spectrometry

The application described below was performed in the Saclay Study Centre [Pilato 1998]. It deals with the measurement of the concentration of radioactive materials. The inspection of nuclear processing units (power plants, reprocessing factories) requires the measurement of concentrations of certain radioactive materials. Concentration measurements are performed on solutions from the water circuits of the plants. One of the techniques used is the X-ray fluorescence, which enables fast, non-destructive analyses to be carried out directly on sampling containers or pipes. X-ray fluorescence consists in exciting the material of interest, and in analyzing the spectra of the photons generated by deactivation.

Figure 3.9 shows an example of a spectrum obtained by X-ray fluorescence on a vessel containing Uranium 235 and Thorium. The peaks denote the presence and concentration of those two elements. In our application, each spectrum is quantized on 4096 energy values. Each value on the vertical axis indicates to the number of photons counted for a given energy level.

Conventional spectrum analysis methods are based on physical models, which assess correlations between the proportion of an element and the integral around the peaks that correspond to certain lines of the element to be analyzed. In that case, the physics is relatively complex: overlapping of peaks, spurious effects or measurement noise. The method is based on a local analysis of the phenomena. Concentrations are estimated using computations on data from the spectrum in the vicinity of the lines.

The CCA approach is different. It is based on a global analysis of the curve. The spectrum is viewed as part of a space with 4096 components. In that $\mathbb{R}^{4096}$ space, the actual dimension of the distribution surfaces of the spectrum points is equal to two: the spectra depend on 2 parameters only, namely, the uranium and thorium concentrations. A reduction in size of $\mathbb{R}^{4096}$ to $\mathbb{R}^2$ was found to be suitable for the problem: the information "lost" by projection is not a discriminating factor for the measurement of concentrations.

The database contains 60 spectra. Each spectrum has 4096 components. The dimension of the matrix of the data sample is thus 60×4096. Reduction by CCA therefore consists in transforming that sample into a matrix of 60×2.

Figure 3.10 shows all the examples in the space reduced to 2 dimensions. We have deliberately meshed the representation by showing the spatial topology of the quantification performed by the investigators on the values of concentrations of uranium and thorium.

The projection obtained by CCA has the same topology as the experimental quantification. The concentrations of uranium and thorium were quantified on the Cartesian product $[(u_1, u_2, \ldots, u_6) \times (t_1, t_2, \ldots, t_{10})]$. Actually, closer inspection shows that a test is missing: the base only contained 59 spectra. Figure 3.10 shows the data missing in the CCA projection.

The example shows the advantages of CCA: despite the nonlinear combinations of several effects on the spectra, dimensionality reduction allowed us to display the inherent size of the data, that of the variation in relation to the concentration of thorium and uranium. Using reduced spectra, the estimation of concentrations in uranium and in thorium becomes: regression with a small neural network, or even simple linear interpolation is more than sufficient.

Applied to more complex problems, when inherent size is not that obvious, one may proceed iteratively by increasing, if necessary, the number of components of the projection space, whilst monitoring the preservation of the local topology on the bisector for short distances.

3.6 The Bootstrap and Neural Networks

The final section describes a new approach that allows automatic design and training of neural networks. It is based on the statistical *bootstrap* method and on the early stopping technique (the latter technique is described in Chap. 2). The approach advocated here consists in starting the design of the model

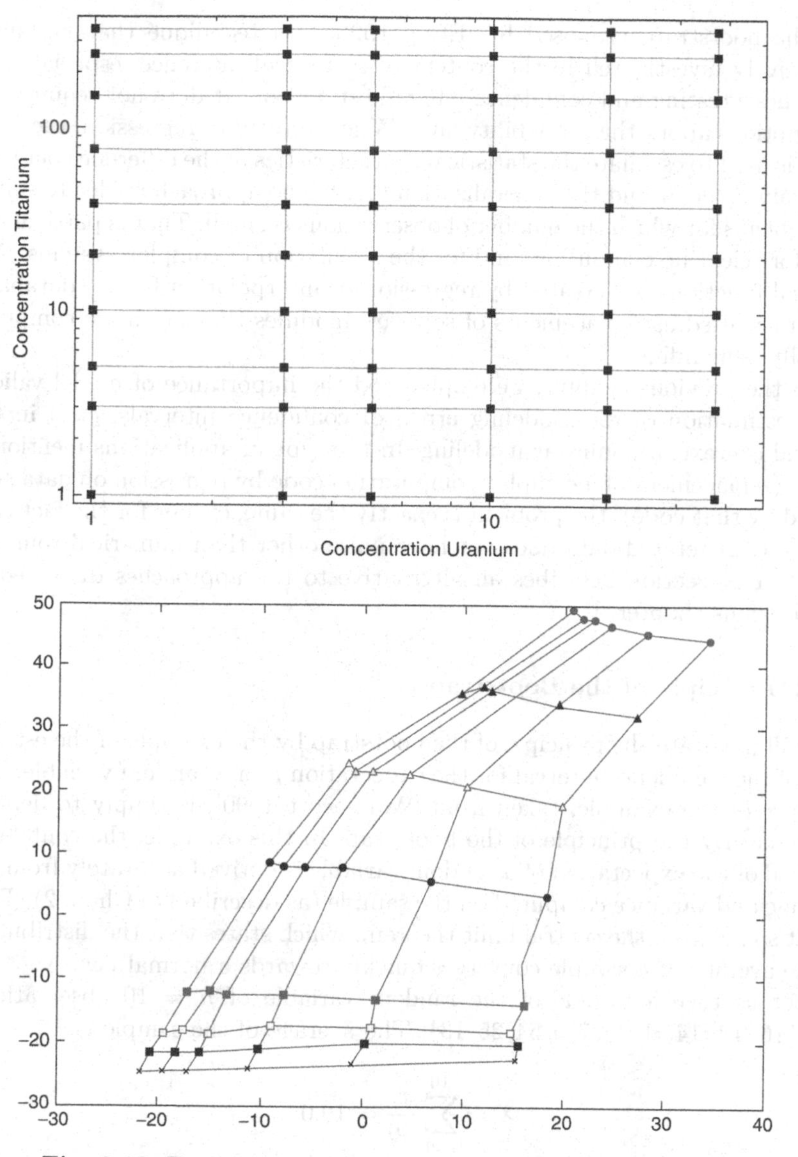

Fig. 3.10. Experimental quantification—Representation by CCA

with a complex network, which is regularized by early stopping of the training process. The bootstrap is used to assess efficiently the variability of the network and its error in relation to the data. Combined with early stopping, it is used to monitor the training process by automatically optimizing the number of training epochs, whilst providing the statistical characteristics of the generalization error.

The bootstrap, proposed by [Efron 1993] is a technique that has been extensively investigated in the context of statistical inference, especially for hypothesis testing and confidence interval estimation. It does not require any assumptions about the probability laws. When applied to regression, the *bootstrap* is used to estimate the statistical characteristics of the difference between the training error and the generalization error. The approach is ideally suited to problems for which the number of observations is small. That is particularly true for scientific computing, and for the simulation of complex systems. Analytical functions are created by regression or interpolation from a database, which are used as replacements of software modules that are more computationally demanding.

In the previous chapter, we emphasized the importance of model validation (estimation of the modeling error, of confidence intervals, etc.) in the general context of nonlinear modeling. In the type of applications mentioned above (replacement of a complex computation code by regression on data generated by that code), the problem is exactly the same, except for the fact that computer-generated-data does not have noise other than numerical roundoff errors. This section describes an alternative to the approaches discussed in the previous chapter.

3.6.1 Principle of the Bootstrap

We will illustrate the principle of the bootstrap by the example of the estimation of the confidence interval for the expectation μ of a random variable. The purpose of the example, taken from [Wonnacott 1990], is simply to demonstrate clearly the principle of the bootstrap. In this example, the confidence interval of the expectation of a random variable is derived accurately from the average and variance computed on the sample (as described in Chap. 2). That result stems from the central limit theorem, which states that the distribution of the average of a sample converges quickly towards a normal law.

Let us take a sample of the random variable of $n = 10$ observations: $x = \{16, 12, 14, 6, 43, 7, 0, 54, 25, 13\}$. The average of the sample is

$$\overline{X} = \sum_{i=1}^{10} \frac{x_i}{10} = 19.0,$$

and its standard deviation is

$$s = \sqrt{\sum_{i=1}^{10} \frac{(x_i - 19)^2}{9}} = 17.09.$$

The 95% confidence interval of the expectation μ is

$$\mu = \overline{X} \pm t_{.025} \frac{s}{\sqrt{n}} = 19.0 \pm 2.26 \frac{17.09}{\sqrt{10}} \approx 19 \pm 12 \Rightarrow 7 < \mu < 31.$$

The confidence interval may be also calculated by the bootstrap, with the following algorithm.

Using the initial sample, we simulate new samples, known as "bootstrapped" bases, the size of which is n, by random selection with replacement. For instance, consider the initial sample defined above $x = \{16, 12, 14, 6, 43, 7, 0, 54, 25, 13\}$. By random selection with replacement, we obtain for example the following bootstrapped base $x^* = \{54, 0, 16, 7, 43, 54, 0, 25, 25, 6\}$, in which some values of the initial sample are missing, whilst others appear several times. Several samples are thus simulated. For each sample, an average is computed. The confidence interval at 95% is defined for that set of averages. The simulation produces the following:

$$9 < \mu < 26.$$

It should be noted that the interval obtained using the *bootstrap* is virtually identical to the 95% confidence interval computed above from the central limit theorem.

Bootstrap—General

The bootstrap does not require any assumption on the underlying statistical distribution.

The bootstrap may therefore be applied to all estimators other than the average, such as the median, the coefficient of correlation between two random variables, or the largest eigenvalue of a variance-covariance matrix, for example. For those estimators, no analytical expression is available for the standard error or the confidence interval. The only applicable methods are the so-called resampling methods, which consist in the simulation of samples such as the *bootstrap* or the *jackknife* [Efron 1993].

3.6.2 Bootstrap Estimation of the Standard Deviation

Consider a random variable X that obeys the probability distribution F. We want to estimate a parameter θ of F. θ is estimated from an n-sample $\boldsymbol{x} = \{x_1, x_2, \ldots, x_n\}$. We denote by $\hat{F}$ the empirical distribution, and by $\hat{\theta} = s(\boldsymbol{x})$ the estimation of θ from sample $\boldsymbol{x}$. The algorithm is as follows:

1. Select B bootstrapped n-samples, $\boldsymbol{x}^{*1}, \boldsymbol{x}^{*2}, \ldots, \boldsymbol{x}^{*B}$, each of them being obtained from the initial sample $\boldsymbol{x}$ by n random selections with replacement
2. For each bootstrapped n-sample, compute a replica of the estimate of θ as

$$\hat{\theta}(b) = s(\boldsymbol{x}^{*b}), \quad b = 1, 2, B.$$

3. Estimate the standard deviation from the standard error computed for all replicas,

$$\hat{\theta}^*(\cdot) = \frac{\sum_{b=1}^{B} \hat{\theta}^*(b)}{B}$$

$$\hat{\sigma}_B^2 = \frac{\sum_{b=1}^{B} \left(\hat{\theta}^*(b) - \hat{\theta}^*(\cdot)\right)^2}{B-1}.$$

One of the theorems proved by Efron deals with the composition of the bootstrap estimator. The estimate $\hat{\sigma}_B$ converges to the standard deviation $\sigma_{\hat{F}}(\hat{\theta}^*)$ of the parameter θ estimated from the sample distribution,

$$\lim_{B \to \infty} \hat{\sigma}_B = \sigma_{\hat{F}}.$$

That algorithm applies to any estimator. For instance, consider the computation of the largest eigenvalue for PCA. It is the largest eigenvalue of the variance-covariance matrix $X_{n \times p} = X^{\mathrm{T}}X$ of the observations. The *bootstrap* consists in simulating the replicas $X_{n \times p}$ obtained by n random selections of the lines of matrix $X_{n \times p}$. The statistics (average and standard deviation) may then be generated easily. That shows the power of the method and its ease of use. However, that method was not widely used in the past, because of the amount of computation required: 50 to 200 bootstrapped bases are enough to estimate an average, but several thousand bootstrapped bases are required to determine the confidence intervals.

3.6.3 The Generalization Error Estimated by the Bootstrap

In the previous chapter, we emphasized the importance of estimating the generalization error, and we described the estimation by ≪*leave-one-out*≫. The *bootstrap* technique can also be used advantageously to estimate that error. The principle is the same: it consists in simulating B "bootstrapped" bases. Each base may contain the same example several times, because of random selection with replacement.

Binomial Distribution of Bootstrapped Bases

For each random selection, all examples have the same probability $p = 1/n$, where n is the number of examples. The number of occurrences of an example in a *bootstrapped* base therefore obeys a binomial distribution $B(n, p = 1/n)$. The probability of an example appearing k times is given by $P(k) = C_n^k p^k (1 - p)^{n-k}$ [Saporta 1990].

The probability of an element not appearing in the *bootstrapped* base is therefore $P(0) = (1 - 1/n)^n$. For sufficiently large values of n, one has $P(0)_{n \to \infty} = e^{-1} \approx 0.368$. On the average, 37% of the examples will not be used for training.

Statistics of the Generalization Error

The difference between the training error computed on the *bootstrapped* base and the testing error evaluated on the initial base is considered to be a random variable that represents the difference between the training error and the generalization error.

Statistics are produced for all those differences (1 per *bootstrapped* base) in order to estimate the probability law of the difference between the training and the generalization error.

We denote by B the initial database and by B_b^*, $b = 1, \ldots, N$ the set of bootstrapped bases. Denote by ε_b^* the training error on bootstrapped base k, and by ε_b the error of the same network computed on the initial base B. The difference $\delta_b = \varepsilon_b - \varepsilon_b^*$ between the errors may be considered as a random variable that arises from overtraining. That difference may also be viewed as the bias that appears when estimating the generalization error by the training error. The expectation value $\overline{\delta}$ and the variance $\sigma_{\overline{\delta}}^2$ of the bias may be estimated on the set of values of δ_b,

$$\delta_b = \varepsilon_b - \varepsilon_{b_k}^*, \quad \overline{\delta} = \frac{1}{B}\sum_{b=1}^{B} \delta_b, \quad \sigma_{\overline{\delta}}^2 = \frac{1}{B-1}\sum_{b=1}^{B} \left(\delta_b - \overline{\delta}\right)^2.$$

3.6.4 The NeMo Method

The algorithm proposed above was programmed in the NeMo software. The bootstrap is associated with early stopping for automatic monitoring of the training of the network.

The NeMo Tool

NeMo is a tool developed by the Systems and Structure Modeling Department of the Study Centre at Saclay using the Stuttgart neural network simulator (SNNS) available on *http://www.ra.informatik.uni-tuebingen.of/SNNS*, which is designed to simplify neural network learning and testing tasks.

The user chooses the number of *training cycles* N_c and the number of bootstrapped bases B. NeMo performs B training cycles and saves the average quadratic training and test errors for each cycle. NeMo analyses the training and test error profiles in order to select the most appropriate value for the number of cycles.

Modeling of Errors

The average quadratic error EQM_r is calculated from the centered and reduced output variables (estimated and measured). Therefore, the analysis of the error deals with the part of the variance that is *not explained* by the model or coefficient of nondetermination that was described in the section on output preprocessing.

We denote by j the rank of the bootstrapped base and by i the iteration on the number of cycles; the average quadratic learning and testing errors are represented by the following two tables:

$$\underbrace{\begin{bmatrix} \varepsilon_1^{*1} & \varepsilon_1^{*2} & \cdots & \varepsilon_1^{*B} \\ \varepsilon_2^{*1} & \varepsilon_2^{*2} & \cdots & \varepsilon_2^{*B} \\ \vdots & \vdots & \ddots & \vdots \\ \varepsilon_{N_c}^{*1} & \varepsilon_{N_c}^{*2} & \cdots & \varepsilon_{N_c}^{*B} \end{bmatrix}}_{\text{learning error}} \quad \underbrace{\begin{bmatrix} \varepsilon_1^{1} & \varepsilon_1^{2} & \cdots & \varepsilon_1^{B} \\ \varepsilon_2^{1} & \varepsilon_2^{2} & \cdots & \varepsilon_2^{B} \\ \vdots & \vdots & \ddots & \vdots \\ \varepsilon_{N_c}^{1} & \varepsilon_{N_c}^{2} & \cdots & \varepsilon_{N_c}^{B} \end{bmatrix}}_{\text{testing error}} .$$

After that phase, NeMo determines the number of cycles by application of heuristics based on game theory. A first pessimistic player considers the worst possible situation on the test error, for each number of cycles value,

$$\varepsilon_i^{\text{Max}} = \text{Max}_b\left(\varepsilon_i^b\right).$$

The second player then determines the number of cycles that corresponding to the worst situation obtained; that is, the number of cycles that corresponds to the maximum testing error,

$$N_{\text{c}}^{\text{optimal}} = \text{Arg}_i\left\{\text{Min}\ \varepsilon_i^{\text{Max}}\right\}.$$

That strategy for the selection of $N_{\text{c}}^{\text{optimal}}$ may be relaxed by adopting only a fraction of the set of B training cycles. To make it more robust, just exclude the outliers, that is, training situations that differ greatly from the average. By default, NeMo determines the optimum number of cycles on the 90th percentile of the test error.

Percentile

The αth percentile corresponds to the interval made up of the values for which the distribution function is smaller than α: a fraction $(1-\alpha)$ of the maximum values is excluded.

The optimum number of cycles may also be estimated by the tri-median method, which is more *stable* but more *risky* since 25% of cases are rejected: the last quartile that corresponds to the largest test errors.

Quartile

If F is the distribution function, the 1st and 3rd quartile Q_1 and Q_3 and the median Q_2 are defined respectively by $F(\text{Q}_1) = 0.25$, $F(\text{Q}_2) = 0.5$, $F(\text{Q}_3) = 0.75$.

Tri-Median

The tri-median corresponds to 0.25 Q_1 (1st quartile) + 0.5 Q_2 (2nd quartile or median) + 0.25 Q_3 (3rd quartile).

After determining the optimum number of cycles by one of the strategies, NeMo starts a new training cycle based on all examples, with the optimized number of cycles N_c^{optimal} defined during the previous phase. For that last training cycle, the same training parameters (initial value and variation law of the gradient step) are used. If ε_a denotes the average error computed on the initial base, and $\overline{\delta}$ the average value of the bias, the generalization error is estimated by

$$\varepsilon_g = \varepsilon_a + \overline{\delta}.$$

More generally, the distribution function of the generalization error is estimated by the empirical distribution function of the shifted bias of the value ε_a. Note the contribution of the bootstrap associated with early stopping in relation to cross-validation,

- to some extent, the automation of the design of the network by adapting the number of early stopping cycles,
- a wider estimate of the variability of the model with respect to the data set,
- estimates the confidence intervals (margins, uncertainty),
- the use of all examples to construct the network.

Finally, it should be noted that NeMo may monitor the suitability of the model to the data: if the optimized number of cycles is too close to user-chosen maximum number of cycles, there is no minimum test error. In that case, the user must increase the complexity of the network (number of hidden neurons) or increase the number of training cycles.

3.6.5 Testing the NeMo Method

In the following, we describe the results of an experiment designed to validate the method. The test consists in comparing the average error estimated by NeMo to the actual error. The actual error is approximated according to the Monte Carlo method, i.e., by making a very large number of computations of the average quadratic error, then by computing its average. We used NeMo for the approximation of two nonlinear analytical functions,

- $\phi_8(x) \quad \mathbb{R}^8 \to \mathbb{R}$,
- $\phi_{12}(x) \quad \mathbb{R}^{12} \to \mathbb{R}$.

We chose those functions in order to evaluate the method on the approximation of sufficiently complex functions (large dimensions of input space).

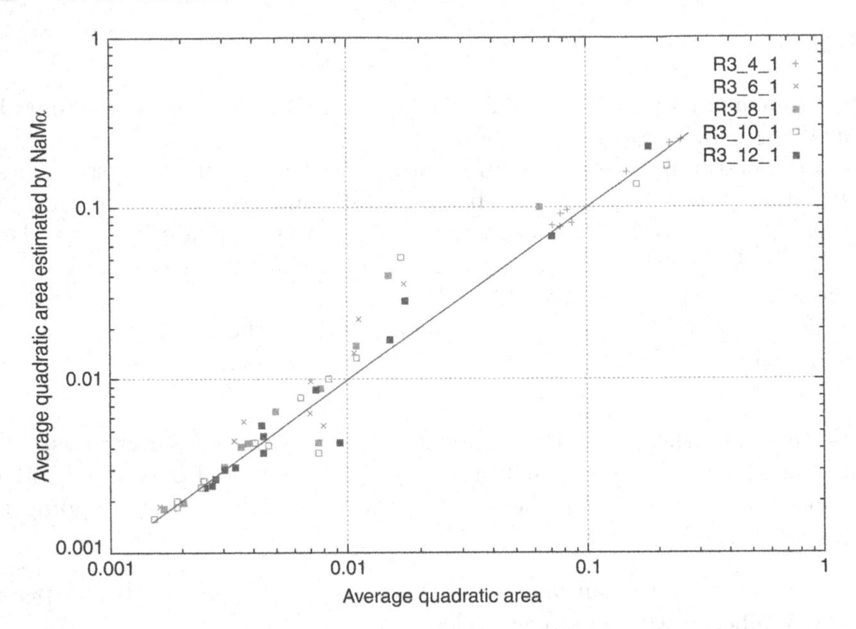

Fig. 3.11. Function ϕ_8

Using those two functions, several bases of examples were generated by varying the number of examples from 100 to 1500 by steps of 100. The inputs were uniformly distributed in $[-1, 1]$.

The model networks adopted are feedforward networks with one hidden layer and a linear output neuron. For the bases generated by the first function ϕ_8, five model networks were proposed to NeMo with 4, 6, 8, 10 and 12 hidden neurons respectively. For the bases generated by the second function ϕ_{12} (a larger input space), six networks were tested, with 10, 14, 18, 22, 26 and 30 hidden neurons respectively.

Large Dimensions

It should be noted that the very low density of points in $\mathbb{R}^{12}$; 1500 points in $\mathbb{R}^{12}$ mean that the average number of points per axis is smaller than two.

The actual error is obtained from 10^6 random selections using the same input generation law (uniform distribution) and by computing the reduced average quadratic error EQM_r between the measured output and the estimated output.

The figures below show (on a log-log scale) the true error EQM_r (horizontal axis) *vs.* the error estimated by NeMo (vertical axis). The points displayed correspond to the different networks created from all bases of examples. Each network was generated 15 times from databases with $100, 200, \ldots, 1500$ examples respectively.

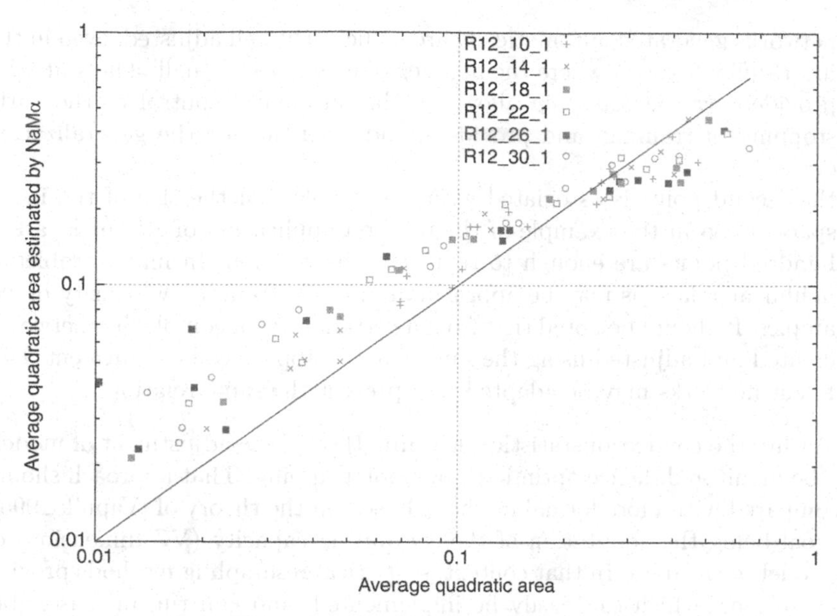

Fig. 3.12. Generator ϕ_{12}EMBED

The analysis of all results given in Figs. 3.11 and 3.12 illustrate the salient features of the NeMo method,

- the generalization error is estimated accurately, even in complex cases (large number of inputs + few examples);
- the bootstrap is used to automate the adjustment of the network to the data by monitoring the termination of training.

Figures 3.11 and 3.12 show estimates of the generalization error very close to the real values. The low error values correspond to training cycles performed from databases with enough examples. For these cases, the estimated error on the Y-axis is virtually equal to the actual error on the X-axis.

A slight overestimate should be noted for 4 cases out of 75 between values 0.01 and 0.02 for ϕ_8EMBED (Fig. 3.11) and less precision for the more complex ϕ_{12}EMBED (Fig. 3.12). In the latter case, regression concerns a relation from $\mathbb{R}^{12}$ to $\mathbb{R}$ with a maximum of 1500 points to represent the relation. There is an overestimate of the error for the low values and a underestimate for values greater than 0.2. Nevertheless, and in spite of the large dimensions of the input spaces, the relation of $\mathbb{R}^{12}$ in R is correctly modeled using a few hundred examples.

3.6.6 Conclusions

The above illustrative example shows that

- networks generated automatically are sufficiently well adjusted, even in the most difficult cases, where the number of examples is small. The statistics provided by the *bootstrap* allow for the automatic control of the early stopping of training, and provide sound statistics for the generalization error;
- the second point is associated with the problem of the size of the input space. Even in the example of the relation application of $\mathbb{R}^{12}$ in $\mathbb{R}$, a few hundred points are enough to represent the relation. In many problems, nonlinear relations may be approximated easily from a low density of examples. It should be noted that from a certain level of complexity, networks created and adjusted using the same sample appear to be equivalent. Different networks may be adapted to represent the same relation.

Within the context of statistical learning theory, the adjustment of models may be monitored, hence optimized, by bootstrapping. That approach should be compared with more formal methods based on the theory of [Vapnik 1995], the goal being the adaptation of the computing capacity (VC dimension) of the model to the data. In that context, statistical resampling methods provide real solutions, which can easily be implemented, and can run in reasonable time on present-day computers.

References

1. A. Cichoki, R. Unbehauen, Neural Networks for Optimization and Signal Processing, Wiley, 1993
2. P. Demartines, Analyse de données par réseaux de neurones auto-organisées, thesis at the Institut National Polytechnique de Grenoble
3. Patrick Davaud, Traitement du Signal Concepts et Applications, Hermès, 1991
4. Bradley Efron, Robert J. Tibshirani, An introduction to the Bootstrap, Chapman & Hall, 1993
5. Jeanny Hérault, Christian Jutten, Réseaux de neurones et traitement du signal, Hermes, 1993
6. Vincent Pilato, Application des réseaux de neurones aux méthodes de mesure basées sur l'interaction rayonnement matière, thesis Université Paris-Sud, 4/11/1998
7. Gilbert Saporta, Probabilités Analyse des données et Statistique, Editions Technip, 1990
8. Vladimr N. Vapnik, The Nature of Statistical Learning Theory, Springer, 1995
9. Vincent Vigneron, Méthodes d'apprentissage statistiques et problèmes inverses—Applications à la spectrographie, thesis for the Université d'Evry Val d'Essonne, 5/5/1997
10. Thomas H. Wonnacott, Ronald J. Wonnacott, Statistique Economie-Gestion-Sciences-Médecine, Economica, 4th issue, 1990

4

Neural Identification of Controlled Dynamical Systems and Recurrent Networks

M. Samuelides

Modeling of controlled dynamical systems or "process identification" is a major application of neural networks. This topic was cursorily addressed in Chap. 2. It is more systematically developed hereafter. Moreover, it is compared to similar statistical methods that are commonly used, especially for linear systems identification.

We start with the presentation of several examples of controlled dynamical systems. We show that the addition of a "state noise" in order to take into account the uncertainty of the model leads to viewing the evolution of the state as a Markov process. Neuronal identification of nonlinear processes is essentially a generalization of well known linear regression. We first recall the elements of linear regression in the section "Regression, a tool for controlled dynamic al system identification." Based on examples, we show how to compute the regression coefficients of an auto-regressive model. Then neural identification is presented as a natural nonlinear regression methodology. Following section is devoted to on-line or adaptive identification of dynamical systems. Our starting point is recursive identification of linear systems, which is a mere generalization of the basic statistical Law of Large Numbers. Furthermore, we develop the recursive prediction error method (RPEM), which is a nonlinear extension thereof. Adaptive identification algorithms, including neural identification algorithms, will be addressed.

In most applications, the state of the system cannot be completely known because some state variables cannot be measured and because one cannot avoid measurement errors. Therefore, filtering techniques are commonly used to reconstruct the state of a dynamical process from the measurement results. The popular technique of Kalman filtering is addressed in the section "Innovation filter in a state model." It is subsequently used for designing a neural learning algorithm that may be used to identify dynamical processes. At the end of the chapter, the sections "Recurrent neural networks" and "Learning for recurrent neural networks" are devoted to recurrent neural networks. The most popular models of recurrent neural networks (Elman networks and

Hopfield networks) are described. Finally, we show how to use that type of networks for the identification of controlled dynamical system.

4.1 Formal Definition and Examples of Discrete-Time Controlled Dynamical Systems

4.1.1 Formal Definition of a Controlled Dynamical System by State Equation

Since all the applications of neural networks to control are implemented on computers, the present chapter, and the next, will be essentially devoted to discrete-time dynamical systems. The sampling techniques of analog signals that are delivered by physical devices will not be addressed.

The mathematical model of a dynamical system is defined by a set $\boldsymbol{E}$, called the state space of the system, and an evolution equation, which describes completely the evolution of the state of the system in state space from the initial state conditions. In most problems, the evolution is said to be autonomous: the evolution law is not time-dependent. We will stick to that hypothesis in order to alleviate the notations. In control problems, the state of the system at time $t + \Delta t$ does not depend on the state of the system at time t only: it also depends on an external signal at time t, which is called input or control of the system. In such a case, the system is termed controlled, in contrast to autonomous. The set of controls will be referred to as $\boldsymbol{F}$. Using classical notations, we will write

- $\boldsymbol{x}(t) \in \boldsymbol{E}$ for the state of the system at time t,
- $\boldsymbol{u}(t) \in \boldsymbol{F}$ for the value of the control at time t.

Thus, in order to define the whole state trajectory of a controlled system from time 0 to time τ, one needs the initial state $\boldsymbol{x}(0)$ of the system and the control trajectory $[u(t)]_{t \in [0,\tau]}$. The control system is designed in order to build a control trajectory that is as close as possible to a reference state trajectory, or that minimizes the cost of the trajectory with respect to a given cost function.

Notice that if a closed loop control law is implemented, i.e., if the control system computes the control as a function of the current state (or the past state trajectory of the system, or the past results of measurements performed on the system), then the whole system (controlled dynamical system+control system) is an autonomous dynamical system. The design of closed-loop control law and their neural implementation will be the main topic of the next chapter.

As mentioned above, we focus here on discrete-time dynamical systems. A discrete-time dynamical system can be derived from a continuous-time dynamical system by sampling the state trajectory of the system. As previously in Chap. 2, the sampling period is denoted by T and we write time k for time

$t = kT$. The controlled dynamical system time evolution is described by the following evolution equation

$$\boldsymbol{x}(k+1) = f[\boldsymbol{x}(k), \boldsymbol{u}(k)],$$

where f is the mapping from $\boldsymbol{E} \times \boldsymbol{F}$ into $\boldsymbol{E}$ that allows us to infer the state at time $(k+1)T$ from the state at time kT. It is possible to make this general set-up more specific for particular systems.

The more classical model is the linear model. In that case, the state space and the control space are vector spaces, $\boldsymbol{A}$ is a linear mapping from $\boldsymbol{E}$ to $\boldsymbol{E}$, $\boldsymbol{B}$ is a linear mapping from $\boldsymbol{F}$ to $\boldsymbol{E}$ and the evolution equation has the following form:

$$\boldsymbol{x}(k+1) = \boldsymbol{A}\boldsymbol{x}(k) + \boldsymbol{B}\boldsymbol{u}(k).$$

Because mathematical models are just an approximation of the real evolution of physical devices, the modeling error is generally represented by a random additional term. This term is often called the state noise.

For instance, in the stationary linear model, the model error is modeled by an additive noise that is generally White and Gaussian. Then the evolution equation takes the form

$$\boldsymbol{x}(k+1) = \boldsymbol{A}\boldsymbol{x}(k) + \boldsymbol{B}\boldsymbol{u}(k) + \boldsymbol{v}(k+1),$$

where the $v(k)$ are gaussian centered (null expectation) independent random vectors with covariance matrix $\boldsymbol{\Gamma}$.

In that case, the state trajectory is partially random, and the process is called a stochastic process. In the following, we provide some examples of controlled dynamical systems, as illustrations for more formal considerations.

4.1.2 An Example of Discrete Dynamical System

First consider an example of a dynamical system with discrete state space. A labyrinth with 18 possible positions is shown on Fig. 4.1.

The state space is an 18-element set {12, 13, 14, 15, 21, 22, 24, 32, 33, 34, 35, 41, 42, 44, 52, 53, 54, 55}. The set of controls may be chosen as the set of four directions (N, W, S, E). The evolution is given by the natural mapping that associates to an initial position and a course order either the resulting position if the order is feasible, or the initial state if it is not:

$$\begin{aligned}
f(12,\mathrm{N}) = 12,\; f(13,\mathrm{N}) = 13,\; \ldots,\; f(21,\mathrm{N}) = 21,\; f(22,\mathrm{N}) = 12,\; \ldots,\\
f(12,\mathrm{W}) = 12,\; f(13,\mathrm{W}) = 12,\; \ldots,\; f(21,\mathrm{W}) = 21,\; f(22,\mathrm{W}) = 21,\; \ldots,\\
f(12,\mathrm{S}) = 22,\; f(13,\mathrm{S}) = 13,\; \ldots,\; f(21,\mathrm{S}) = 21,\; f(22,\mathrm{S}) = 32,\; \ldots,\\
f(12,\mathrm{E}) = 13,\; f(13,\mathrm{E}) = 14,\; \ldots,\; f(21,\mathrm{E}) = 22,\; f(22,\mathrm{E}) = 22,\; \ldots.
\end{aligned}$$

Other modeling rules may be chosen, corresponding to other state representations of the same problem. For instance, one may prefer characterizing

Fig. 4.1. A labyrinth

the state of a robot by the couple of the actual position and of the direction. In our example, the state space would have then $18 \times 4 = 72$ elements and it will be completed by a control set of three elements (A for move ahead, L for move towards left, R for move towards right).

Software products that are used to exploit large database or to perform data mining on the Web are facing a lot of problems that are formalized as navigation problems in a graph. The state space is the set of the nodes of the graph.

4.1.3 Example: The Linear Oscillator

Let us consider now the harmonic oscillator that is governed by the second-order linear differential equation,

$$\frac{\mathrm{d}^2 x}{\mathrm{d}t^2} = -x.$$

First, notice that the differential equation does not provide a genuine state representation since it is a second-order equation. The associated continuous state representation is

$$\frac{\mathrm{d}}{\mathrm{d}t}\begin{pmatrix} x_1 \\ x_2 \end{pmatrix} = \begin{pmatrix} x_2 \\ -x_1 \end{pmatrix},$$

where the state incorporates the mobile current position x_1 and its speed x_2. In order to derive a discrete time evolution, we have to solve the differential equation on the sampling period T. In that trivial example, the solution is obvious, and the function that maps the state at time t to the state at time $t+T$ can be written explicitly. Generally, it will not be the case, neither for other models in the following nor for most applications. Therefore, one has to obtain a numerical approximation with a differential equation solver (Runge-Kutta algorithm for instance [Demailly 1991]).

To control the mobile, we consider a scalar additive control on the speed, denoted by u.

For instance, in our example, the second time derivative is easily obtained,

$$\frac{\mathrm{d}^2}{\mathrm{d}t^2}\begin{pmatrix} x_1 \\ x_2 \end{pmatrix} = \begin{pmatrix} -x_1 \\ -x_2 \end{pmatrix}.$$

From that expression, one can derive a second-order Taylor approximation of the state evolution

$$\begin{pmatrix} x_1 \\ x_2 \end{pmatrix}(t+T) = \begin{pmatrix} x_1 \\ x_2 \end{pmatrix}(t) + T\frac{\mathrm{d}}{\mathrm{d}t}\begin{pmatrix} x_1 \\ x_2 \end{pmatrix}(t) + \frac{T^2}{2}\frac{\mathrm{d}^2}{\mathrm{d}t^2}\begin{pmatrix} x_1 \\ x_2 \end{pmatrix}(t) + \begin{pmatrix} 0 \\ u(t) \end{pmatrix}.$$

Thus, the following dynamical discrete-time controlled system is obtained as

$$\begin{pmatrix} x_1(k+1) \\ x_2(k+1) \end{pmatrix} = f\begin{pmatrix} x_1(k) \\ x_2(k) \end{pmatrix} = \begin{pmatrix} x_1(k) + Tx_2(k) - \frac{T^2}{2}x_1(k) \\ x_2(k) - Tx_1(k) - \frac{T^2}{2}x_2(k) + u(k) \end{pmatrix},$$

such that the trajectories of that system are a close approximation of the sampled trajectories of the continuous-time dynamical system.

4.1.4 Example: The Inverted Pendulum

We consider now the nonlinear dynamical system called inverted pendulum because its unstable equilibrium is considered as the Reference State. The device diagram is represented on Fig. 4.2.

The differential equation of the controlled system is

$$\frac{\mathrm{d}^2\theta}{\mathrm{d}t^2} = g\sin(\theta) - k\frac{\mathrm{d}\theta}{\mathrm{d}t} + u.$$

Its continuous-time state representation is

$$\frac{\mathrm{d}}{\mathrm{d}t}\begin{pmatrix} x_1 \\ x_2 \end{pmatrix} = \begin{pmatrix} x_2 \\ g\sin x_1 - kx_2 \end{pmatrix} + \begin{pmatrix} 0 \\ u \end{pmatrix}.$$

Notice that the state space is not really a vector state, since the angle θ is only defined up to a multiple of 2π. Actually the physical problem makes sense only if the angle is constrained within a given viability domain. The differential equation solver is not detailed. Simulations that are used to illustrate the present chapter are performed using Matlab™ software.

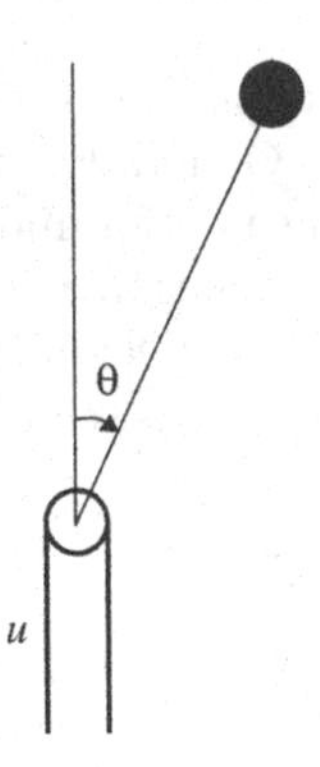

Fig. 4.2. Diagram of the inverted pendulum

4.1.5 Example of Nonlinear Oscillator: The Van Der Pol Oscillator

Stable oscillations in uncontrolled operating mode are another example of adverse oscillations in physical devices. They arise very frequently from nonlinearities. A typical example is provided by the following Van der Pol differential equation:

$$\frac{\mathrm{d}^2x}{\mathrm{d}t^2} - 2z\omega_0\frac{\mathrm{d}x}{\mathrm{d}t} + \omega_0^2 x + 3kx^2\frac{\mathrm{d}x}{\mathrm{d}t} = u.$$

The parameter z is the damping rate of the system and ω_0 is the eigenfrequency of the oscillator. The state representation is 2-dimensional, i.e.,

$$\frac{\mathrm{d}}{\mathrm{d}t}\begin{pmatrix} x_1 \\ x_2 \end{pmatrix} = \begin{pmatrix} x_2 \\ 2z\omega_0 x_2 - \omega_0^2 x_1 - 3kx_1^2 x_2 \end{pmatrix} + \begin{pmatrix} 0 \\ u \end{pmatrix}.$$

Note that the system is linear with respect to the control. The dynamics of the uncontrolled system ($u = 0$) in the 2-dimensional sate space features a limit cycle as an attractor. That means that, whatever the initial state, the state trajectory winds around a specific periodic trajectory when times is going on. This phenomenon is illustrated on Fig. 4.3:

4.1.6 Markov Chain as a Model for Discrete-Time Dynamical Systems with Noise

Let us now return to discrete-time dynamical systems. Consider first the following very simple dynamical system: the random walk on a triangle. The state space has three elements a, b and c. The dynamics is defined by the following evolution function f:

$$f(a) = b, \quad f(b) = c, \quad f(c) = a.$$

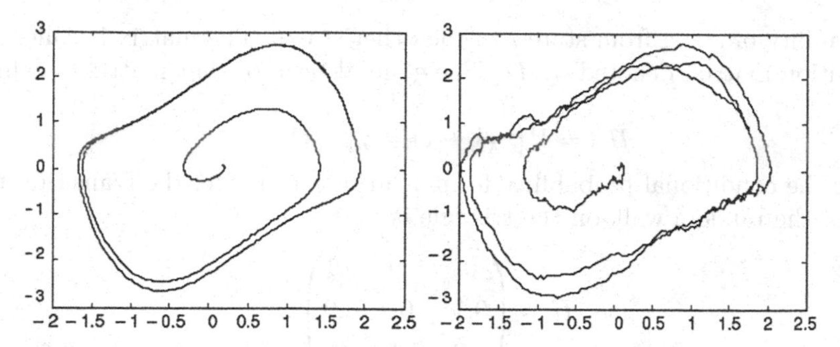

Fig. 4.3. Trajectory of a Van der Pol oscillator. In figure (**a**) a limit cycle is observed. In figure (**b**), the trajectory is perturbed by an additive random input in the equation

Let us introduce now uncertainty in the model. We assume that, at any time, there is a probability of 0.1 of heading into the wrong direction,

$$P[f(a) = b] = 0.9, \quad P[f(a) = c] = 0.1,$$

and so on.

The picture of that random dynamics is outlined on Fig. 4.4.

The state trajectory is no longer deterministic. That random dynamical system, or stochastic process, is called a Markov chain. In the long time limit, the behavior of a Markov chain is quite different from that of a deterministic dynamical system. In that simple example, the state does not depend on the initial state and it is straightforward to show that it is distributed according to the uniform law in the limit of large time. That probability law is called the stationary distribution of the Markov chain.

The dynamics of a Markov chain can be conveniently described by a matrix representation. The state set is ordered, and a matrix is built, whose rows are the transition probabilities: the elements of row i are the values of the

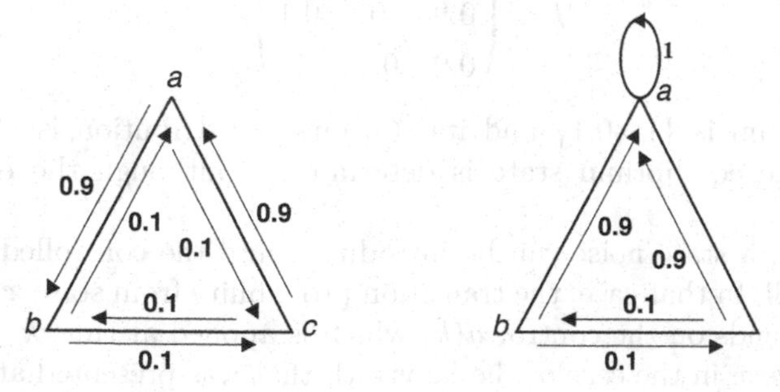

Fig. 4.4. Diagrams of random dynamical evolutions on the triangle. (**a**) Periodic dynamics with random disturbance (**b**) Point attractor dynamics with random disturbance

probability of going from state i to the other states. The matrix is called the transition matrix, denoted by $\boldsymbol{\Pi}$. The general term of that matrix is defined as

$$\Pi_{ij} = P\left[x\left(k+1\right) = j \,|x\left(k\right) = i\right]$$

using the conditional probability formalism. For instance, the transition matrix of the random walk on the triangle is

$$\Pi = \begin{pmatrix} 0 & 0.9 & 0.1 \\ 0.1 & 0 & 0.9 \\ 0.9 & 0.1 & 0 \end{pmatrix}.$$

One can check that the stationary distribution is invariant under the application of the transition matrix. Actually, 1 is an eigenvalue of any transition matrix. It can be shown that the magnitude of any eigenvalue is smaller than or equal to 1. For instance, in our example, the eigenvalues of the transition matrix Π are (approximately) 1, $-0,5+0.6928\mathrm{i}$ and $-0,\ 5-0,\ 6928\mathrm{i}$. The uniform distribution can readily be checked to be invariant.

$$\begin{pmatrix} 1/3 & 1/3 & 1/3 \end{pmatrix} \begin{pmatrix} 0 & 0.9 & 0.1 \\ 0.1 & 0 & 0.9 \\ 0.9 & 0.1 & 0 \end{pmatrix} = \begin{pmatrix} 1/3 & 1/3 & 1/3 \end{pmatrix}.$$

The invariant distribution plays the same role as the equilibrium state of deterministic dynamics. In statistical physics it is termed precisely equilibrium state (see, for instance, Gibbs state in statistical physics).

Here is another example of dynamics on the triangle that exhibits symmetry breaking. The evolution function f is defined by

$$f(a) = a, \quad f(b) = a, \quad f(c) = a.$$

Then, the transition matrix is

$$\Pi = \begin{pmatrix} 1 & 0 & 0 \\ 0.9 & 0 & 0.1 \\ 0.9 & 0.1 & 0 \end{pmatrix}.$$

Its spectrum is $\{1,\ 0.1\}$ and its stationary distribution is (1, 0, 0). In that case, the equilibrium state is deterministic although the dynamics is stochastic.

Of course, a state noise can be introduced into the controlled dynamical system as well. In that case, the transition probability from state $\boldsymbol{x}(k)$ to state $\boldsymbol{x}(k+1)$ depends on the control $\boldsymbol{u}(k)$ which is applied at time k.

For instance, in the case of the labyrinth that was presented at the beginning of this section, $f(13, \mathrm{N}) = 13$. If we introduce a uniformly distributed error probability of 0.1 for the control system, then $f(13, \mathrm{N})$ is a random variable that takes the values 13, 12 and 14 with probabilities 0.9, 0.05, 0.05 respectively.

4.1.7 Linear Gaussian Model as an Example of a Continuous-State Dynamical System with Noise

Engineers are commonly dealing with state noise in continuous state dynamical systems. In that case, probability calculus is more complex and cannot be solved analytically except for the case of linear models with additive gaussian noise. We will describe that model because it is frequently used for Kalman filtering.

Let us consider the linear controlled dynamical system with state equation

$$\boldsymbol{x}(k+1) = \boldsymbol{A}\boldsymbol{x}(k) + \boldsymbol{B}\boldsymbol{u}(k) + \boldsymbol{C}\boldsymbol{v}(k+1),$$

where $(\boldsymbol{v}(k))$ is a centered reduced white gaussian noise, i.e., a sequence of random independent identically distributed random vectors which follow a gaussian distribution with 0 mean and identity covariance matrix.

If $\boldsymbol{x}(k)$ is a gaussian vector with expectation equal to $\boldsymbol{m}(k)$ and covariance matrix equal to $\boldsymbol{P}(k)$, then, from the stability of gaussian law under linear transform, $\boldsymbol{x}(k+1)$ is a gaussian vector with mean equal to

$$\boldsymbol{m}(k+1) = \boldsymbol{A}\boldsymbol{m}(k) + \boldsymbol{B}\boldsymbol{u}(k)$$

and with covariance matrix equal to

$$\boldsymbol{P}(k+1) = \boldsymbol{A}\boldsymbol{P}(k)\boldsymbol{A}^{\mathrm{T}} + \boldsymbol{C}\boldsymbol{C}^{\mathrm{T}},$$

where $\boldsymbol{A}^{\mathrm{T}}$ and $\boldsymbol{C}^{\mathrm{T}}$ are transposed matrices $\boldsymbol{A}$ and $\boldsymbol{C}$.

We recall that if $\boldsymbol{P}$ is the covariance matrix of a random vector $\boldsymbol{x}$, that takes its values in a finite-dimensional vector space $\boldsymbol{E}$, and if $\boldsymbol{A}$ is a linear mapping defined on $\boldsymbol{E}$, then the covariance matrix of the random vector $\boldsymbol{A}\boldsymbol{x}$ is $\boldsymbol{A}\boldsymbol{P}\boldsymbol{A}^{\mathrm{T}}$. (We merge here the notations for linear mapping $\boldsymbol{A}$ and its matrix representation in the reference basis). That result will be crucial for the computation of the Kalman filter.

The above equation is termed the propagation equation of covariance. Then we can determine the asymptotic behavior of gaussian stochastic process $(\boldsymbol{x}(k))$ for long times. If matrix $\boldsymbol{A}$ is stable (i.e., if the magnitude of all its eigenvalues is smaller than 1), the gaussian process converges when times goes to infinity towards a gaussian law with 0 mean and with covariance matrix $\boldsymbol{P}_\infty$ which is the solution of the following equation:

$$\boldsymbol{P}_\infty = \boldsymbol{A}\boldsymbol{P}_\infty\boldsymbol{A}^{\mathrm{T}} + \boldsymbol{C}\boldsymbol{C}^{\mathrm{T}}.$$

Conversely, if $\boldsymbol{A}$ is not stable (i.e., if there exists an eigenvalue whose modulus is larger than or equal to 1) then there does not exist a stationary regime and the process diverges for long time. The linear model is said to be unstable.

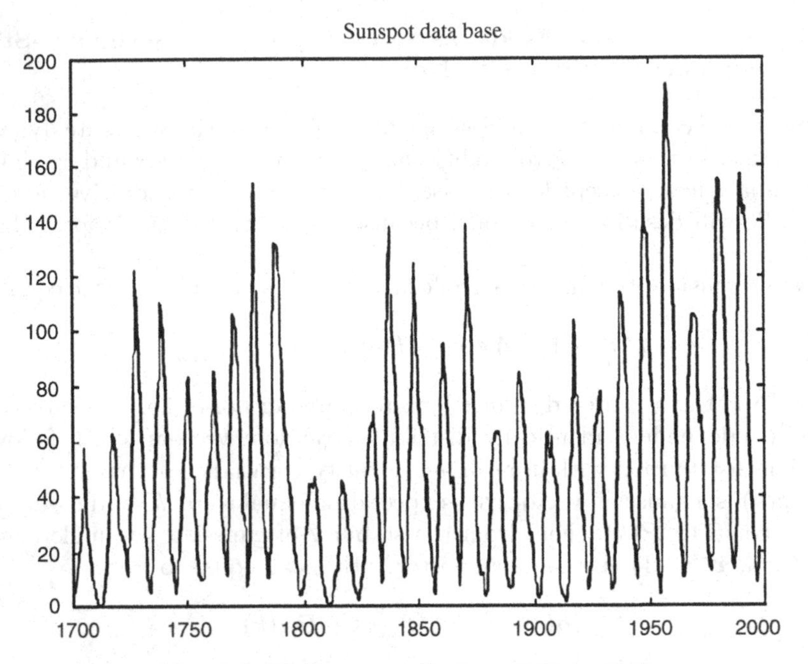

Fig. 4.5. Wolf sunspot file from 1700 to 1997

4.1.8 Auto-Regressive Models

The Wolf sunspot file is an example of database that is commonly used as a benchmark for identification and prediction algorithms. It is maintained since 1700; its variations are shown on Fig. 4.5.

The diagram exhibits some regularity with obvious cycles with approximate period of 11 years. Therefore, it is natural to look for a law that predicts the evolution of the phenomenon [Tong 1995]. There is a wealth of papers on that topic; we consider, for instance, the following model, built up in 1984 by Subba and Gabr (the original data were first centered):

$$\begin{aligned} x(k+1) = 1.22x(k) - 0.47x(k-1) - 0.14x(k-2) + 0.17x(k-3) \\ -0.15x(k-4) + 0.05x(k-5) - 0.05x(k-6)\ldots \\ +0.07x(k-7) + 0.011x(k-8) + v(k+1), \end{aligned}$$

where $(v(k))$ is a gaussian white noise with variance equal to 199. Such a model is called an auto-regressive (AR) model.

Thus, an AR(p) model is defined by the following regression equation

$$x(k+1) = a_1 x(k) + \cdots + a_p x(k-p+1) + v(k+1),$$

where $(v(k))$ is a gaussian white noise. Note that the relevant signal may be interpreted as the response of a linear filter [infinite impulse response (IIR)] to white noise [Duvaut 1994].

Remark. An infinite impulse response filter (IIR filter or "recursive filter") is characterized by the fact that its response at time $k+1$ depends on its input at time k and on its response at previous times. On the other hand, a finite impulse response filter (FIR filter or "transverse filter") is characterized by the fact that its response at time $k+1$ does not depend on its response at previous instants but solely on the input signal at the same instant and at previous instants.

In addition, the model of the response of FIR filters to white noise input, such as

$$x(k+1) = b_0\, v(k+1) + b_1\, v(k) + \cdots + b_q\, v(k-q+1),$$

are usually called moving average process MA(q). The natural generalization of these two models is the auto-regressive moving average model of order (p,q), or ARMA (p,q), model.

Although ARMA models enjoy universal approximation property, it is more efficient to build nonlinear evolution equations to model phenomena or signals that admit parsimonious nonlinear representations [Tong 1995]. So NARMA models are introduced with the following regression equation

$$x(k+1) = f[x(k), \ldots, x(k-p+1), v(k+1), v(k) \ldots, v(k-q+1)].$$

We point out that these models are particular examples of dynamical systems that have been addressed in previous paragraphs. Their state representations are obvious but quite voluminous. For instance in the previous order (p,q) NARMA model, the state of the system at time k is the vector $\boldsymbol{x}(k)$, which has $p+q$ components, namely,

$$\begin{aligned}[\boldsymbol{x}_1(k) = x(k), \ldots, \boldsymbol{X}_P(k) = x(k-p+1), \boldsymbol{x}_{p+1}(k) \\ = v(k) \ldots, \boldsymbol{x}_{p+q}(k) = v(k-q+1)],\end{aligned}$$

and the state equation is

$$\begin{aligned}
\boldsymbol{x}_2(k+1) &= \boldsymbol{x}_1(k) \\
&\ldots\ldots\ldots\ldots\ldots\ldots \\
\boldsymbol{x}_p(k+1) &= \boldsymbol{x}_{p-1}(k) \\
\boldsymbol{x}_{p+1}(k+1) &= v(k+1) \\
\boldsymbol{x}_{p+2}(k+1) &= \boldsymbol{x}_{p+1}(k) \\
&\ldots\ldots\ldots\ldots\ldots\ldots \\
\boldsymbol{x}_{p+q}(k+1) &= \boldsymbol{x}_{p+q-1}(k) \\
\boldsymbol{x}_1(k+1) &= f[\boldsymbol{x}_1(k), \ldots, \boldsymbol{x}_p(k), v(k+1), \boldsymbol{x}_{p+1}(k) \ldots,), \boldsymbol{x}_{p+q}(k)].
\end{aligned}$$

In the same way we considered controlled dynamical systems built from autonomous dynamical systems by introducing an input, time series theory considers autoregressive models with exogenous inputs which are called

ARMAX models or NARMAX models (with X for exogenous). In these models, the evolution equation takes into account exogenous variables at current instant or in the past. These exogenous variables are known and are the exact equivalent of the control signal. So we get the ARMAX (p, q, r) model,

$$x(k+1) = a_1 x(k) + \ldots + a_p x(k-p+1) + b_0 v(k+1) + b_1 v(k) + \ldots \\ + b_q v(k-q+1) + c_1 u(k) + \cdots + c_r u(k-r+1),$$

and the NARMAX (p, q, r) model,

$$x(k+1) = f[x(k), \ldots, x(k-p+1), v(k+1), v(k), \ldots, \\ v(k-q+1), u(k), \ldots, u(k-r+1)].$$

4.1.9 Limits of Modeling Uncertainties Using State Noise

We introduced in the previous sections the state noise $(v(k))$, which models uncertainty on the state variables using random variable and the probabilistic framework. Of course this type of model is relevant if the uncertainty is subject to statistical regularity that enables to identify some knowledge about this uncertainty and to improve prediction and the quality of control. Yet, it is not always the case and the occurrence of uncertainties or unknown that are ill represented by random variables is an intrinsic limitation of any statistical algorithm. A good example of this situation is the example of a non-cooperative target tracking: if we model the unknown control of the target by a stochastic process, the intention of the pilot is badly represented by such a statistical modeling.

In that case, when there is no other specific knowledge, probabilistic framework is just a less evil. Then it is important to use all the available information rather than to represent all the ill-identified variables in a large-dimensional stochastic process. The number of unknown parameters to be identified has to be reduced. These considerations support the use of parsimonious models and among them neural networks as it was explained in Chap. 2.

4.2 Regression Modeling of Controlled Dynamical Systems

4.2.1 Linear Regression for Controlled Dynamical Systems

4.2.1.1 Outline of the Algorithm

In Chap. 2, linear regression was described as the task of finding the $(n, 1)$ column vector $\boldsymbol{w} = (w_1; \ldots; w_n)$ that minimizes the sum of the squared errors (SSE)

$$J = \sum_{k=1}^{N} (y_k - \boldsymbol{x}_k \boldsymbol{w})^2$$

from $N(1, n)$ input vectors $(\boldsymbol{x}_1, \ldots, \boldsymbol{x}_k, \ldots, \boldsymbol{x}_N)$ and N output scalars $(y_1, \ldots, y_k, \ldots, y_N)$, or, equivalently that minimizes the half mean squared error (MSE),

$$\phi_N(\boldsymbol{w}) = \frac{1}{2N} \sum_{k=1}^{N} (y_k - \boldsymbol{x}_k \boldsymbol{w})^2.$$

We restrict ourselves to the classical case of a scalar output. The extension to the case of a vector output is trivial. It is well known (see Chap. 2) that a quadratic cost function has a single minimum, which can be derived through the following matrix formula:

$$\hat{\boldsymbol{w}} = \left(\boldsymbol{X}^{\mathrm{T}} \boldsymbol{X}\right)^{-1} \boldsymbol{X}^{\mathrm{T}} \boldsymbol{Y},$$

where the (N, n) observation matrix $\boldsymbol{X} = (\boldsymbol{x}_1; \ldots; \boldsymbol{x}_k; \ldots \boldsymbol{x}_N)$ and the $(N, 1)$ column vector $\boldsymbol{y} = (y_1; \ldots; y_k; \ldots; y_N)$ are constructed from the input and the output data. This result is available if the quadratic minimization problem is well-posed, i.e., the matrix $(\boldsymbol{X}^{\mathrm{T}} \boldsymbol{X})$ is invertible.

That algorithm may be used for autoregressive model identification. That type of model (ARX model) was introduced in the previous section.

$$x(k+1) = a_1 x(k) + \cdots + a_p x(k-p+1) + b_0 v(k+1) + c_1 u(k) + \cdots + c_r u(k-r+1).$$

Note that there is a correlation of the input and the output. Here the regression coefficient vector is $\boldsymbol{w} = [a_1, \ldots, a_p, b_0, c_1, \ldots, c_r]^{\mathrm{T}}$.

When an input trajectory $[u(1), \ldots, u(k), \ldots, u(N)]$ and an output trajectory $[x(1, \ldots, x(k), \ldots, x(N)]$ are available, the $(1, p+r)$ input vectors of the regression are constructed as follows: $\boldsymbol{x}_k = [x(k); \ldots; x(k-p+1); u(k); \ldots; u(k-r+1)]$ for k varying from $\max(p, r) + 1$ to $(N-1)$ and the associated output is $y_k = x(k+1)$.

High quality results are obtained provided that the linear model of the estimator is relevant. This assertion is supported by the following example.

4.2.1.2 Example of Application

Let us consider the (2, 2) order ARX model,

$$x(k+1) = a_1 x(k) + a_2 x(k-1) + b_0 v(k+1) + c_1 u(k) + \cdots + c_2 u(k-1),$$

with the following real values for the parameters:

$$a_1 = 1.2728, \quad a_2 = -0.81, \quad b_0 = 0.5, \quad c_1 = 0.5, \quad c_2 = -0.5,$$

and where the operator-designed input trajectory (u_k) is a white noise.

As requested by the method, we build the input vectors of the regression $\boldsymbol{x}(k) = [x(k); x(k-1); u(k); u(k-1)]$ for k varying from 2 to $N-1$.

The computation of regression coefficients yields the following numerical results:

$$\hat{a}_1 = 1.29, \quad \hat{a}_2 = -0.83, \quad \hat{c}_1 = 0.49, \quad \hat{c}_2 = -0.51.$$

Assume now that we ignore the input data; then the regression input vectors are two-dimensional $\boldsymbol{x}(k) = [x(k); x(k-1)]$. We have a simple AR model, and the computation gives the following, inaccurate estimations:

$$\hat{a}_1 = 1.17, \quad \hat{a}_2 = -0.71.$$

The model was not relevant: since the input trajectory was white noise, an AR model was used to process data that were actually generated by an ARMA model with vector noise (u_k, v_k).

Now, assume that a measurement noise is introduced into the simulator, so that our observation of the state is inaccurate while the process dynamics is unaffected (that point will be developed further in the section devoted to filtering). Then the data-generating process is the following:

$$\begin{cases} x(k+1) = a_1 x(k) + a_2 x(k-1) + c_1 u(k) + c_2 u(k-1) \\ y(k) = x(k) + b_0 w(k) \end{cases}.$$

In that case we have poor results if we use the ARX regression, even when the input trajectory is taken into account. We get

$$\hat{a}_1 = 0.61, \quad \hat{a}_2 = -0.36, \quad \hat{c}_1 = 0.49, \quad \hat{c}_2 = -0.11.$$

That numerical simulated example shows how it is important to get a relevant model for the noises to achieve linear regression. This problem was already addressed in Chap. 2 in the framework of dynamical neural modeling. We shall give a more detailed account further in this chapter: the occurrence of measurement noise creates a new problem, namely the filtering problem.

4.2.1.3 Statistical Background

Statistical analysis of time series is well known and will not be detailed hereafter. One can consult [Chatfield 1994] for a classical and practical handbook and [Gouriéroux 1995; Azencott 1984] for more mathematical details. We shall just outline here the explanation of the least-squares methodology in the simplest case of a stable autoregressive model with a gaussian centered noise. The variables are written with capital letters because they are considered as random variables.

Consider the gaussian stationary stochastic process associated to the AR(p) model,

$$X(k+1) = a_1 X(k) + \cdots + a_p X(k-p+1) + b_0 V(k+1).$$

Assume that the model is stable (i.e., the roots of the polynomial $P(z) = 1 - a_1 z - \cdots - a_p z^p$ are outside the unit ball) and that the white noise (V_k) is gaussian and centered. Then set $r_j = \mathrm{Cov}(X_k, X_{k-j})$, and take the covariance of the two members of the previous equation binding the variables $(X_{k-i})_{i=0\ldots p-1}$. Then the classical Yule-Walker relations are obtained

$$\begin{cases} r_1 = a_1 r_0 + \cdots + a_p r_p \\ \dots\dots\dots\dots\dots\dots\dots\dots \\ r_p = a_1 r_{p-1} + \cdots + a_p r_0 \end{cases}$$

The same relations are obeyed approximately by the empirical estimators $\hat{r}_i = 1/(N-p) \sum_{k=i+1}^{k=N} x(k)x(k-i)$ of the covariance coefficients r_j and by the mean-squared estimators $\hat{a}_i$ of the regression coefficients a_j (for higher accuracy, one should consider the truncation errors that vanish as the ratio p/N).

The estimators $\hat{r}_i$, however, are consistent, without bias and asymptotically gaussian with a variance of order $1/N$. Then, it may be proved that the estimators $\hat{a}_i$ are consistent, asymptotically without bias, asymptotically gaussian with a variance of order $1/N$ too. That result allows us to build statistical tests in order to validate the model.

Remark. An estimator is said to be consistent if the mean squared estimation error goes to 0 when the sample size goes to infinity.

We have just provided here a cursory introductory outline of linear regression. Actually, statisticians and control engineers have improved those methods to a great extent. Spectral representation is a key tool of linear modeling, and the transfer function of linear filters associated to ARMA models are generally the object of identification process. Those basic techniques are addressed in the literature (see the references) and are not within the scope of this book. Neural networks are a methodology that is relevant in the nonlinear framework.

4.2.1.4 Application to a Linear System: The Harmonic Oscillator

Let us use the previous algorithm to identify the harmonic oscillator that was described in the previous section. Suppose we know only the input trajectory and the angle trajectory (oscillator position). If a hundred step data file is available, ARX(2, 2) model-based identification gives the correct coefficients with high accuracy. Note that the order of the model is 2. If we use an ARX(2, 1) model to perform the identification, the results are significantly corrupted. That can easily be explained: since the control is implemented on speed increment, its order is 2.

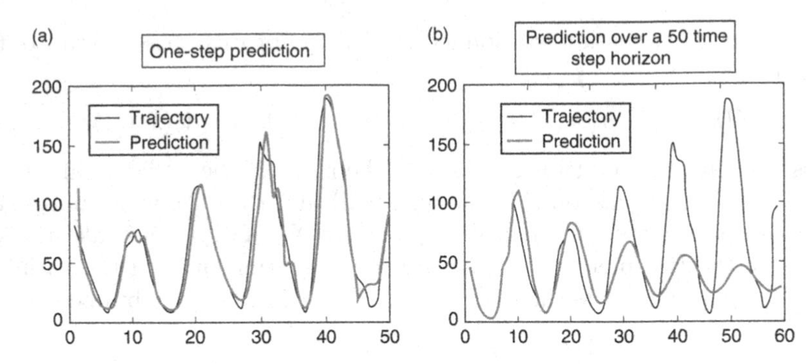

Fig. 4.6. Prediction of sunspot using AR(9) model based linear regression (**a**) One step prediction (**b**) 50 step horizon prediction

4.2.1.5 Application to Sunspot Data Modeling

If we use an AR(9) model based on linear regression to the sunspot series, omitting any data preprocessing, from a 150-step data file, we obtain a predictive model. Its performances on a 50-step test set are shown on Fig. 4.6. The difference between the real observation and the predicted value computed from the 9 closest past observations is represented on picture (a). That prediction is quite accurate. In picture (b), we represented the difference between the real observation and a predicted value, which uses solely the initial 9 first values of the data file. Of course, oscillations are damped. The damping phenomenon is normal because the autoregressive model is stable and the predictions do not use new measurements, but initial values only. Nevertheless, the regression process captured the basic frequency of the phenomenon.

4.2.2 Nonlinear Identification Using Feedforward Neural Networks

4.2.2.1 Limitations of Linear Regression

When the linearity assumption on the state equation is not obeyed, linear-regression-based modeling of controlled dynamical systems is very inaccurate and uses very heavy models with too many parameters. That is illustrated on Fig. 4.7.

In that example, which was described in previous section, there is no possible linear model, which exhibits both a stable equilibrium and an unstable equilibrium. Yet, the linear regression captured the right frequency of the oscillator.

4.2.2.2 Network with Delay (NARX Model)

The simplest example of neural identification of a controlled dynamical system is based on regression algorithms. The model is an autoregressive model with

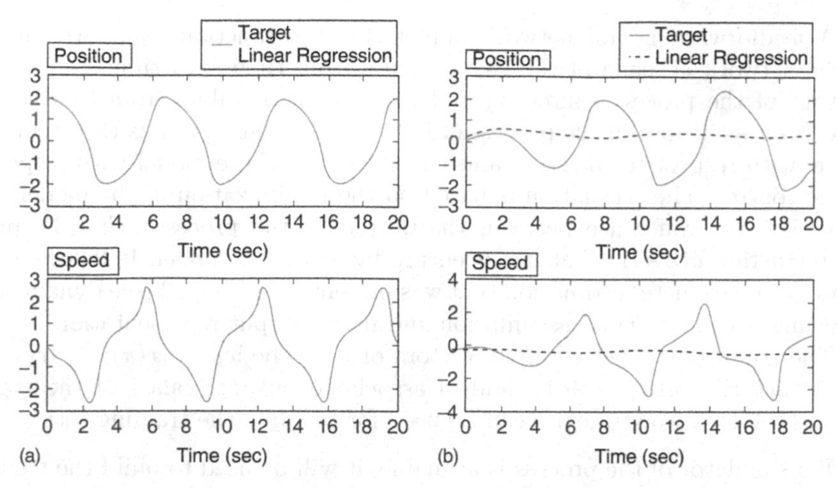

Fig. 4.7. Identification of Van der Pol oscillator using linear regression (1000 step regression) (**a**) Initialization on the attractor (limit cycle) (**b**) Initialization far from the attractor

exogenous input (NARX model). The stochastic NARX(p, r) model equation is

$$X(k+1) = f[X(k), \dots, X(k-p+1), V(k+1), u(k), \dots, u(k-r+1)].$$

Regression order is p for the state and r for the control. The diagram of the network, which is used for that purpose, is displayed on Fig. 4.8.

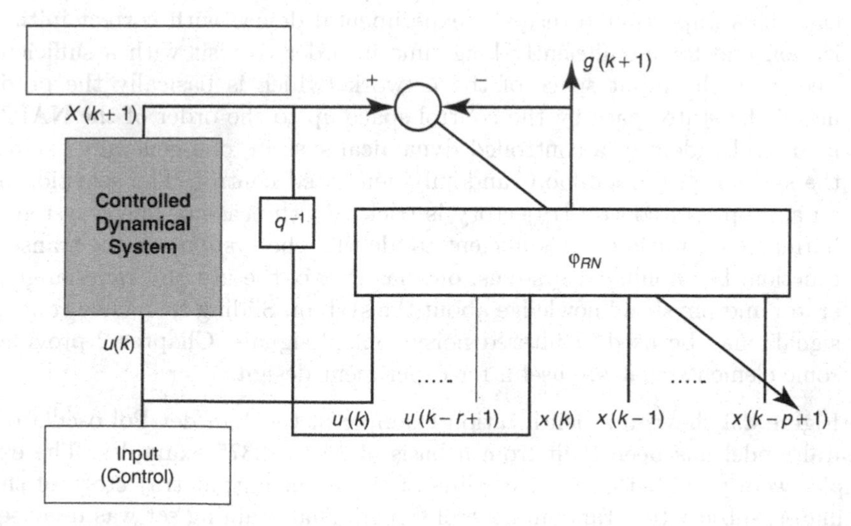

Fig. 4.8. Learning diagram of a NARX based neural model to identify a controlled dynamical system (see also Chap. 2, Fig. 2.31)

A feedforward neural network implements the function φ_{RN}. An input of the network is made of the signal values from time k to time $k-p+1$ (output of the process f interest) and of the control values from time k to time $k-r+1$ (input of the process of interest). In that case, p is the order of model with respect to the state and r is the order of the model with respect to the control. The estimation is based on the minimization of the modeling error, i.e., the difference between the output of the process $x(k+1)$ and the prediction $g(k+1)$ that has been produced by the model. It follows the strategy of parameter estimation that was presented in Chap. 2 (see dynamical modeling with state noise assumption and input-output representation).

The training is a set of input vectors of the type $\boldsymbol{x}_k = [x(k);\ldots;x(k-p+1);u(k);\ldots;u(k-r+1)]$ and of associated output scalars of the type $g_k = x(k+1)$. Two strategies can be used for building the training set:

- If a simulator of the process is available, it will be used to build the training set. In that case, one has the freedom of choosing a representative sampling of the network output. To that effect, one can select either a regular sampling of the input space, or select the input samples according to a probability law, which favors the usual operating region of input space. Sometimes, on the contrary, the limit operating points and the boundary of the safety domain are favored to ensure security and accuracy of the representation on the entire operation domain. That situation, where a simulator is available, is common when one is looking for a semi-physical representation or "grey-box model" (see Chap. 2).
- By contrast, if training is performed from actual experimental data, a sampling of input space cannot be chosen at will: the training set is obtained from the sampling of the input-output experimental trajectory. In that case, it is important to use the experimental device with correct initialization and for a sufficiently long time in order to visit with a sufficient frequency the input space of the network (which is basically the product of the state space by the control space up to the order of the NARX model). To identify a controlled dynamical system, one generally excites the system with open-loop randomly generated control. The selection of an appropriate control trajectory is tricky. In the case of linear systems, harmonic excitations are sufficient to identify the system via the transfer function. For nonlinear systems, one has to mix the use of a random generator and physical knowledge about the system. Sliding frequency control signals may be used or filtered noisy control signals. Chapter 2 provides some elements that are useful for experiment design.

Figure 4.9 shows an identification example of the Van der Pol oscillator. Neural model has been built from a basis of $15^3 = 3375$ examples. The examples were provided by the sampling of the input-output trajectory of the oscillator, subject to a random control signal. That training set was used before, for linear regression, as shown on Fig. 4.7. The results are far better here.

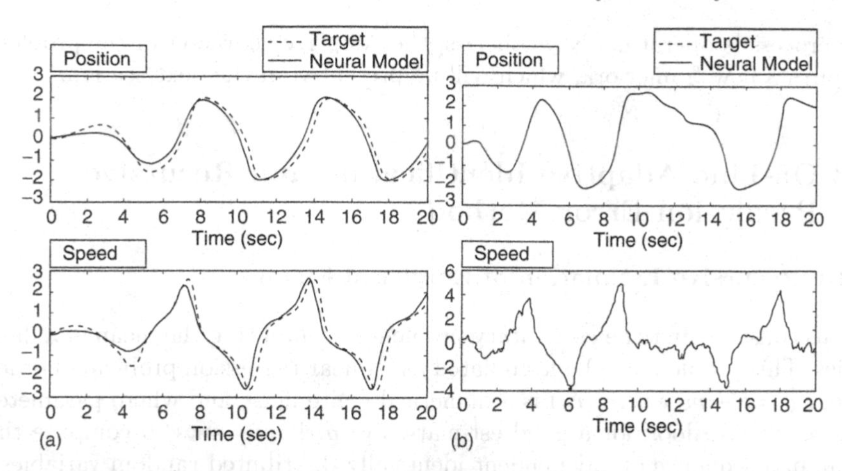

Fig. 4.9. Comparison of controlled Van Der Pol oscillator with its neural model: (**a**) No control has been used for learning (**b**) A random control signal has been provided to the system during learning

Those results have been obtained with a classical neural architecture with three inputs, ten hidden neurons and two output neurons. If training is performed from a training set that has the same size, but that was built from a regular sampling of the state space and of the feasible control set, the results are poor (if there is no data preprocessing). That result shows the crucial importance of the selection of the training set. Actually, as emphasized in Chap. 2, it is important to build the training set according to the probability distribution of visit on the input space in experimental conditions. That will be further elaborated when on-line training will be explained. Note the importance of the implemented open-loop control signal to visit the full input space, especially when the system is pushed on a single attractor (like the limit cycle of Van der Pol oscillator. In the next chapter, we will come back to that "*exploration policy*" in the neuro-dynamical programming framework.

The choice of the order of the model is important because it has a direct impact on the number of parameters to be estimated. It is more critical than in the linear case. Actually, the problem of selecting the model order is not fully solved in nonlinear regression theory. In practice, one combines an empirical approach (estimation of the generalization error) and theoretical criteria that were designed for linear models [Gouriéroux 1995]. Moreover, the model may be validated ex-post using hypothesis testing [Urbani 1993]. Nonadaptive identification from a representative training set is not especially troublesome when neural networks with supervised training are used, provided a cautious methodology, and efficient training algorithms are used.

Similar considerations apply in the framework of adaptive identification, where one has to use a flow of experimental data in an adaptive way, i.e., as the process is operating. Nevertheless, the adaptive character of the problem requires a new framework, which will be presented in the next section.

4.3 On-Line Adaptive Identification and Recursive Prediction Error Method

4.3.1 Recursive Estimation of Empirical Mean

Let us consider first the elementary problem of computing the mean of a data series. This problem can be formulated as a linear regression problem of order zero, $x_k = a + v_k$ where (v_k) is a numerical white noise and where parameter a is scalar. We look for a good estimation of a. It amounts to compute the mean of a sequence of independent identically distributed random variables.

The minimization of the cost function $J_N(a) = 1/2N \sum_{k=1}^{N} (x_k - a)^2$ with respect to a is a well-known problem. Its solution is the empirical mean $\hat{a}_N = \sum_{k=1}^{N} x_k / N$.

That estimate has all the desirable properties of current linear regression estimators such as consistency, bias free, minimal variance among bias free estimates. Its consistency (i.e., its convergence towards a when the sample size goes to infinity) is called the law of large numbers. It intuitively expresses that the arithmetic mean of a sequence of independent random measurements provides an accurate estimate of the expectation value of the random variable that models the phenomenon of interest.

A simple rewriting of the previous definition gives

$$(N+1)\hat{a}_{N+1} = \sum_{k=1}^{N} x_k + x_{N+1} = N\hat{a}_N + x_{N+1}.$$

The following recursive definition follows immediately:

$$\hat{a}_{N+1} = \hat{a}_N + \frac{1}{N+1}(x_{k+1} - a_N).$$

This recursive formulation of the definition of the empirical mean allows an adaptive estimation. A single observation is sufficient to initialize the algorithm. To update the estimate, it is not necessary that all the observations be available. The previous estimate and the current observation are sufficient to perform the update. The coefficient $\gamma_{\kappa+1} = 1/(N+1)$ is called the learning rate.

Another advantage of the recursive estimation is that it allows tracking slow variations of the parameter, which is currently estimated if the model is not stationary. The estimation is adaptive. In that case, one has to replace the slowly decreasing learning rate by a small constant learning rate. Then, estimation amounts to filtering (in that case, a first order filter). In order to

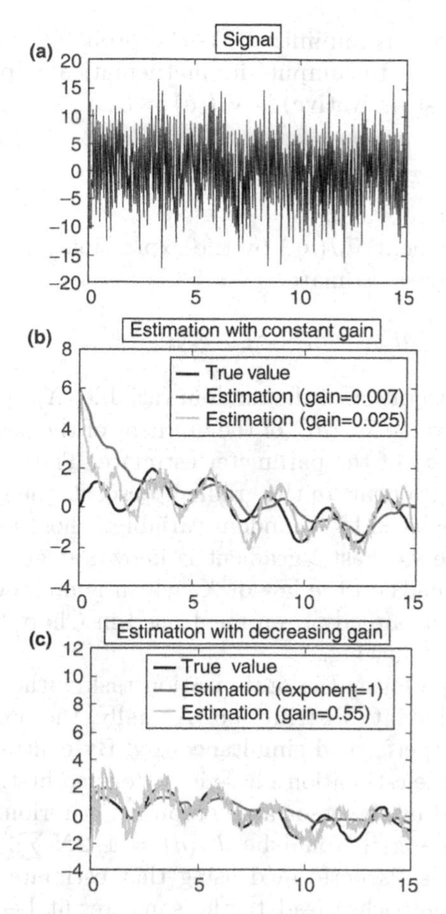

Fig. 4.10. Behavior of the empirical mean estimate. (**a**) Original signal (**b**) estimate of the parameter using a constant-gain filter (**c**) Estimate of the parameter using decreasing gains

compare a first order filter and a recursive estimation of the mean, the behavior of such an estimator has been shown on Fig. 4.10. The task is to track the quasi-periodic variation of the deterministic component of a random signal with a signal-to-noise ratio of 1/5. The original signal is shown on picture (a). On picture (b), the results for various values of the gain are compared. On picture (c), the performances of slowly decreasing gain estimates are compared. It is shown that the ability of the estimate to track the slow variations of the parameter in that case are poor.

One can notice that the empirical mean estimation is based on the minimization of a quadratic cost function using gradient descent. Actually, in the case of the stationary model, the data are a sample of the probability distribution of a random variable X. The quadratic cost function

$J(a) = 1/2\,E[(X-a)^2]$ is minimized. As the probability distribution is unknown, it is not possible to compute its mathematical expectation. The gradient of J (i.e., its first derivative) is $\nabla J(a) = E(X-a)$. A gradient descent algorithm is

$$a_{k+1} = a_k - \gamma_{k+1} \nabla J(a_k),$$

where γ_{k+1} is a positive scalar.

Replacing the gradient $\nabla J(a_k)$ by the expression $(X_{k+1} - a_k)$ yields the empirical mean recursive estimate,

$$a_{k+1} = a_k - \gamma_{k+1}\left(X_{k+1} - a_k\right).$$

Note that the expectation of the random variable $(X_{k+1} - a_k)$ is $E(X) - a_k$, which is exactly the value $\nabla J(a_k)$ of the gradient of the cost function J taken at the current value a_k of the parameter estimate. Therefore, this algorithm is termed stochastic gradient. In the gradient descent, the gradient of the cost function has been replaced by a random variable, whose expectation is equal to this gradient. The stochastic gradient is known at any time, whereas the "total" gradient depends on the law of X, which is unknown. The stochastic gradient algorithm has already been mentioned in Chap. 2. We will study it more in detail in the following.

That algorithm performs the optimization task without prior estimation of the unknown probability distribution. Actually, the optimization and the estimation tasks are performed simultaneously. By contrast, in the classical estimation process, the estimation phase is performed first. In that phase, the criterion is estimated by the associated empirical criterion, here the estimate of $J(a) = 1/2\,E[(X-a)^2]$ would be $J_N(a) = 1/2N \sum_{k=1}^{N} (x_k - a)^2$; then, the optimization phase is performed using that estimate. Actually, on that example, the two approaches lead to the same result because the model is linear with respect to the parameter of interest, namely a. Nevertheless, the implementations of the two algorithms are different: the implementation of the stochastic gradient is recursive.

4.3.2 Recursive Estimation of Linear Regression

The stochastic gradient approach that we have just used to estimate the empirical mean is frequently used for linear and nonlinear regression. In the context of linear regression, the algorithm is named recursive least mean squares (LMS) or Widrow-Hoff algorithm. This algorithm is well known in signal theory, where it is used to compute a linear regression in an adaptive way (see also Chap. 2).

Consider the problem of the minimization of $J(\boldsymbol{w}) = 1/2\,E[(Y - \boldsymbol{X}\boldsymbol{a} - b)^2]$ with respect to $\boldsymbol{w}$, where $\boldsymbol{X}$ is a second-order random vector $(1, n)$ (i.e., its expectation and covariance matrices are well defined). The vector parameter

$\boldsymbol{w}$ is the concatenation of the vector $(n, 1)$ of the vector parameter $\boldsymbol{a}$ and of the scalar parameter b; Y is a second-order real random variable.

We have

$$\nabla J(\boldsymbol{a}, b) = -E[(Y - \boldsymbol{X}\boldsymbol{a} - b)\boldsymbol{X}, (Y - \boldsymbol{X}\boldsymbol{a} - b)].$$

The data samples $(\boldsymbol{X}_1, Y_1), \ldots, (\boldsymbol{X}_k, Y_k), \ldots$ are available on-line to solve the estimation problem. They are independent. Then the stochastic gradient approach may be used. The recursive stochastic gradient estimate is defined by the following formula:

$$\begin{cases} \boldsymbol{a}_{k+1} = \boldsymbol{a}_k + \gamma_{k+1}\left(Y_{k+1} - \boldsymbol{X}_{k+1}\boldsymbol{a}_k - b_k\right)\boldsymbol{X}_{k+1} \\ b_{k+1} = b_k + \gamma_{k+1}\left(Y_{k+1} - \boldsymbol{X}_{k+1}\boldsymbol{a}_k - b_k\right) \end{cases}.$$

We have the following convergence statement:

- If the gain of the algorithm obeys the following conditions $\sum_{k=1}^{\infty} \gamma_k = \infty$, $\sum_{k=1}^{\infty} \gamma k^2 < \infty$, then the algorithm converges almost certainly to the linear regression coefficients of Y with respect to $\boldsymbol{X}$.

The conditions on the gain that have just been stated are general. Hereinafter, they will be referred to as the stochastic approximation conditions for the gain. In particular, the sequence $\gamma_k = 1/k$ obeys those conditions.

4.3.3 Recursive Identification of an AR Model

Consider the identification problem of the AR(p) model

$$X(k+1) = a_1 X(k) + \cdots + a_p X(k-p+1) + V(k+1).$$

We assume that the data are collected under a stationary regime, and we are looking for a recursive estimate that minimizes the least square criterion

$$J(\boldsymbol{w}) = \frac{1}{2} E[(X(k+1) - a_1 X(k) - \cdots - a_p X(k-p+1))^2].$$

The gradient of the cost function is: $\nabla J(\boldsymbol{w}) = -E\{[X(k+1) - a_1 X(k) - \cdots - a_p X(k-p+1)] \cdot [X(k; \ldots; X(k-p+1)]\}$ Thus, the stochastic gradient recursive estimate is defined by the algorithm

$$\hat{\boldsymbol{w}}(k+1) = \hat{\boldsymbol{w}}(k) + \gamma_{k+1}\vartheta(k+1)[X(k); \ldots; X(k-p+1)],$$

with $\vartheta(k+1) = X(k+1) - a_1 X(k) - \cdots - a_p X(k-p+1)$.

This rule was encountered previously and has been long known as the delta rule or Widrow rule. If the gain sequence obeys the stochastic approximation conditions, the algorithm converges, so that the estimate is consistent.

In the case of AR models, the input-output data are no longer independent. Therefore, the classical assumptions of the elementary law of large numbers are not fulfilled. The following Markov linear model produces the data:

$$\boldsymbol{X}(k+1) = \boldsymbol{A}[\boldsymbol{w}]\boldsymbol{X}(k) + \boldsymbol{V}(k+1),$$

where $\boldsymbol{A}[\boldsymbol{w}]$ depends linearly on $\boldsymbol{w}$, $(\boldsymbol{V}_k)$ is a white vector noise and where

$$\boldsymbol{X}(k) = [X(k); \ldots; X(k-p+1)] \quad \text{and} \quad \boldsymbol{A}(\boldsymbol{w}) = \begin{pmatrix} a_1 & a_2 & . & . & a_k \\ 1 & 0 & . & . & 0 \\ 0 & 1 & . & . & 0 \\ . & . & . & . & 0 \\ 0 & 0 & . & 1 & 0 \end{pmatrix}.$$

The stochastic approximation theory is valid in that general Markov frame and provides the consistency statement about the almost sure convergence of the recursive estimate.

There exist as well recursive versions of second-order optimization algorithms (Newton rule). The estimates are consistent too. Their convergence may be proved in the general stochastic approximation framework [Ljung 1983]. They are of special interest for linear models, because they provide a powerful way to speed up convergence. Recall (see Chap. 2) the Newton formula,

$$\hat{\boldsymbol{w}} = \boldsymbol{w}^* - \boldsymbol{H}J[\boldsymbol{w}^*]^{-1}\nabla J[\boldsymbol{w}^*],$$

where $\boldsymbol{H}J[\boldsymbol{w}^*]$ is the Hessian matrix of the cost function. The elements of the Hessian matrix of a function of several variables are the second partial derivatives, and its symmetry is guaranteed by the inversion Schwarz rule. The formulation of Newton formula leads to the recursive relation

$$\hat{\boldsymbol{w}}(k+1) = \hat{\boldsymbol{w}}(k) - \boldsymbol{H}\Phi[\hat{\boldsymbol{w}}(k)]^{-1}\nabla\Phi[\hat{\boldsymbol{w}}(k)].$$

In the case of a strictly convex function (e.g., for a quadratic criterion), this matrix is definite positive thus invertible. In the previous example of the AR(p) model, it is equal to the covariance matrix of the stationary random vector $\boldsymbol{X}_k$.

The recursive second-order algorithm combines the second-order optimization of $\boldsymbol{J}$ and the recursive estimation of covariance matrix $\boldsymbol{R}$,

$$\begin{aligned} \hat{\boldsymbol{w}}(k+1) &= \hat{\boldsymbol{w}}(k) + \gamma_{k+1}\vartheta(k+1)\hat{\boldsymbol{R}}(k)^{-1}\boldsymbol{X}(k) \\ \hat{\boldsymbol{R}}(k+1) &= \hat{\boldsymbol{R}}(k) + \gamma_{k+1}\boldsymbol{X}(k+1)\boldsymbol{X}(k+1)^{\mathrm{T}}. \end{aligned}$$

That method is called recursive prediction error method (RPEM). It is fully described in [Ljung 1983], with emphasis on the applications to identification. The RPEM method may be extended to nonlinear models. It may be used for neural network adaptive learning when the learning data are provided on-line by an experimental process or by a simulation.

4.3.4 General Recursive Prediction Error Method (RPEM)

The general recursive prediction error method is an application to the estimation of stochastic approximation. We have just provided some examples for linear identification. The general theory has been developed since the fifties (Robbins and Monroe have done some pioneering work). A detailed presentation is provided in [Kushner 1978]. It has been used for adaptive neural network learning. Its advantage is to be recursive, so that the storage of a large amount of data is not necessary. Its main drawback is its slow convergence. To apply with full security the general method one has to check a number of non-trivial assumptions. Precise convergence statements are given in [Ljung 1983; Benveniste et al. 1987; Duflo 1996]. We will give a detailed treatment of the particular case of the NARX(p, r) model identification $X(k+1) = f[X(k), \ldots, X(k-p+1), V(k+1), u(k), \ldots, u(k-r+1)]$ This model is relevant for neural networks. It is a Markov model when its state representation is given as

$$\boldsymbol{X}(k+1) = f[\boldsymbol{X}(k), V(k+1), \boldsymbol{u}(k)].$$

We assume that the model is stable and converges towards a unique stationary regime.

Function f, as well as the state noise $\{V(k)\}$, are unknown. Conversely, we assume that the state $\boldsymbol{X}(k)$ is accurately determined at time k. We are looking for an adaptive nonlinear parametric identification scheme of the type $\boldsymbol{X}(k+1) = g[\boldsymbol{X}(k), \boldsymbol{u}(k), \boldsymbol{w}]$, by minimization of the quadratic prediction error. The prediction error is defined for the input-ouput data $(\boldsymbol{x}, \boldsymbol{u}, \boldsymbol{y})$ and for a given value $\boldsymbol{w}$ of the vector parameter by: $\boldsymbol{\vartheta}(\boldsymbol{y}, \boldsymbol{x}, \boldsymbol{u}, \boldsymbol{w}) = \boldsymbol{y} - g(\boldsymbol{x}, \boldsymbol{u}, \boldsymbol{w})$.

We must compute the value of the parameter $\boldsymbol{w}$ that minimizes the mean quadratic prediction error,

$$J(\boldsymbol{w}) = \frac{1}{2} E[\| f(\boldsymbol{x}, V, \boldsymbol{u}) - g(\boldsymbol{x}, \boldsymbol{u}, \boldsymbol{w})\|^2],$$

where the mathematical expectation is taken over the probability law of the state noise, and is then averaged over the stationary regime of the input vector variable (state-control).

In order to apply the stochastic gradient method, one has to compute the gradient of the function $1/2\,[\|\vartheta(\boldsymbol{x}, \boldsymbol{y}, \boldsymbol{u}, \boldsymbol{w})\|^2]$ with respect to $\boldsymbol{w}$. That gradient is equal to $-\partial g / \partial \boldsymbol{w}\,(\boldsymbol{y}, \boldsymbol{x}, \boldsymbol{u}, \boldsymbol{w}) \vartheta\,(\boldsymbol{x}, \boldsymbol{y}, \boldsymbol{u}, \boldsymbol{w})$. It will be denoted below as $\boldsymbol{G}(\boldsymbol{y}, \boldsymbol{x}, \boldsymbol{u}, \boldsymbol{w})$. Similarly, we will denote $\boldsymbol{G}(k+1) = \boldsymbol{G}[\boldsymbol{X}(k+1), \boldsymbol{X}(k), \boldsymbol{u}(k), \boldsymbol{w}(k)]$.

Consider the following algorithms:

Stochastic gradient algorithm

$$\begin{aligned}\boldsymbol{w}(k+1) &= \boldsymbol{w}(k) - \gamma_{k+1}\boldsymbol{G}(k+1)\\ &= \boldsymbol{w}(k) + \gamma_{k+1}\frac{\partial g}{\partial \boldsymbol{w}}(\boldsymbol{X}(k+1), \boldsymbol{X}(k), \boldsymbol{u}(k), \boldsymbol{w}(k))\vartheta(k+1)\end{aligned}$$

Gauss-Newton stochastic algorithm

$$\begin{aligned}\boldsymbol{R}(k+1) &= \boldsymbol{R}(k) + \gamma_{k+1}\frac{\partial g}{\partial \boldsymbol{w}}(\boldsymbol{X}(k+1), \boldsymbol{X}(k), \boldsymbol{u}(k), \boldsymbol{w}(k))\\ &\quad\times\left[\frac{\partial g}{\partial \boldsymbol{w}}(\boldsymbol{X}(k+1), \boldsymbol{X}(k), \boldsymbol{u}(k), \boldsymbol{w}(k))\right]^T\\ \boldsymbol{w}(k+1) &= \boldsymbol{w}(k) - \gamma_{k+1}\boldsymbol{R}(k+1)^{-1}\boldsymbol{G}(k+1).\end{aligned}$$

Under usual stochastic approximation assumptions on the gain sequence, and if the current estimate is bounded during the processing, those algorithms are converging towards a local minimum of the quadratic criterion.

The boundedness assumption is impossible to guarantee a priori as far as practical applications are concerned (analogical noises are generally gaussian). So according to [Ljung 1983], one has to add a nonlinear projection onto a safety domain. That projection respects convergence if the true value is inside the safety domain.

It is possible to approximate Hessian matrix inversion, which is necessary in Gauss-Newton formula by other classical second-order algorithms (quasi-Newton algorithms as Levenberg-Marquardt, conjugate gradients, etc.). This was detailed in Chap. 2. A good empirical review of application of prediction error method to neural network learning is given in [Norgaard 2000].

If the gain is small and constant, the tracking abilities of the algorithm are similar to sliding regime technique performance [Benveniste et al. 1987].

4.3.5 Application to the Linear Identification of a Controlled Dynamical System

Figure 4.11 shows schematically the application of RPEM to on-line identification.

At that stage, let us overlook the measurement noise. The dynamic system is described in Fig. 4.11 by

- the evolution block, the inputs of which are the current state and the control, and whose output is the next state,

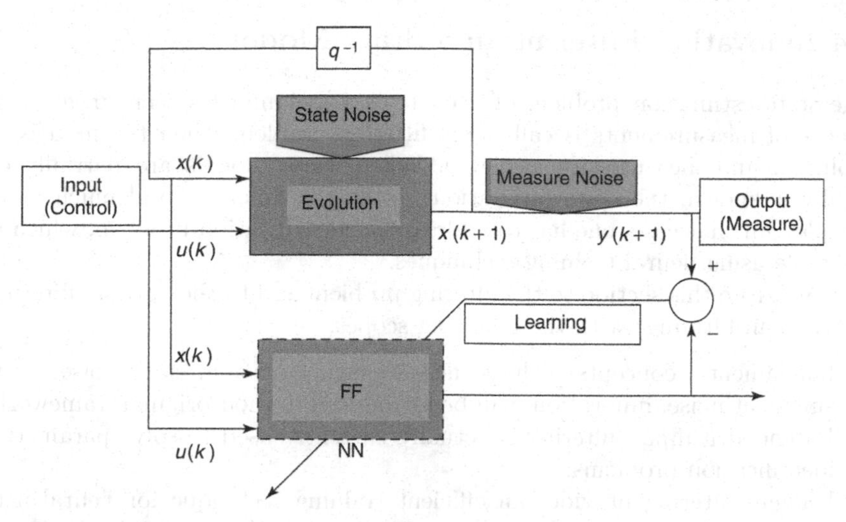

Fig. 4.11. Identification of the internal neural model of a controlled dynamic system (teacher forcing learning)

- the closed loop, which is ensured by the delay operator, and which maintains dynamics.

The current state and the control are sent as input to the neural net in its current configuration.

The state is supposed to be fully measured. In the case of an auto-regressive model, the current signal and the current control are processed to reconstruct the current state using delay lines that are shown in Fig. 4.8. The net computes its own prediction of the next state, which is compared to the state of the system. The computed prediction error is sent back to the network and is used to compute the gradient of the criterion, using the back-propagation algorithm. That supervised control algorithm (teacher forcing algorithm in control theory) has been described in detail in Chap. 2, in the framework of nonlinear dynamic system learning.

4.3.5.1 Addressing Measurement Inaccuracy

If measurement noise must be considered, regression identification using feed-forward neural networks and teacher forcing learning provides poor results. Some examples of that situation have been provided in this chapter for linear models, and a numerical demonstration was given in Chap. 2 in the section dedicated to dynamic systems.

When the state of the system cannot be completely known, the current state must be estimated. It is not a usual statistical problem, since, at a given instant, a single data is available. One has to take advantage of past knowledge provided by previous data. That is the purpose of filtering algorithms, which are the topics of next section.

4.4 Innovation Filtering in a State Model

The state estimation problem of a controlled dynamical system from a sequence of measurements is called the filtering problem when the models of evolution and measure are available. When those models are partially or totally unknown, the state estimation problem is different, and much more complex. In that case, one has to address the identification problem, which is tractable using neural training techniques.

We devote this section to the filtering problem and to the optimal filtering (or Kalman filtering) with the following scopes:

- Fundamental concepts such as measurement equation, state noise, measurement noise, innovation, will be introduced in their original framework;
- Extended Kalman filtering is actually currently used to solve parametric identification problems;
- Kalman filtering provides an efficient training technique for neural networks.

4.4.1 Introduction of a Measurement Equation

4.4.1.1 Observing Dynamic Linear Systems

We recall here the form of the state equation, which was previously stated in the section "Regression modeling of controlled dynamical systems" in a deterministic version,

$$\boldsymbol{x}(k+1) = f[\boldsymbol{x}(k), \boldsymbol{u}(k)].$$

For simplicity, the system is assumed to be time-invariant. In the linear case, the previous equation has the form:

$$\boldsymbol{x}(k+1) = \boldsymbol{A}\boldsymbol{x}(k) + \boldsymbol{B}\boldsymbol{u}(k).$$

Assume that the system is not completely observed. Then a measurement equation (or output equation) is introduced by the following expression:

$$\boldsymbol{y}(k) = h[\boldsymbol{x}(k)]$$

or, in the framework of linear time-invariant models:

$$\boldsymbol{y}(k) = \boldsymbol{H}\boldsymbol{x}(k).$$

In order to identify the state trajectory from the measurement sequence, we just have to guess the initial state $\boldsymbol{x}(0)$ since the evolution is deterministic. The following equations is obtained

$$\boldsymbol{y}(k) = \sum_{j=0}^{k-1} \boldsymbol{H}\boldsymbol{A}^{k-1-j}\boldsymbol{B}\boldsymbol{u}(j) + \boldsymbol{H}\boldsymbol{A}^{k}\boldsymbol{x}(0)$$

where the control sequence is known; we derive a linear system of equations, where the unknown vector is $\boldsymbol{x}(0)$ when k varies from 0 to n (n is the dimension of the state vector):

$$\boldsymbol{H}\boldsymbol{A}^k\boldsymbol{x}(0) = \boldsymbol{y}(k) - \sum_{j=0}^{k-1} \boldsymbol{H}\boldsymbol{A}^{k-1-j}\boldsymbol{B}\boldsymbol{u}(j).$$

The solution of that linear system is unique if and only if the rank of the matrix $[\boldsymbol{H}; \ldots; \boldsymbol{H}\boldsymbol{A}^n]$ is n. In that case, the system $(\boldsymbol{H}, \boldsymbol{A})$ is said to be *completely observable.*

That concept may be extended to nonlinear dynamical systems [Sontag 1990; Slotine et al. 1991] using differential geometry tools such as Lie brackets, which are not in the scope of this book.

4.4.1.2 Filtering State Noise and Reconstructing the State Trajectory

When the evolution is not deterministic, uncertainty at time k is modeled by a random vector $\boldsymbol{v}(k)$. Therefore, the state equation takes the following form:

$$\boldsymbol{x}(k+1) = f[\boldsymbol{x}(k), \boldsymbol{u}(k), \boldsymbol{v}(k+1)].$$

In the linear additive model, it has the particular form

$$\boldsymbol{x}(k+1) = \boldsymbol{A}\boldsymbol{x}(k) + \boldsymbol{B}\boldsymbol{u}(k) + \boldsymbol{v}(k+1).$$

In the section "Regression modeling of controlled dynamical systems", we mentioned that, in that case, the model of the state evolution is a particular stochastic process, namely a Markov chain. Assume that the state is not completely observed. Then we define the measurement process through the following measurement equation:

$$\boldsymbol{y}(k) = h[\boldsymbol{x}(k)],$$

which takes on the particular form for linear system

$$\boldsymbol{y}(k) = \boldsymbol{H}\boldsymbol{x}(k).$$

In the following, we assume, in a first approach, that the model is linear. Further, when the nonlinear extension will be considered, we will specify it explicitly.

To reconstruct the state trajectory, it would be necessary to solve recursively the following linear equation where $\boldsymbol{v}(k+1)$ is the unknown variable:

$$\boldsymbol{H}\boldsymbol{v}(k+1) = \boldsymbol{y}(k+1) - \boldsymbol{H}\boldsymbol{A}x(k) - \boldsymbol{H}\boldsymbol{B}\boldsymbol{u}(k).$$

Actually, it is not possible to find exact and well-defined solutions for this system. The right-hand side of that equation,

$$\varphi(k+1) = \boldsymbol{y}(k+1) - \boldsymbol{HAx}(k) - \boldsymbol{HBu}(k),$$

is called the innovation at time k. It is the error of the prediction of the new observation $\boldsymbol{y}(k+1)$ from the previous estimate of the state. That error provides us with new information that can be used for estimating a posteriori the state $\boldsymbol{x}(k+1)$ in a Bayesian framework.

If the system is completely observable, it can be shown that it is possible to select a matrix gain sequence $(\boldsymbol{K}_k)$ such that the following recursive estimate converges:

$$\hat{\boldsymbol{x}}(k+1) = \boldsymbol{A}\hat{\boldsymbol{x}}(k) + \boldsymbol{Bu}(k) + \boldsymbol{K}_{k+1}\varphi(k+1).$$

The $(\boldsymbol{K}_k)$, are called innovation gains. This model is called the Luenberger state observer. The innovation gain sequence is constrained by a *stability condition* in order to avoid the *divergence* of the filter. For instance, if we want to take a constant innovation gain $\boldsymbol{K}$ n order to get a steady filter, the spectrum of the matrix $\boldsymbol{A} - \boldsymbol{KHA}$ must be embedded in the unit disc (all the eigenvalue modules must be smaller than 1).

4.4.1.3 Variational Approach of Optimal Filtering

The computation of the innovation gain sequence is performed by minimizing a cost function. One can take the sum over k of the $\|\boldsymbol{v}_k\|^2$. However, in most applications, one has to take into account the measurement errors as well. Then, the cost function will be the sum over k of the following instantaneous cost:

$$j(\boldsymbol{v}_{k+1}) = \lambda\|v_{k+1}\|^2 + \mu\|\boldsymbol{y}_{k+1} - \boldsymbol{HAx}(k) - \boldsymbol{HBu}(k) - \boldsymbol{Hv}(k+1)\|^2.$$

That least squares criterion is a balance between the model uncertainty, which is weighted by the parameter λ, and the measurement uncertainty, which is weighted by the parameter μ. The tuning of those two hyperparameters requires some prior knowledge of the system.

Then the innovation gain is computed by solving the quadratic optimization problem. The solution is straightforward by canceling the gradient of the cost function

$$0 = 2(\lambda\boldsymbol{I} + \mu\boldsymbol{H}^{\mathrm{T}}\boldsymbol{H})v_{k+1} - 2\mu\boldsymbol{H}^{\mathrm{T}}[y(k) - \boldsymbol{HAx}(k-1) - \boldsymbol{HBu}(k-1)].$$

Therefore, the innovation gain is equal to

$$\boldsymbol{K}_{k+1} = (\lambda\boldsymbol{I} + \mu\boldsymbol{H}^{\mathrm{T}}\boldsymbol{H})^{-1}\mu\boldsymbol{H}^{\mathrm{T}} = \mu\boldsymbol{H}^{\mathrm{T}}(\lambda\boldsymbol{I} + \mu\boldsymbol{H}^{\mathrm{T}}\boldsymbol{H})^{-1}.$$

Note that the values of hyperparameters λ and μ may be time-dependent, or may be in matrix form. A fine tuning of the hyperparameters is only possible when sufficient prior knowledge of the system is available. Moreover, one has to check that the solution obeys the stability constraint. The probabilistic interpretation of optimal filtering gives insight into those issues, which will be further considered in the next section.

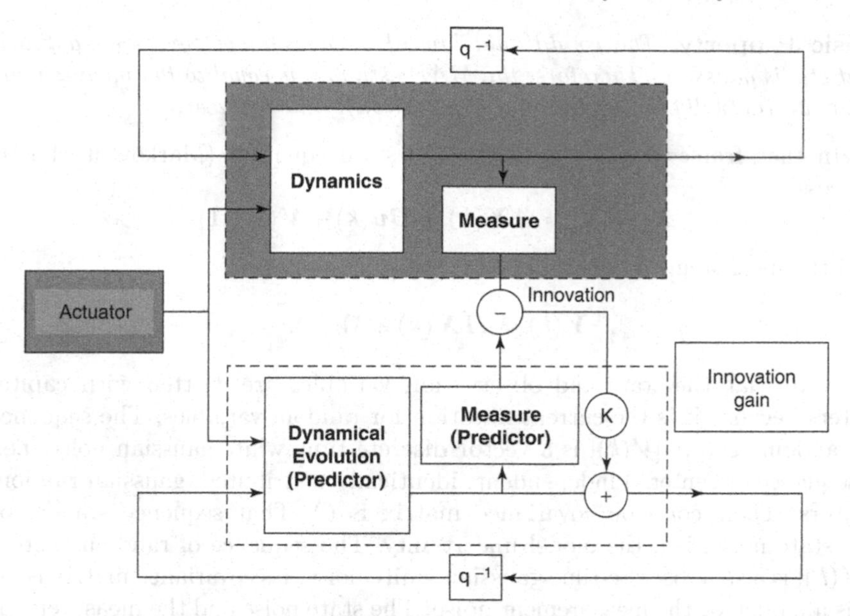

Fig. 4.12. Diagram of an innovation filter. The innovation filter is a predictor-corrector algorithm. The correction is brought by the innovation, which is computed from the measurement. The filter is recursive, and the estimate is fed back to the filter, which sets a stability problem

4.4.2 Kalman Filtering

4.4.2.1 Definition of the Kalman Filter for a Linear Stationary System

The algorithms that are currently used to estimate the state from the measurements are called filters. Actually, those algorithms cancel the noises in order to supply the true value for the current state. The filters that were described in the previous section are based on predictor-corrector schemes: they use current information to revise previous estimate. That is shown diagrammatically in Fig. 4.12. Those filters are called innovation filters.

The principle of Kalman filtering [Anderson 1979; Haykin 1996] consists in using probabilistic models of both model and measurement uncertainties for computing the innovation gain. The reconstruction of the state from the measurements is just a Bayesian estimation problem: the posterior probability law of the state is computed from the available measurements, and the decision is made using a least squares estimate or a maximum likelihood estimate (MAP estimate). However, such a computation may be very difficult in the general case. In the framework of linear model with additive gaussian noise, the solution is a recursive filter, which is just the optimal filter that was designed in the previous section. That simple solution results from the following basic property of the gaussian law, which is well known in probability theory:

Basic Property. *The conditional law of a gaussian vector given a linear statistic is gaussian. Therefore, the MAP estimate is equal to the mean-square estimate (actually the conditional expectation), and is linear.*

In that framework, let us write the state equation (Markov stochastic process)

$$\boldsymbol{X}(k+1) = \boldsymbol{A}\boldsymbol{X}(k) + \boldsymbol{B}\boldsymbol{u}(k) + \boldsymbol{V}(k+1)$$

and the measurement equation

$$\boldsymbol{Y}(k) = \boldsymbol{H}\boldsymbol{X}(k) + \boldsymbol{W}(k).$$

Note that the state and observation variables are written with capital letters because it is the current notation for random variables. The sequence of random vectors $[\boldsymbol{V}(k)]$ is a vector discrete time white gaussian noise, i.e., a sequence of centered independent, identically distributed, gaussian random vectors. Their common covariance matrix is $\boldsymbol{Q}$. That sequence stands for the state noise, i.e., the model uncertainty. The sequence of random vectors $[\boldsymbol{W}(k)]$ is also a discrete-time gaussian white noise. Its covariance matrix is $\boldsymbol{R}$. It is a model for the measurement noise. The state noise and the measurement noise are independent.

The filtering problem consists in reconstructing, at time $k+1$, the current state given the past or present measurements. The available information is gathered in the vector $\mathbf{y}(\mathbf{k}+\mathbf{1}) = [\boldsymbol{y}(\mathbf{1}), \ldots, \boldsymbol{y}(\mathbf{k}+\mathbf{1})]$. The criterion is the quadratic difference between the estimate $\hat{X}(k+1)$ and the true value of the state $\boldsymbol{X}(k+1)$.

It is a classical estimation problem in the linear gaussian model. It has been stated that the optimal solution $\hat{\boldsymbol{X}}(k+1)$ is the linear regression of the random state $\boldsymbol{X}(k+1)$ onto the random vector $\mathbf{Y}(k+1) = [\boldsymbol{Y}(1); \ldots; \boldsymbol{Y}(k+1)]$, which stands for the available information.

In order to compute the linear regression, let us split the vector $\boldsymbol{Y}(k+1)$ into the sum of two uncorrelated random vectors, the vector $\boldsymbol{Y}(k)$ and the residual of $\boldsymbol{Y}(k+1)$ onto $\mathbf{Y}(k)$. Then, the linear regression onto the vector $\boldsymbol{Y}(k+1)$ will be the sum of the two linear regressions onto its uncorrelated components (from the orthogonal projection theorem). Therefore, we can first compute the regression of the current measurement $\boldsymbol{Y}(k+1)$ onto $\mathbf{Y}(k)$. We start from

$$\begin{aligned}\boldsymbol{Y}(k+1) &= \boldsymbol{H}\boldsymbol{X}(k+1) + \boldsymbol{W}(k+1)\\ &= \boldsymbol{H}\boldsymbol{A}\boldsymbol{X}(k) + \boldsymbol{H}\boldsymbol{B}\boldsymbol{u}(k) + \boldsymbol{H}\boldsymbol{V}(k+1) + \boldsymbol{W}(k+1).\end{aligned}$$

Because $\boldsymbol{V}(k+1)$ and $\boldsymbol{W}(k+1)$ are independent from the past (from the white noise assumption), the regression is equal to $\boldsymbol{H}\boldsymbol{A}\hat{\boldsymbol{X}}(k) + \boldsymbol{H}\boldsymbol{B}\boldsymbol{u}(k)$ where $\hat{\boldsymbol{X}}(k)$ is the optimal estimate of $\boldsymbol{X}(k)$ given $\mathbf{Y}(k)$.

The residual of the regression of $\boldsymbol{Y}(k+1)$ onto $\mathbf{Y}(k)$ is

$$\begin{aligned}&\boldsymbol{Y}(k+1)-\boldsymbol{HA}\hat{\boldsymbol{X}}(k)-\boldsymbol{HBu}(k)\\&\quad=\boldsymbol{HA}[\boldsymbol{X}(k)-\hat{\boldsymbol{X}}(k)]+\boldsymbol{HV}(k+1)+\boldsymbol{W}(k+1).\end{aligned}$$

It is exactly the innovation term, which was defined in the previous section in the variational formulation of the filtering problem. From now on, the innovation will be written

$$\boldsymbol{\vartheta}(k+1)=\boldsymbol{Y}(k+1)-\boldsymbol{HA}\hat{\boldsymbol{X}}(k)-\boldsymbol{HBu}(k).$$

Note that the innovation at time $k+1$ in independent from $\boldsymbol{Y}(k)$.

The optimal estimate of the state at time $k+1$ can be split into the sum of two terms:

- A prediction term, which depends on the previous available information, equal to $\boldsymbol{A}\hat{\boldsymbol{X}}(k)+\boldsymbol{Bu}(k)$;
- A correction term, which depends linearly on the innovation $\boldsymbol{\vartheta}(k+1)$ at time $k+1$ and is equal to

$$\boldsymbol{K}_{k+1}\boldsymbol{\vartheta}(k+1)=\boldsymbol{K}_{k+1}[\boldsymbol{Y}(k+1)-\boldsymbol{HAX}(k)-\boldsymbol{HBu}(k)],$$

where $\boldsymbol{K}_{k+1}$ is called the Kalman gain of the filter at time $k+1$.

Thus, the filter is recursive and defined by the following formula

$$\hat{\boldsymbol{X}}(k+1)=\boldsymbol{A}\hat{\boldsymbol{X}}(k)+\boldsymbol{Bu}(k)+\boldsymbol{K}_{k+1}\boldsymbol{\vartheta}(k+1).$$

That computation shows that the Bayesian estimate is an innovation filter. The Kalman gain is the matrix coefficient of the linear regression of the state $\boldsymbol{X}(k+1)$ at time $k+1$ onto the innovation $\boldsymbol{\vartheta}(k+1)$. That coefficient is computed from the covariance matrix according to the following formula (linear regression has been recalled in Chap. 2):

$$\boldsymbol{K}_{k+1}=\mathrm{Cov}[\boldsymbol{X}(k+1),\boldsymbol{\vartheta}(k+1)]\mathrm{Var}[\boldsymbol{\vartheta}(k+1)]^{-1}.$$

To compute the Kalman gain, the covariance matrix of the errors must be computed. The details are given in the appendix of this chapter. We just state here the results.

If P_k stands for the covariance matrix of the estimation error $\boldsymbol{X}(k)-\hat{\boldsymbol{X}}(k)$ and $\boldsymbol{P}^{\circ}_{k+1}$ for the covariance matrix of the prediction error $\boldsymbol{X}(k+1)-\boldsymbol{A}\hat{\boldsymbol{X}}(k)-\boldsymbol{Bu}(k)$, then the Kalman gain is given by the following formula:

$$\boldsymbol{K}_{k+1}=\boldsymbol{P}^{\circ}_{k+1}\boldsymbol{H}^{\mathrm{T}}[\boldsymbol{HP}^{\circ}_{k+1}\boldsymbol{H}^{\mathrm{T}}+\boldsymbol{R}]^{-1},$$

where the dynamics of matrices P_k and $\boldsymbol{P}^{\circ}_{k+1}$ are determined by the following updating equations, the so-called covariance propagation equations:

$$\begin{aligned}&\boldsymbol{P}^{\circ}_{k+1}=\boldsymbol{AP}_k\boldsymbol{A}^{\mathrm{T}}+\boldsymbol{Q}\\&\boldsymbol{P}_{k+1}=(\boldsymbol{I}-\boldsymbol{K}_{k+1}\boldsymbol{H})(\boldsymbol{AP}_k\boldsymbol{A}^{\mathrm{T}}+\boldsymbol{Q})(\boldsymbol{I}-\boldsymbol{K}_{k+1}\boldsymbol{H})^{\mathrm{T}}+\boldsymbol{K}_{k+1}\boldsymbol{RK}^{\mathrm{T}}_{k+1}.\end{aligned}$$

Note that the evolution of the covariance matrix does not depend on the measurements. That remark is of practical importance for real-time applications of the Kalman filter (on-board navigation devices, for instance) because the computation of the Kalman gain sequence may be performed once and for all, from the model equations and the initial covariance error.

4.4.2.2 Properties of the Kalman Filter

The consequences of the previous paragraph are very important. Some of them may be extended to more general models. Let us summarize the main properties of the Kalman filter:

- When we compare the results of the computation of the innovation gain in the variational framework and in the probabilistic framework, the Kalman filter appears to be an optimal innovation filter in the sense of the variational framework. Penalties are time-dependent, in matrix form, and may be pre-computed. They are interpreted as the covariance of the prediction error (penalizing the model uncertainty) and the covariance of the measurement error (penalizing the measurement uncertainty).
- It was shown that the Kalman filter is unconditionally stable and is a consistent estimate of the state: the dynamics of the error converges towards an optimal steady regime even when the dynamical system itself is unstable (for details and proofs see [Anderson 1979; Haykin 1996]).
- The innovation sequence is the result of successive linear regressions. Therefore, it is uncorrelated and independent in the gaussian model. Whitening of innovation is an optimality characteristics, which may be computationally observed and tested.

4.4.2.3 Kalman Filtering for a Time-Varying Linear System

Kalman filtering may be applied in a straightforward way for linear nonstationary models. Let the state equation be

$$\boldsymbol{X}(k+1) = \boldsymbol{A}(k)\boldsymbol{X}(k) + \boldsymbol{B}(k)\boldsymbol{u}(k) + \boldsymbol{V}(k+1)$$

and the measurement equation

$$\boldsymbol{Y}(k) = \boldsymbol{H}(k)\boldsymbol{X}(k) + \boldsymbol{W}(k),$$

where the state noise sequence $\boldsymbol{V}(k)$ and the measurement noise sequence $\boldsymbol{W}(k)$ have time-varying covariance matrices $\boldsymbol{Q}(k)$ and $\boldsymbol{R}(k)$. The filter equation is

$$\hat{\boldsymbol{X}}(k+1) = \boldsymbol{A}(k)\hat{\boldsymbol{X}}(k) + \boldsymbol{B}(k)\boldsymbol{u}(k) + \boldsymbol{K}_{k+1}\boldsymbol{\vartheta}(k+1),$$

with

$$\boldsymbol{\vartheta}(k+1) = \boldsymbol{Y}(k+1) - \boldsymbol{H}(k+1)\boldsymbol{A}(k)\hat{\boldsymbol{X}}(k) - \boldsymbol{H}(k+1)\boldsymbol{B}(k)\boldsymbol{u}(k).$$

An iteration of the covariance and Kalman gain updates is

$$
\begin{aligned}
\boldsymbol{P}^{\circ}_{k+1} &= \boldsymbol{A}(k)\boldsymbol{P}_k\boldsymbol{A}(k)^{\mathrm{T}} + \boldsymbol{Q}(k+1) \\
\boldsymbol{K}_{k+1} &= \boldsymbol{P}^{\circ}_{k+1}\boldsymbol{H}(k+1)^{\mathrm{T}}[\boldsymbol{H}(k+1)\boldsymbol{P}^{\circ}_{k+1}\boldsymbol{H}(k+1)^{\mathrm{T}} + \boldsymbol{R}(k+1)]^{-1} \\
\boldsymbol{P}_{k+1} &= [\boldsymbol{I} - \boldsymbol{K}_{k+1}\boldsymbol{H}(k+1)][\boldsymbol{A}(k)\boldsymbol{P}_k\boldsymbol{A}(k)^{\mathrm{T}} + \boldsymbol{Q}(k+1)] \\
&\quad \times [\boldsymbol{I} - \boldsymbol{K}_{k+1}\boldsymbol{H}(k+1)]^{\mathrm{T}} + \boldsymbol{K}_{k+1}\boldsymbol{R}(k+1)\boldsymbol{K}^{\mathrm{T}}_{k+1}.
\end{aligned}
$$

The innovation sequence is uncorrelated as before. Conversely, there is no steady regime, and the stability of the filter is no longer guaranteed.

We just give here the principle of the algorithm. In practical cases, one can face problems if the dimension of the state space is too large. Such problems occur if the computation is too expensive, if the covariance matrix inversion fails, or if the positivity constraint of the covariance matrix is violated. Some special care allows overcoming these difficulties. For more details see [Anderson 1979; Haykin 1996].

4.4.3 Extension of the Kalman Filter

4.4.3.1 Case of Nonlinear Systems

Filtering nonlinear dynamic systems is a difficult issue. It is a field of active research. Neural networks are one of the tools that allow performing that task. For an introduction to nonlinear filtering which is both rigorous and application-oriented, one may consult the old textbook [Jazwinsky 1970]. It is nice and clear but it does not deal with numerical filtering. The paper [Levin 1997] gives a much shorter introduction. Moreover, it is written to introduce neural filtering. We will not address the general subject of nonlinear filtering here. Specifically, we will not address the observability problems, which deserve a special development.

The scope of this section is just to the presentation of a convenient formal framework for extended Kalman filtering, which is a common technique. That technique will be used below for the training of neural networks. Consider a time-invariant, nonlinear, controlled dynamical system with additive state noise and measurement noise. Its state equation is

$$\boldsymbol{X}(k+1) = f[\boldsymbol{X}(k), \boldsymbol{u}(k)] + \boldsymbol{V}(k+1)$$

and its measurement equation is

$$\boldsymbol{Y}(k) = h[\boldsymbol{X}(k)] + \boldsymbol{W}(k),$$

where the covariance matrices of gaussian white noise are denoted by $\boldsymbol{Q}(\boldsymbol{x})$ and $\boldsymbol{R}(\boldsymbol{x})$, for the state noise and the measurement noise respectively. That means that the noise laws are defined as conditional gaussian probability distributions given the current state. That is a Markov model.

To apply the Kalman filter algorithm, the nonlinear evolution model is replaced by its linearization around the current estimate $\hat{\boldsymbol{X}}(k)$, and the nonlinear measurement model is replaced by its linearization around the predicted state $f[\hat{\boldsymbol{X}}(k), \boldsymbol{u}(k)]$ in order to compute the covariance propagation.

Thus, $\boldsymbol{A}(k)$ stands for the gradient of f with respect to x at the point $[\hat{\boldsymbol{X}}(k), \boldsymbol{u}(k)]$, and $\boldsymbol{H}(k+1)$ stands for the gradient of h at the point $f[\hat{\boldsymbol{X}}(k), \boldsymbol{u}(k)]$.

The filter equation is determined by the predictor-corrector scheme,

$$\hat{\boldsymbol{X}}(k+1) = f[\hat{\boldsymbol{X}}(k), \boldsymbol{u}(k)] + \boldsymbol{K}_{k+1}\boldsymbol{\vartheta}(k+1),$$

with

$$\boldsymbol{\vartheta}(k+1) = \boldsymbol{Y}(k) - h\{f[\hat{\boldsymbol{X}}(k), \boldsymbol{u}(k)]\}.$$

The iteration of covariance update is (using convenient linearizations, see [Anderson 1979])

$$\begin{aligned}
\boldsymbol{P}^{\circ}_{k+1} &= \boldsymbol{A}(k)\boldsymbol{P}_k\boldsymbol{A}(k)^{\mathrm{T}} + \boldsymbol{Q}(k+1) \\
\boldsymbol{K}_{k+1} &= \boldsymbol{P}^{\circ}_{k+1}\boldsymbol{H}(k+1)^{\mathrm{T}}[\boldsymbol{H}(k+1)\boldsymbol{P}^{\circ}_{k+1}\boldsymbol{H}(k+1)^{\mathrm{T}} + \boldsymbol{R}(k+1)]^{-1} \\
\boldsymbol{P}_{k+1} &= [\boldsymbol{I} - \boldsymbol{K}_{k+1}\boldsymbol{H}(k+1)][\boldsymbol{A}(k)\boldsymbol{P}_k\boldsymbol{A}(k)^{\mathrm{T}} + \boldsymbol{Q}(k+1)] \\
&\quad \times [\boldsymbol{I} - \boldsymbol{K}_{k+1}\boldsymbol{H}(k+1)]^{\mathrm{T}} + \boldsymbol{K}_{k+1}\boldsymbol{R}(k+1)\boldsymbol{K}^{\mathrm{T}}_{k+1}.
\end{aligned}$$

The computation of the gain is subject here to an approximation. Therefore, no optimality property can be provided any longer. If the approximation is valid, that algorithm can provide a sub-optimal solution (the quality of the solution is near optimal). The stability of the linearized Kalman filter is much more difficult to prove than for time-varying linear Kalman filter. Moreover, the gain computation must be performed on-line. That is very hard for real-time applications and on-board computers. For that type of applications, tracking a reference trajectory, and computing a filter for a linearization of the model along that trajectory, are usually preferred. In that case, the algorithm of the previous section is used. Nevertheless, the extended Kalman filter is currently used, especially for identification problems. In the following section, we address that issue, using a state extension.

4.4.3.2 Using Extended Kalman Filter for Parametric Identification

Consider the following model for an observed controlled dynamical system

$$\begin{aligned}
\boldsymbol{X}(k+1) &= \boldsymbol{A}(\boldsymbol{\theta})X(k) + \boldsymbol{B}(k)\boldsymbol{u}(k) + \boldsymbol{V}(k+1) \\
\boldsymbol{Y}(k) &= \boldsymbol{H}(\boldsymbol{\theta})\boldsymbol{X}(k) + \boldsymbol{W}(k),
\end{aligned}$$

where the model depends on an unknown parameter $\boldsymbol{\theta}$. One has to estimate the unknown parameter. Depending on the application, $\boldsymbol{\theta}$ may be constant or may change slowly. Several methods have been proposed to estimate on-line both the current state $\boldsymbol{X}(\mathrm{k})$ and the parameter $\boldsymbol{\theta}$. In the method of extended Kalman filter, the parameter $\boldsymbol{\theta}$ is incorporated into the state. Then the extended model equations become

$$\begin{aligned} \boldsymbol{X}(k+1) &= \boldsymbol{A}[\boldsymbol{\theta}(k)]\boldsymbol{X}(k) + \boldsymbol{B}(k)\boldsymbol{u}(k) + \boldsymbol{V}_1(k+1) \\ \boldsymbol{\theta}(k+1) &= \boldsymbol{\theta}(k) + \boldsymbol{V}_2(k+1) \\ \boldsymbol{Y}(k) &= \boldsymbol{H}[\boldsymbol{\theta}(k)]\boldsymbol{X}(k) + \boldsymbol{W}(k). \end{aligned}$$

The state noise $[\boldsymbol{V}_2(k)]$, which allows the parameter to vary, is artificial for a steady model: nevertheless, it improves the operation of the filter because it helps to stabilize the algorithm [Haykin 1999]. Independence and stationarity of $[\boldsymbol{V}_1(k)]$, and of $[\boldsymbol{V}_2(k)]$ are assumed here for the sake of simplicity. Those assumptions must sometimes be dropped. From the previous paragraph, application of linearization techniques for extended Kalman filter provides the following equations:

$$\begin{aligned} \hat{\boldsymbol{X}}(k+1) &= \boldsymbol{A}[\hat{\boldsymbol{\theta}}(k)]\hat{\boldsymbol{X}}(k) + \boldsymbol{B}(k)\boldsymbol{u}(k) + \boldsymbol{K}_{1,k+1}\boldsymbol{\vartheta}(k+1) \\ \hat{\boldsymbol{\theta}}(k+1) &= \hat{\boldsymbol{\theta}}(k) + \boldsymbol{K}_{2,k+1}\boldsymbol{\vartheta}(k+1), \end{aligned}$$

with the same notation for innovation as in the linear case,

$$\boldsymbol{\vartheta}(k+1) = \boldsymbol{Y}(k+1) - \boldsymbol{H}[\hat{\boldsymbol{\theta}}\,(k)]\{\boldsymbol{A}[\hat{\boldsymbol{\theta}}\,(k)]\hat{\boldsymbol{X}}\,(k) + \boldsymbol{B}(k)\boldsymbol{u}(k)\}.$$

Note that the parameter and state estimates are simultaneously updated using the same innovation and different Kalman gains. Iterations of computation for covariance updates and Kalman gain computations are derived from the previous section.

Though its computer implementation is quite simple when the state dimension is not too large, using the extended Kalman filter to estimate both parameter and state presents major drawbacks: stability issue, and dependence on initialization. Therefore, whenever possible, trickier methods are preferred because they are more reliable. Such methods generally combine Kalman filtering techniques for state estimation and Bayesian techniques or maximum likelihood estimation techniques for parameter estimation.

4.4.3.3 Adaptive Training of Neural Networks Using Kalman Filtering

Figure 4.13 provides the diagram of a neural network training algorithm using the extended Kalman filter.

The system under estimation is the neural network itself. It is supposed to be a model of the process that generated the training set. Actually, its state is the configuration of the network (i.e., the set of all the parameters of the

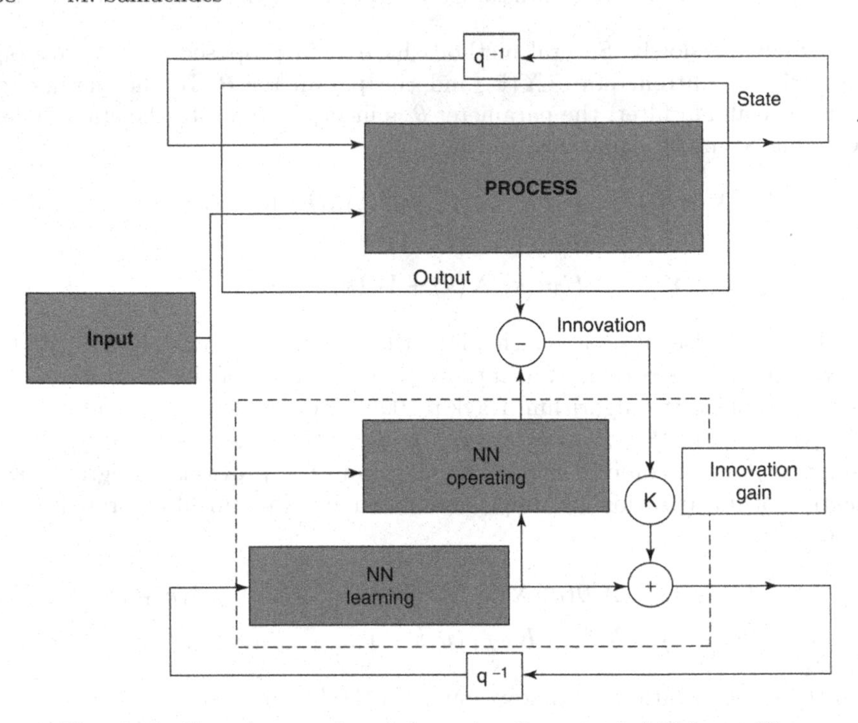

Fig. 4.13. Neural network training using the extended Kalman filter

network). The input-output pair sequence of the neural network is the measurement process that provides information on the evolution of the configuration. Thus, that adaptive algorithm is well suited to tracking slow variations of the environment, which is a typical task for an adaptive algorithm.

In that context, training amounts to estimating the state. Operating the network amounts to measuring the current state. Therefore, the innovation is the classical error for supervised learning i.e., the difference between the desired network output and the computed network output when an input is presented to the network.

The linear state equation $\boldsymbol{X}(k+1) = \boldsymbol{AX}(k) + \boldsymbol{Bu}(k) + \boldsymbol{V}(k+1)$ becomes the following expression:

$$\boldsymbol{\mathcal{W}}(k+1) = \boldsymbol{\mathcal{W}}(k) + \boldsymbol{V}(k+1),$$

where $\boldsymbol{\mathcal{W}}(k)$ is the configuration vector of the network (weights+biases) at time k.

The nonlinear measurement equation $\boldsymbol{Y}(k) = h[\boldsymbol{X}(k)] + \boldsymbol{W}(k)$ becomes

$$\boldsymbol{y}(k) = g[\boldsymbol{x}(k), \boldsymbol{w}(k)] + \boldsymbol{W}(k).$$

Therefore, the innovation error of the model is: $\boldsymbol{\vartheta}(k+1) = \boldsymbol{y}(k+1) - g[\boldsymbol{x}(k+1), \boldsymbol{w}(k)]$. It is indeed the expression of the learning error that was stated in Chap. 2. The measurement equation must be linearized to update the covariance and to recursively compute the Kalman gain as described in the previous section. Actually, the state evolution is just a random walk and the covariance of noise is constant, so that the equations of the filter take a simpler form.

If $\boldsymbol{H}(k+1)$ stands for the gradient of the network output g with respect to the weight vector $\boldsymbol{w}$ at the point $[\boldsymbol{x}(k+1), \boldsymbol{w}(k)]$, we get

$$\begin{aligned}
\boldsymbol{P}_{k+1}^{\circ} &= \boldsymbol{P}_k + \boldsymbol{Q} \\
\boldsymbol{K}_{k+1} &= \boldsymbol{P}_{k+1}^{\circ}\boldsymbol{H}(k+1)^{\mathrm{T}}[\boldsymbol{H}(k+1)\boldsymbol{P}_{k+1}^{\circ}\boldsymbol{H}(k+1)^{\mathrm{T}} + \boldsymbol{R}]^{-1} \\
\boldsymbol{P}_{k+1} &= [\boldsymbol{I} - \boldsymbol{K}_{k+1}\boldsymbol{H}(k+1)]\boldsymbol{P}_{k+1}^{\circ}[\boldsymbol{I} - \boldsymbol{K}_{k+1}\boldsymbol{H}(k+1)]^{\mathrm{T}} + \boldsymbol{K}_{k+1}\boldsymbol{R}\boldsymbol{K}_{k+1}^{\mathrm{T}},
\end{aligned}$$

where $\boldsymbol{Q}$ and $\boldsymbol{R}$ are the classical notations for covariance matrices of the state noise and of the measurement noise in Kalman filtering theory. The equation of the filter is

$$\ddot{\boldsymbol{w}}(k+1) = \ddot{\boldsymbol{w}}(k)\boldsymbol{K}_{k+1}\vartheta(k+1),$$

with

$$\boldsymbol{\vartheta}(k+1) = \boldsymbol{y}(k+1) - g[\boldsymbol{x}(k+1), \ddot{\boldsymbol{w}}(k)].$$

It should be emphasized that the neural network under identification is a virtual object: the only existing configuration is the current configuration $\ddot{\boldsymbol{w}}(k)$ under estimation. The ideal configuration that we try to identify or to track has no actual existence; it is an approximate representation of the real process.

The equation of the filter is a form of nonadaptive optimization algorithms that was reviewed in Chap. 2. In that algorithm, the descent direction is not the gradient of quadratic error that is equal to $\boldsymbol{H}(k+1)^{\mathrm{T}}\boldsymbol{\vartheta}(\mathrm{k}+1)$. The gradient may be computed using the backpropagation algorithm. Actually, the Kalman filter training algorithm is a second-order method, but it is an adaptive method, by contrasts to the methods that were presented in Chap. 2. The estimation of error surface curvature is performed by updating the covariance matrices. The implementation problems are similar to other second-order algorithms (inversion of a large matrix, positivity constraint) and are overcome by similar algorithmic techniques.

In order to reduce the complexity of the covariance matrix update, a decoupled extended Kalman filter technique (DEKF) was proposed in the literature. The parameters are grouped into clusters. The clusters are supposed to be uncorrelated. For instance a cluster may be the set of weights afferent to a single neuron and the associated bias. Then the covariance matrix acquires a block structure, so that it is easier to update and to invert ([Puskorius et al. 1994; Haykin 1999]).

The Kalman filter training method is not commonly used because it is relatively complex to implement. Nevertheless, it is potentially very interesting, because it is a second-order *adaptive* method. The choice of covariance matrices may seem arbitrary. That can be used advantageously to express empirical knowledge on perturbations and system noise. Thus, the tracking abilities of the algorithm may be tuned. We will apply that method to neural control in the next chapter.

4.5 Recurrent Neural Networks

4.5.1 Neural Simulator of an Open-Loop Controlled Dynamical System

In the section that was dedicated to neural identification of a controlled dynamical system, a feedforward neural network was designed as a one-step-ahead prediction model. We presented on Fig. 4.11 the diagram of the learning process of an input-output model according to the NARX hypothesis. We showed in Chap. 2, in the section entitled "black box modeling" and in the section "state noise hypothesis, input-output representation", that that approach is relevant when a state noise is present: the output of the model at time k may be reconstructed from the past values of the process output and the past values of the control signal. After completion of training, the network output may be plugged into the state input, with a delay operator in the feedback loop. That generates a recurrent neural network. Actually, the network graph exhibits a closed circuit. One may use that recurrent network, which models the function ψ_{RN}, to predict the process output within a finite horizon.

Figure 4.14 shows an input-output recurrent neural network: the network state input consists in past values of the output. If the network parameters have been estimated using open-loop training as in Fig. 4.11, and if the network is used to predict the process output beyond one time step, then it was shown in Chap. 2 that such an approach is not optimal: the prediction is corrupted by the iteration of the state noise. Conversely, it was shown theoretically, and illustrated through examples, that if the noise is a measurement noise (output noise), if training was performed by a semidirected algorithm, and if, during learning, the model outputs were used as input states, then the accuracy of the prediction is optimal.

In that context, we assumed that the control signal $u(k)$ did not depend on the state (actually the output of the network). It was therefore an open-loop controlled dynamical system. We shall now model a closed-loop controlled dynamical system using a combination of neural networks.

4.5.2 Neural Simulator of a Closed Loop Controlled Dynamical System

Just as a feedforward neural network whose inputs were the (state+control) signals, and whose output was a state output, was used to model a controlled

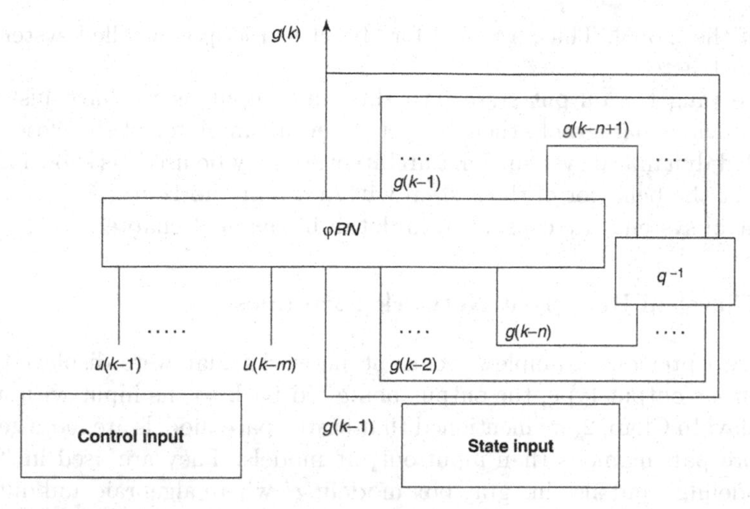

Fig. 4.14. Recurrent input-output neural network for an open loop controlled dynamical system modeling

dynamical system, a controller may be modeled as an application of the state space onto the control space, which associates the appropriate control to the current state. The diagram of Fig. 4.15 represents the connection of these two neural networks.

In that diagram, the state input feeds both the process model and the controller, which computes the control signal. That control signal is the second

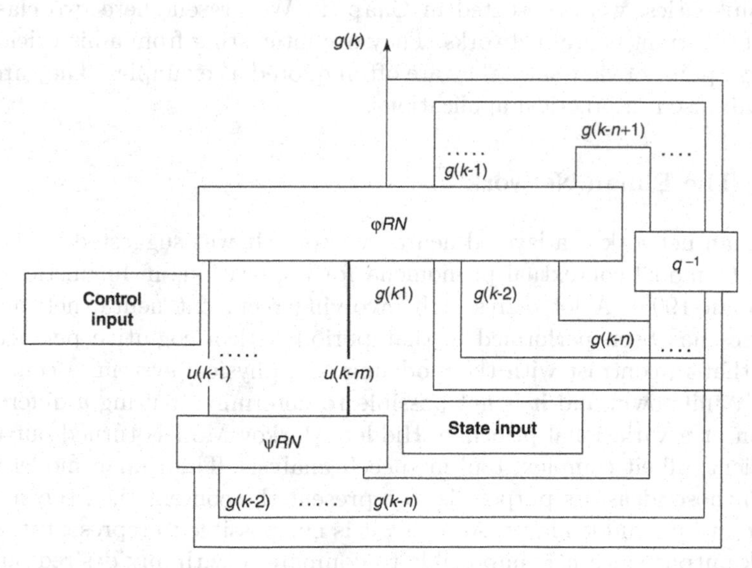

Fig. 4.15. Recurrent neural network modeling a closed loop controlled dynamical system. The model is the combination of network φ_{RN} and network ψ_{RN}

input of the model. Thus, a model for the closed loop controlled system has been constructed.

If we plug the output state into the state input as we have just done in the previous paragraph, then, we get a neural simulator of the closed loop controlled dynamical system. That architecture may be used, as shown above, to predict the behavior of the system within a finite horizon.

Control systems are considered in detail in the next chapter.

4.5.3 Classical Recurrent Network Examples

In the two previous examples, recurrent networks that were displayed were of the input-output type, the output being fed back to the input with a unit time delay. In Chap. 2, we mentioned that state-space models are more general and more parsimonious than input-output models. They are used in "black box modeling" and also in "gray box modeling" where algebraic and differential equations that express domain knowledge are taken into account in the structure of the network.

It should be remembered that, in recurrent networks, the delays must be fully specified in order to avoid ambiguity in the networks dynamics. In the appendix of the chapter we emphasize the importance of the delay distribution with an example. The following should also be remembered (Chap. 2):

Rule. *For a recurrent neural network to be causal, any cycle of the network graph must have a nonzero delay.*

Other examples of recurrent neural networks, having structures of different complexities, were presented in Chap. 2. We present here two classical types of recurrent neural networks. They are interesting from a historical and didactical point of view since they are often quoted as examples. They are not commonly used in practical applications.

4.5.3.1 The Elman Network

The Elman network is a layered neural network. It was suggested in the late eighties to model contextual phenomena for applications in linguistic analysis. [Elman 1990]. A lot of research involving recurrent neural networks in linguistics has been performed in that period with a cognitive perspective. Notice that in contrast with the modeling of a physical system, a context is generally unknown, and it is not possible to determine it using a differential equation or a variational principle. Hidden Markov Models turned out to be an efficient, albeit complex, tool in speech analysis. The Elman model is related to those ideas: its purpose is to represent the context (i.e., the state of the system) in a hidden layer. Actually, it is not possible to represent it as the network output since it is impossible to compare it with any desired output. A diagram of the Elman recurrent network is displayed on Fig. 4.16.

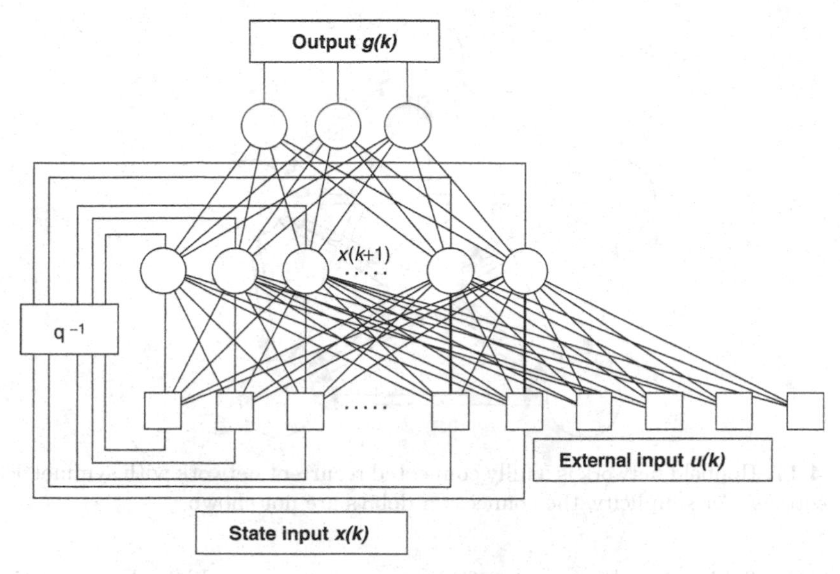

Fig. 4.16. Application of Elman network for dynamical system modeling

The *Elman network* is a network with one hidden layer. The outputs of the neurons of that layer are fed back to the network input with a unit time delay. Therefore, the order of the model is equal to the number of hidden units. Elman calls the hidden units the contextual units. The network output at a given time is a nonlinear function of the external input at that time and of the output of hidden units at the previous time step.

Note that the basic components of an observed dynamical system are clearly disclosed in the Elman model: the network input layer stands for the control of the system, the contextual hidden layer stands for the state of the system, the output layer stands for the measurement. The connection of the input layer to the hidden layer stands for the influence of the control on the state evolution.

4.5.3.2 The Hopfield Network

Hopfield networks played an important historical part for several years from 1982. They were motivated by the progress of the statistical physics of disordered media and its applications to complex systems. In 1982, Hopfield proposed [Hopfield 1982] a neural network model that was a decisive step away from the popular perceptron. (The perceptron is studied in detail in Chap. 6 of this book). He emphasized the dynamical characteristics of biological networks that stem from the recurrent connectivity: a recurrent neural network, being a dynamical system, exhibits attractors that are steady states of the dynamics.

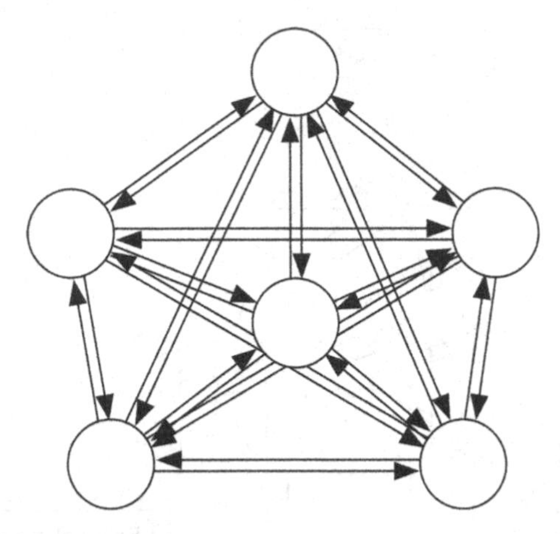

Fig. 4.17. Hopfield network is a fully connected recurrent network with symmetrical connections. For simplicity, the connection delays are not shown

A Hopfield network is made of binary neurons, of which the activation function is a step function. The output y of a neuron is given by

$$y = H\left(\sum_j w_{ij}x_j\right) \quad \text{where } H(x) = 1 \text{ if } \sum_j w_{ij}x_j \geq 0 \text{ and } H(x) = 0 \text{ else,}$$

and where x_j are the values of the inputs of neuron i; the latter are the outputs of all other neurons. Thus, each neuron encodes binary information and the state of the network is the vector of the neuron outputs. It is a binary vector that encodes information.

A Hopfield network has no external inputs. Its behavior is autonomous and depends only on its own dynamics. To guarantee the global stability of such a network, (i.e., to guarantee that the network state converges to a steady state, irrespective of the initial state) and to be able to compute those equilibria, Hopfield introduced a symmetry in the connections rule: the connection weight from neuron i to neuron j is equal to the connection weight of neuron j to neuron i. Moreover, each connection has a unit delay (those assumptions are not plausible from a biological standpoint). Figure 4.17 represents a Hopfield network with six binary neurons. The symbols for the connection delays have been omitted. The dynamics of the Hopfield net can be viewed as follows: the steady states of the dynamics encode codes of memory contents, and the dynamical process, which starts from an initial sate and proceeds autonomously towards an equilibrium state, can be interpreted as the recall process of an associative memory or a content addressable memory (CAM): the initial state is the binary code of a corrupted information (as delivered by a sensor), and the final state is the binary code of the correct information

The network training process consists in computing the network parameters in such a way that the items to be stored be coded by equilibrium states of the network dynamics. To obtain that behavior, Hopfield proposed that the connection matrix should be equal to the correlation matrix of the stored items. More precisely, assume that the network has N neurons and that p items should be encoded and stored. Those items are encoded by binary vectors $\boldsymbol{\xi}_i = (\xi_i^j)$. The weight matrix is denoted by $\boldsymbol{w} = (w_{jl})$ with $w_{jl} = (1/p)\sum_{i=1}^{p} \xi_i^j \xi_i^j$ if $j \neq l$ and $w_{jj} = 0$. Note that the connection matrix is symmetric. That learning rule is a simplistic version of Hebb's rule, which was first proposed by Hebb to model some biological learning processes. Later on, other learning rules have been proposed to ensure that any set of vectors (with a cardinal smaller than $N/2$) or any state sequence can be stored as a fixed point or as a cycle of the network dynamics.

To conclude, 20 years after J. J. Hopfield's seminal paper, the following conclusions may be drawn:

- From the point of view of biological modeling, the advantage of Hopfield model is to emphasize the role of dynamics in the cognitive functions of biological neural networks, and to show the connection between learning and correlation, which is set by Hebb's rule. Some older models, which were less popular, already emphasized those points. Later on, more biologically plausible models integrated new properties: information temporal coding using action potential or spikes, sparseness and asymmetry of the connections. These new properties outdated the Hopfield model, however rich and innovative.
- As content addressable memories (CAM), the performance of Hopfield models is rather poor. Improvements have been developed in the eighties (mean-field Hopfield networks with continuous activation functions, stochastic Hopfield networks and Boltzmann machines). There was a lot of published literature. Nevertheless, applied research on those topics was progressively abandoned, especially in the fields of pattern recognition and error correction.
- A close connection was soon established between the Hopfield model and the simulated annealing algorithm. Simulated annealing was discovered at about the same period by Kirkpatrick, Gelatt and Vecchi [Kirkpatrick 1983] from statistical physic inspiration. A new research direction originated from that connection: the application of neural networks to optimization. That approach is the scope of Chap. 8 of this book.

4.5.4 Canonical Form for Recurrent Networks

The examples of recurrent neural networks that were provided in previous paragraphs show that those networks are inherently dynamical systems. Therefore, if those networks are considered as dynamical systems, they are by input signals and provide output signals. Consequently, it is convenient to express them in a state-space representation. That state representation provides a unified way of handling those models, whatever their special architecture may be (delay distribution, etc.). That state representation was called canonical form and was fully described in Chap. 2.

Any *recurrent neural network*, however complex, has a minimal state representation, called "canonical form." The algorithms that are described in the previous section may be applied to the canonical form in a straightforward way.

In Chap. 2, the paragraph that is entitled "Canonical form of dynamical models" and complementary sections address that issue. Several examples are presented there to illustrate the approach.

4.6 Learning for Recurrent Networks

E. Sontag has proven in [Sontag 1996] that recurrent neural networks are universal approximators of controlled, observable, deterministic dynamical systems. Note that, just as Hornik's universal approximation theorem for function approximation, the present theorem is not constructive, and provides no indication either on the architecture or on the learning algorithm.

The main problem with recurrent neural network learning using a descent method (first order gradient method or second-order method) comes from the time range of the consequences of changing a weight value. The influence of a weight value on the cost function is not limited to the current time: it propagates through the computing horizon, which is theoretically unbounded. In a rigorous mathematical treatment, the computation of the gradient of the cost function requires propagating the computation for each example on the full computational horizon, compute the weight correction and iterate as necessary. The training process for recurrent networks would be then a very expensive procedure for very long training sequences. It would be difficult to implement on real-time applications. Therefore, when recurrent architectures for neural networks were suggested for dynamical system identification and control, approximate solutions were used. The basic paper [Williams 1989] presents an interesting approach.

When the state of the system is completely known because it is measured at each time step, there is no particular problem. A teacher-forcing algorithm can readily be implemented, although (see Chap. 2) that technique is appropriate only in applications where the relevant uncertainty is modeled by a state noise. That approach was shown to be poor when a measurement noise must be taken into account, a very frequent situation in industrial applications.

In the general case, where the knowledge of the state of the process is corrupted by a measurement noise, or is not fully measured, one must make a choice between two approximations:

- Either compute the true gradient with respect to the current weights but change the cost function by truncating the computation period to a sliding window: that is called back-propagation through time (BPTT)

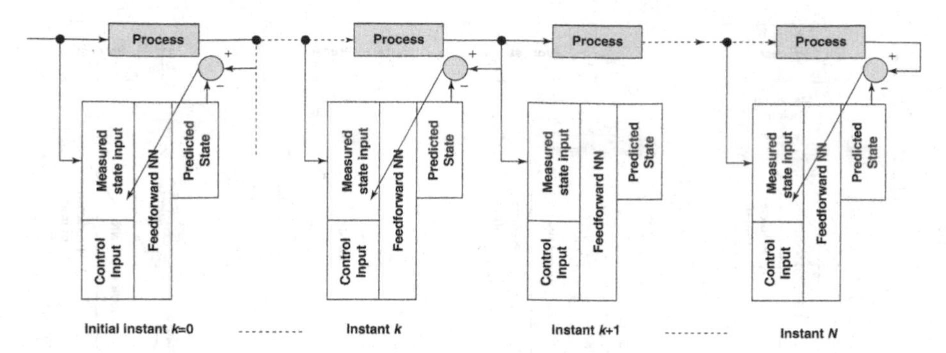

Fig. 4.18. Teacher-forced learning of a recurrent network

- Or approximate the gradient of previous states with respect to the current weights by the values of those gradients with respect to the previous weights: that is called real-time recurrent learning (RTRL).

More details are provided in the next section.

4.6.1 Teacher-Forced Learning

In the teacher-forced learning method, all the inputs of the canonical form of the network are known during the training process. The name of the algorithm is inspired from the teacher's behavior: the teacher corrects the student's behavior at each step instead of first observing the behavior for a while and correcting it afterwards. The engineer just takes into account the fact that experimental data are available to set the model at any time-step. Then the network learning process amounts to a nonlinear regression of the network on its input (NARX) as shown in Chap. 2. That learning process is depicted graphically in Fig. 4.18.

An input-state trajectory (set of N input-state pairs) is used as the database for the training process. The intermediate states (time k) are used both as outputs to assess the evolution of the current network computing evolution from time $k-1$ to time k and as inputs to feed the network and compute evolution from time k to time $k+1$. Of course to apply this simple method, one has to know the full input of the process at each time-step. It cannot be implemented in the general case.

4.6.2 Unfolding of the Canonical Form and Backpropagation Through Time (BPTT)

In this method, the recurrent characteristics of the network are considered by building a feedforward network whose outputs are identical to the sequence of outputs of the recurrent network. As mentioned in Chap. 2, that network is obtained by copying the feedforward part of the canonical form N times if

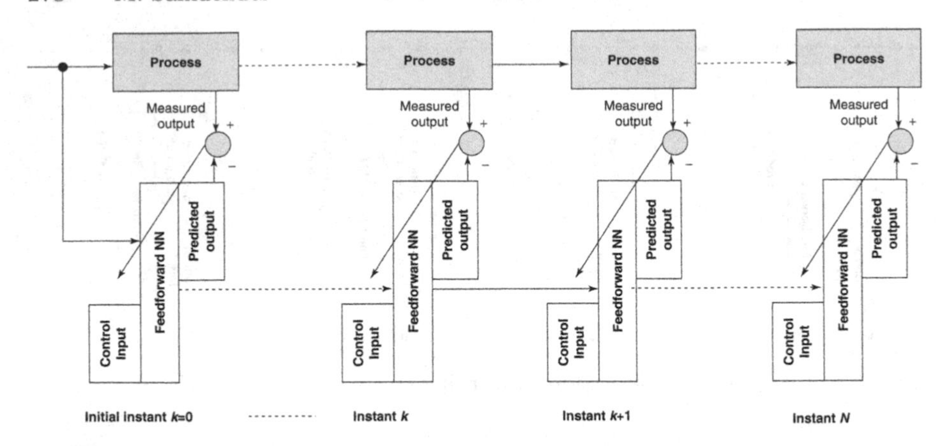

Fig. 4.19. Time unwrapping of the canonical form of a recurrent network along the whole learning sequence

the training sequence is of length N. The state trajectory of the process is not measured , hence cannot be used as state inputs during training. The state of the network is set initially to the state of the process if the latter is available (semidirected learning). If it is not the case, the input state of the network is initialized to a likely value (a priori initialization). The unfolding of the network canonical form is shown schematically on Fig. 4.19. That generates a feedforward network. Any training algorithm of feedforward networks can be used, subject to the constraint that all the weights of the copies are identical. Therefore, the shared-weight technique must be used (see Chap. 2).

If the training sequences are too long, or if an *adaptive* training is required (on-line training), the training sequences must be truncated to a finite duration, so that, for each training step, only a limited portion of past information will be used. Let p be the duration of that period. Thus, at time n, only information pertaining to time $n-p+1$ to time n is taken into account. That leads to a new notation: from now on, k will stand for the number of the copy at step n; k varies from 1 to p. The training scheme is similar to Fig. 4.19 with the following modifications:

- The length of the sequence is not n time steps, but p time steps;
- The state inputs at the first of those p time-steps can be set in to two different ways:
  1. If the state of the process is fully measured, then the measured values of the process may be assigned to these state inputs (at the first time-step): then the algorithm is called *semidirected*;
  2. If the state of the process is not fully measured, those inputs must be fed with the previously computed values for that specific copy (those quantities were computed at learning step $n-1$). Then, the algorithm is said "*undirected*" because the true state of the process is never taken into account during the training process. In that case, that assignment

integrates all the available information coming from the past up to time $n-p+1$, through p successive revisions. It can be trusted. Nevertheless, it introduces both error and instability. It was shown in [Lion 2000] that it is possible to control this approximation by introducing a projection to provide reasonable boundaries to the results. Then, using stochastic approximation theory, it is shown that the algorithm converges towards a local minimum. (The minimum is local since the framework is not linear and not necessarily convex).

Thus, it is important to discriminate, in the computation, two different times indices, the learning step index n and the time-step index of the unfolded network, which is denoted by $k, k = 1$ to p. A copy of the network is defined by the two functions $\boldsymbol{g}$ and $\boldsymbol{h}$ that respectively determine the state and the output of the network at step k as a function of the network state, its input and its previous parameter values. We are now able to describe in detail the operations that are necessary to compute the gradient using backpropagation through time during the training step $n+1$. All the current values of the network parameters are stored in the parameter vector $\mathcal{W}$.

For the nth learning step, the following components of the input vector be used:

$$\boldsymbol{u}_{n+1}^{k-1} = \boldsymbol{u}_{n-p+k}, k = 1 \text{ to } p,$$

and the following output data:

$$\boldsymbol{\psi}_{n+1}^{k} = \boldsymbol{\psi}_{n-p+k+1}, k = 1 \text{ to } p.$$

If the state of the network cannot be measured (undirected learning), the previous learning step state estimate is used as the initial state of the unfolded network

$$\boldsymbol{x}_{n+1}^{0} = \hat{\boldsymbol{x}}_{n-p+1} = \boldsymbol{x}_{n}^{1}.$$

At the training step $n+1$, the following computations are performed on the unfolded network that was obtained at the previous step:

- computation of the state and of the output for $k = 1$ to p,

$$\begin{aligned} \boldsymbol{x}_{n+1}^{k} &= g(\boldsymbol{u}_{n+1}^{k-1}, \boldsymbol{x}_{n+1}^{k-1}, \boldsymbol{w}) \\ \boldsymbol{y}_{n+1}^{k} &= h(\boldsymbol{u}_{n+1}^{k-1}, \boldsymbol{x}_{n+1}^{k-1}, \boldsymbol{w}); \end{aligned}$$

- comparison with the desired outputs for $k = 1$ to p,

$$\boldsymbol{\varepsilon}_{n+1}^{k} = \boldsymbol{\psi}_{n+1}^{k} - y_{n+1}^{k};$$

- computation of the adjoint unfolded network (the adjoint unfolded network is built by inverting the signal propagation direction, by replacing nodes by adders, adders by nodes, and activation functions by their derivatives)

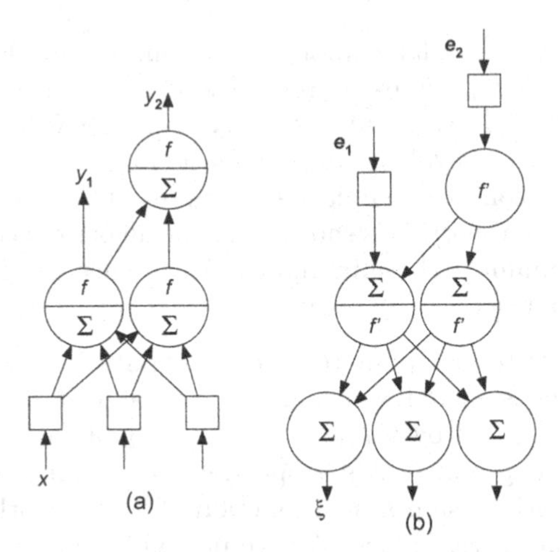

Fig. 4.20. Adjoint of a layered feedforward network (**a**) Original network, f stands for the transition operator through the nonlinear activation function (**b**) adjoint network, notation f stands for the linear product by the derivative of function f taken at the current point of the original network

- backpropagation of the error through the adjoint unfolded network for $k = 1$ to p,

$$\boldsymbol{\xi}_{n+1}^{k-1} = g^*(\boldsymbol{\varepsilon}_{n+1}^k, \boldsymbol{\xi}_{n+1}^k, \boldsymbol{w}).$$

Figure 4.20 illustrates the construction of the adjoint network in a simple case.

We showed on picture (a) a layered network with three inputs, a first layer with a hidden neuron and an output neuron and a second layer with a single output neuron. Thus, the network performs a nonlinear application of R^3 into R^2. On picture (b), the adjoint network performs a linear mapping of R^2 into R^3. The inputs of the adjoint network are the error signals that are associated to the original network outputs. The mathematical definition is simple: the adjoint of the nonlinear application $\boldsymbol{y} = g(\boldsymbol{x})$ is the linear application $\boldsymbol{\xi} = [Dg(\boldsymbol{x})]^{\mathrm{T}}\boldsymbol{\varepsilon}$, where $[Dg(\boldsymbol{x})]^{\mathrm{T}}$ is the transposed matrix of the Jacobian matrix of $\boldsymbol{g}$ at $\boldsymbol{x}$, i.e., the matrix of the partial derivatives. It is just a graphical representation of the backpropagation algorithm, which is frequently used to compute the gradient of the cost function with respect to the parameters.

Once the error signals of the adjoint network are computed, the computation of the quadratic error gradient is achieved through implementing the classical backpropagation rule. However, one has to consider that the network is an unfolded network. Actually, the network has been duplicated p times where p is the width of the time window. Therefore, the numerical value of one connection weight is shared by several connections that have different locations in the unfolded network.

Gradient Computation

The component of the gradient of the quadratic error with respect to a connection weight of the recurrent network is the sum of computed values of the components of the gradient with respect to all the connections of the unwrapped network that share the same value of the connection weight.

That result was shown in Chap. 2 in the paragraph dedicated to the "shared weights" technique.

The reader wishing to implement on a computer one of the foregoing algorithms will find all the needed formulas in a synthetic framework in the Yacine Oussar's Ph.D. entitled "Wavelet networks and neural networks for static and dynamic process modeling" (Chap. 3, pages 64 to 69 for input-output models and pages 72 to 81 for state models). This thesis is available in pdf format at the following URL http://www.neurones.espci.fr. A full technical discussion of the algorithms is developed there.

4.6.3 Real-Time Learning Algorithms for Recurrent Network (RTRL)

The real time recurrent learning method (RTRL) relies on another approximation, different from time truncation. Let us write again the recurrent network evolution equation from time n to time $n+1$ under its canonical form,

$$\begin{aligned} \boldsymbol{x}(n+1) &= g[\boldsymbol{u}(n), \boldsymbol{x}(n), \boldsymbol{w}(n)] \\ \boldsymbol{y}(n+1) &= h[\boldsymbol{u}(n), \boldsymbol{x}(n), \boldsymbol{w}(n)]. \end{aligned}$$

We want to compute, with the weights $\boldsymbol{w}(n)$, the gradient of the application Ψ_1^{n+1} that takes $\boldsymbol{w}$ as input and delivers $\boldsymbol{y} = \Psi_1^{n+1}(\boldsymbol{w})$. The computation will be performed from an initial state $\boldsymbol{x}(0)$ by using the following sequence of equations. For $k = 0, \ldots, n$,

$$\boldsymbol{x}(k+1) = \boldsymbol{g}[\boldsymbol{u}(k), \boldsymbol{x}(k), \boldsymbol{w}] \quad \text{and} \quad \boldsymbol{y} = \boldsymbol{h}[\boldsymbol{u}(n), \boldsymbol{x}(n), \boldsymbol{w}].$$

Differentiating those expressions, we obtain

$$\begin{aligned} \nabla_w \Psi_1^{n+1}[\boldsymbol{w}(n)] = {} & \nabla_w h[\boldsymbol{u}(n), \boldsymbol{x}(n), \boldsymbol{w}(n)] \\ & + \nabla_x h[\boldsymbol{u}(n), \boldsymbol{x}(n), \boldsymbol{w}(n)] \cdot \nabla_w \Phi_1^n[\boldsymbol{w}(n)], \end{aligned}$$

where Φ_1^n is defined as the application that takes $\boldsymbol{w}$ as input and delivers $\boldsymbol{x} = \Phi_1^n(\boldsymbol{w})$ using the following recursive computation sequence: for $k = 0, \ldots, n-1$,

$$\boldsymbol{x}(k+1) = \boldsymbol{g}[\boldsymbol{u}(k), \boldsymbol{x}(k), \boldsymbol{w}] \quad \text{and} \quad \boldsymbol{x} = \boldsymbol{x}(n).$$

The problem is the computation of $\nabla_w \Phi_1^n[\boldsymbol{w}(n)]$ though the value $\boldsymbol{w}(n)$ was not available at the past time steps. Since we are operating in real time, we cannot go back to the past as in the BPTT algorithm. For instance at step $n-1$, we perform the computation

$$\boldsymbol{x}(n) = g[\boldsymbol{u}(n-1), \boldsymbol{x}(n-1), \boldsymbol{w}(n-1)]$$

instead of the computation

$$\boldsymbol{x}(n) = g[\boldsymbol{u}(n-1), \boldsymbol{x}(n-1), \boldsymbol{w}(n)]$$

using a different state trajectory that was computed in real-time with the time-varying weight trajectory $\boldsymbol{w}(k)$, instead of computing again a new state trajectory with the single constant configuration $\boldsymbol{w}(n)$.

The idea is to update an approximation of $\nabla_w \Phi_1^n[\boldsymbol{w}(n)]$ that is noted $\hat{\nabla}_w \Phi_1^n$ by the following recursive equation:

$$\hat{\nabla}_w \Phi_1^n = \nabla_g[\boldsymbol{u}(n-1), \boldsymbol{x}(n-1), \boldsymbol{w}(n-1)] \cdot \hat{\nabla}_w \Phi_1^{n-1}.$$

That approximation is proven mathematically using stochastic approximation theory in the framework of controlled Markov chains subject to assumptions that will not be detailed here (see [Benveniste et al. 1987]).

For computational issues, it should be emphasized that RTRL does not use the adjoint network. Indeed, in contrast to backpropagation, the full gradient has to be computed explicitly. The computation must be performed from inputs to outputs, instead of being performed backwards.

4.6.4 Application of Recurrent Networks to Measured Controlled Dynamical System Identification

The applications of recurrent neural networks to identification, using undirected or hybrid algorithms are not very common. Generally, academic examples are presented in the literature. The stability of the algorithms is more difficult to guarantee than for linear models [Ljung 1996].

In practice, for the identification of nonlinear models, directed learning algorithms should be tested first. In [Haykin 1999], it is shown that the identification using a NARX model of the time series $\sin(n+\sin(n^2))$ outperforms the identification using a semidirected algorithm with the same number of parameters. However, numerous examples of applications advocate for the opposite conclusions. Generally, noise is essentially output noise and one has to use semidirected or undirected learning algorithms (for examples, see Chap. 2). It should be emphasized that, in a lot of published examples, the success of directed algorithms rely essentially on the regularity of the functions to be approximated and that the dumb predictor (as defined in Chap. 2) outperforms any directed learning algorithms.

For feed-forward networks, to the key problems that have to be faced are

- the selection of inputs,

- the selection of architecture (generally, the size of the hidden layer).

For recurrent networks, three additional questions must be addressed:

- the selection of the representation (input-output or state representation),
- the choice of the model order,
- the length of the sliding time window for backpropagation through time.

It may be quite useful to perform a linear identification (where structural tests are better understood) to select the order of the model. The selection of the truncation horizon for BPTT is also a tricky issue: theoretically the order of observability of the model is sufficient as a truncation horizon, practically computing time limits the size of the sliding time window.

Another problem for recurrent network learning is to capture long-range time dependency going backwards to the past. That issue is investigated in [Bengio 1994]. Nevertheless, long-range time dependencies are seldom considered in practical applications because the true physical processes are not steady along very long epochs: there exist slow drifts to cope with; adaptive methods (that have been developed here) are then used to update the model.

When one is facing big difficulties, it is advised to use directed and evolutive learning strategy, progressively increasing the time-depth of the learning process and using robust optimization methods. An efficient methodology for practical applications is the use of "gray box" modeling (presented in Chap. 2) to take advantage of all the available knowledge on the process to be modeled, the mathematical form of the model equations, the order of the model and so on. Thus, the designer has to address a smaller number of issues.

Of course data preprocessing, using first linear regression methods then using the residues to feed nonlinear learning algorithms, often improves the accuracy of nonlinear identification methods, since approximation problems are correctly decoupled and scaled.

Recurrent neural networks may be used to design controllers. That question will be addressed in the following chapter.

4.7 Appendix (Algorithms and Theoretical Developments)

4.7.1 Computation of the Kalman Gain and Covariance Propagation

Let us consider the Markov stochastic state model,

$$\boldsymbol{X}(k+1) = \boldsymbol{A}\boldsymbol{X}(k) + \boldsymbol{B}\boldsymbol{u}(k) + \boldsymbol{V}(k+1),$$

with the following measurement equation:

$$\boldsymbol{Y}(k) = \boldsymbol{H}\boldsymbol{X}(k) + \boldsymbol{W}(k).$$

We define $\hat{\boldsymbol{X}}(k)$ as the least squares optimal estimate, i.e., the linear regression of the random state vector $\boldsymbol{X}(k)$ onto the random vector of past measurements up to time k : $\boldsymbol{Y}(k) = [\boldsymbol{Y}(1); \ldots; \boldsymbol{Y}(k)]$. Let $\boldsymbol{\vartheta}(k+1)$ be the innovation at time $k+1$. It is defined by

$$\boldsymbol{\vartheta}(k+1) = \boldsymbol{Y}(k+1) - \boldsymbol{HA}\hat{\boldsymbol{X}}(k) - \boldsymbol{HBu}(k).$$

The innovation filter recursive equation is

$$\hat{\boldsymbol{X}}(k+1) = \boldsymbol{A}\hat{\boldsymbol{X}}(k) + \boldsymbol{Bu}(k) + \boldsymbol{K}_{k+1}\boldsymbol{\vartheta}(k+1)$$

where the innovation gain is inferred from the computation formula of linear regression:

$$\boldsymbol{K}_{k+1} = \mathrm{Cov}[\boldsymbol{X}(k+1), \boldsymbol{\vartheta}(k+1)]\mathrm{Var}[\boldsymbol{\vartheta}(k+1)]^{-1}.$$

P_k stands for the covariance matrix of the estimation error $\boldsymbol{X}(k) - \hat{\boldsymbol{X}}(k)$ and $\boldsymbol{P}^{\circ}_{k+1}$ stands for the covariance matrix of the prediction error

$$\boldsymbol{X}(k+1) - \boldsymbol{A}\hat{\boldsymbol{X}}(k) - \boldsymbol{Bu}(k).$$

Let us compute covariance of the prediction error. One obtains

$$\boldsymbol{X}(k+1) - \boldsymbol{A}\hat{\boldsymbol{X}}(k) - \boldsymbol{Bu}(k) = \boldsymbol{A}[\boldsymbol{X}(k) - \hat{\boldsymbol{X}}(k)] + \boldsymbol{V}(k+1).$$

Because $\boldsymbol{V}(k+1)$ is uncorrelated to $\boldsymbol{X}(k) - \hat{\boldsymbol{X}}(k)$, the prediction error covariance propagation equation is easily computed using a quadratic expansion,

$$\boldsymbol{P}^{\circ}_{k+1} = \boldsymbol{AP}_k\boldsymbol{A}^{\mathrm{T}} + \boldsymbol{Q}.$$

From the definition of innovation error,

$$\begin{aligned}\boldsymbol{\vartheta}(k+1) &= \boldsymbol{Y}(k+1) - \boldsymbol{HA}\hat{\boldsymbol{X}}(k) - \boldsymbol{HBu}(k)\\ &= \boldsymbol{H}\{\boldsymbol{A}[\boldsymbol{X}(k) - \hat{\boldsymbol{X}}(k)] + \boldsymbol{V}(k+1)\} + \boldsymbol{W}(k+1).\end{aligned}$$

The value of its covariance matrix is deduced in a similar way, expressed as a function of the prediction error at time k

$$\mathrm{Var}[\boldsymbol{\vartheta}(k+1)] = \boldsymbol{HP}^{\circ}_{k+1}\mathrm{H}^{\mathrm{T}} + \boldsymbol{R}.$$

Let us compute the covariance between the state $\boldsymbol{X}(k+1)$ and the innovation $\boldsymbol{\vartheta}(k+1)$,

$$\begin{aligned}&\mathrm{Cov}[\boldsymbol{X}(k+1), \boldsymbol{Y}(k+1) - \boldsymbol{HA}\hat{\boldsymbol{X}}(k) - \boldsymbol{HBu}(k)]\\ &\quad= \mathrm{Cov}\{\boldsymbol{AX}(k) + \boldsymbol{V}(k+1), \boldsymbol{HA}[\boldsymbol{X}(k) - \hat{\boldsymbol{X}}(k)]\\ &\qquad\quad + \boldsymbol{HV}(k+1) + \boldsymbol{W}(k+1)\}\\ &\quad= \mathrm{Cov}\{\boldsymbol{AX}(k), \boldsymbol{HA}[\boldsymbol{X}(k) - \hat{\boldsymbol{X}}(k)]\}\end{aligned}$$

$$+\,\mathrm{Cov}[\boldsymbol{V}(k+1), \boldsymbol{H}\boldsymbol{V}(k+1)+\boldsymbol{W}(k+1)]$$
$$= \boldsymbol{A}\mathrm{Cov}[\boldsymbol{X}(k), \boldsymbol{X}(k)-\hat{\boldsymbol{X}}(k)]\boldsymbol{A}^{\mathrm{T}}\boldsymbol{H}^{\mathrm{T}} + \mathrm{Var}[\boldsymbol{V}(k+1)]\boldsymbol{H}^{\mathrm{T}}.$$

Yet, from the correlation of $\hat{\boldsymbol{X}}(k)$ with $\boldsymbol{X}(k)-\hat{\boldsymbol{X}}(k)$, one gets

$$\mathrm{Cov}[\boldsymbol{X}(k), \boldsymbol{X}(k)-\hat{\boldsymbol{X}}(k)] = \mathrm{Var}[\boldsymbol{X}(k)-\hat{\boldsymbol{X}}(k)] = \boldsymbol{P}_k.$$

Therefore,

$$\mathrm{Cov}[\boldsymbol{Y}(k+1)-\boldsymbol{H}\boldsymbol{A}\hat{\boldsymbol{X}}(k)-\boldsymbol{H}\boldsymbol{B}\boldsymbol{u}(k), \boldsymbol{X}(k+1)]$$
$$= (\boldsymbol{A}\boldsymbol{P}_k\boldsymbol{A}^{\mathrm{T}}+\boldsymbol{Q})\boldsymbol{H}^{\mathrm{T}} = \boldsymbol{P}^{\circ}_{k+1}\boldsymbol{H}^{\mathrm{T}}.$$

Finally, we obtain,

$$\boldsymbol{K}_{k+1} = \boldsymbol{P}^{\circ}_{k+1}\boldsymbol{H}^{\mathrm{T}}[\boldsymbol{H}\boldsymbol{P}^{\circ}_{k+1}\boldsymbol{H}^{\mathrm{T}}+\boldsymbol{R}]^{-1}.$$

In order to iterate the algorithm, which is recursive, let us compute finally the covariance matrix of the estimation error at time $k+1$. From its value,

$$\boldsymbol{X}(k+1)-\hat{\boldsymbol{X}}(k+1) = \boldsymbol{A}[\boldsymbol{X}(k)-\hat{\boldsymbol{X}}(k)] + \boldsymbol{V}(k+1) - \boldsymbol{K}_{k+1}[\boldsymbol{Y}(k+1) - \boldsymbol{H}\boldsymbol{A}\hat{\boldsymbol{X}}(k) - \boldsymbol{H}\boldsymbol{B}\boldsymbol{u}(k)]$$
$$\boldsymbol{X}(k+1)-\hat{\boldsymbol{X}}(k+1) = (\boldsymbol{I}-\boldsymbol{K}_{k+1}\boldsymbol{H})\{\boldsymbol{A}[\boldsymbol{X}(k)-\hat{\boldsymbol{X}}(k)]+\boldsymbol{V}(k+1)\} - \boldsymbol{K}_{k+1}\boldsymbol{W}(k+1),$$

the expression of the covariance matrix can be derived,

$$\boldsymbol{P}_{k+1} = (\boldsymbol{I}-\boldsymbol{K}_{k+1}\boldsymbol{H})(\boldsymbol{A}\boldsymbol{P}_k\boldsymbol{A}^{\mathrm{T}}+\boldsymbol{Q})(\boldsymbol{I}-\boldsymbol{K}_{k+1}\boldsymbol{H})^{\mathrm{T}} + \boldsymbol{K}_{k+1}\boldsymbol{R}\boldsymbol{K}^{\mathrm{T}}_{k+1}.$$

4.7.2 The Delay Distribution Is Crucial for Recurrent Network Dynamics

In this chapter, we provided examples of recurrent neural networks. Most of them were of the input-output type, i.e., they were built from a feedforward neural network whose outputs are fed back to the input with a unit time delay. Other recurrent network models have been shown in this chapter (Hopfield and Elman models) and in Chap. 2 ("gray box" modeling taking into account algebraic and differential equations from prior knowledge for the network architecture).

Let us emphasize that for a recurrent neural networks, delay distribution has to be specified. If it is neglected, the network behavior is not properly defined. To illustrate this, Fig. 4.21 shows a comparison of the delay specification for a network without any closed circuit in the connection graph (feed-forward network) and a recurrent network with circuits in the connection graph.

Pictures (a) and (b) show the graph of an elementary feedforward network with four connections. In pictures (c) and (d), feedback was added. The

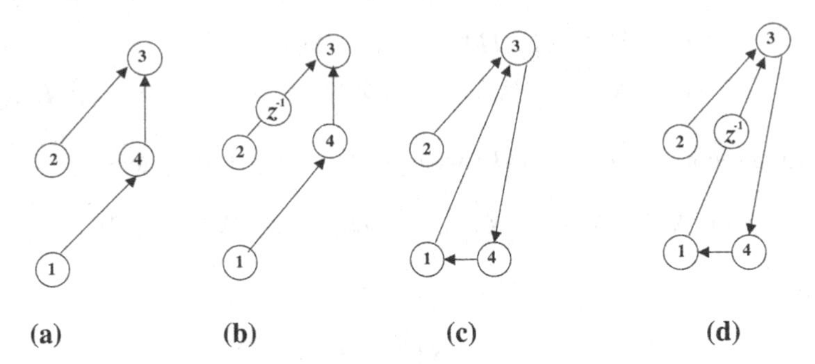

Fig. 4.21. The delays are crucial to update a neural network

connection weights are the same for cases (a) and (b) and for cases (c) and (d). Those two pairs of networks are not simple replication because a delay operator was added in the networks (b) and (d). We shall now investigate the consequences of that feedback, assuming that the network inputs are constant.

In case (a), at time 1, the state of unit 3 depends on the initial states of units 2 and 4, as the state of unit 4 is determined by the state of unit 1. At time 2, the state of unit 3 is determined by the state of units 2 and 4, hence by the states of units 1 and 2. In case (b), the state of unit 3 is only determined at time 2. At that time, its state value is similar to the state value of unit 3 in case (a).

Remark. In fact, open-loop feedforward neural networks that are fed by static inputs stabilize onto a final state, which depends only on the initial input state, whatever the distribution delay. Therefore, it does not depend on the update order; the units are assumed to be updated synchronously.

Moreover, the update order and the delay distribution are not taken into account in a layered network with feedforward information propagation and connections only from each layer to the following layers. In Fig. 4.14, even if the network is operated in an open-loop fashion with a connection of the state at time k towards the controller and the internal model, there is an ambiguity for the update order. In that case, the relevant rule to update units is to update synchronously the units of a same layer and sequentially along the information propagation direction. Thus, the units of the first hidden layer of the internal model have to wait, to be updated, for the controller to deliver the control signal that is an input of the internal model. That rule is even more important if the inputs are time-dependent.

In that case, one has to discriminate between the simulation time representation (one time step for operating the whole simulation block including the controller and the internal model) and the update steps of the different layers of the whole simulation block that are nested in one time step of the algorithm.

Consider now cases (c) and (d) of Fig. 4.21. Pictures are representing recurrent network architectures. In the two cases, the architectures are similar with respect to static characteristics and differ only by one delay, which has been added in case (d). At time 2, the state of unit 3 is different: in case (c), it depends on initial states of units 2 and 4, whereas, in case (d), it depends on initial states of units 2 and 1. This difference is propagated at the following time step to unit 4 then to unit 1 and so on. Finally, the network states are completely different.

For recurrent networks, the state of units do not stabilize across time even when the network is fed by static inputs. The state dynamics depend strongly on the delay distribution. The update order plays a key role in the dynamics.

References

A list of references is given at the end of Chap. 5 for Chaps. 4 and 5.

5

Closed-Loop Control Learning

M. Samuelides

In the previous chapter, we showed how to use training in order to model controlled dynamical systems, with emphasis on neural modeling. This chapter extends that presentation to the problem of designing a closed-loop control law by training. Nonlinear control has been a growing field during the past twenty years. However, there is no methodology based on first principles, in contrast to linear control. A number of practical methods have been proposed, starting from various points of view. Some results are essentially theoretical, addressing the problems of controllability, existence of stabilizing control law, and validity of linearization techniques. Such results are beyond the scope of this book.

However, we will recall some elements of control theory in the following section, emphasizing the connections between linear and nonlinear control laws. Actually, "neural" techniques extend "classical" techniques of nonlinear control to systems, which has been previously modeled using neuronal identification and training. Those techniques are described in the section "Design of Neural Control by Inverse Model", where several techniques are successively studied: straightforward inversion, simple but too often inefficient, reference model control, which is more frequently used, and recurrent models whose implementation may be more difficult.

Further sections are devoted to optimal decision problems in the classical framework of dynamic programming (section "Dynamic Programming and Optimal Control") and to its counterpart in learning theory (section "Reinforcement Learning and Neuro-Dynamic Programming"). Those techniques were in existence long before neural networks became popular; they addressed the problems of control in discrete spaces. Neural networks provided good approximations of those methods. Meanwhile, reinforcement learning can be applied now to continuous state spaces avoiding "combinatorial explosion" which was a drastic limit to the field of application of classical reinforcement learning. That set of more modern techniques was termed recently "Neuro-Dynamic Programming".

5.1 Generic Issues in Closed-Loop Control of Nonlinear Systems

5.1.1 Basic Model of Closed-Loop Control

The principle of *closed-loop* control or *feedback* control is to cancel the effects of disturbances on the reference dynamics of the system by *closing the control loop*, i.e. by establishing a functional dependence of the command signal on the state of the system. That is achieved by implementing a control law into a controller. A controller is a device whose input is the state of the system to be controlled (or more generally the output of that system if its state is not completely observed). Then the controller determines the value of the control signal that will be used to control the original system at the next instant. Let us consider a dynamical system defined in Chap. 4,

$$\boldsymbol{x}(k+1) = f[\boldsymbol{x}(k), \boldsymbol{u}(k)],$$

where $\boldsymbol{x}(k)$ is the state vector of the model at instant k, and $\boldsymbol{u}(k)$ is the vector of control signals at instant k. The controller determines the value of that control signal vector from the state vector according to a function ψ,

$$\boldsymbol{u}(k) = \psi[\boldsymbol{x}(k)].$$

That function is called the control law.

The simpler purpose that can be assigned to a control system is to keep the system in a desired state in spite of any disturbance (then the control is said to reject disturbances): a servo-system is thus designed. Another possible purpose is to keep the state trajectory of the controlled system as closed as possible from a desired state trajectory: a tracking system is thus designed. In those cases, which are very common in applications, the desired state is called setpoint in servo control, and reference trajectory in tracking; naturally enough, the control law is based on the difference between the setpoint or reference trajectory and the actual state.

Such a closed-loop system is shown in Fig. 5.1.

When the state is not completely known, the control can only be a function of the observations. Therefore, for such a system, the relevant equations are the state equation, the measurement equation and the control law,

$$\begin{aligned} \boldsymbol{x}(k+1) &= f[\boldsymbol{x}(k), \boldsymbol{u}(k)] \\ \boldsymbol{y}(k) &= g[\boldsymbol{x}(k)] \\ \boldsymbol{u}(k) &= \psi[\boldsymbol{y}(k)]. \end{aligned}$$

Clearly, a controlled dynamical system with its control law is actually equivalent to an autonomous dynamical system. Therefore, its stability must be investigated. If some stochastic process is added in the equations to model

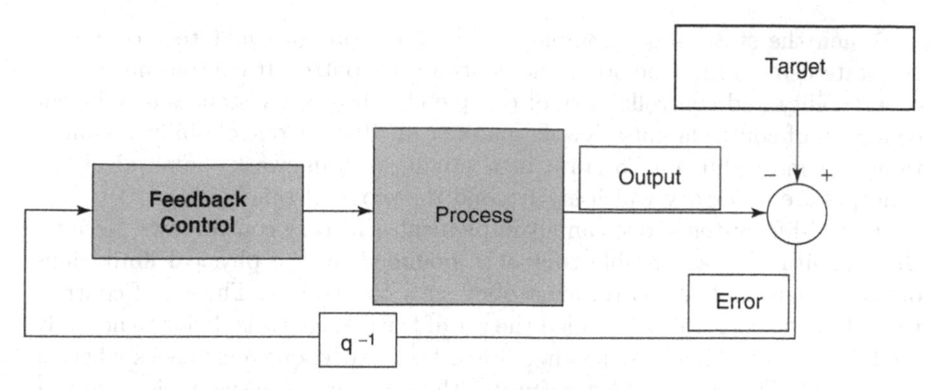

Fig. 5.1. Principle of a closed-loop control

state noise or observation noise, then the stability of the controlled system is again a crucial issue.

In the previous paragraphs, we just mentioned control laws that depend on the current state or on the current observation only. Such a control law is called a static control law. Actually, all the past information can be used for implementing the current control. Such a control law is called a dynamic control law. In practice, however, the complexity of the computation of the control law is must obey stringent time constraints. The computation must be performed during the sampling period of the controller to implement the control in real time.

5.1.2 Controllability

The purpose of a control law cannot always be achieved. *Controllability* is the property of the system whereby it can reach any target if it is provided with an appropriate control law. Note that the simplest dynamic controlled systems, such as linear systems, are not necessarily controllable when the state dimensionality is larger than 1.

Consider for instance the following linear system:

$$\boldsymbol{x}(k+1) = \boldsymbol{x}(k) + \begin{pmatrix} 1 \\ 0 \end{pmatrix} u(k).$$

Its order is 2 and the control is scalar. It is not possible to change the second component of the state with the scalar control. On the other hand, it is very easy to show that the following linear system enjoys the controllability property,

$$\boldsymbol{x}(k+1) = \begin{pmatrix} 1 & 1 \\ 0 & 1 \end{pmatrix} \boldsymbol{x}(k) + \begin{pmatrix} 1 \\ 0 \end{pmatrix} u(k).$$

Controllability can readily be expressed for a linear system: in order to reach any state, it is necessary and sufficient to be able to reach the state zero from any initial state [Kwakernaak et al. 1972].

When the state is not completely observed, one has first to reconstruct the state using a filter before implementing the control. It can be shown that observability and controllability of completely observed system is a sufficient condition of controllability [Kwakernaak et al. 1972]. Controllability assumptions are more difficult to state in a nonlinear framework. Some algebraic concepts are necessary which are beyond the scope of this book.

In real-life systems, one cannot implement arbitrary control laws, because the magnitude of acceptable control is bounded by the physical limitations of the actuators. The control must obey such constraints. The set of controls that obey the constraints is called the set of feasible controls. Prior to actually applying a control law designed in a linear framework, one must check whether it is feasible. If the control law saturates the actuators, the system is no longer linear.

5.1.3 Stability of Controlled Dynamical Systems

The most important property of a control law is that it guarantees the stability of the controlled dynamical system. We explained in the previous chapter that a controlled dynamical system with a closed-loop control law behaves just like a usual dynamical system without any control. Let us recall some definitions about stability of discrete-time nonlinear dynamical systems. In this section, discrete-time dynamical systems are considered, with the following state equation:

$$\boldsymbol{x}(k+1) = f[\boldsymbol{x}(k)].$$

A state $\boldsymbol{x}^*$ such that $f(\boldsymbol{x}^*) = \boldsymbol{x}^*$ is called an *equilibrium state*. $\boldsymbol{x}^*$ is also said to be a *fixed point* of f.

An equilibrium $\boldsymbol{x}^*$ is said *stable* if

$$\forall\, \varepsilon,\ \exists\, \eta,\ \|x\,(0) - x^*\| \leq \eta \Rightarrow \forall\, k,\ \|x\,(k) - x^*\| \leq \varepsilon.$$

An equilibrium $\boldsymbol{x}^*$ is said *asymptotically stable*, with an attraction basin Ω, if for any initial condition in Ω, the state trajectory originating from that initial condition reaches the fixed point $\boldsymbol{x}^*$.

The stability properties of dynamical linear systems $\boldsymbol{x}(k+1) = \boldsymbol{A} \cdot \boldsymbol{x}(k)$ can be easily derived from the spectral properties of matrix $\boldsymbol{A}$. The point 0 is a fixed point of the linear system. If the eigenvalues of $\boldsymbol{A}$ are strictly included in the open unit disc, the equilibrium is stable and asymptotically stable. If there exists an eigenvalue, whose module is strictly larger than 1, then 0 equilibrium is neither stable nor asymptotically stable. Critical cases of eigenvalues of module equal to 1 deserve a specific analysis.

That simple characterization of linear dynamical systems is the basis of the methodology to build control laws for linear controlled dynamical systems by locating the poles of transfer functions [Kwakernaak et al. 1972]. That methodology is traditional in control theory, and is very popular in real world applications. It was first invented in the framework of single variable systems with the intensive use of Laplace transform and it was easily extended to multivariable systems. Nevertheless, those basic methodologies have no straightforward extensions for nonlinear controlled dynamical systems. We just mention them here for the sake of completeness.

In the asymptotically stable case, the stability of equilibrium of nonlinear dynamical systems can be derived from the stability of the linearized system around that equilibrium. If $\boldsymbol{x}^*$ is a fixed point of the dynamical system $\boldsymbol{x}(k+1) = f[\boldsymbol{x}(k)]$, the «linearized dynamical system around $\boldsymbol{x}^*$» is the following dynamical system $\boldsymbol{x}(k+1) = \nabla f_{x^*}[\boldsymbol{x}(k) - \boldsymbol{x}^*] + \boldsymbol{x}^*$. That system is linear with respect to the fixed point $\boldsymbol{x}^*$; ∇f_{x^*} is the Jacobian matrix of the partial derivatives of f in $\boldsymbol{x}^*$. Then the following result holds:

Linearization Theorem. Provided that the linearized system around $\boldsymbol{x}^*$ is asymptotically stable, $\boldsymbol{x}^*$ is a stable and asymptotically stable fixed point of the nonlinear dynamical system.

With the linearization, the transfer function of the linearized system becomes a convenient tool for the analysis and synthesis of control laws for nonlinear systems [Slotine et al. 1991]. Specifically, a linearization theorem of controlled dynamical system states that if the linearized system is controllable, that system is locally stabilized when the closed loop control law of the linearized system is applied to the original nonlinear system [Sontag 1990].

The *Liapunov function* method [Slotine et al. 1991], which is a straightforward generalization of stability concept of dissipative physical systems, is a general method of investigation of the stability of equilibrium of nonlinear dynamical systems.

In spite of the important linearization theorem we have just mentioned, numerous difficulties must be overcome when studying the stability of nonlinear systems:

- The dynamics of a nonlinear system may exhibit several fixed points with different stabilities: the linearization theorem is a local theorem which gives no information on the size of attraction basins of asymptotically stable fixed points;
- Dynamical attractors may exist, which confer a global stability to the nonlinear system even if there is no stable fixed point: the simpler example of such attractors is the stable limit cycle. Such an attractor exists in the Van der Pol oscillator, which was described in the previous chapter.

When noise is considered, the study of stability of systems is completely changed. In the previous chapter, in the section devoted to the modeling of dynamical systems, that the stochastic equivalent of a dynamical deterministic system is a Markov process, and that the stochastic equivalent of equilibrium is the invariant measure of this process (see definition in Chap. 4). When a linear dynamical system is disturbed by gaussian white noise, that probability describes the asymptotic fluctuations of the state around the zero fixed point of the dynamical system in the absence of disturbances. In the general case of a nonlinear dynamical system with several attractors, the situation is still more intricate. Actually, fluctuations occur "almost surely", which enable the system to pass from an attraction basin of the deterministic dynamical system to another. The theory of "Large deviations" allows provides a tool for estimating the transition probability of those events [Benveniste et al. 1987; Duflo 1996].

However, in this chapter (and in most applications), we are interested in stabilizing a fixed point or in tracking a reference trajectory; therefore the investigation of the coexistence of several dynamical attractors is not really relevant.

5.2 Design of a Neural Control with an Inverse Model

5.2.1 Straightforward Inversion

The simplest method, to design a neural control law from a neural model of a controlled dynamical system that was identified as an open-loop neural network, is the straightforward inversion of that model. The control system is just the inverse of the model of the process. If that model is nonlinear, its inverse is nonlinear too, hence can be implemented as a neural network. The training and operation of such neural control are demonstrated in Fig. 5.2.

In that figure, a neural network that computes the control signal is added to the neural model of the process. That neural controller is a feedforward network whose inputs are the state and, optionally, the desired state (at the next time) if the task is the tracking of a state trajectory. Otherwise, the only input of the controller is the current state of the system (at time k). The output of the neural controller is the control signal at time k. That control is fed to the control input of the model of the process during the training phase, and to the process input during the operation phase.

The set (controller + model) is a feedforward neural network whose input is the state at the next time step. Training is performed by minimizing the difference between the reference state or setpoint and the network output. The only parameters subject to change are the controller's parameters (weights and bias). The model parameters stay unchanged during the training process.

The cost function is usually the squared distance between the desired output and the measured output. If constraints are imposed to the control signal, they can be embedded into the controller. For instance, if the admissible control is bounded, those bounds can be embedded into the activation sigmoid function of the output neuron of the controller. Alternatively, a penalization that grows drastically when the constraint is violated may be added to the error cost function.

That straightforward methodology gives good results for simple problems, where the objective is a static function of the current state. If the objective

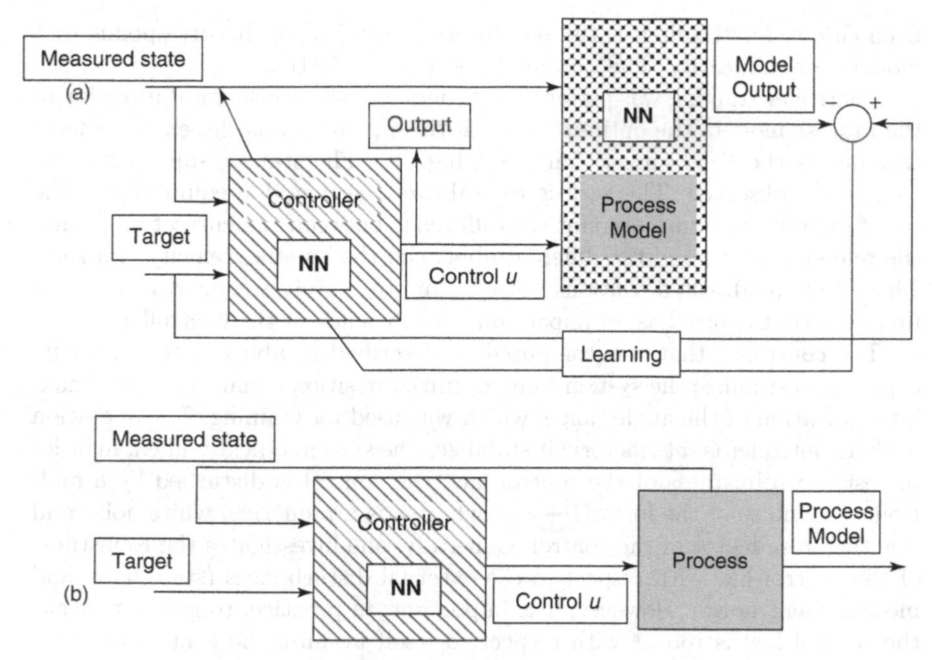

Fig. 5.2. (a) Principle of training and (b) operating a closed loop-neural control which was designed by inverting the model

is a function of the final state in a finite horizon problem, or if it is related to the whole state trajectory, the above straightforward method cannot be used. Time unfolding of the global network (controller + model) can be considered and time delayed back-propagation can be used for training. That methodology will be discussed below. Even if the objective is a static function of the current state, straightforward static training of a neural controller does not always provide satisfactory results.

In addition, that method is not robust with respect to the modeling errors the control is computed from the model, so that it can be inaccurate if the model itself is not an accurate approximation of the process. The internal model control method that will be considered below can overcome that problem.

5.2.1.1 Illustrative Example: The Inverted Pendulum

The following numerical results show the limits of straightforward inversion of the process model to build the control law on an elementary example. The problem is stabilization of the inverted pendulum. That controlled dynamical system was introduced in the previous chapter, in section "Example: the inverted pendulum". The neural model has been easily identified with a good accuracy from the state equation. The range of angle variation, which has been chosen for training, is the real interval $[-\pi/5, \pi/5]$. It corresponds to a moderate nonlinearity. The sampling frequency is 50 Hz.

Controller training was performed through the straightforward inversion of the process model. The optimization algorithm, which was chosen to perform training, is the BFGS algorithm (see Chap. 2). The state is supposed to be completely observed. The task is to stabilize the unstable equilibrium. The cost function takes into account the difference between the current angle and the reference angle and the difference between the angular velocity and zero. Those two quadratic deviations must be appropriately weighted in the cost function; that choice has an impact on the efficiency of the controller.

The controller that was computed as described as above is tested for its capacity to stabilize the system from an initial position equal to half the maximal deviation of the angle range, which was used for training. The operation of that controller is satisfactory: it stabilizes the system quickly. Then, in order to test the robustness of the control law, the control is disturbed by a multiplicative noise of the form $(1 + \kappa\varepsilon)$, where ϵ is a numerical white noise and κ is the noise factor of the control. Generally, one investigates the robustness of the control law with respect to the external disturbances (state noise and measurement noise). However, it is important, in practice, to guarantee that the control law is robust with respect to itself because the control law may be implemented with errors (numerical roundoff errors, electro-mechanical errors for the servomotors, etc.). The efficiency of the controller depends on the choice of the cost function, as shown in the following figures.

In the first experiment, the weight of the velocity deviation in the cost function is larger to the angle deviation weight. Figure 5.3 shows a typical trajectory of such a controlled system. The system is stabilized only if the noise factor is smaller than 0.5. When the noise factor is larger, generally the trajectory leaves its viability domain during the experiment (20 seconds). The velocity is stabilized around the reference as shown on Fig. 5.3. The stabilization of the position is slower and equilibrium is not reached within the allotted time.

In the second experiment, the weight of the velocity deviation is smaller than the weight of the angle deviation. The controller is more robust with respect to the control noise ($\kappa = 3$) as shown in Fig. 5.4.

If the weight of the angle deviation is chosen smaller than the weight of the velocity deviation in the cost function, the system is ill-stabilized as soon as a control noise exists in the controller. Thus, straightforward model inversion method assumes that empirical knowledge of the system is available in order to choose a relevant cost function. The robustness of the optimal controller relies heavily on that choice.

To summarize, straightforward model inversion method is a simple solution. However, a deep knowledge of the system to be controlled may be needed to implement it efficiently. It is necessary to check its robustness versus various disturbances and modeling errors. Quite often, improvements of that method

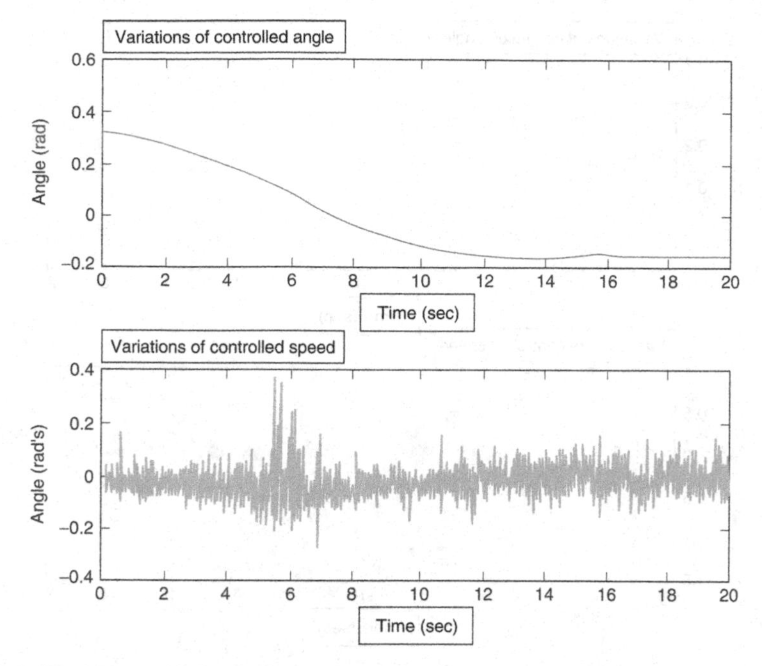

Fig. 5.3. Trajectory of the stabilized system with a noise factor of 0.5. (Straightforward model inversion control learning with larger weighting of the speed deviation in the cost function)

are necessary in order to obtain satisfactory results. We are going to review some of those improvements.

5.2.2 Model Reference Adaptive Control

In the method called MRAC, for model reference adaptive control, prior knowledge of the system is used to build the control law [Rivals et al. 2000]. In that method, the cost function does not push the system directly to the desired target, but it is designed to force the closed-loop controlled system to track a reference trajectory. The choice of that reference trajectory is made from prior knowledge of the controlled system, especially on the actuator abilities.

Actually, there is always an implicit reference model; in the straightforward inverted control of the previous section, the reference model is just a single time lag.

Figure 5.5 shows the general organization of the training process of a neural controller with a reference model.

The Reference Model method turns out to be valuable in numerous applications to real world problems. It is generally used to improve the performance of classically controlled dynamical systems. Whenever possible, the reference trajectory is chosen as the trajectory of a linear system with a critical damping

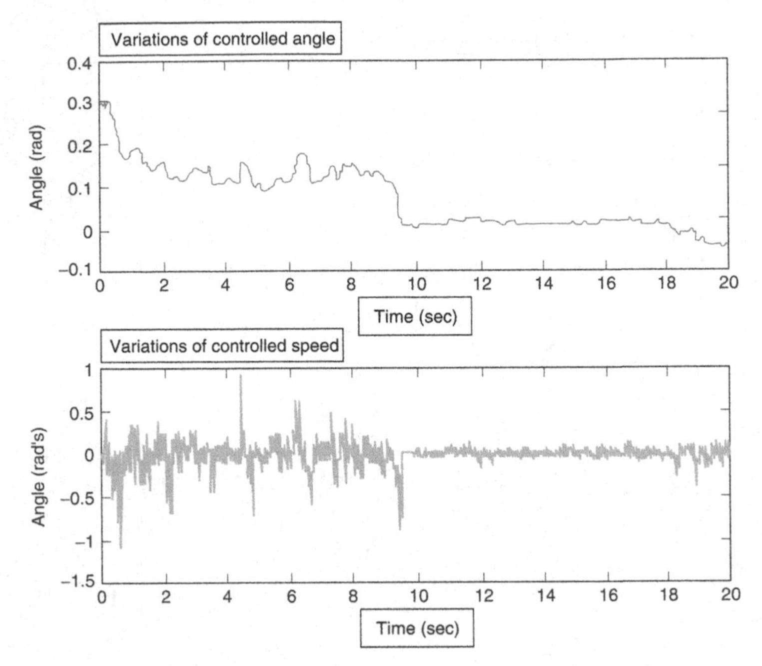

Fig. 5.4. Trajectory of the stabilized system with a noise factor of 3. (Straightforward model inversion control learning with larger weighting of the angle deviation in the cost function)

with the desired time constant. On the previous example of inverted pendulum, that method gives much better results in the neighborhood of equilibrium with the same model. A typical trajectory of the closed-loop controlled system is shown on Fig. 5.6 with a control noise factor equal to 3.

Another neural controller design method is proposed in [Levin 1993]. That method is somewhat similar to the previous method; it consists in choos-

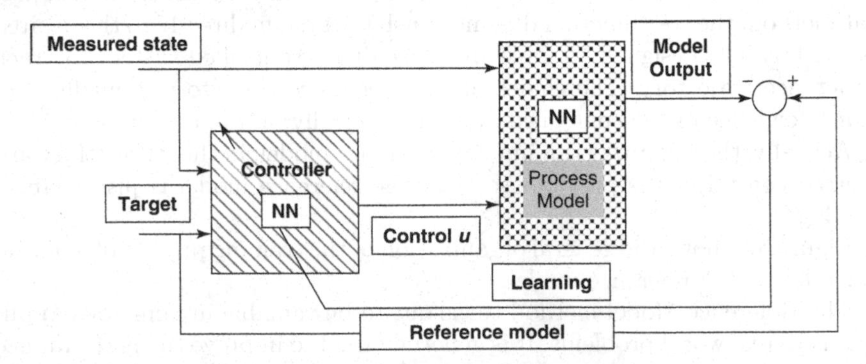

Fig. 5.5. Learning process of a reference model controller

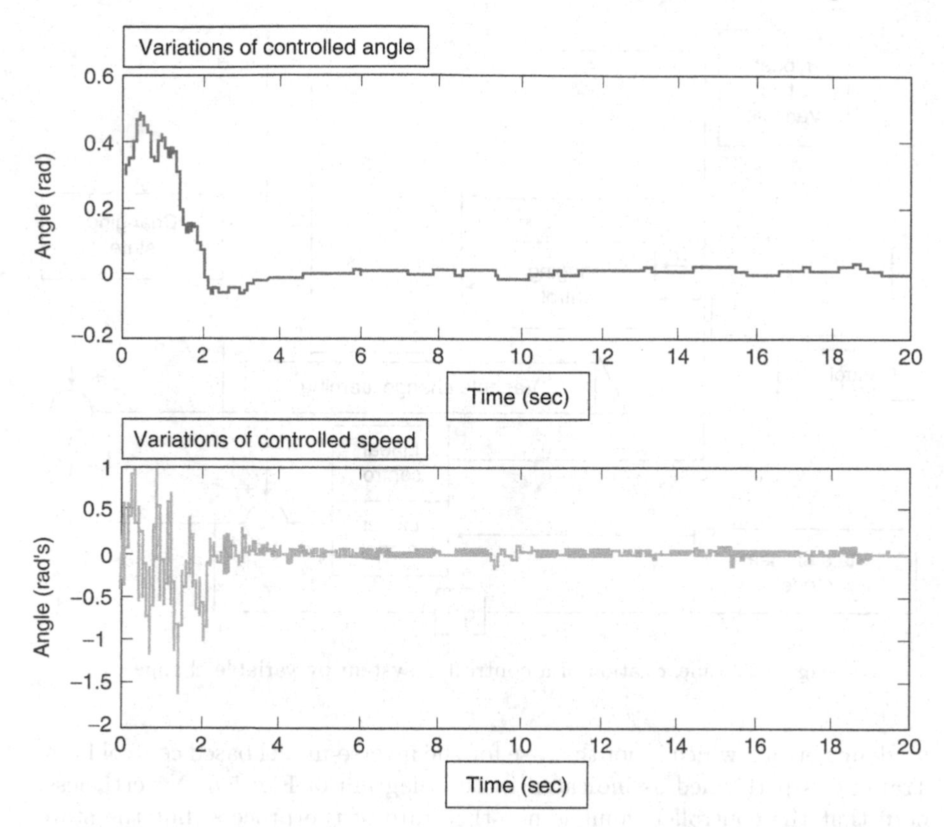

Fig. 5.6. Trajectory of the controlled system with a control noise factor equal to 3 (Reference Model method)

ing, as the reference, the linearization of the controlled dynamical system in the neighborhood of the stabilization target. Neural training is then used to compute nonlinear variable change of state and control to reduce the nonlinear system to its linearization. Such design is displayed on Fig. 5.7.

We point out that the state change deals only with current state but that the control change deals with control *and* state.

5.2.3 Internal Model Based Control

As emphasized earlier, designing nonlinear methods as extensions of methods that are efficient in linear modeling and control is often fruitful. «Neural» control with internal model is a good instance of such an approach. Figure 5.8 is the diagram of a control with internal model (and explicit reference model). In addition to the controller, a model of the process (called «internal model») is involved in the internal model control law. The modeling error is used to change the reference, so the controlled system is robust with respect to the

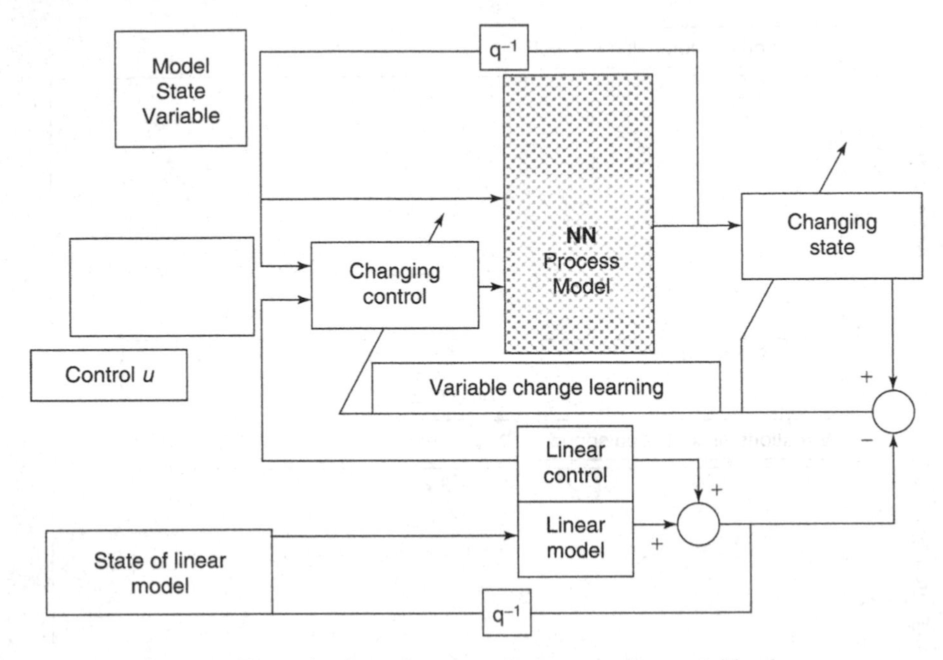

Fig. 5.7. Linearization of a controlled system by variable change

modeling errors, which is not the case for the inverse-model based control laws. Training is performed as indicated in the diagram of Fig. 5.5. Nevertheless, note that the controller input is not the state of the process, but the state of the *internal model*. Therefore, during training, it is advantageous to use relevant values of the state variable of that model.

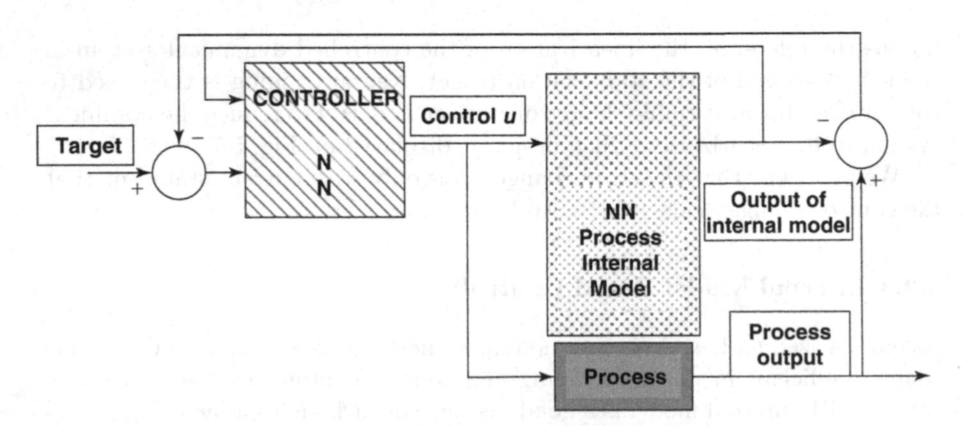

Fig. 5.8. Internal Model control law

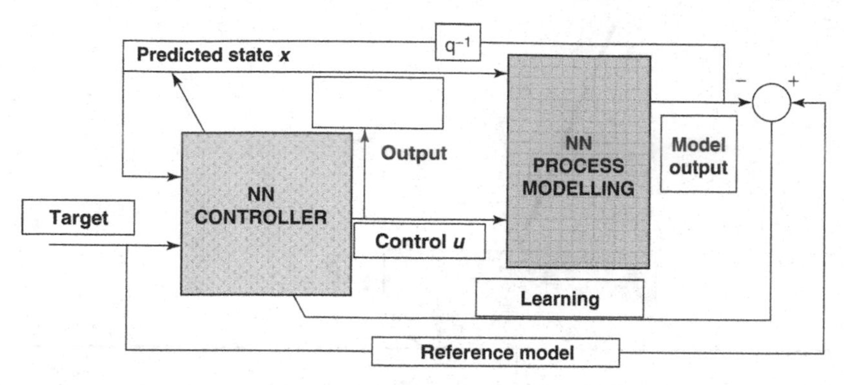

Fig. 5.9. Learning a closed-loop control law by the dynamics model inversion using recurrent back-propagation

That control methodology spurred the development of numerous practical applications. A detailed description of an application of that technique in order to perform the automatic piloting of a land vehicle is displayed in [Rivals 1995].

In the linear domain, predictive control is also very powerful and is commonly used in process technology (oil industry, etc.). It has a natural neural extension that gave very good results in process control [Grondin 1994].

5.2.4 Using Recurrent Neural Networks

We just saw the limits of building a control law by a straightforward inversion of the system model. Model Reference method allows taking advantage of available heuristic or analytical knowledge about system dynamics. When no such knowledge is available, the dynamics of the system must be taken into account during training. That leads to using recurrent neural networks and dynamical backpropagation as described in Chaps. 2 and 4. The diagram that is represented in Fig. 5.9 shows how to design such a recurrent neural.

In that case, the global network is made of the neural controller and of the system model. The input of the global network is the current state, and its output is the state at the next time. The global network is turned into a recurrent network by adding the state feedback to close the loop.

5.2.4.1 Using a Recurrent Neural Network to Control a Partially Observed Dynamical System

We will end this section on neural control design by describing a major work by industry researchers that was mentioned in previously in the context of Kalman filters and recurrent networks [Puskorius et al. 1994]. The authors study several examples of stabilization of nonlinear controlled dynamical systems, where the state is not completely known (partial and noisy observation). That paper presents several examples selected from real industry problems (biochemical reactor with limit cycles, rotation speed of a car engine).

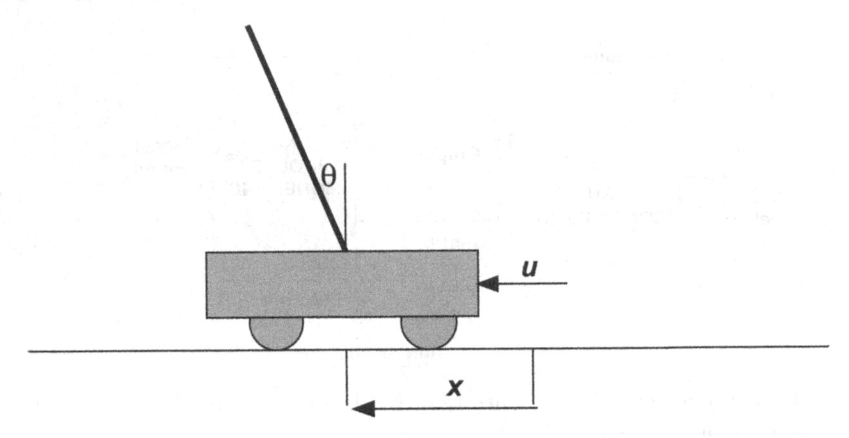

Fig. 5.10. The pole-cart system

Moreover, a difficult variant of the inverted pendulum stabilization problem is studied, where the inverted pendulum is set on a cart and is controlled by moving the cart. That problem is known as the pole-cart problem. It is a classical benchmark of nonlinear dynamic system stabilization. A diagram is shown on Fig. 5.10.

The control is a force, which is applied to the cart. That actuator changes the state of the cart. Then the inertial force changes also the state of the pendulum, which is nonlinearly coupled to the state of the cart. There are 4 state variables: the cart position $\boldsymbol{x}$, the pendulum angle $\boldsymbol{\theta}$ and the associated velocities. The observable variables are usually the two position variables. The objective consists in stabilizing the pole. Meanwhile, the cart has to be kept within a given range around its central position.

The state is not completely known. Therefore, the authors use an Elman recurrent network to identify the system (Elman recurrent networks are described in Chap. 4 in the section devoted to recurrent networks). The diagram of that kind of network is shown on Fig. 5.11.

The global network is made of the controller and the system model. Two closed loops are embedded in that network: an external feedback, feeding back the measurements to the controller, and an internal recurrence of hidden layer that models the unknown state dynamics. The neural controller itself is a recurrent network. Its input is the reference and the two observed variables. Its hidden layer has with six neurons and its output is a self-recurrent neuron. The training of the recurrent networks is performed by an adaptive Decoupled Extended Kalman Filter (DEKF) that was presented in the previous chapter, in the section devoted to Kalman filtering.

The problem is made more difficult, adding control noise shown previously in the stabilization of inverted pendulum. The above technique allows the stabilization of the system in various experimental conditions.

5.3 Dynamic Programming and Optimal Control

5.3.1 Example of a Deterministic Problem in a Discrete State Space

Let us return to the simple example of controlled dynamical system that is shown on Fig. 4.1 of previous chapter. That example was described at the beginning of the section ≪Formal definition and examples of discrete time controlled dynamical systems≫. In order to define a control problem, we have to define the criterion as a cost function to minimize. In the considered example, it is possible to choose a location in the labyrinth as a target to reach as soon as possible. In that case, we will associate to each triple (current state, current action, next state) a unit cost, except for the triple whose next state is state 35 (the target): that triple will enjoy a high negative cost $-A$ (≪reward≫).

The problem of optimal control consists in designing a closed-loop control law. In the context of operational research for discrete time and discrete state space, the terms policy, or strategy are preferred. It is a function from state space $\boldsymbol{E}$ to the control set (or action set) $\boldsymbol{A}$, which associates an action to each current state. A couple, which consists in one state and one action that can be carried out from that state, is called a feasible (state-action) couple.

Actually, for finite horizon problems, it is natural to consider nonstationary policies: if we are traveling in a dangerous country at the beginning of the day, we surely choose to advance as quickly as possible. Conversely, at the end of the day, we rather choose to move towards a safe place to spend the night. In a given location, the two directions are generally not the same. Therefore, in finite horizon problems, nonstationary policies must be considered, which are functions of the current time and of the current state and which take their values in the set of feasible actions.

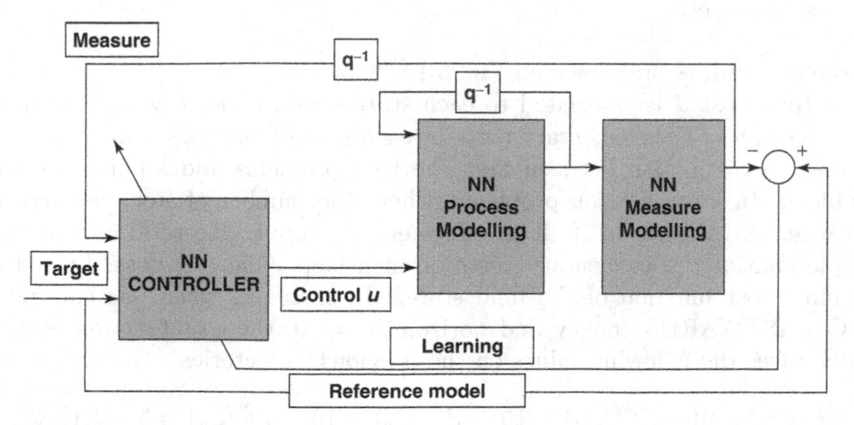

Fig. 5.11. Building up a closed-loop control law using recurrent back-propagation through an Elman network

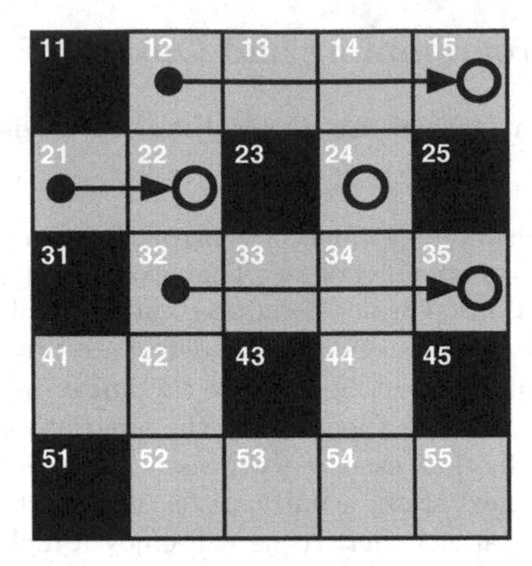

Fig. 5.12. Diagram of labyrinth from Fig. 4.1 with trajectories that are associated to constant policy GO EASTWARDS

In our simple example, the set of feasible actions is always N, S, E, W (north, south, east, west) and does not depend on the current state. A dynamical system is associated to a given policy. If this policy is stationary, the dynamical system is autonomous. Thus, in our example, consider the stationary constant policy «GO EASTWARDS», which associates E action to any current state. State trajectories of the associated dynamical system are

$\omega_1 =$ ((12, E), (13, E), (14, E), (15, E), (15, E) ...) trajectory coming from initial state 12,
$\omega_2 =$ ((21, E), (22, E), (22, E), ...) trajectory coming from initial state 21,
$\omega_3 =$ ((24, E), (24, E),...) trajectory coming from initial state 24,
$\omega_4 =$ ((32, E), (33, E), (34, E), (35, E),(35) ...) trajectory coming from initial state 32, etc.

Those trajectories are shown on Fig. 5.12.

A total cost J is associated to each state-action trajectory. In principle, it is the sum of the elementary costs of each step of the trajectory. One has to make a distinction between finite horizon problems and infinite horizon problems. In finite horizon problems, where the number of steps is fixed in advance, for instance to N, it is sufficient to compute the total cost as the simple sum of the elementary costs of each step. One can possibly add a terminal cost function of the final state. For instance, when one considers "GO EASTWARDS" policy and horizon $N = 10$, the cost function is J^N, which takes the following values on the previous trajectories

$$J^N(\omega_1) = 10, \quad J^N(\omega_2) = 10, \quad J^N(\omega_3) = 10, \quad J^N(\omega_4) = 3 - 7A \ldots$$

in the case without terminal cost.

When modeling that example, it is more natural to take a unit cost for each (state-action) couple and to choose a terminal cost equal to $-A$ for the target 35 and to A for any other terminal state. Thus, one gets the following values for the previous trajectories:

$$J^N(\omega_1) = 10 + A, \quad J^N(\omega_2) = 10 + A,$$
$$J^N(\omega_3) = 10 + A, \quad J^N(\omega_4) = 10 - A \ldots.$$

Unfortunately, the horizon is not precisely known in most applications. Therefore, infinite horizon problems are often considered. In those problems, it is not always possible to define the total cost as the sum of elementary costs for each step. Actually, the sequence of elementary costs may diverge for an infinite number of steps. There are several possible ways of defining the total cost of an infinite length trajectory.

One can define, when it exists, the limit of the average cost on the N first steps of the trajectory when N goes to infinity. In our simple problem, that solution does not make sense: it amounts to assign the cost $-A$ to each trajectory ending on the desired target, and the cost 1 to any other trajectory. The trajectories that reach the desired target quickly are not favored.

When the problem is to reach a specified target within a finite number of steps, it is possible to take the effective sum of the elementary costs as the total cost just as in finite horizon problems. It is the case in our example.

In the general case, we suggest taking the *discounted cost* as a global criterion to minimize. That cost function is inspired from economics, where future costs are discounted by a discount rate α here $0 < \alpha < 1$.

Thus, in our problem, for an infinite horizon model, we have

$$J^\alpha(\omega_1) = J^\alpha(\omega_2) = J^\alpha(\omega_3) = 1 + \alpha + \alpha^2 + \cdots = \frac{1}{1-\alpha}$$
$$J^\alpha(\omega 4) = 1 + \alpha + \alpha^2 - A\alpha^3 - A\alpha^4 - \cdots = 1 + \alpha + \alpha^2 - \frac{A\alpha^3}{1-\alpha}.$$

Thus, shorter trajectories are favored.

The problem consists in finding an optimal policy π_* such that the total cost of the state-action trajectory associated to that policy is minimal for any initial state.

5.3.2 Example of a Markov Decision Problem

A Markov decision problem (MDP) is the generalization to a stochastic framework of the previous problem. Randomness is introduced in the state evolution model and in costs. The total cost is then a random variable $\mathcal{J}$. We have to minimize a functional, which is just the mathematical expectation[1] of that random variable $J = E(\mathcal{J})$.

[1] Let us recall that the mathematical expectation of a random variable is the average of this random variable for its probability distribution. As dynamics is considered here, the probability law is defined on the set of trajectories.

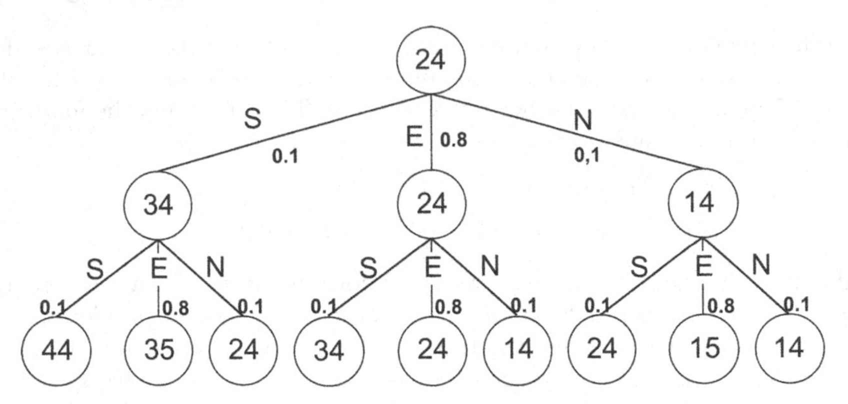

Fig. 5.13. Tree of the 2-step trajectories obtained from initial state 24 in the labyrinth of Fig. 4.1 with constant policy GO EASTWARDS

For instance, in our example, we can consider that there is some control noise: the evolution does not associate to each (state-action) couple a single state, but it associates a random variable that takes its value in the state space. For instance, we select the nominal state at next time with probability 0.8 and the two neighbor states with probability 0.1. For each transition, the random variables are independent.

Therefore, when the constant policy GO EASTWARDS is applied, a Markov chain is generated. Figure 5.13 shows the trajectories of horizon 2 of that Markov chain from the initial state 24 on a ternary tree.

That tree consists in $3^2 = 9$ trajectory sections. The result of the first transition that is associated to the (state-action) couple (24, E) is a random variable that takes the 3 values 14, 24, 34 in the state space with the respective probabilities 0.1, 0.8 and 0.1. At the next step, according to the current state, a transition is performed from one of the (state-action) couples (14, E), (24, E) or (34, E). For instance the result of the transition, which is associated to (14, E) is a random variable. It is independent from the previous one and takes values 24, 15, 14 with probabilities 0.1, 0.8 and 0.1 respectively.

The probability of an N-horizon trajectory (i.e. with N transitions) is computed by multiplying the elementary probabilities of the transitions for each step. For instance, the probability of the 2-trajectory ((24, E), (14, E), (24)) is equal to $0.1 \times 0.1 = 0.01$. The probability of the interesting 2-trajectory ((24, E), (34, E) (35)) is equal to $0.1 \times 0.8 = 0.08$. The valuation of a policy involves the computations of the probability of every trajectory.

Therefore, even more clearly than in the deterministic problem, it is impossible to enumerate all the trajectories in order to compute the policy cost, because combinatorial explosion makes it quickly intractable. Appropriate methods are needed to value a policy. Those methods will be developed in a further section, which will be dedicated to reinforcement learning and neuro-dynamic programming. We are looking for an algorithm that enables to value each (state-action) couple but not each feasible (state-action) trajectory; otherwise, the size of the data will grow exponentially with time.

One may also choose to randomize the elementary costs for each transition. This generalization is very easy to address when the criterion is the expected cost because the random cost of a transition is immediately replaced by its expectation.

5.3.3 Definition of a Decision Markov Problem

5.3.3.1 Controlled Markov Chain

The previous example is formalized with the following definition, which is limited to the case of finite state space and action set for the sake of simplicity. A Markov decision problem (MDP) consists in the following ingredients: a control Markov chain, an elementary cost function, a horizon length and, possibly, either a terminal cost (if the problem is a finite horizon problem), or a discount rate (if the problem is an infinite horizon problem).

We met previously the concept of controlled Markov process, which is the stochastic analog of a controlled dynamical system. Let us give a precise definition.

A *controlled Markov chain* consists in the following ingredients: a state space E, an action set A, a subset $\mathcal{A} \subset E \times A$ of the feasible (state-action) set and an application p from $\mathcal{A}$ into the set of probabilities upon E. That application takes as input any feasible (state-action) couple (x, u) and returns the probability denoted $P_u(x, y)$ of going to the state y when action u is performed from state x.

Remark. P_u is a probability law and not a probability density; it is a transition probability kernel.

Thus, from the initial (state-action) couple, the probability of the (state-action) N-trajectory

$$\omega = ((x_0, a_0), (x_1, a_1), \ldots, (x_{N-1}, a_{N-1}), (x_N))$$

is equal to

$$P(\omega) = P_{a_0}(x_0, x_1)\, P_{a1}(x_1, x_2) \ldots P_{a_{N-1}}(x_{N-1}, x_N)\,.$$

A (*feasible*) *policy* of the controlled Markov chain is an application π from $E \times N$ into A such that, for any state x and for any time k, the (state-action) couple $(x, \pi(x, k))$ is feasible.

If policy π does not depend on time it is called a *stationary policy*. In order to simplify notations, a stationary policy will be also denoted by π as a function of the state. Any stationary policy π defines a Markov chain, whose transition probability P_π is defined by:

$$P_\pi(x,y) = P_{\pi(x)}(x,y).$$

The *elementary cost* is an application c from $A \times E$ into R; the *terminal cost* is an application C from E into R.

5.3.3.2 Finite Horizon Markov Decision Problem

A strictly positive integer N called horizon is given. It is the number of authorized transitions. The problem is to find a minimal cost trajectory from time 0 to time N.

To any policy π and any horizon N, we associate the cost function $J_\pi^{0,N}$ from E into R. That cost function takes the state x as input and returns the expected cost of the random N-trajectory from initial state x according to the law of the π-controlled Markov chain. $J_\pi^{0,N}$ is computed by the following relation:

$$\begin{aligned} J_\pi^{0,N}(x) = & \sum_{(x_{k+1},\dots x_N)\in E^{N-k}} P_{\pi(x,0)}(x,x_1)P_{\pi(x_1,1)}(x_1,x_2)\cdots P_{\pi(x_{N-1},N-1)} \\ & \times(x_{N-1},x_N)c(x,\pi(x,0),x_1) \\ & + \sum_{k=1}^{N-1} c(x_k,\pi(x_k,k),x_{k+1}) + C(x_N). \end{aligned}$$

More generally, the cost $J_\pi^{k,N}$ is defined from time k according to

$$\begin{aligned} J_\pi^{k,N}(x) = & \sum_{(x_{k+1},\dots x_N)\in E^{N-k}} P_{\pi(x,k)}(x,x_{k+1})P_{\pi(x_{k+1},k+1)}(x_{k+1},x_{k+2})\cdots \\ & \times P_{\pi(x_{N-1},N-1)}(x_{N-1},x_N)c(x,\pi(x,k),x_{k+1}) \\ & + \sum_{k'=k+1}^{N-1} c(x_{k'},\pi(x_{k'},k'),x_{k'+1}) + C(x_N). \end{aligned}$$

Finite Horizon Markov Decision Problem

The N-horizon *Markov decision problem* consists in finding the optimal policy π^* that minimizes the cost function $J_\pi^{0,N}$.

5.3.3.3 Shortest Stochastic Path Problem

The problem that consists in reaching a target state x^* as fast as possible is called the shortest stochastic path problem [Bertsekas et al. 1996]. In that kind of problem, there is a single terminal state, denoted by x^*, such that, for any feasible action, the only possible transition from that state is the identical transition $x^* \to x^*$. In addition, it is assumed that there exists at least one stationary policy, which enables to reach, with non-zero probability, the terminal state from any state. Such stationary policies are called proper stationary policies. Therefore, the terminal state is the equilibrium state (actually deterministic) of the controlled Markov chain that is associated to any proper stationary policy.

For infinite horizon problems, because elementary costs are stationary and without any terminal cost, there is no point in looking for an optimal non-stationary policy. For a given state, the optimal action does not depend on current time.

We set that elementary cost of the identical transition from terminal state to zero, and the elementary cost of any other transition to strictly positive values; therefore, the latter is lower bounded by a strictly positive constant since state space is finite.

The expected total cost of a stationary policy π is defined by

$$J_\pi(x) = \lim_{N\to\infty} \sum_{(x_1,\dots x_N)\in E^N} P_\pi(x,x_1)P_\pi(x_1,x_2)\dots P_\pi(x_{N-1},x_N) \times \left[c(x,\pi(x),x_1) + \sum_{k=1}^{N-1} c(x_k,\pi(x_k),x_{k+1})\right].$$

This may be written more formally using probability theory notation

$$J_\pi(x) = E_{P_{\pi,x}}\left[c(x,\pi(x),X_1) + \sum_{k=1}^{\infty} c(X_k,\pi(X_k),X_{k+1})\right],$$

where $P_{\pi,x}$ is the probability distribution of the Markov chain that is associated to the stationary policy π and the initial state x.

One infers that if a stationary policy is not proper, there exists at least one initial state such that the expected total cost from that state is infinite.

The *shortest stochastic path problem* consists in finding optimal policy π^* that minimizes the cost function J_π.

5.3.3.4 Infinite Horizon Problem with Discounted Cost

A real number α strictly between 0 and 1 is given. It is called the discount rate.

To any stationary policy π and any discount rate α we associate a cost function J_π^α from $\boldsymbol{E}$ into R. It takes the state x as input and returns the expected cost from initial state x for the Markov chain with transition probability kernel P_π.

$$J_\pi^\alpha(x) = \lim_{N\to\infty} \sum_{(x_1,\dots x_N)\in E^N} P_\pi(x,x_1)P_\pi(x_1,x_2)\dots P_\pi(x_{N-1},x_N) \times \left[c(x,\pi(x),x_1) + \sum_{k=1}^{N-1} \alpha^k c(x_k,\pi(x_k),x_{k+1})\right].$$

We may rewrite it as in the case of the shortest stochastic path:

$$J_{\pi}^{\alpha}(x) = E_{P_{\pi,x}}\left[c(x,\pi(x),X_1) + \sum_{k=1}^{\infty}\alpha^k c(X_k,\pi(X_k),X_{k+1})\right].$$

The *Infinite horizon Markov decision problem with discount rate* α consists in finding an optimal stationary policy, which minimizes the cost function J_{π}^{α}.

In the next sections, when no ambiguity can arise, we will just denote by N the finite horizon or by α the discount rate without further explanation. We will omit the superscript in the cost function in order to alleviate notations.

A discount rate infinite horizon problem may be transformed into a shortest stochastic path problem as follows: every elementary cost is shifted by an appropriate constant to be made positive. Then an artificial terminal state is added, and transitions are changed: any probability transition of the initial problem is multiplied by α and the other possibility is to be sent into the terminal state (cemetery state). All stationary policies of the initial problem are proper stationary policies of the shortest stochastic path problem thus defined. The costs of the two problems (infinite horizon with discount rate and shortest stochastic path) are equal. This transformation is rather formal. Thus, methods that are used to solve shortest stochastic path problems can be used to solve more general infinite horizon problems with discount rate.

Similarly, given a shortest stochastic path problem, it can be transformed, in practice, into an infinite horizon problem with average cost and discount rate by random restart of the state as soon as the terminal state is reached.

5.3.4 Finite Horizon Dynamic Programming

5.3.4.1 Bellman's Optimality Principle

The previous definition of $J_{\pi}^{0,N}$ can be transformed into the following:

$$\begin{aligned} J_{\pi}^{0,N}(x) = \sum_{x_1 \in E} & P_{\pi(x,0)}(x,x_1) \\ & \times \left[c(x,\pi(x,0),x_1) + \sum_{(x_2,\ldots x_N)} P_{\pi(x_1,1)}(x_1,x_2)\ldots P_{\pi(x_{N-1},N-1)}\right. \\ & \left. \times (x_{N-1},x_N)\sum_{k=1}^{N-1} c(x_k,\pi(x_k,k),x_{k+1}) + C(x_N)\right], \end{aligned}$$

which amounts to

$$J_{\pi}^{0,1}(x) = \sum_{x_1 \in E} P_{\pi(x,0)}(x,x_1)[c(x,\pi(x,1),x_1) + J_{\pi}^{1,N}(x_1)]$$

$$= E_{P_{\pi(x1,1)}}(x,0)[c(x,\pi(x,1),x_1) + J_\pi^{1,N}(X_1)].$$

It is a very simple consequence of the summation of elementary costs step by step along a trajectory. This form shows that the optimal policy $\pi*$, which minimizes $J_\pi^{0,N}$, also minimizes $J_\pi^{k,N}$. Thus, the following basic statement is proven:

$$J_{\pi_*}^{0,N}(x) = \min_{u/(x,u)\in \boldsymbol{A}} E_{p_{u(x_1)}}\left[c\left((x,u),X_1\right) + J_{\pi_*}^{1,N}(X_1)\right].$$

That equation, which is verified by optimal policy, is called Bellman's optimality principle.

5.3.4.2 Dynamic Programming Algorithm under Finite Horizon Assumption

Bellman's optimality principle enables to deduce an algorithm which solves the finite horizon Markov decision problem: it is the celebrated «Dynamic Programming algorithm». Its principle is to determine the optimal policy for the last instant and then to go back in time by sequentially optimizing $J_\pi^{N-1,N}, \ldots, J_\pi^{k,k+1}, \ldots, J_\pi^{0,1}$.

For k varying sequentially from $N-1$ to 1, one solves problem

$$\pi_*(x,k) = \operatorname*{Arg\,min}_{u/(x,u)\in \boldsymbol{A}} \left[\sum_{y\in E} P_u(x,y)c(x,u,y) + J_\pi^{k+1,N}(y)\right].$$

Then the optimal cost is updated

$$J_{\pi_*}^{k,N}(x) = \sum_{y\in E} P_{\pi_*(x,k)}(x,y)[c(x,\pi_*(x,k),y) + J_{\pi_*}^{k+1,N}(y)].$$

It is convenient to introduce a new quantity: the value function $Q^{k,N}$, which is defined on the set of feasible (state-action) couples

$$Q^{k,N}(x,u) = \sum_{y\in E} P_u(x,y)\left[c(x,u,y) + J_{\pi_*}^{k+1,N}(y)\right].$$

Then the dynamic programming algorithm takes the following form:

$$\begin{aligned}
Q^{k,N}(x,u) &= \sum_{y\in E} P_u(x,y)\left[c(x,u,y) + J_{\pi_*}^{k+1,N}(y)\right] \\
\pi_*(x,k) &= \operatorname*{Arg\,min}_{u/(x,u)\in \boldsymbol{A}} \left[Q^{k,N}(x,u)\right] \\
J_{\pi_*}^{k,N}(x) &= Q^{k,N}(x,\pi_*(x,k)).
\end{aligned}$$

5.3.5 Infinite-Horizon Dynamic Programming with Discounted Cost

5.3.5.1 Bellman's Optimality Principle

Similarly, the definition of J_π^α may be modified to get the following relation:

$$J_\alpha^\pi(x) = E_{p_{\pi,x}}[c(x,\pi(x),X_1) + \alpha J_\pi^\alpha(X_1)].$$

We express now the fact that the optimal policy $\pi*$ is better then the nonstationary policy which consists in first implementing action a from initial state x, and next keep the application of stationary optimal policy. Thus, we obtain the following relation:

$$J_{\pi_*}^\alpha(x) = \min_{u/(x,u)\in \boldsymbol{A}} E_{P_u(x,.)}[c((x,u),X_1) + \alpha J_{\pi_*}^\alpha(X_1)],$$

which expresses Bellman's optimality principle in the infinite horizon with discount rate framework. We introduce a value function Q, *which* is deduced from the cost function J, just as in the previously considered case of finite horizon problem. The value function is defined on the set of feasible (state-action) couples,

$$Q_J^\alpha(x,u) = \sum_{y\in E} P_u(x,y)[c(x,u,y) + \alpha J(y)].$$

Retaining that definition of the value function, Bellman's optimality principle may be written in the simpler way: $J_*^\alpha(x) = \min_{u/(x,u)\in \boldsymbol{A}} Q_J^\alpha \times (x,u)$. That Bellman variational equation is a fixed-point equation that characterizes the optimal cost function J_{π_*}. It does not provide an immediate algorithm to compute the optimal policy within a finite number of computation steps, as it was the case for finite horizon problems and dynamic programming. On the other hand, the following characterization theorem is easy to prove [Bertsekas et al. 1996].

Theorem. *The Bellman variational equation has a single solution. This unique solution is the optimal cost function* J_{π_*}.

This theorem is proven using the mathematical technique of «contraction». That technique not only provides an existence and uniqueness mathematical proof: it can be applied to prove the convergence towards solution of practical algorithms. Those algorithms are iterative algorithms that will be described in following sections. In order to alleviate notations, from now on, we omit writing α as an index.

5.3.5.2 Policy Iteration Method

This algorithm focuses on displaying a sequence of stationary policies, which undergo step-by-step improvements. Hereafter, we describe an iteration $n+1$ from the policy π_n that was obtained at iteration n:

J_n is computed as the expected cost of policy π_n,

Q_n is computed as the associated value function:

$$\pi_{n+1}(x) = \underset{u/(x,u)\in \boldsymbol{A}}{\text{Arg min}}\ Q_n(x,u).$$

That algorithm provides an explicit sequence of feasible policies that improve monotonically step by step. Their cost may be controlled, so that the algorithm can be stopped as soon as an acceptable cost is reached. In "actor-critics" methods, a policy is first applied (computation of the cost function J_n) and then criticized (here by minimization of the value function) to obtain a new policy. Of course, here, the policy application is rather heavy since it needs the exhaustive computation of the values of the cost function. Moreover it is a virtual application, which is just simulated.

That computation is performed by solving the following linear system:

$$\forall\, x \in E, \quad J_n(x) = \sum_{y\in E} P_{\pi_n(x)}(x,y)[c(x,\pi(x),y) + \alpha J_n(y)].$$

That algorithm can be shown to converge "linearly" towards the optimal policy π_*. In other words, the deviation between the current cost and the optimal cost vanishes and is bounded by a geometric sequence, whose ratio is strictly smaller than 1. For some classical problems, the algorithm may end up within a finite number of iterations.

5.3.5.3 Value-Function Iteration Method

This algorithm focuses on generating a sequence of cost functions that are not necessarily associated to feasible policies, but which converges to the optimal cost function. We will also write down the iteration $n+1$ of this algorithm from the cost function J_n, which is the algorithm output at iteration n:

Q_n is computed as the value function from cost function J_n

$$J_{n+1}(x) = \min_{u/(x,u)\in \boldsymbol{A}} Q_n(x,u).$$

It can be also shown, by a contraction method, that this algorithm converges linearly (i.e. at exponential speed) towards the value function that is associated to the optimal policy. The optimal policy is recovered from this limit value function by the classical variational Bellman equation:

$$\pi_*(x) = \underset{u/(x,u)\in \boldsymbol{A}}{\text{Arg min}}\ Q_*(x,u).$$

5.3.6 Partially Observed Markov Decision Problems

In actual applications, the state is not completely observed in general. For instance, an autonomous robot's perception is limited to its environment. Actually there is a close connection between dynamic programming in discrete state space and optimal control in continuous state space. Therefore, in a way that is similar to the reconstruction of an estimation of the state by filtering in continuous problems, we have to face partially observed Markov decision problems (POMDP). In those problems, the state is not completely observed, because it is impossible to measure some state variables and because the observation is corrupted by measurement noise.

Basic principles of dynamic programming can be applied, but the framework is far more complex because the policies are not defined on the state space but on the belief state space, which is continuous. Actually a belief state is a probability distribution over the state space.

From an observation trajectory, the belief state is defined as the conditional probability on the state space with respect to the observations that were obtained in the past. That probability is updated by Bayes rule. Therefore, it is possible to obtain an optimal policy among the functions that are defined on the belief state space.

Unfortunately, for the belief state to be updated, the model must be known. Therefore, we cannot exploit the method we have just outlined to build up learning strategies. For this reason, Partially Observed Markov Decision Problems will not be developed in the following sections, which are dedicated to training. We will just mention empirical approaches, which have been implemented in practical problems. This question is an open research topic.

5.4 Reinforcement Learning and Neuro-Dynamic Programming

5.4.1 Policy Evaluation Using Monte Carlo Method and Reinforcement Learning

Dynamic programming methods that have just been developed in previous section face practical implementation difficulties in real-world problems. In particular, the policy cost may be difficult to compute by solving a linear system:

- The goal is to evaluate exactly the expected policy cost. If the cardinality of state space is very large, the solution of the linear system for each iteration may have a prohibitive computational cost.
- To write down the linear system, one has to know exactly all the transition probabilities from one state to the other according to the different feasible

actions. Very often now, in complex practical applications, the knowledge of the process is gathered in a computer simulator. The simulator program models real events, so that the knowledge of the transition probabilities is not straightforward. Sometimes, one has to estimate them by performing a first complex set of simulations.

These considerations lead searchers to use simulation straightforwardly to define the optimal policy without identifying the model by estimating the transition probabilities.

The simplest use of a Monte Carlo method to value a policy consists in simulating a large number of trajectories from each initial state and then in computing the average cost over the trajectory set. In the same way, one can estimate the value function by averaging the trajectory cost over a generated trajectory set for every initial feasible (state-action) couple, applying the current policy after the first transition.

The advantage of the Monte Carlo method over exact methods is that it may be implemented even when the mathematical model is not known, provided it is possible to perform intensive experiments or simulations. The optimal policy is no longer determined from the model, but from experiments and environment response. The environment response is called the «reinforcement signal». When this signal is positive, the current modifications of policy that are under testing phase are validated, when it is negative they are discarded. This type of learning process is called reinforcement learning. This terminology is relative to the same conceptual line as Actor-critics methodology that was examined previously. Reinforcement learning always attracted researchers, especially in Artificial Intelligence, because it is generally considered that adaptation mechanisms of living systems are ruled by such principles (in particular Pavlov's work on reflex and other psychologists' work of last century may be addressed). Reinforcement learning was first developed independently from neural learning [Barto et al. 1983]. In this section, we will describe usual methods that widely outperform simple Monte Carlo simulations, which were just explained above as an introduction.

Actually, the complexity of the straightforward Monte Carlo method that has just been mentioned here is generally too large for usual applications. When the model is known, its complexity may be larger than the complexity of direct linear system inversion and, even when the model is not known, it may be impossible to get a result within reasonable time. Moreover, this method throws away useful information. Indeed, transition costs of a given trajectory gives us information not only on the value of the cost function on the initial state of the trajectory but also on the values of the cost function on all the intermediate states of the trajectory. In the following section, we describe a method that takes advantage of the information provided by experiments or simulations in a more efficient way.

We will describe it in the framework of Markov decision problems with infinite horizon and discounted cost. As previously, the discounted cost is denoted by α. In that framework, the algorithm that is described below is the most commonly used. The method can be used as well in finite horizon problems or in shortest stochastic path problems.

5.4.2 TD Algorithm of Policy Evaluation

5.4.2.1 TD(1) Algorithm and Temporal Difference Definition

The "temporal difference" method (TD method in abbreviated form) is based on the following additivity equation of cost

$$J_\pi(x) = E_{P_{\pi,x}}[c(x, \pi(x), X_1) + \alpha J_\pi(X_1)],$$

which was stated in the previous section, and which is rewritten in a simpler form without superscripts.

When we implement a transition $(x \to y)$ according to the admissible state-action couple, the corresponding cost $c(x, \pi(x), y)$ must be used for updating the estimation of $J_\pi(x)$. That update is performed by the recursive computation of the average using a filtering technique with gain (or learning rate) γ. Thus, we take into account the new information about the average total cost $c(x, \pi(x), y) + \alpha \hat{J}_\pi(y)$ and the previous information $\hat{J}_\pi(x)$ according to the following relation:

$$\hat{J}_\pi^+(x) = \hat{J}_\pi(x) + \gamma[c(x, \pi(x), y) + \alpha \hat{J}_\pi(y) - \hat{J}_\pi(x)].$$

The properties of decreasing gain and constant gain filtering techniques were reviewed in Chap. 4. If the gain decreases linearly with the number of updates, the filter will (slowly) converge to the desired value (consistent estimation). If a small, constant gain is used in the stationary regime, the filter undergoes small fluctuations around the desired value. However, the filter is able (subject to an appropriate tuning of the gain) to track slow variations of the environment. For practical use, one generally implements first a decreasing gain to get close to the stationary regime, and then a small constant gain filter to track the stationary regime. Note that the gain tuning may be specific here to each state.

Yet, updates of values of J_π for different states are coupled. Actually, the update of J_π after a transition $(x \to y)$ will use the previous update of $J_\pi(y)$. That method is called a "temporal difference method" (TD method) and is extended to a trajectory of length N.

Given a policy π, a current estimation $\hat{J}_\pi$ of J_π and a state trajectory denoted $(x_0, \ldots, x_N)$, whose initial state is x_0, having N transitions, which is obtained by the application of the policy, the temporal difference of order k is the quantity d_k, which is defined by the following:

$$d_k = c(x_k, \pi(x_k), x_{k+1}) + \alpha \hat{J}_\pi(x_{k+1}) - \hat{J}_\pi(x_k).$$

Then the estimation of J_π is updated for each state x_k of the trajectory by the formula

$$\hat{J}^+_\pi(x_k) = \hat{J}_\pi(x_k) + \gamma[d_k + \alpha d_{k+1} + \cdots + \alpha^{N-k-1} d_{N-1}],$$

for all $k \in \{0, \ldots, N-1\}$. Note that the incremental implementation of that update amounts to the batch rule if the updates are performed backwards in time, as for the backpropagation of the error signal.

5.4.2.2 TD(λ) Algorithm and Eligibility Trace Method

In the previous section, we presented an algorithm that, in order to update the cost function at state x, takes into account either an immediate transition from x, or following transitions on a given horizon N. All those algorithms converge. However, their convergence speed relies on the way the information provided by the trajectory is taken into account. Indeed, it may seem inaccurate to update $\hat{J}_\pi(x)$ by assigning the same weight to the contributions coming from the consecutive transition and from remote future transitions that are more unlikely to be observed. It has been proposed in a basic on reinforcement learning [Barto et al. 1983], to weight the information brought by future transitions with a discount rate $\lambda \in [0, 1]$. That suggests the following updating algorithm, which is called TD(λ):

$$\hat{J}^+_\pi(x_k) = \hat{J}_\pi(x_k) + \gamma[d_k + \alpha\lambda d_{k+1} + \ldots + (\alpha\lambda)^{N-k-1} d_{N-1}],$$

for all $k \in \{1, \ldots, N-1\}$. Historically, the discount rate λ was first intended to deal with finite horizon problems or shortest stochastic path problems where the cost was not discounted by the discount rate α. The introduction of the discount rate λ in that context was a new idea, in infinite horizon problems with discounted costs; it is simply equivalent to a change of the discount rate.

The convergence of the algorithms TD(λ) is just guaranteed by the usual hypotheses of the stochastic approximation [Sutton 1988]. In particular, all the states should be visited infinitely often, i.e., in practice, sufficiently frequently. It is especially important for the states that are relevant for the optimal policy. But those states cannot be determined at the initialization of the algorithm. In following sections, we will emphasize the "exploration policy" in reinforcement learning algorithms. If a simulator is used to provide information, one has to make sure that the hypothesis is approximately true by randomly initializing the trajectory from time to time, preventing it form being confined to a specific region of state space. When a real experiment is performed, one has to make sure that the feasible states are sufficiently explored given the experimental constraints, and to settle a tradeoff between the need of complete exploration and the cost of the experiment. Otherwise, the algorithm may converge towards local minima of the cost function that are not optimal.

Various algorithms have been suggested for applying the methodology of temporal difference to various problems ion such areas as games, optimal planning, and combinatorial optimization. Convergence of those algorithms has been proved, and the following general framework of eligibility traces was first described in [Bertsekas et al. 1996].

In that framework, k is an integer that labels the steps of the algorithm. At step k, an initial state x_0^k is chosen. That choice depends on the past history of the algorithm implementation, and guarantees that each state is visited frequently enough. Then, the current policy is applied for N instants and a trajectory of length $N\omega_k = (x_0^k, x_1^k, \ldots, x_m^k, \ldots, x_N^k)$ generated. The associated costs are observed and the temporal differences d_m^k are computed.

Then a finite sequence of positive state functions z_m^k is chosen. That sequence is called an eligibility trace; it has the following properties (index m labels time along the trajectory; δ *is* the Kronecker delta function):

- $z_0^k(x) = \delta_{x_0^k}(x)$, and, in addition, $z_m^k(x) = 1$ when m is the first visiting time of state x in trajectory ω_k.
- $z_{m+1}^k(x) \leq z_m^k(x) + \delta_{x_{m+1}^k}(x)$.

In addition, let us consider a decreasing sequence $\{\gamma_k\}$ of state functions that take their values in [0, 1]. That sequence is the sequence of gains or learning rates, and it obeys the classical assumptions of stochastic approximation theory

- $\sum_k \gamma_k(x) = \infty$.
- $\sum_k \gamma_k(x)^2 < \infty$.

Then, the generalized TD algorithm updates the estimation of the cost according to

$$\hat{J}_{k+1}(x) = \hat{J}_k(x) + \gamma_k(x) \sum_{m=0}^{N-1} z_m^k(x)\, d_m^k.$$

It converges almost surely towards J_π. Therefore, the estimation is consistent.

For instance, TD(λ) algorithm, which has been previously described, is a particular eligibility trace algorithm where the trace decreases according to the constant multiplicative rate $\alpha\lambda$.

5.4.2.3 Back to Actor-Critics Methodology and Optimistic Iteration of Policy

On many real-life problems, the assessment of a given policy may be a question per se. The previous algorithms are designed for that purpose. However, we introduced the valuation of a policy as a step in the computation loop aimed at finding the optimal policy. The valuation algorithms are iterative so that running them as a part of an optimization loop may too time-consuming. Therefore, it is natural to improve the current policy before completing the valuation process. That improvement makes use of partial results that are provided by one or a small number of iterations of the valuation algorithm.

In the subsection "Value-function Iteration method" of the previous section, we outlined a control design algorithm whereby the policy was updated according to an intermediate approximation of the cost function. That method is called "*Actor-Critics*" method or "*Optimistic Iteration of the Policy*", because the current policy is computed on the basis of the current estimation of the cost function that is optimistically supposed to be the optimal cost.

More precisely, as described above, the following steps are implemented at iteration n, given the cost function J_n.

Q_n is computed as the associated value function of J_n,

$$Q_n(x,u) = \sum_{y \in E} p_u(x,y)[c(x,u,y) + \alpha J_{n(y)}].$$

The policy π_n is defined as the solution of the minimization problem,

$$\pi_{n(x)} = \operatorname*{Arg\,min}_{u/(x,u)\in \boldsymbol{A}} Q_{n(x,u)}.$$

One or several iterations of a temporal difference valuation algorithm are performed, using simulation results or experimental measurements, with policy π_n as the current exploration policy. Thus, a new approximation J_{n+1} is obtained.

5.4.3 Reinforcement Learning: Q-Learning Method

5.4.3.1 Description of the Q-Learning Algorithm

The algorithmic variants, which have just been stated above, use the value function Q as a key ingredient in the determination of the optimal policy. Watkins and Dayan suggested an adaptive version of the value function iteration method in [Watkins et al. 1992]. It was called "*Q-learning*" by the authors, since it focused on the learning of the value function Q. It quickly became one of the most popular reinforcement learning algorithms, especially for infinite horizon problems.

The previous value function iteration algorithm consisted in the following:

- Q_n was computed as the value function for the cost function J_n
- $J_{n+1}(x) = \min_{u/(x,u)\in \boldsymbol{A}} Q_n(x,u)$.

In its adaptive version, a key change is performed:

- The update of Q_n is performed from new experimental or simulation results according to an exploration policy of the feasible state-action couples.

Thus a new algorithm is available. Its definition relies on a random exploration policy π, which associates to each state x a probability on the set of actions u such that the state-action couple (x, u) is feasible. Thus, the exploration policy generates a Markov transition on the set of feasible (state-action) couples. Actually, the theoretically mild assumption that each feasible couple is visited infinitely often suffices to guarantee the convergence of the algorithm. We will discuss below the practical consequences of the choice of the exploration policy. Each transition according to the exploration policy is followed by the implementation of the following Q-learning update rule:

$$
\begin{aligned}
Q_{k+1}(x,u) &= Q_k(x,u) \text{if} (x,u) \neq (x_k, \pi_{k+1}(x_k)) \\
Q_{k+1}(x_k, \pi_{k+1}(x_k)) &= (1-\gamma_{k+1}) Q_k(x_k, \pi_{k+1}(x_k)) \\
&\quad + \gamma_{k+1}[c(x_k, \pi_{k+1}(x_k), x_{k+1})] + \alpha J_k(x_{k+1}) \\
J_{k+1}(x_{k+1}) &= \min_{u/(x_{k+1},u)\in \boldsymbol{A}} Q_{k+1}(x_{k+1}, u)
\end{aligned}
$$

Q-Learning Convergence Theorem. *The Q-learning algorithm converges to the value function Q that is associated to the optimal policy π once all the feasible state-action couples are visited infinitely often and the sequence of the learning rates decreases according to the basic assumption of approximation theory (for instance linear decrease with respect to the number of visits).*

When convergence is stopped, providing a satisfactory estimation of the value function Q_*, it is obvious to determine the associated policy as the solution of the minimization problem

$$
\pi_*(x) = \operatorname*{Arg\,min}_{u/(x,u)\in \boldsymbol{A}} Q_*(x,u)
$$

exactly as in the value function iteration algorithm. There is no necessary connection between the exploration policy and the optimal policy. Of course, a blind exploration policy is rather costly. In practice, one tries to track an optimal policy, by applying a sub-optimal policy. We will elaborate on that point in the following section.

5.4.3.2 The Choice of an Exploration Policy

The problem of choosing an exploration policy is commonly addressed in sequential statistics [Thrun 1992]. If a lot of time is dedicated to the exploration, using for instance a blind policy (random selection of a feasible action), the resulting estimation is accurate but exploration is costly

- in computational time if a simulator is used,

- in experimental costs if experiments are performed (destructive testing is not a good solution).

However, if an optimistic exploration policy is performed, which relies on the current estimation of the cost function, the estimation of the value function may be strongly biased. Thus, neither extreme strategies, blind exploration policy or greedy (fully optimistic) policy, are relevant.

Several mixed exploration schemes have been successfully tested. All of them are intertwining phases of greedy policy most of the time and phases of exploration policy which allow to explore new state-action couples and to check the convergence hypothesis of stochastic approximation (see above):

- The iterative exploration-optimization scheme alternates iteration sequences of greedy optimistic policy and iteration sequences of blind exploration policy.
- According to the randomized scheme, a random selection of the policy is performed independently for each step (blind with probability ε and greedy with probability $1-\varepsilon$).
- The *annealed scheme* is inspired from simulated annealing in combinatorial optimization (which is described in detail in Chap. 8). According to that scheme, a random policy is performed subject to the Gibbs distribution,

$$P(\pi_k(x_k)=u)=\frac{\exp\left(-\dfrac{Q_k(x_k,u)}{T_k}\right)}{\displaystyle\sum_{u/(x_k,u)\in A}\exp\left(-\dfrac{Q_k(x_k,u)}{T_k}\right)},$$

  where the temperature sequence $\{T_k\}$ obeys a cooling schedule. The cooling schedule must be tuned according to the problem of interest. Several cooling schedules are described in Chap. 8.

5.4.3.3 Application of Q-Learning to Partially Observed Problems

It is easy to implement Q-learning algorithm for partially observed Markov decision problems if the selection is performed among feasible policies that depend solely on observations and that do not depend on an unknown state. It is currently implemented, especially in autonomous robotics. In that field, perception is limited; it is provided by sensors and does not allow determining exactly the current state. Q-learning *may* provide satisfactory sub-optimal policies but the success is not guaranteed. In that case, Q-learning is just a heuristic. Its success relies on the relevance of the sensors to determine the key features of the environment with respect to feasible optimal policy whenever it exists (as a deterministic policy). It has been shown [Singh et al. 1995] that the limit of Q-learning algorithms (which stabilizes under common mild hypothesis) depends on the exploration policy, and is generally not the optimal feasible policy. That is not the case in Markov decision problems. It is important to emphasize that the feasible optimal policy in partially observed Markov decision problems is generally neither deterministic nor Markov.

Another way to cope with these problems, when the immediate perception is not sufficient for determining a good policy, consists in taking advantage of past observations to estimate the current state. When the model is unknown, that reconstruction step may be time-consuming, and its completion may be checked using statistical tests [Dutech 1999]. No general solution to those problems is available, so that specific applications require the design of specific solutions.

5.4.4 Reinforcement Learning and Neuronal Approximation

5.4.4.1 Approximate Reinforcement Learning

It is often difficult to use reinforcement learning to deal with large size problems, since those algorithms are relatively complex. The algorithms that have been outlined here are based on the iterative updating of a value table. Stochastic approximation allows implementing the algorithm on an adaptive basis. Thus, simultaneous updating of all the values is not necessary, so that the cost information that is provided by a time step can be used efficiently for updating all the relevant values of the cost function. Nevertheless, when the cardinality of state space or of the feasible state-action couple set is too large, the visit of a generic couple is scarce: thus, the updates of a value do not occur often enough, and the convergence of the algorithm slows down. Reliable results cannot be obtained within reasonable computational times.

An alternative solution consists in using the methodology of supervised learning to maintain a current approximation of the cost function or of the value function. A linear approximation or a neural network approximation may be used. The input is the state (when value iteration of the optimistic policy is performed) or the value function (if Q-learning is performed) and the output is the desired approximation of the updated function.

Many algorithms along those lines have been published. One will select one of them, taking into account the condition of the learning process (using a simulator or an experimental device) and the relevant exploration policy.

Here is a description of one iteration of a commonly used approximate Q-learning algorithm.

A current value function Q_n is available, which determines the exploration policy π_n.

A subset E_n of states is randomly selected.

For each state x_k in E_n, a feasible action $\pi_n(x_k)$ is selected according to the current exploration policy π_n. A transition occurs to the state y_k. The elementary cost $c(x_k, \pi_n(x_k), y_k)$ is taken into account.

Then, the construction of a current element of the training set is possible, which associates to the input $(x_k, \pi_n(x_k))$ the output $Q_k^n = c(x_k, \pi_n(x_k), y_k) + \min_{u/(y_k,u)\in \mathbf{A}} Q_n(y_k, u)$.

A supervised learning epoch is implemented to update the approximate value function Q_n, providing a new approximation Q_{n+1}.

The previous iteration is repeated either from the current state set $E_{n+1} = y_k$ or from a new random selection of E_{n+1}.

The diagram of that algorithm is shown on Fig. 5.14.

Note first there is no available general convergence proof for that algorithm, which is quite similar to Q-learning. For practical purposes, the algorithm is computationally economical, and the approximation is accurate if one uses a relevant topology on the set of feasible state-action couples in order to obtain an efficient numerical encoding of the approximation input. The value function must be regular with respect to that encoding in order to decrease the time complexity of the supervised learning process.

To summarize, a good knowledge of the application context must be available, in order to compensate for the lack of generality of the algorithm. Those

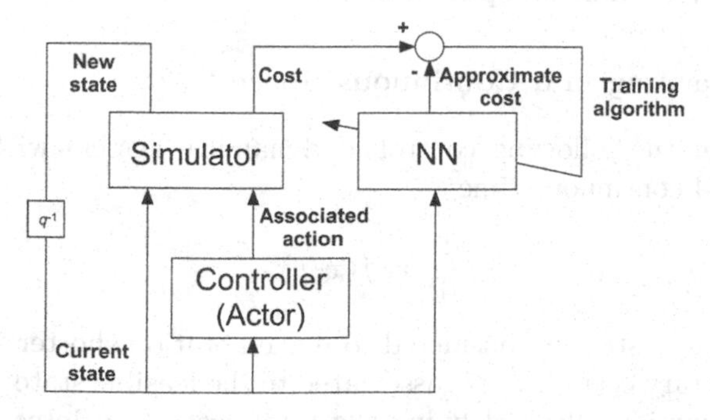

Fig. 5.14. Approximate assessment of a policy using neural network

algorithms have been used efficiently to solve large-size problems such as the backgammon game, the lift planning problem or the dynamic assignment of radio frequencies.

5.4.4.2 Reinforcement Learning with Sampling of a Continuous State-Space

The implementation of approximate reinforcement learning when the value functions are regular suggests that it could be useful for building an approximate optimal control law of a nonlinear continuous system; that topic was addressed at the beginning of this chapter as a direct application of supervised learning and model inversion. Actually, Bellman equation is just a discrete version of the Hamilton-Bellman-Jacobi equation (HBJ equation), which is known to be the variational equation of optimal control with continuous state space and time.

We just saw that the implementation of reinforcement learning in large-size discrete problems is computationally demanding. Therefore, when using that methodology for continuous control problems, one faces the following dilemma:

- A coarse sampling of state space or of the feasible state-action set leads to an inaccurate approximation of the value function, to losing the Markov property of the problem, and possibly to designing a control law that is far from the optimal.
- A fine sampling leads to the combinatorial explosion of the computation complexity.

In order to overcome that difficulty, specific sampling schemes are proposed in the literature. One can use variable sampling steps. In autonomous robotics for instance, space sampling will be fine in key locations (crossings, ambiguous perceptions), where immediate reactions are necessary (new obstacle avoidance), but space sampling will be rough in most regions where optimal navigation is just routine. If the problem allows a multiscale sampling, it may be efficient to determine an optimal policy.

5.4.4.3 Q-Learning in a Continuous Space

Let us consider the following controlled dynamical system with continuous state space and continuous time.

$$\frac{\mathrm{d}x}{\mathrm{d}t} = f(x, u).$$

(A deterministic system is considered to make notation shorter and simpler.)

The elementary cost $c(x, u)$ is associated to the feasible state-action couple (x, u). That function allows defining the total cost as an integral functional that depends on the state-action trajectory,

$$J = \int_0^\infty \mathrm{e}^{-\alpha t} c[x(t), u(t)]\mathrm{d}t.$$

A stationary policy π defines an autonomous dynamical system $\mathrm{d}x/\mathrm{d}t = f(x, \pi(x))$.

To value policy π, one must compute the state function

$$J_\pi(x) = \int_0^\infty \mathrm{e}^{-\alpha t} c[x(t), \pi(x(t))]\mathrm{d}t;$$

the integral is computed on the trajectory of the autonomous dynamical system originating from the initial state x.

Therefore, a stationary optimal policy π_* follows the variational equation:

$$\begin{aligned}\pi_*(x) &= \operatorname*{Arg\,min}_{u/(x,u)\in \boldsymbol{A}} \left[c(x,u) + \nabla_x(J_{\pi_*}) \frac{\mathrm{d}x}{\mathrm{d}t} \right] \\ &= \operatorname*{Arg\,min}_{u/(x,u)\in \boldsymbol{A}} \left[c(x,u) + \nabla_x(J_{\pi_*}) f(x,t) \right].\end{aligned}$$

That equation is exactly the HBJ equation of the control problem. When a neural network approximates the total cost of a policy π, the latter may compute the gradient of the cost function $\nabla_x(J_{\pi_*})$, which can be plugged into the previous formula. Thus, it is possible to infer a training algorithm of the continuous value function Q that is defined by

$$Q(x,u) = c(x,u) + \nabla_x(J_{\pi_*}) f(x,t)$$

and to use it within a generalized continuous Q-learning algorithm.

Recent publications investigate systematically the implementation of reinforcement learning to learn an optimal control law when the model is not known. See for instance [Bertsekas et al. 1996] for a general introduction. More recently, [Doya 2000] presents a nice derivation of several reinforcement learning algorithms in the continuous framework and test them using the inverted pendulum problem as a benchmark.

References

1. Anderson B.D.O., Moore J.B. [1979], *Optimal Filtering*, Prentice Hall
2. Azencott R., Dacunha-Castelle D. [1984], *Séries d'observations irrégulières. Modélisation et prévision*, Masson
3. Barto A.G., Sutton R.S., Anderson C.W. [1983], Neuron-like elements than can solve difficult learning control problemes, *IEEE Trans. On Systems, Man and Cybernetics*, 13, pp 835–846
4. Benveniste A., Métivier M., Priouret P. [1987], *Algorithmes adaptatifs et approximations stochastiques. Théorie et application à l'identification, au traitement du signal et à la reconnaissance des formes*, Masson

5. Bengio Y., Simard P., Frasconi F. [1994], Learning long term dependencies with gradient descent is difficult, *IEEE Trans. on Neural Networks*, 5, pp 157–166
6. Bertsekas D.P., Tsitsiklis J.N. [1996], *Neuro-dynamic programming*, Athena Scientific, Belmont, MA
7. Chatfield C. [1994], *The Analysis of Time series, an Introduction*, Chapman & Hall
8. Demailly J.-P. [1991], *Analyse numérique et équations différentielles*, Presses universitaires de Grenoble
9. Doya K. [2000], Reinforcement learning in continuous time and space, *Neural computation*, pp 219–244
10. Duflo M. [1996], *Algorithmes stochastiques*, Springer
11. Dutech A. [1999], Apprentissage d'environnements: approches cognitive et comportementale, thèse de doctorat de l'École nationale supérieure de l'aéronautique et de l'espace
12. Duvaut P. [1994], *Traitement du signal: concepts et applications*, Hermès
13. Elman J.L. [1990], Finding structure in time, *Cognitive Science*, 14, pp 1179–211
14. Grondin B. [1994], Les réseaux de neurones pour la modélisation et la conduite des réacteurs chimiques: simulations et expérimentations, thèse de doctorat de l'Université de Bordeaux I
15. Haykin S. [1996], *Adaptive Filter Theory*, Prentice Hall
16. Haykin S. [1999], *Neural Networks: a comprehensive foundation*, Prentice Hall
17. Hopfield J.J. [1982], Neural networks and physical systems with emergent collective computational abilities, *Proceedings of the National Academy of Sciences*, États-Unis, 79, pp 2554–2558
18. Isermann R., Lachmann K.H., Matko D. [1992], *Adaptive Control Systems*, Prentice Hall
19. Jazwinsky A.H. [1970], *Stochastic Processes and Filtering Theory*, Academic Press
20. Kirkpatrick S., Gelatt C.D., Vecchi M.P. [1983], Optimization by simulated annealing, *Science*, 220, pp 671–680
21. Kushner K.H.J., Clark D.S. [1978] *Stochastic Approximation Method for constrained and unconstrained Systems*, Applied Mathematical Sciences, 26, Springer-Verlag
22. Kwakernaak H., Sivan R. [1972], *Linear Optimal Control Systems*, Wiley
23. Gouriéroux C., Monfort A. [1995], *Séries temporelles et modèles dynamiques*, Economica
24. Landau I.D., Dugard L. [1986], *Commande adaptative, aspects pratiques et théoriques*, Masson
25. Landau I.D. [1993], *Identification et commande des systèmes*, Hermès
26. Levin A.U., Narendra K.S. [1993], Control of nonlinear dynamical systems using neural networks, *IEEE Transactions on neural networks*, 4.2, pp 192–207
27. Levin A.U., Narendra K.S. [1997], Identification of nonlinear dynamical systems using neural networks *in Neural Systems for Control*, O. Omivar, D.L. Elliott, éd., Academic Press, pp 129–160
28. Lion M. [2000], Filtrage adaptatif par réseaux neuronaux, application à la trajectographie, thèse de doctorat de l'École nationale supérieure de l'aéronautique et de l'espace
29. Ljung L., Söderstrom T. [1983], *Theory and Practice of Recursive Identification*, MIT Press

30. Ljung L., Sjoberg J., Hjalmarsson H. [1996], On neural network model structures in system identification, *in Identification, Adaptation, Learning. The science of learning models from data*, S. Bittanti, G. Pici, éd., NATO ASI Series, Springer
31. Nerrand O., Roussel-Ragot P., Personnaz L., Dreyfus G. [1993], Neural networks and nonlinear adaptive filtering: unifying concepts and new algorithms, *Neural Computation*, 5, pp 165–199
32. Nerrand O., Roussel-Ragot P., Urbani D., Personnaz L., Dreyfus G. [1994], Training recurrent neural networks: why and how ? An illustration in dynamical processes modeling, *IEEE Transactions on neural networks*, 5.2, pp 178–184
33. Norgaard M., Ravn O., Poulsen N.K., Hansen L.K. [2000], *Neural Networks for Modelling and Control of Dynamical Systems*, Springer
34. Puskorius G.V., Feldkamp L.A. [1994], Neurocontrol of nonlinear dynamical systems with Kalman filter-trained recurrent networks, *IEEE Transactions on Neural Networks*, vol. 5, pp 279–297
35. Rivals I. [1995], Modélisation et commande de processus par réseaux de neurones; application au pilotage d'un véhicule autonome, thèse de doctorat de l'Université Pierre et Marie-Curie, Paris VI
36. Rivals I., Personnaz L. [2000], Nonlinear Internal Model Control Using Neural Networks, *IEEE Transactions on Neural Networks*, vol. 11, pp 80–90
37. Singh S.P., Jaakkola T., Jordan M. [1995], Learning without state estimation in a partially observable Markov decision problems, *Proceedings of the 11th Machine Learning conference*
38. Slotine J.J.E., Li W. [1991], *Applied Nonlinear Control*, Prentice Hall
39. Slotine J.J.E., Sanner R.M. [1993], Neural Networks for Adaptive Control and Recursive Identification: A Theoretical Framework, in *Essays on Control*, H.L. Trentelman, J.C. Willems, éd., Birkhauser, pp 381–435
40. Sontag E.D. [1990], *Mathematic Control Theory. Deterministic finite dimensional systems*, Springer Verlag
41. Sontag E.D. [1996], *Recurrent Neural Networks: Some Systems-Theoretic Aspects*, Dept. of Mathematics, Rutgers University, NB, États-Unis
42. Sutton R.S. [1988], Learning to predict by the method of temporal differences, *Machine Learning*, 3, pp 9–44
43. Thrun S.B. [1992], The role of exploration in learning control, *in Handbook of intelligent control*, D.A. White, D.A. Sofge, éd., pp 527–559, Van Nostrand
44. Tong H. [1995], *Nonlinear Time Series, a dynamical system approach*, Clarendon Press
45. Urbani D., Roussel-Ragot P., Personnaz L., Dreyfus G. [1993], The selection of nonlinear dynamical systems by statistical tests, *Neural Networks for Signal Processing*, 4, pp 229–237
46. Watkins C.J.C.H., Dayan P. [1992] Q-learning, *Machine Learning*, 8, pp 279–292
47. Williams, R.J., Zipser, D. [1989], "A learning algorithm for continully runnig fully recurrent neural networks", *Neural Computation*, pp. 270–280

6

Discrimination

M. B. Gordon

The task of assigning patterns to classes based on their characteristics is called discrimination. For example, medical diagnosis, handwritten character recognition, non-destructive tests of defects, are particular cases of pattern discrimination.

In Chap. 1, a general introduction to the problem of discrimination was provided. A general methodology for the design of statistical classifiers was described, and was illustrated by detailed presentations of actual applications. That methodology is based on considerations developed in the present chapter. We have already pointed out that the problem of automatic classification may be considered from different viewpoints, depending on the application. We may consider the classifier training problem as a regression problem, and view the continuous output as an estimate of the probability that the patterns belong to a given class. Conversely, in other applications, we may just need the frontiers between classes, called discriminant surfaces; those may be obtained using neural networks of binary neurons, as was suggested already in the sixties, and further developed from the eighties up to the present.

This chapter is mainly devoted to the second approach: we provide detailed explanations of the modern techniques allowing linear separations between classes using binary neurons, and, if necessary, how to go beyond and determine more complex separations. A probabilistic interpretation of the results is also presented.

We also introduce many theoretical justifications, stemming mainly from work due to physicists, as well as from recent developments in learning theory. However, it should be borne in mind that in any application, the time devoted to the following tasks should not be underestimated:

- The choice of the data representation requires a careful analysis, because the quality of the results depends critically on that issue. An appropriate representation is both compact (the dimension of the input vectors should be as small as possible) and discriminant (it allows efficient separation of patterns belonging to different classes).

- It may be essential to define a class of rejected patterns (that the classifier is unable to discriminate).

As in the previous chapters, we will consider problems where the data are represented by vectors. Their components are characteristics that are relevant to the discrimination task. For example, in the case of medical diagnosis, these are the patient age, blood pressure, etc.; in the case of pattern recognition, the pixels of the image... The classes may be encoded as integer numbers, representing the kind of disease, or the image type. We will mainly consider problems where the data can only belong to one of two classes. As will be discussed in the corresponding section, problems with more than two classes may be reduced to a set of two-class problems.

The chapter is divided into five sections. After general considerations, we describe several training algorithms for linear separation. Then, we present various cases where the discriminant surfaces are more complex. In the fourth section, we consider the discrimination in problems with more than two classes. At the end of the chapter we describe theoretical concepts, such as the Vapnik-Chervonenkis dimension and the capacity of a classifier, which are important in applications.

6.1 Training for Pattern Discrimination

Can we learn to classify new patterns using the information contained in a set of examples previously classified by an expert? This is a variant of the general problem already considered in previous chapters, where we tried to predict the behavior of a process on fresh data, not used to adjust the model's parameters. As explained in Chap. 1, regression and discrimination are ill-posed problems.

Remark. Some authors use the term "discrimination" to refer to the classification task when the classes are known a priori. That is the case of the so-called *supervised learning*, in contrast to *non-supervised* learning, whose goal is to organize data not previously classified. In this chapter, we consider supervised learning of classification tasks, that we will loosely call either classification or discrimination.

As in other training problems, the parameters of the classifier are estimated from a training set of M examples L_{M}, where each example is an input vector and its class,

$$L_M = \{(\boldsymbol{x}^1, y^1), (\boldsymbol{x}^2, y^2), \ldots, (\boldsymbol{x}^M, y^M)\},$$

where the input

$$\boldsymbol{x}^k = [x_1^k, x_2^k, \ldots, x_M^k]^{\mathrm{T}}$$

is a vector of N discrete, binary or real-valued components, describing the example k $(k = 1, 2, \ldots, M)$ and $y^k \in \{-1, +1\}$ is its class.

Remark. It is possible to encode the two classes using $z \in \{0, 1\}$. Codes $\{-1, +1\}$ and $\{0, 1\}$ are formally equivalent. They are related by the transformation $y = 2z - 1$. The ± 1 encoding adopted in this Chapter is elegant and presents advantages in programming. However, in electronic implementations, it may be useful to use the $\{0, 1\}$ code.

The output $\sigma(\boldsymbol{x}; \boldsymbol{w})$ of the classifier, neural network or any other classifier depends on its input $\boldsymbol{x}$ and on its parameters, hereafter denoted by $\boldsymbol{w}$ even if the classifier is not a neural network. The output corresponding to input $\boldsymbol{x}^k \in L_M$ will be denoted by $\sigma^k(\boldsymbol{x}; \boldsymbol{w})$ or simply $\sigma^k (\sigma^k \in \{-1, +1\})$. The classifier is able to classify correctly the example $\boldsymbol{x}^k$ if $\sigma^k = y^k$, i.e., if the following *condition of correct classification* is obeyed:

$$\sigma^k y^k > 0.$$

Otherwise, $\sigma^k \neq y^k$, so that $\sigma^k y^k < 0$.

6.1.1 Training and Generalization Errors

The quality of training may be assessed through the training (or learning) error $\varepsilon_t(\boldsymbol{w})$, which is the fraction of misclassified examples of L_M. From the condition of correct classification, we have,

$$\varepsilon_t(\boldsymbol{w}) = \frac{1}{M} \sum_{k=1}^{M} \Theta(-y^k \sigma^k(\boldsymbol{x}; \boldsymbol{w})),$$

where $\Theta(u)$ is the Heaviside function, which takes on the value 1 if its argument is positive or zero, and 0 otherwise,

$$\Theta(u) = \begin{cases} 1 & \text{if } u \geq 0 \\ 0 & \text{if } u < 0. \end{cases}$$

In fact, the goal of learning a classification task using the examples in L_M is mainly to determine the classifier parameters that will correctly classify new inputs, that is, *generalize.* Obviously, the patterns to be classified are unknown, but we will assume that they present the same regularities which are as those used for training. Mathematically, we consider that the input vectors $\boldsymbol{x}$ are realizations of a real-valued random vector $\boldsymbol{X}$. Similarly, the output y (that is the code given to the class of $\boldsymbol{x}$) is the realization of a discrete random variable Y. We thus assume that there is an unknown probability density $p_{XY}(\boldsymbol{x}, y) \equiv p_X(\boldsymbol{x})\, P_Y(y \mid \boldsymbol{x})$ from which are drawn

- the inputs and outputs of the training set,
- the new inputs, whose class, given by $P_Y(y \mid \boldsymbol{x})$, is unknown.

The quantity that we would like to minimize is the generalization error $\varepsilon_g(\boldsymbol{w})$, defined by

$$\varepsilon_g(\boldsymbol{w}) = \sum_{y=\pm 1} \int \Theta(-y\sigma(\boldsymbol{x};\boldsymbol{w}))p_{x,y}(\boldsymbol{x},y)\mathrm{d}\boldsymbol{x}$$

where σ is the class assigned by the classifier to the input $\boldsymbol{x}$. The generalization error is thus the probability that the classifier with parameters $\boldsymbol{w}$ makes a classification error on an input $\boldsymbol{x}$ drawn with probability $p_X(\boldsymbol{x})$, whose class y has probability $P_Y(y \mid \boldsymbol{x})$. Clearly, the generalization error cannot be computed in actual applications because $p_X(\boldsymbol{x})$ and $P_Y(y|\boldsymbol{x})$ are unknown. In practice, ε_g is estimated by statistical methods such as cross-validation, as discussed in Chap. 2. Later in the present chapter, we will come back to that probabilistic formulation, because it is one of the foundations of the statistical learning theory. It allows the determination of bounds to training and generalization errors, or the estimation of their typical values. Clearly, training from examples raises the following fundamental questions:

1. What are the properties of the classifier designed through learning, and more specifically, what is its generalization error?
2. What is the minimal number of examples needed to catch the regularities in the data?
3. What are the properties of different training algorithms?
4. Given a training set, are the classifier parameters $\boldsymbol{w}$ unique? If multiple solutions are possible, is there an optimal one?

6.1.2 Discriminant Surfaces

Assume that the inputs are vectors $\boldsymbol{x} \in R^N$ (the assumption of real-valued components is not essential: the results presented in this chapter are also valid for discrete-valued components, unless explicitly stated). We can represent them as colored points in an N dimensional space, each color indicating the class of the corresponding point. The surface that separates the points of different class is termed discriminant surface. As shown on Fig. 6.1, that surface is not necessarily unique, and can possibly be a combination of parts of surfaces. Training aims at determining the equation of an appropriate discriminant surface.

As indicated in Chap. 1, classification may be considered as a particular case of regression, where we search for a continuous surface $g(\boldsymbol{x})$ whose values are close to the desired output, i.e., a function that is equal to $+1$ for the examples $\boldsymbol{x}^k$ of class $y^k = 1$, and to -1 for those of class $y^k = -1$, as shown on Fig. 6.2. The techniques presented in Chap. 2 can be used to find this function. The discriminant surface is the set of points $\boldsymbol{x}$ where the sign of $g(\boldsymbol{x})$ changes.

Two situations may arise in an application:

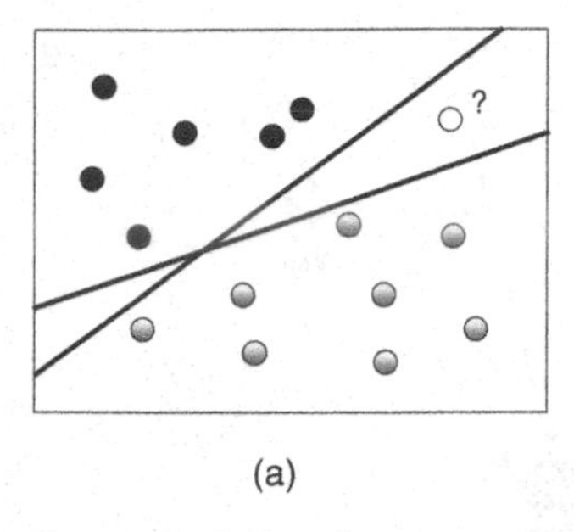

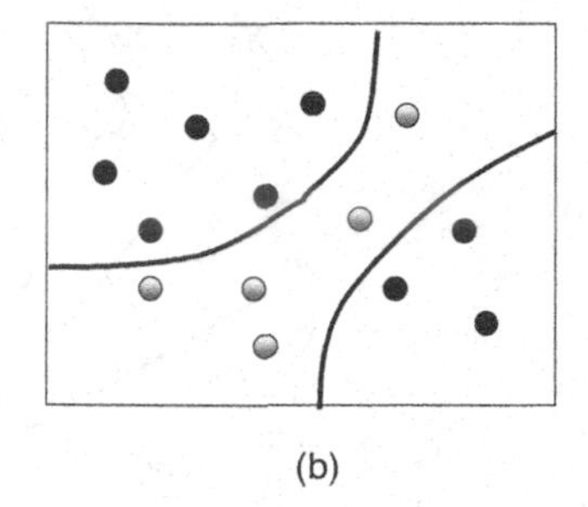

Fig. 6.1. Examples belonging to two classes in dimension $N = 2$. The lines represent the discriminant surfaces. (**a**) linearly separable training set, showing two separations without training errors but giving different answers for the class of a new input (*empty circle*). (**b**) A general case

- if one just needs to classify the patterns $\boldsymbol{x}$, the discriminant surface only is needed. Since the classifier performs a binary function of its inputs, in this chapter we will show how the discriminant surface may be represented using binary neurons only. That cannot be done if we transform the problem into a regression problem;
- if the probability that the input pattern belongs to a given class is necessary, in order to make a posterior decision (such is the case for instance when the classification is performed after comparison of the outputs of different classifiers) we may use either continuous output units (e.g., sigmoidal neurons), but also binary neurons. In this chapter we show that the latter can also be assigned a probabilistic interpretation.

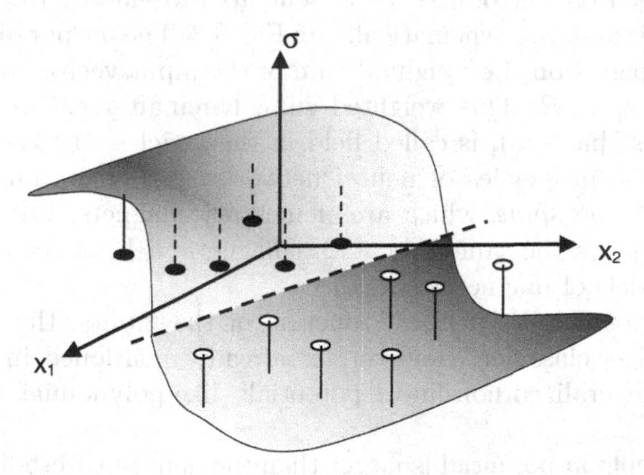

Fig. 6.2. Examples in dimension 2. *Black points* stand for patterns of class +1; *white* denote those of class −1. The *shadowed surface* is the regression; the discriminant surface (a line in that case) is shown as a *dotted line*

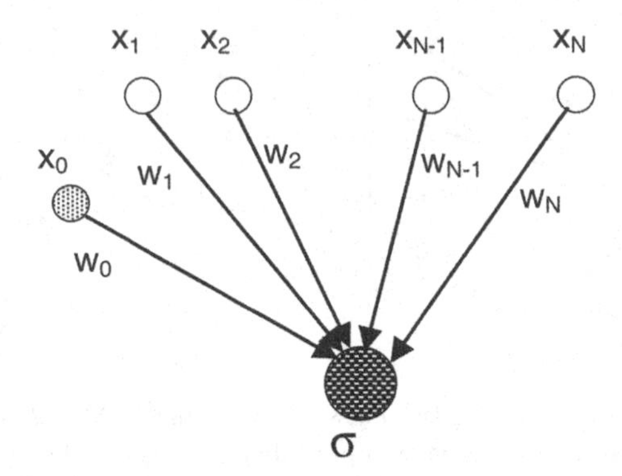

Fig. 6.3. Schematic diagram of a perceptron

One element of choice between those approaches is the actual implementation of the application. Estimating probabilities on a digital computer is not a problem. If the classifier will be implemented on a special-purpose device, binary units can considerably improve the computation times and the complexity of the circuits. A comparison between neural networks using binary and real-valued neurons on an application of radar signals recognition may be found in the Ph.D. thesis of Christelle Godin [Godin 2000].

6.2 Linear Separation: The Perceptron

The simplest network allowing the classification of data into two classes is a single binary neuron. Introduced by Rosenblatt [Rosenblatt 1958], who called it perceptron, it is shown schematically on Fig. 6.3. The output of Rosenblatt's perceptron depends on the weighted sum of the input vector components x_i, with weights $w_i \in R$. This weighted sum, hereinafter called potential (as in the previous chapters), is called field in the articles written by physicists that studied the properties of neural networks. A Perceptron shares many properties with the *spins*, which are elementary magnets. In particular, the weighted sum plays the same role as the magnetic field in the context of the physicists' models of magnetism.

Since the potential is a linear function of the inputs, the perceptron is also called linear classifier. However, as already mentioned in Chap. 1, we will consider generalized non-linear potentials, like polynomials (higher-order neurons).

If the perceptron potential is larger than the neuron threshold s_0, the output is $\sigma = +1$, otherwise $\sigma = -1$. Thus, the perceptron is a neuron whose activation function is a threshold function. In Fig. 6.3 we followed the convention used in the previous chapters, and we included a constant input $x_0 = 1$ with a weight w_0. Its role is to shift the threshold of the activation function. Now the input has an additional component $x_0 = 1$, so that we consider a

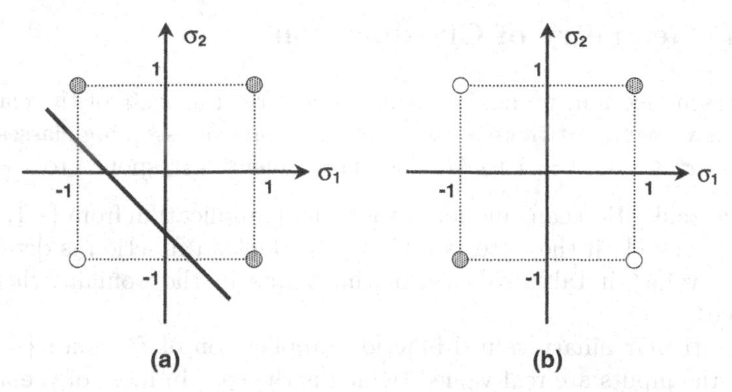

Fig. 6.4. Two Boolean functions; *left*: function OR, realizable with a perceptron; *right*: the complement of the XOR (the non-XOR), not realizable by a perceptron

perceptron without threshold in an *extended input space* of dimension $N+1$. In that space, we define the *linear potential* as follows:

$$v_L = \sum_{i=0}^{N} w_i x_i = \boldsymbol{w} \cdot \boldsymbol{x},$$

where the subscript stands for linear (to distinguish it from the generalized potentials introduced in the following sections). The perceptron's output is given by

$$\sigma_L = \mathrm{sign}(v_L).$$

The perceptron separates the inputs $\boldsymbol{x}$ into two subsets, depending on the output value σ. From the definitions of the (linear) potential and the output of the perceptron, the weight vector $\boldsymbol{w}$ is normal to a hyperplane in R^{N+1} (a hyperplane is the generalization to dimension N of a plane in three-dimensional space). That hyperplane contains the origin and separates the examples with $\sigma = 1$ from those with $\sigma = -1$. The former obey $\boldsymbol{w} \cdot \boldsymbol{x} > 0$, the latter $\boldsymbol{w} \cdot \boldsymbol{x} < 0$. Thus, the perceptron performs linear separations in input space.

If we restrict the inputs to take only binary values, a perceptron performs a Boolean function of its inputs, that is, an application from $\{0,1\}^{N+1}$ onto $\{0,1\}$ (or from $\{-1,+1\}^{N+1}$ onto $\{-1,+1\}$). Clearly, it is unable to perform some functions, like the exclusive-or (XOR) or its complement, as shown on Fig. 6.4b.

Given the inputs $\boldsymbol{x}^k$ of the training set, which are M points in a space of N dimensions, there are 2^M possible Boolean functions of those points. That number increases exponentially with M. Although it is not known how many Boolean functions are linearly separable, it has been shown that their number increases with M like a power law ($\approx M^n$, with $n > 1$), much slower than exponentially. For large M, the linearly separable functions are a very small fraction of all the possible Boolean functions. Thus, the perceptron can perform a very small number of Boolean functions of its inputs.

6.3 The Geometry of Classification

In the present section, we analyze some geometrical aspects of the classification. We have seen that there are 2^M different ways of assigning classes to the M vectors $\boldsymbol{x}^k \in L_M$, $k = 1$ to M. Each assignment corresponds to

- one particular Boolean function of the inputs (application from $\{-1,+1\}^{N+1}$ onto $\{-1,+1\}$) if the latter are binary (a Boolean function is determined by the values it takes on each of the points in the domain where it is defined),
- one particular binary valued function (application of R^N onto $\{-1,+1\}$) when the inputs are real-valued (which is the case in most of the applications).

The selection of a weight vector is actually a choice of one particular Boolean function that, in the case of error-free training, takes exactly the values y^k on the M patterns in L_M. Thus, training amounts to selecting a single function, determined by the network weights. Even restricting the weights to those that discriminate correctly the examples, the corresponding functions generalize differently on the points $\boldsymbol{x}$ that do not belong to the training set L_M.

6.3.1 Separating Hyperplane

As shown by the definition of the (linear) potential, the latter it is the scalar product of the weight vector $\boldsymbol{w}$ and the input vector $\boldsymbol{x}$. That product vanishes when the points $\mathbf{x}$ lie exactly on the *separating hyperplane.*

The points $\boldsymbol{x}^H$ that obey the equation

$$\boldsymbol{w} \cdot \boldsymbol{x}^H = 0$$

belong to the separating hyperplane; it is normal to $\boldsymbol{w}$ in the extended input space, of dimension $N + 1$. In that space, the hyperplane contains the origin.

Remark. If the examples of the training set are linearly separable, there is a continuum, i.e., an infinite number, of hyperplanes that classify correctly the training patterns.

Consider an arbitrary input $\boldsymbol{x}$. If $\boldsymbol{w} \cdot \boldsymbol{x} > 0$, the perceptron assigns it to the class +1; if $\boldsymbol{w} \cdot \boldsymbol{x} < 0$, the class is -1. In both cases, the vector $\boldsymbol{x}$ is at a distance $|d|$ of the separating hyperplane; d is given by

$$d = \frac{\boldsymbol{w} \cdot \boldsymbol{x}}{\|\boldsymbol{w}\|},$$

where $\|\boldsymbol{w}\| = \sqrt{\sum_{j=0}^{N} w_j^2}$ is the norm of the weight vector.

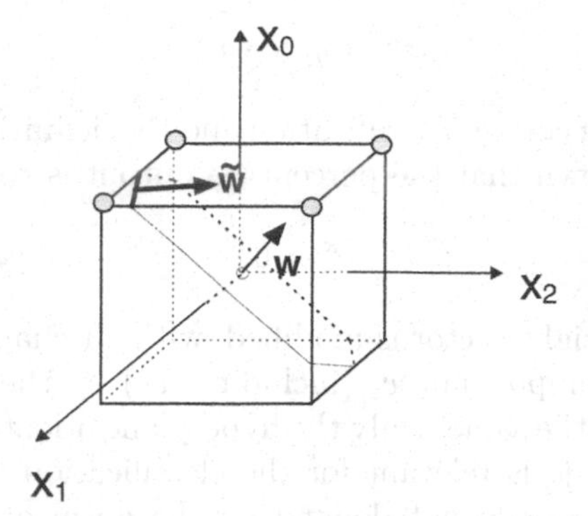

Fig. 6.5. Examples (*grey circles*) in two dimensions, represented in the extended three-dimensional space. The separating hyperplane and the corresponding separating *straight line* in the original input space are shown

Remark. It may be useful to come back to the original N-dimensional input space. The points $\tilde{x}^H \in R^N$ that satisfy

$$\sum_{j=1}^{N} w_j \tilde{x}_j^H = -w_0$$

lie on the hyperplane that is normal to the vector $\tilde{\boldsymbol{w}} = [w_1, w_2, \ldots, w_N]^T$, whose distance to the origin is the absolute value of

$$d_0^H = \frac{w_0}{\sqrt{\sum_{j=0}^{N} w_j^2}}.$$

The relation between $\tilde{\boldsymbol{w}}$ and $\boldsymbol{w}$ is illustrated on Fig. 6.5. The separating hyperplane in the N-dimensional space is the intersection of the hyperplane in extended space with the sub-space $x_0 = 1$. Clearly, the distances of the examples to the hyperplanes are different, depending on which space is considered.

To summarize, each weight vector $\boldsymbol{w}$ determines one hyperplane that separates the input space in two regions. Inputs $\boldsymbol{x}$ with positive projection onto $\boldsymbol{w}$ have outputs equal to $+1$, whereas the others have outputs equal to -1. A perceptron performs linear separations (or discriminations), since the equation of the separating (discriminant) surface is a linear function of the inputs. The implementations of more complex separations need either neural networks with hidden units or perceptrons with higher order potentials, like spherical perceptrons or Support Vector Machines. These are described in the Sect. 6.5 "beyond linear separation" .

6.3.2 Aligned Field

Consider an input $\boldsymbol{x}^k$ belonging to L_M, of class y^k. The *aligned field* z^k with respect to the perceptron of weights $\boldsymbol{w}$, is defined by

$$z^k = y^k \boldsymbol{w} \cdot \boldsymbol{x}^k.$$

From the condition of correct classification and the definition of the perceptron output, it can be shown that the perceptron output is correct if

$$z^k > 0.$$

If the norm of the weight vector is modified, without changing its orientation, by multiplying all components w_i (including w_0) by the same constant, the perceptron output is the same. Only the hyperplane orientation, defined by the unitary vector $\boldsymbol{w}/\|\boldsymbol{w}\|$, is relevant for the classification task. The properties of the linear separation do not depend on the norm of $\boldsymbol{w}$, but only on its orientation.

6.3.3 Stability of an Example

In order to investigate the training properties of the neurons, we introduce the concept of *stability* γ of an example

$$\begin{aligned}\gamma^k &= \frac{y^k \boldsymbol{w} \cdot \boldsymbol{x}^k}{\|\boldsymbol{w}\|} \\ &= \frac{z^k}{\|\boldsymbol{w}\|}.\end{aligned}$$

Comparing with the distance to the separating hyperplane, and given that $|y^k| = 1$, the magnitude of the stability $|\gamma^k|$ is nothing but the distance of the example k to the separating hyperplane. That is illustrated on Fig. 6.6 for the case of real-valued inputs. In terms of stability, the condition for correct classification can be written as

$$\gamma^k > 0.$$

Remark. The stabilities of the examples are a measure of our confidence in the classification. We will see in the last part of this chapter that the probabilistic interpretation of the classification is a function of those stabilities.

Some examples have interesting properties. The distance κ to the separating hyperplane of the example of L_M that is closest to the hyperplane is called the *margin*. The region in input space on both sides of the hyperplane, of width 2κ, centered on the latter, does not contain any example.

Among all possible separating hyperplanes, the hyperplane with *maximal margin*, also called *optimal stability perceptron*, has interesting properties. In particular, it is robust with respect to small perturbations in the inputs or the weights. The support vector machines, introduced later in this chapter, are based on the concept of maximal margin.

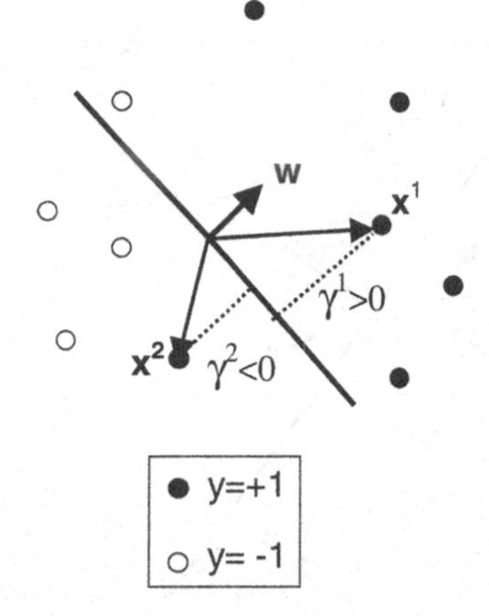

Fig. 6.6. Input space, with two examples x^1 and x^2, of class +; the hyperplane corresponding to a perceptron of weights w (with $\|w\| = 1$) is shown, together with the stabilities γ^1 and γ^2 of the examples

6.4 Training Algorithms for the Perceptron

There are many learning algorithms that allow to determine the perceptron weights based on the training set $L_M = \{(x^k, y^k)\}$, $k = 1$ to M. Historically, the oldest one is the "perceptron algorithm". Although it is seldom used in practice, it has interesting properties. We will see that the alternative training algorithms may be considered as generalizations of it.

Remark. If the examples of the training set are linearly separable, a perceptron should be able to learn the classification.

6.4.1 Perceptron Algorithm

The following algorithm was proposed by Rosenblatt to train the perceptron:

Perceptron Algorithm

- **Initialization**
  1. $t = 0$ (counter of updates)
  2. either $w(0) = 0$ (tabula rasa initialization) or each component of $w(0)$ is initialized at random.
- **Test**
  1. *if* $z^k \equiv y^k w(t) \cdot x^k > 0$ for all examples $k = 1, 2, \ldots, M$ (they are correctly learned) *then* **stop**.
  2. *else go to* **learning**
- ***Learning***
  1. select an example k of the training set L_M, either at random or following a pre-established order.

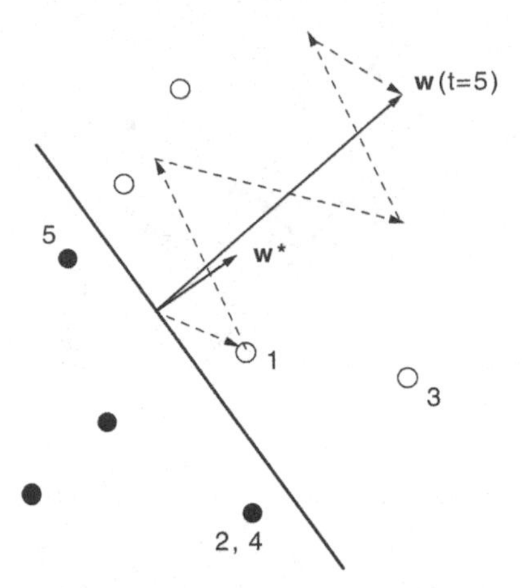

Fig. 6.7. Vector $\boldsymbol{w}$ after 5 iterations of the perceptron algorithm: the examples used for learning (*black circles*: class −1, *white circles*: class +1) are numbered following the order in which they have been used. The weight vector $\boldsymbol{w}^*$ is one solution. The weight vector $\boldsymbol{w}(t=5)$ correctly separates all the examples

2. *if* $z^k \equiv y^k \boldsymbol{w}(t) \cdot \boldsymbol{x}^k > 0$ (the example k is correctly classified) *then go to* **learning**
3. *else*
   (a) update the weights' components ($i = 1, \ldots, N+1$): $w_i(t+1) = w_i(t) + y^k x_i^k$
   (b) increment the counter of updates: $t = t + 1$,
   (c) *go to* **test**

The perceptron algorithm iterates the weights modifications as long as there are classification errors of the training patterns, that is, examples with negative aligned fields z^k. Figure 6.7 presents an example of the application of the Perceptron Algorithm in two dimensions. Clearly, if the training set is not linearly separable, the algorithm will never stop (in contrast with the algorithm of Ho and Kashyap presented in Chap. 1).

Remark. The perceptron algorithm cannot determine whether a training set is linearly separable or not, because it is not easy to distinguish in a reasonable training time whether the algorithm entered an infinite loop because the examples are not linearly separable, or whether the examples are separable but convergence is slow. The algorithm of Ho and Kashyap [Ho 1965] provides an answer to that question: if the examples are linearly separable, it obtains a solution (which should not be expected to be optimal); if they are not linearly separable, it signals that after a finite number of iterations.

If the examples are linearly separable, the perceptron algorithm converges, as proved in the following theorem.

6.4.2 Convergence Theorem for the Perceptron Algorithm

Theorem. *If the examples of the training set are linearly separable, the perceptron algorithm finds a solution in a finite number of iterations.*

In order to prove the theorem, we assume that the weights are initialized to zero, i.e., using the option of *tabula rasa.* Strictly, that assumption is not necessary. It just makes the proof easier than starting with arbitrary weight values.

Since the examples of the training set L_M are assumed to be linearly separable, there exists a weight vector $\boldsymbol{w}_*$, hereinafter called *reference perceptron,* that classifies correctly the examples. Without any loss of generality, we assume that $\boldsymbol{w}_*$ is unitary. Otherwise, it would suffice to divide its components by its norm. The stabilities of the examples in L_M with respect to the reference hyperplane are all positive. Since $\boldsymbol{w}_*$ is unitary, they are identical to the corresponding aligned fields,

$$\gamma_*^k = y^k \boldsymbol{x}^k \cdot \boldsymbol{w}_* = z_*^k.$$

To prove the theorem, we derive upper and lower bounds to the norm of the weight vector generated by the algorithm. Those bounds (see the Sect. 6.8 "Additional material" at the end of the chapter) are increasing functions of the number of iterations t, but they increase at different rates. The lower bound increases linearly with the number of iterations t, whereas the upper bound increases more slowly, as $\sqrt{t}$ (see Fig. 6.8). The bounds cross each other, which is absurd, after a number of iterations T given by

$$T = \left(\frac{\|\boldsymbol{x}^{\max}\|}{\gamma_*^{\min}} \right)^2,$$

where $\|\boldsymbol{x}^{\max}\|$ is the norm of the example of L_M with the largest norm, and $\gamma_*^{\min}$ is the stability of the example of L_M that has the smallest stability with respect to the reference perceptron hyperplane. Thus, the perceptron algorithm must converge, since the number of iterations cannot be larger than T. Notice that the learning time may be very long, especially if there are examples very close to the reference hyperplane ($\gamma_*^{\min}$ small relative to $\|\boldsymbol{x}^{\max}\|$). However, the algorithm may converge in times much smaller than that given by the above relation for two reasons:

- first, because the reference hyperplane $\boldsymbol{w}_*$ is arbitrary, and the value of $\gamma_*^{\min}$ may be particularly small;
- on the other hand, the learning time is a random variable that depends on the particular sequence of examples selected for the successive updates.

Remark 1. The expression of the number of iterations T has a simple intuitive interpretation. The correction to the weights performed at each iteration is bounded, because its norm cannot be larger than the norm of the example

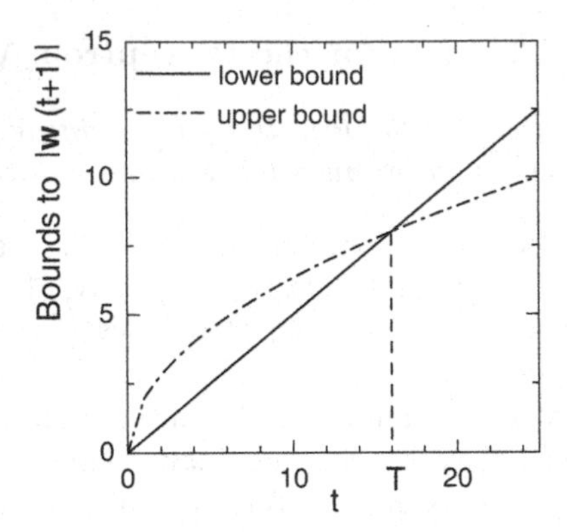

Fig. 6.8. Behavior of the *upper* and *lower bounds* as a function of the number of iterations, for a case with $\gamma_*^{\min} = 0.5$ and $||\boldsymbol{x}^{\max}|| = 2$

used at that iteration, $\|\boldsymbol{x}^k\|$. On the other hand, the successive learning steps increase the norm of the weights. Therefore, the relative correction produced by an example, whenever its presentation generates a weight update, decreases during training. As a consequence the hyperplane undergoes successive reorientations of decreasing amplitude. If some examples are very close to the hyperplane (small $\gamma_*^{\min}$), the corrections must become small enough to reach the necessary precision. This explains why the convergence time is inversely proportional to $\gamma_*^{\min}$.

Remark 2. Instead of considering the inputs $\boldsymbol{x}^k$ of classes y^k, it is equivalent to consider input vectors $\boldsymbol{x}'^k \equiv y^k \boldsymbol{x}^k$, all of the same class $y'^k = +1$. If $\boldsymbol{w}$ classifies correctly the set of $\boldsymbol{x}^k$, it will do so with the $\boldsymbol{x}'^k$, since the sign of the aligned fields are unaltered by that transformation: $y'^k \boldsymbol{w} \cdot \boldsymbol{x}'^k \equiv y^k \boldsymbol{w} \cdot \boldsymbol{x}^k > 0$. The computation time of training algorithms may be shortened if that transformation is applied to the training set.

6.4.3 Training by Minimization of a Cost Function

Most training algorithms compute $\boldsymbol{w}$ through the minimization of a differentiable cost function, which is the sum of partial costs per example. We have seen that, for correct classification, the examples should have positive aligned fields. It is therefore reasonable to consider partial costs that are functions of z^k: $V(z^k)$. The cost function to be minimized is

$$C(\boldsymbol{w}) = \frac{1}{M} \sum_{k=1}^{M} V\left(z^k\right) .$$

It depends on $\boldsymbol{w}$ through the aligned fields of the examples. We will see later that the assumption of an additive cost over the examples is consistent with the hypothesis that the examples are independent random variables.

Remark. Of course, the factor $1/M$ in front of the sum does not play any role in the minimization of the cost function. It allows the definitions of the *average cost per example*, a quantity that allows easy comparisons between results obtained with training sets of different sizes.

The partial cost $V(\boldsymbol{z}^k)$ must satisfy some conditions in order that the minimum of the cost function corresponds to appropriate weights. Weights $\boldsymbol{w}$ that produce negative aligned fields must have a higher cost than weights producing positive aligned fields. Thus, $V(z)$ must be a non-increasing function of the aligned field z. However, that condition on V is not sufficient, at least in the case of a linearly separable training set: if $\boldsymbol{w}_*$ separates correctly L_M, then any weight vector of the form $a\boldsymbol{w}_*$ with $a > 1$ is also a solution, with a *lower* cost. Hence, a minimization algorithm would never converge, since the cost can decrease without bounds by increasing the norm of $\boldsymbol{w}$ without modifying the hyperplane orientation. To avoid this, we impose the constraint that $\|\boldsymbol{w}\|$ be constant. Normalizations $\|\boldsymbol{w}\| = 1$ and $\|\boldsymbol{w}\| = N+1$ in the extended space (or $\|\boldsymbol{w}\| = N$ in input space) are the most popular ones.

The simplest method of minimizing $C(\boldsymbol{w})$ is to use the algorithm of *gradient descent,* as described in Chap. 2, which modifies iteratively the weights following

$$\boldsymbol{w}(t+1) = \boldsymbol{w}(t) + \Delta\boldsymbol{w}(t),$$

with

$$\begin{aligned}\Delta\boldsymbol{w}(t) &= -\mu\frac{\partial C(\boldsymbol{w})}{\partial \boldsymbol{w}}(t) \\ &= -\mu\frac{1}{M}\sum_{k=1}^{M}\frac{\partial V\left(z^k\right)}{\partial z^k}(t)y^k\boldsymbol{x}^k \\ &= \sum_{k=1}^{M} c^k(t)y^k\boldsymbol{x}^k,\end{aligned}$$

where μ is the learning rate, and we introduced the relation $\partial z^k/\partial \boldsymbol{w} = y^k\boldsymbol{x}^k$. It is convenient to normalize the weights after each iteration.The last relation shows that the weights can be written under the general form:

$$\boldsymbol{w} = \sum_{k=1}^{M} c^k y^k \boldsymbol{x}^k.$$

The parameters c^k, which are the sum of the $c^k(t)$ over all the iterations, depend on the algorithm. If $c^k = 1$ in the expresion of $\boldsymbol{w}$, the mathematical expression of Hebb's rule is retrieved. That learning rule states that the information used for modifying the synaptic efficacies in the nervous system is the correlation between the activity of the pre-synaptic neuron (neuron excitation) and of the post-synaptic neuron (neuron firing rate). It is worth pointing out that Hebb's rule has very poor performances in Machine Learning. However, even if its interest for applications is mainly historical, we will show that it is possible to accelerate the convergence of some algorithms by initializing the weights with Hebb's rule.

Remark. If the weights were not normalized after each iteration to satisfy the constraint $\|\boldsymbol{w}\|$ = constant, it would be possible to control the convergence, and stop the algorithm as soon as the corrections become parallel to the weights, that is, when $\boldsymbol{w}(t+1)\cdot\boldsymbol{w}(t) = \|\boldsymbol{w}(t+1)\|\|\boldsymbol{w}(t)\|$ (within the accuracy limits imposed by the considered application).

In the rest of this section we review some partial costs proposed in the literature.

6.4.4 Cost Functions for the Perceptron

The cost function that seems most appropriate intuitively is the number of training errors. The corresponding partial cost is shown on Fig. 6.9, and can be written as

$$V(z) = \Theta(-z),$$

where $\Theta(u)$ is the Heaviside function defined at the beginning of this chapter. It takes the value 1 if the example is incorrectly classified, and 0 otherwise. At its minimum, the total cost is the smallest fraction of examples incorrectly classified. This cost function is not differentiable, and cannot be minimized using a gradient descent. Its minimization is performed by combinatorial optimization techniques, or using simulated annealing, described in Chap. 8.

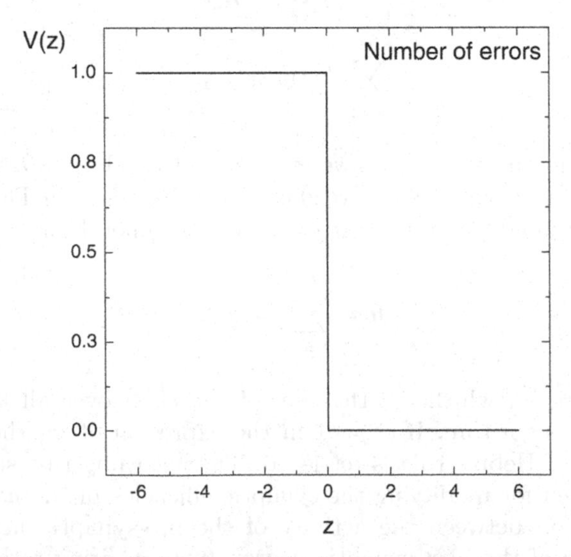

Fig. 6.9. Partial cost corresponding to the number of training errors

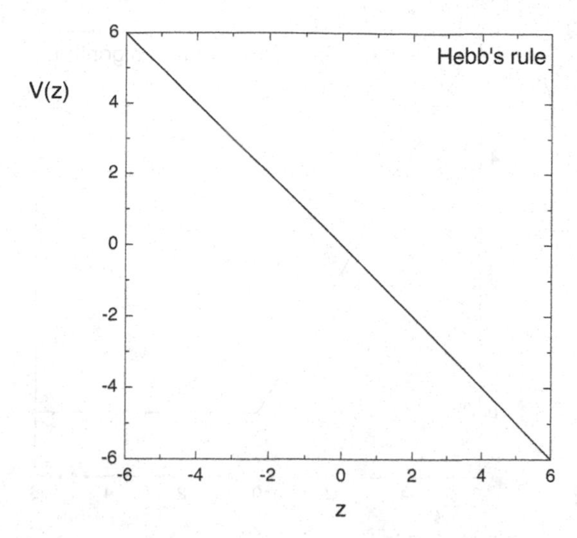

Fig. 6.10. Partial cost corresponding to Hebb's rule

We now turn to differential costs. The following partial cost, represented on Fig. 6.10,

$$V(z) = -z$$

is the simplest monotonic decreasing function.

After introducing its derivative in $\Delta \boldsymbol{w}(t)$, we find

$$\Delta \boldsymbol{w} = \mu \frac{1}{M} \sum_{k=1}^{M} y^k \boldsymbol{x}^k,$$

which is nothing but Hebb's rule. As was already discussed, since the partial cost is a monotonic decreasing function, it is necessary to introduce the normalization constraint on the weights to guarantee that the algorithm will stop. Then a single iteration suffices to find the cost minimum. We will take advantage of that result to initialize the Minimerror algorithm.

The perceptron algorithm may be derived from the following partial cost:

$$V(z) = -z\Theta(-z)$$

shown on Fig. 6.11. The weight updates at each iteration are

$$\Delta \boldsymbol{w} = \mu \frac{1}{M} \sum_{k=1}^{M} \Theta(-z^k) y^k \boldsymbol{x}^k.$$

This is equivalent to a nonadaptive ("batch") version of the perceptron algorithm. Here the weights are updated using all the incorrectly classified examples at each iteration (thanks to the Θ function in $\Delta \boldsymbol{w}$, which eliminates

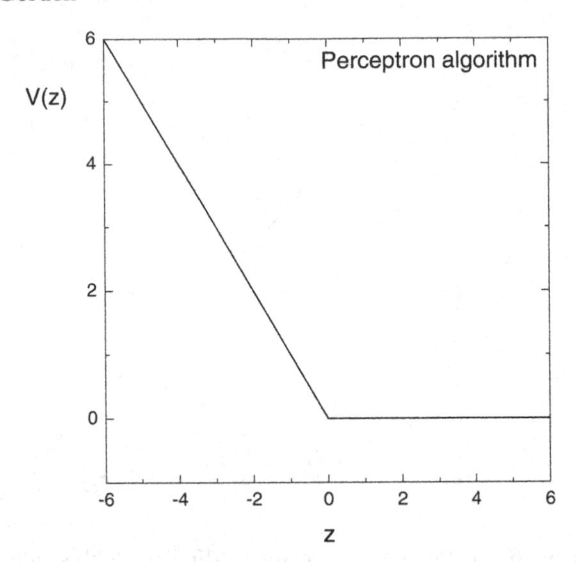

Fig. 6.11. Partial cost corresponding to a nonadaptive ("batch") version of the perceptron algorithm

all correctly classified patterns from the sum), instead of considering only one pattern at each update, as the perceptron algorithm does.

The Adaline algorithm, also called delta rule, Widrow-Hoff algorithm, or relaxation algorithm, derives from the following partial cost:

$$V(z) = \frac{1}{2} z^2 \Theta(-z)$$

shown on Fig. 6.12. The weights updates at successive iterations are given by

$$\Delta \boldsymbol{w} = \mu \frac{1}{M} \sum_{k=1}^{M} z^k \Theta(-z^k) y^k \boldsymbol{x}^k.$$

Remark. If a solution without training errors exists, i.e., if the training set is linearly separable, and if the training set is equilibrated, that is, contains positive and negative examples in similar proportions, then most of the presented algorithms may find the separating hyperplane after more or less iterations. However, it is worth to remind that the learning rate μ must be small enough.

The above algorithms penalize the weights that produce training errors, since the partial costs corresponding to negative aligned fields have positive values. Correctly classified patterns have vanishing cost (except for Hebb's rule), independently of their position in input space. However, intuitively we are more confident about the classification of patterns far from the separating hyperplane, generally surrounded by patterns of the same class, than about that of patterns very close to it. Thus, hyperplanes too close to the patterns should

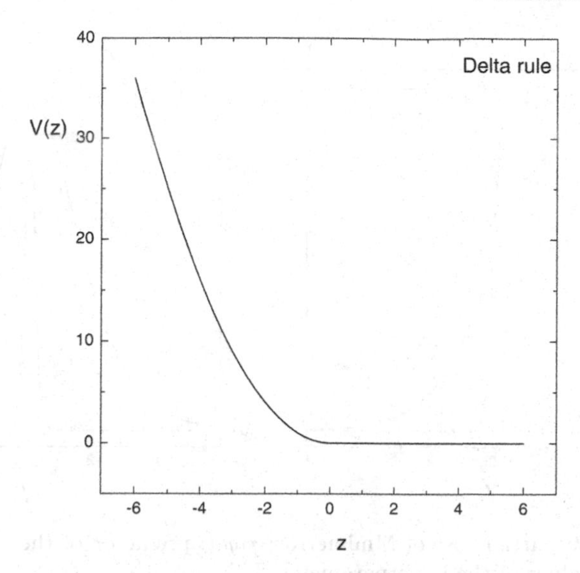

Fig. 6.12. Partial cost corresponding to the Delta rule

be avoided, even if they classify them correctly. This is the aim of algorithms that look for the hyperplane of a given margin $\boldsymbol{\kappa}$, that is, the weights $\boldsymbol{w}(\kappa)$ such that, for all examples k,

$$\gamma^k \equiv \frac{z^k}{\|\boldsymbol{w}\|} \geq \kappa.$$

The hyperplanes closer to the examples than the margin κ can be penalized through a simple modification of the costs $V(z)$, replacing everywhere the aligned field z^k by $z^k - \|\boldsymbol{w}\|\kappa$. In that case, if the training set is linearly separable, the solutions of vanishing cost satisfy the above relation for all the examples. The largest value of $\boldsymbol{\kappa}$ for which a solution with zero cost exists defines the *maximal stability perceptron.* We should point out that in practice, the procedure that consists in maximizing κ may be complex and time consuming.

Other cost functions have adjustable parameters more or less equivalent to κ, generically called hyperparameters. Those allow finding solutions that have better generalization properties than the above costs.

In general, when the training examples are not linearly separable, the discriminant surface may be represented with hidden neurons. In that case the hyperplane defined by each neuron should separate correctly the examples, at least in a limited neighborhood of the hyperplane. However, when the examples are not separable, the cost functions presented above have many local minima. Generally, the solution found by minimizing those costs does not exhibit the property of local separation. The following partial cost (used by the algorithm Minimerror described later) allows finding such a solution,

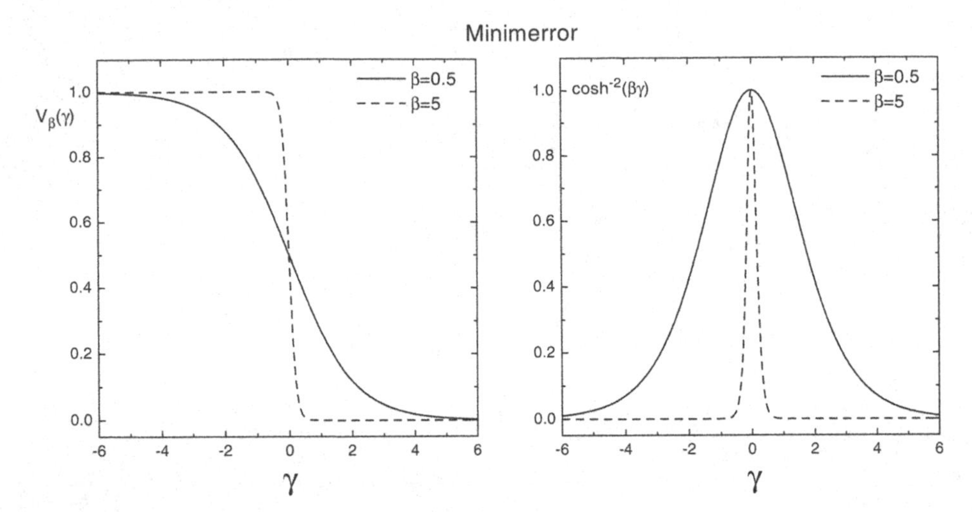

Fig. 6.13. *Left*: partial cost of Minimerror; *right*: prefactor of the weights corrections, for two values of the hyperparameter β

through the use of a hyperparameter β. This cost is not a function of the aligned field z, but of the stability γ:

$$V_\beta(\gamma) = \frac{1}{2}[1 - \tanh(\beta\gamma)].$$

The contribution of each example to learning is proportional to

$$-\frac{\partial V_\beta(\gamma)}{\partial \boldsymbol{w}} \propto \frac{b}{2\cosh^2(\beta\gamma)}[y\boldsymbol{x} - \gamma\boldsymbol{w}].$$

The partial cost, as well as the prefactor $\cosh^{-2}(\beta\gamma)$, are shown on Fig. 6.13 as a function of γ, for two different values of β. The hyperparameter β has a simple intuitive meaning: the larger β, the narrower the region on both sides of the hyperplane where the examples contribute significantly to learning. The examples that contribute effectively to the learning process are those within a *virtual window* of width proportional to $1/\beta$ centered on the hyperplane. Due to the factor $\cosh^{-2}(\beta\gamma)$, the contribution of the examples outside this window is vanishingly small.

Remark 1. With reference to the algorithms with partial costs depending on the aligned field, the derivative of $V_\beta(\gamma)$ with respect to the weights exhibits an extra term that is proportional to $\gamma\boldsymbol{w}$. That quantity, subtracted from the term between square brackets in the gradient of the cost function, is the component of $y\boldsymbol{x}$ parallel to $\boldsymbol{w}$. It only contributes to changing the norm of the weight vector, without modifying its orientation. If the weights are normalized at each iteration, that term can be neglected.

Remark 2. Even the examples that are correctly classified, i.e., with $\gamma > 0$, contribute to the cost function. the closer they lie to the hyperplane, the larger their contribution.

Remark 3. If β is small enough that $\beta\gamma^k \ll 1$ for all k, then all the examples contribute with almost the same prefactor, like in Hebb's rule discussed before. Moreover, in the limit $\beta \to 0$, the stabilities of all the examples are in the region where the cost function is linear (in the neighborhood of $\gamma = 0$), and the prefactor in the gradient of the cost function is the same for all examples.

Remark 4. For intermediate values of β, the examples with large stabilities with respect to the virtual window width $1/\beta$ ($\beta|\gamma| \gg 1$) do not contribute significantly to training, since their prefactor in the gradient of cost function is exponentially small (in the limit $\beta|\gamma| \gg 1$, one has $1/(\cosh^2(\beta\gamma)) < 4\exp(-2\beta|\gamma|)$). For example, if $\beta|\gamma| > 5$, the prefactor is of order 10^{-4}. Loosely speaking, the algorithm uses for learning only the examples lying inside a virtual window of width $\beta|\gamma|$ on both sides of the hyperplane.

The above remarks are at the basis of the Minimerror algorithm. The hyperparameter β, which increases throughout the iterations to optimize the solution, allows one to obtain a linear separation with large margin if it exists, or finds surfaces that are locally discriminant (with large margins) otherwise. The weights are initialized using Hebb's rule, which corresponds to $\beta = 0$. The iterations start with β sufficiently small for all the patterns to be inside the virtual window. If $\|\boldsymbol{x}\|^{\max}$ is the norm of the example of largest norm, one can use $\beta_{ini} = 10^{-2}/\|\boldsymbol{x}\|^{\max}$. Then, at each training step (iteration) the weights are updated and β is increased by a small amount $\delta\beta$. This procedure is known in the literature as *deterministic annealing*, a concept close to that of *simulated annealing*, used in optimization problems (see the Chap. 8 on optimization).

A heuristic improvement consists in considering two different values of β, β_+ for the examples with positive stability and β_- for those with negative stability. In order to keep a small number of parameters, the ratio β_+/β_- does not change during training. Thus, the Minimerror algorithm has three parameters: the learning rate μ, the annealing step $\delta\beta$ and the asymmetry $\beta_\pm \equiv \beta_+/\beta_-$. It proceeds as follows:

Minimerror Algorithm

- **Parameter Settings**
  1. learning rate μ (suggested value: 10^{-2}),
  2. ratio $\beta_\pm$ (suggested value: 6)
  3. annealing step $\delta\beta_+$ (suggested value: 10^{-2})
- **Initialization**
  1. iteration counter: $t = 0$
  2. weights: $\boldsymbol{w}(0)$ (suggested initialization: apply Hebb's rule and then normalize the weights to $\|\boldsymbol{w}\| = N + 1$)

3. hyperparameter β_+ (suggested value: $10^{-2}/\|\boldsymbol{x}\|_{\max}$)

- **Training**

1. update and normalize the weights according to

$$\boldsymbol{w}(t+1) = \frac{\boldsymbol{w}(t) + \Delta\boldsymbol{w}}{\|\boldsymbol{w}(t) + \Delta\boldsymbol{w}\|}$$

$$\Delta\boldsymbol{w} = -\frac{\mu}{M}(\delta\boldsymbol{w}_+ + \delta\boldsymbol{w}_-)$$

where $\mu > 0$ is the learning rate, and

$$\delta\boldsymbol{w}_\pm = \sum_{k/\gamma^k \in \gamma_\pm} \frac{\beta_\pm}{\cosh^2 \beta_\pm \gamma^k} y^k \boldsymbol{x}^k$$

where $\gamma_\pm$ denotes the subset of examples with positive (γ_+) and negative (γ_-) stabilities, respectively.

2. update the iteration counter and the hyperparameters: $t \leftarrow t+1, \beta_+ \leftarrow \beta_+ + \delta\beta_+, \beta_- = \beta_+/\beta_\pm$.
3. *if* β_+ and β_- are sufficiently large that $\beta\gamma^k \gg 1$ for all k, no example can significantly contribute to modify the weights (within the accuracy limits of the problem),
then **stop**.
else, go to **training**.

It is possible, and often useful, to modify the learning rate and adapt it at each iteration, as discussed in Chap. 2.

Remark. The Minimerror algorithm combines a gradient descent with the adaptation of the hyperparameter. It converges towards a local minimum. It has been shown [Gordon 1995] that if the training patterns are linearly separable, the minimization of the cost function for increasing values of β allows finding the hyperplane with maximal margin. If the examples are not linearly separable, the algorithm converges to weights that minimize locally (in the neighborhood of the hyperplane) the number of training errors. These properties are very useful for constructive training algorithms, as explained later in this chapter.

The hyperparameter β may be interpreted as the inverse of a noise, or a temperature, $T = 1/\beta$ [Gordon 1995]. We will come back to that interpretation below. Further details, and examples of applications of Minimerror can be found in [Raffin et al. 1995], [Torres Moreno et al. 1998], [Torres Moreno 1997] and [Godin 2000].

Remark. The least squares partial cost is of particular interest when applied to a network without hidden units, i.e., to a single neuron with sigmoidal activation function. Since $y^k = \pm 1$, one has

$$\begin{aligned}V(z^k) &= \frac{1}{2}\left(y^k - \tanh(\boldsymbol{w}\cdot\boldsymbol{x}^k)\right)^2\\ &= \frac{1}{2}\left(1 - y^k\tanh(\boldsymbol{w}\cdot\boldsymbol{x}^k)\right)^2\\ &= \frac{1}{2}\left(1 - \tanh(z^k)\right)^2 .\end{aligned}$$

The corresponding modification of the weights with the algorithm of gradient descent has the form given previously with

$$\begin{aligned}c^{k(t)} &= \frac{\mu}{M}\frac{1-\tanh(z^k)}{\cosh^2(z^k)}\\ &= \frac{\mu}{M}\frac{1-\tanh(\|\boldsymbol{w}\|\gamma^k)}{\cosh^2(\|\boldsymbol{w}\|\gamma^k)} .\end{aligned}$$

The latter relation is similar to the relation of Minimerror. Here, $\|\boldsymbol{w}\|$ plays the same role as β. The essential difference between both algorithms is that β is a controllable parameter of Minimerror, while $\|\boldsymbol{w}\|$ cannot be controlled when minimizing the Least Squares cost.

6.4.5 Example of Application: The Classification of Sonar Signals

The data of this application is available at the address http://www.ics.uci.edu/mlearn/MLRepository.html [Blake 1998]. The problem is the discrimination of sonar signals generated by cylindrical mines from those generated by rocks with the same shape. The benchmark has a training set of 208 preprocessed signals, defined by $N = 60$ real values $x_i \in [0,1] (i = 1, \ldots, N)$, and their corresponding classes. The first 104 signals are traditionally used as training examples, the last 104 ones are used for estimating the generalization error. Despite the fact that this benchmark has been used to test many learning algorithms with many different network architectures, we discovered using Minimerror that not only were the training set and the test set linearly separable, but that the complete set of 208 signals is also linearly separable [Torres Moreno et al. 1998]. That result was subsequently confirmed by the algorithm of Ho and Kashyap (see Chap. 2). The left part of Fig. 6.14 shows the distances of the patterns to the separating hyperplane found with Minimerror, with a sign corresponding to the class assigned by the perceptron trained with the first 104 patterns. The solution has a margin $\kappa = 0,1226$: none of the training examples lie at a distance to the hyperplane smaller than κ. On the contrary, among the 104 test patterns, that hyperplane makes 23 classification errors. The right part of Fig. 6.14 shows the distances (with a sign corresponding to the class assigned by the trained perceptron) after learning with all the database of 208 signals. In that case, the margin is smaller ($\kappa = 0,0028$). The histogram of the pattern stabilities with respect to that hyperplane is shown on Fig. 6.15. We will see later that, if we assume that

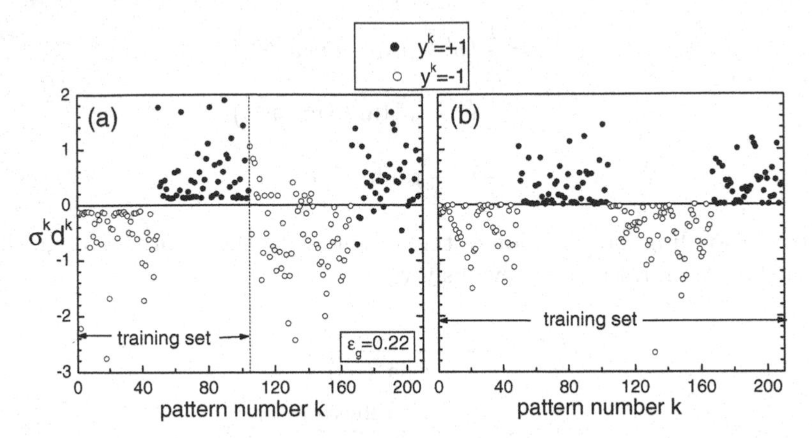

Fig. 6.14. Distance of the patterns to the separating hyperplane, with different colors for the different classes. The sign of d^μ represents the class assigned by the perceptron after learning. *Left*: learning with the $M = 104$ first patterns. The last $G = 104$ examples belong to the test set. *Right*: distances to the hyperplane determined with all the patterns in the database, showing that they are linearly separable

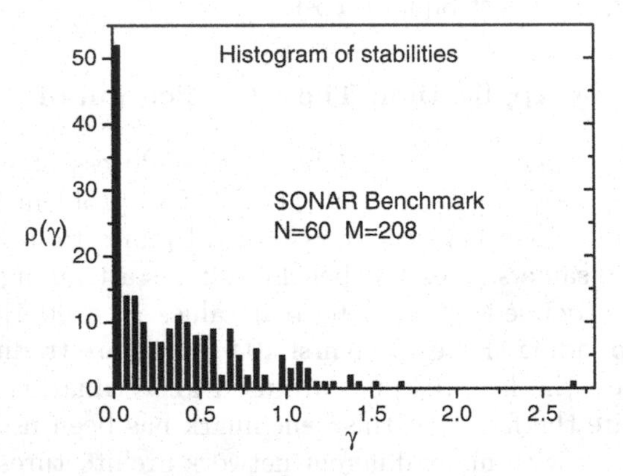

Fig. 6.15. Histogram of the stabilities of the examples with respect to the hyperplane that separates the complete set of patterns

the training examples stem from noisy measurements of the corresponding physical signals, those distances allow assigning a degree of plausibility (or a probability density) to the perceptron output.

Remark 1. The fact that the 208 patterns of the sonar database turned out to be linearly separable is not surprising. A theorem due to Cover [Cover 1965], and later generalized by E. Gardner (see [Gardner 1989]) to the case of correlated data [Engel et al. 2001], states that the probability that a set of points in general position (that is, such that no subset of more than N points lie on one hyperplane) is linearly separable only depends on the ratio M/N, where M is the number of points and N the space dimension. In particular, for $N = 60$ and $M = 208$, and if the patterns are correlated, as is the case of the sonar signals, that probability is not negligible. Those theorems, whose importance has already been mentioned in the first chapter, are further discussed at the end of this chapter.

Remark 2. One may wonder why the fact that the sonar database is linearly separable was not discovered earlier, since we have already shown in Chap. 1 that the algorithm of Ho and Kashyap [Ho 1965] provides the answer in a few minutes. That is a consequence of the multidisciplinary character of the field of neural networks: important results are frequently rediscovered. The authors of this book hope that it will contribute to overcoming such problems.

6.4.6 Adaptive (On-Line) Training Algorithms

Adaptive algorithms update the weights after the presentation of each example, just as the perceptron algorithm does. As already pointed out in previous chapters, adaptive training is useful when the training set is too large to be stored in the computer memory—as requested by the optimization algorithms described above—or in problems where the examples are available one at a time, as is the case when a robot explores its environment.

As mentioned in Chap. 2 and 4, adaptive training can be performed by updating the weights proportionally to the derivative of the partial costs defined in the previous section. Such implementations are called methods of *stochastic gradient descent* since the true gradient is replaced by a stochastic term whose average is equal to the gradient. Stochasticity is due to the more or less arbitrary order of presentation of the examples. Different orderings may end up with different, statistically equivalent results.

Among the on-line learning algorithms for the perceptron, we mention Minover [Krauth et al. 1987] and Adatron [Anlauf et al. 1989], which achieve better performances than the perceptron algorithm. In fact, Adatron is an adaptive version of the relaxation algorithm described above.

6.4.7 An Interpretation of Training in Terms of Forces

In this section, we provide an interpretation of training in terms of forces produced by the examples on the hyperplane, which provides insight into the non-convergence of some algorithms when the training set is not linearly separable.

Given the orientation of the hyperplane at iteration t, the contribution of an example k to the weight update may be interpreted as a force

$$\begin{aligned} \boldsymbol{F}^k(t) &= -\frac{\mu}{M}\frac{\partial V(z^k)}{\partial \boldsymbol{w}}(t) \\ &= -\frac{\mu}{M}\frac{\partial V(z^k)}{\partial z^k}(t) y^k \boldsymbol{x}^k \\ &= c^k(t) y^k \boldsymbol{x}^k \end{aligned}$$

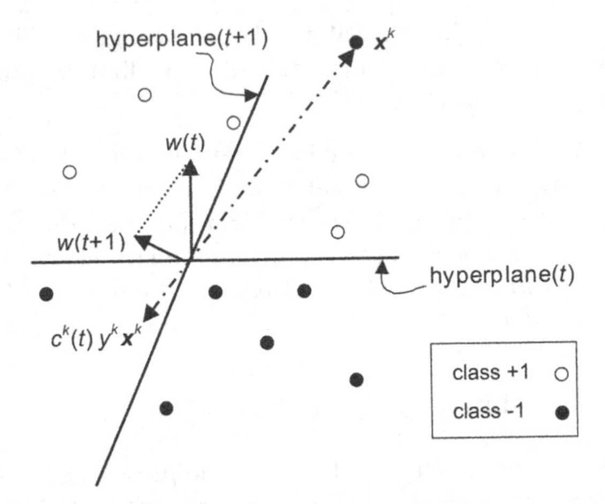

Fig. 6.16. Forces on the hyperplane. At iteration t, the example k incorrectly classified attracts the hyperplane. Its contribution to the weight update is indicated by the vector $c^k(t)y^k\boldsymbol{x}^k$, which is added to $\boldsymbol{w}(t)$ to give $\boldsymbol{w}(t+1)$

that acts on the hyperplane. Note that that force derives from a potential, which is nothing but the partial cost V (this is why the latter is usually called potential in the literature written by physicists working in this field; we will not use this term here in order to avoid confusions with the neuron potential). Since the hyperplane contains the origin of the extended input space, this force will make it turn around the origin. If $V(z)$ is a non-increasing function of its argument, then $c^k \geq 0$. As we can see on Fig. 6.16, if the stability of $\boldsymbol{x}^k$ is negative, then the force attracts the hyperplane towards x^k, like if the example tried to pass to the other side of the hyperplane. Conversely, if the stability of $\boldsymbol{x}^k$ is positive, the example repels the hyperplane.

Remark. The rotation angle is proportional to the learning rate μ. If it is too large, the effect of the force may be excessive and produce oscillations upon successive iterations.

The hyperplane orientation is stabilized, and the algorithm converges, when a balance between the forces produced by examples on both sides of the hyperplane is reached. If the partial cost V is zero for positive stabilities, only the wrongly classified patterns produce forces, which are attractive, on the hyperplane. If $V \neq 0$ for positive stabilities, as in the Minimerror algorithm, the correctly classified patterns also produce (repulsive) forces on the hyperplane.

If the examples of the training set are not linearly separable, training with partial costs that diverge at negative stabilities may exhibit convergence problems. In non-separable problems, the misclassified patterns attract the hyperplane with forces that are stronger the farther the examples. The hyperplane orientation may thus oscillate during successive iterations and never stabilize.

To avoid this problem, the learning rate μ may be decreased throughout the learning procedure. The same remark applies to on-line (adaptive) training: in that case, the solution not only depends on the rate at which μ was modified, but also on the order of presentation of the patterns.

6.5 Beyond Linear Separation

We have seen that a perceptron can only implement linear separation of the inputs. To overcome that limitation, multilayered networks were introduced in Chap. 1. However, other possibilities exist. One of them is to use non-linear potentials. An example is presented in the next section. It allows the implementation of spherical separation surfaces using a generalized perceptron with the same number of parameters as a linear perceptron. Clearly, restricting to hyperspheres is still an important limitation that can be overcome by using two very different approaches. Either the network complexity is increased through the addition of hidden neurons using incremental methods, or we increase the perceptron complexity is increased, as in "support vector machines". The latter can find discriminant surfaces of arbitrary shapes at the price of learning a larger number of parameters.

6.5.1 Spherical Perceptron

A hyperspherical discriminant surface may be implemented through a simple generalization of the linear perceptron. Let us define a spherical potential or activity,

$$v_S = \sum_{i=1}^{N} (x_i - \tilde{w}_i)^2 - w_0^2.$$

The sum over i is the square of the distance between the input $\boldsymbol{x}$ and the weight vector in input space $\tilde{\boldsymbol{w}} = [\boldsymbol{w}_1, \boldsymbol{w}_2, \ldots, \boldsymbol{w}_N]^T$, which is the center of a hypersphere of radius w_0. The perceptron output is

$$\sigma_S = \text{sign}(v_S).$$

Thus, $\sigma_S = +1$ if the point $\boldsymbol{x}$ is outside the hypersphere, and $\sigma_S = -1$ if it is inside (see Fig. 6.17). Notice that the spherical perceptron has the same number of parameters as the linear perceptron. Only the definition of the potential is different. All the training algorithms for the linear perceptron can thus be easily transposed to the spherical perceptron, by introducing the following expression:

$$z_S^k = y^k v_S$$

for the aligned field of example k. It is important to emphasize that, in this case, the weights must not be normalized, since that would impose that the hypersphere center be at a fixed distance of the origin, given by the normalization constant.

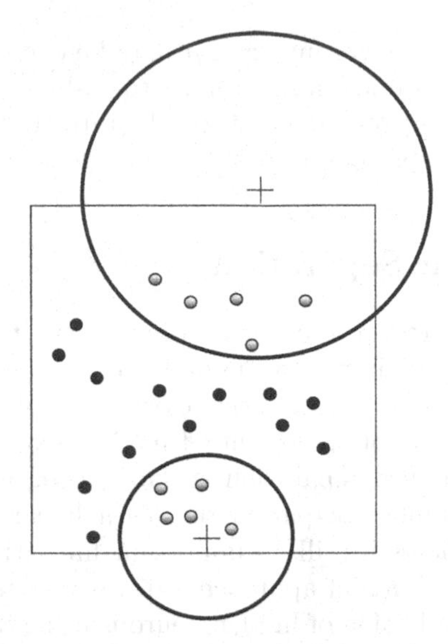

Fig. 6.17. Two discriminant hyperspherical surfaces in dimension $N = 2$. The center of the sphere, indicated by a cross, may be outside the region occupied by the examples

6.5.2 Constructive Heuristics

As mentioned above, if the discriminant surface is neither linear nor spherical, the classification problem may be turned into a regression problem, and the techniques of training and model selection of Chap. 2 may be applied. In that case, the neurons must have differentiable activation functions. The number of hidden units has to be postulated a priori, and is generally adjusted by comparing results obtained with different network sizes, at the cost of time and resources. An alternative solution, presented in this chapter, is to determine the discriminant surfaces by combining linear and spherical separations using binary hidden neurons. Those are included in the network sequentially, following constructive heuristics that use different criteria to associate binary *internal representations* to the input patterns in the training set. The hidden units states corresponding to an input pattern constitute its *internal representation.* Its dimension is the number of hidden units. If those internal representations are linearly separable, an output perceptron connected to the hidden units can learn the discrimination. The probability that the pattern belongs to the class assigned by the classifier may be estimated using the results described in the section "Probabilistic formulation of learning" of the present chapter. In any case, whether the neurons used are binary or continuous, it is important to use the techniques for model selection through cross validation, or statistical tests, explained in Chap. 2, to avoid overfitting: a classifier that classifies correctly all training patterns may be unable to generalize satisfactorily.

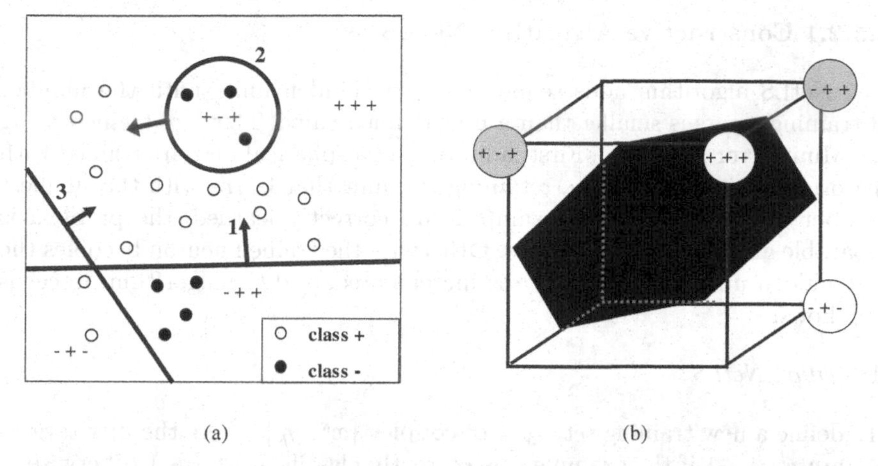

Fig. 6.18. (**a**) Discriminant surfaces generated with the algorithm NetLS. (**b**) Internal representations corresponding to the regions of figure (**a**). The surface shown corresponds to the linear separation of the internal representations, implemented by the output neuron

Remark. The internal representations of the training set are said to be *faithful* if the examples of different class are given different representations. Note that different examples may have the same representation, provided they belong to the same class. This is even desirable, since it allows information compression.

The different constructive or incremental algorithms in the literature generate the internal representations through the sequential addition of hidden units. Especially adapted to learn discrimination tasks with neural networks, they differ by the heuristics (what has to be learnt to the successive hidden units), the final network architecture (trees, layered, etc.) and the algorithm used to train the individual neurons. In particular, the *number* of hidden neurons, which determines the dimension of the internal representations, depends crucially on the efficiency of the training algorithm.

In the following we briefly describe the constructive algorithm NetLS. It generates separations such as those shown on Fig. 6.18(a). In this example, the first hidden neuron (numbered 1 on the figure) implements a linear separation in input space. The second and third ones perform a spherical and a linear separation respectively. The regions thus defined in input space are mapped to faithful internal representations, shown on Fig. 6.18(b). Since they are binary vectors (in dimension 3 in this case), they lie at the vertex of the hypercube in the space of hidden neuron outputs. A separating hyperplane is shown on the same figure. The internal representations are linearly separable: an output perceptron connected to the hidden units can correctly implement the discrimination. It is important to emphasize that the only way to obtain a neural network with binary hidden units is to build it by adding binary perceptrons sequentially.

6.5.2.1 Constructive Algorithm NetLS

The NetLS algorithm adds sequentially the hidden units until the number of training errors is smaller than a user-defined value. They are trained using the Minimerror algorithm. First, a linear and a spherical unit are trained with the original training set L_M, retaining the unit that learns with the smallest number of errors. If all the examples are correctly learned, the problem is separable and the algorithm stops. Otherwise the trained neuron becomes the first hidden unit, $h = 1$. Then, h is incremented, and the algorithm proceeds as follows:

Algorithm NetLS

1. define a new training set $L_{M,h}$ of couples $\{\boldsymbol{x}^k, y_h^k\}$ where the new targets are $y_h^k = +1$ if the example was correctly classified, $y_h^k = -1$ otherwise.
2. train two perceptrons (one linear, one spherical) with the set $L_{M,h}$. Keep the perceptron that makes the smallest number of errors as hidden neuron h.
3. connect an output neuron to the set of h hidden units, and train it to assign the original outputs y^k to the internal representations of the corresponding input patterns $\boldsymbol{x}^k$. If the classification of all the patterns is correct, the algorithm stops. Otherwise, the output neuron is deleted, and the counter of hidden units is incremented.
4. go to 1.

Figure 6.19 shows the algorithm schematically. At iteration $t = 1$, two perceptrons are trained with the original training set L_M. If an error-free solution is found, the algorithm stops. Otherwise, the training set is modified. At $t = 2$, two new perceptrons are trained, and again, the perceptron that generates the smallest number of errors is retained. An output neuron is then connected to the hidden ones, and is trained to discriminate the classes of the examples, based on their internal representations. If an error-free solution is reached, the algorithm stops. Otherwise, the output neuron is eliminated and new targets are defined for each pattern, depending on whether the output neuron classified it correctly or incorrectly. The process is iterated until all examples are classified correctly.

There are several variations of NetLS, which improve the computation time of the algorithm. The interested reader can read the Ph.D. thesis of Juan Manuel Torres Moreno and Christelle Godin referenced above, where applications of the algorithm to different problems are described in details, (see also [Torres Moreno et al. 1998]).

Remark 1. One advantage of the constructive algorithms is their computation time. At each stage, only one neuron is trained. The weights of previously trained hidden neurons are unchanged.

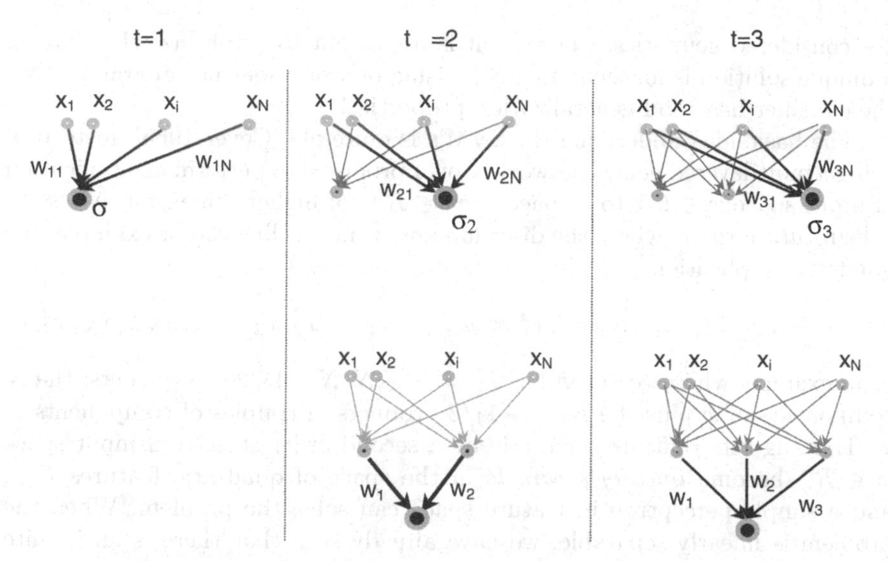

Fig. 6.19. Schematic representation of the NetLS training algorithm

Remark 2. The network has a single hidden layer because it has been shown [Cybenko 1989] that, under mild conditions, a single hidden layer is sufficient to represent any function of the inputs.

Remark 3. The main drawback of constructive algorithms is that the result depends critically on the separation obtained with the first hidden unit. In some cases, keeping the neuron that makes the smallest number of errors may not be the best strategy. Since the neurons added sequentially learn to improve the classification quality by making the internal representations faithful, a poor choice of the first hidden unit may have important consequences on the classifier performance. To avoid this problem, it may be wise to begin with different initial separations, and use the techniques of model selection described in Chap. 2.

6.5.3 Support Vector Machines (SVM)

Support vector machines (SVMs) are generalized perceptrons with high-order potentials. They allow finding, at least in principle, discriminant surfaces of any shape. Recently, their applications have been extensively developed. In this section we introduce the SVM's, and describe the principles of the learning algorithm.

One of the assets of SVMs is the fact that their cost function is convex (i.e., it has a single minimum), in contrast with the algorithms described thus far. Both the least squares cost function usually used in regression or the cross-entropy used for classification problems in Chap. 1, may have local minima. The constructive algorithms find different solutions depending on the considered heuristics. However, it is important to emphasize that having a unique solution is mathematically satisfactory, but does not guarantee that the classifier has good generalization properties!

The basic idea underlying the SVM's is quite old [Cover 1965]: instead of using a multilayered neural network, Cover proposed to perform an application of input space $\boldsymbol{x} \in R^N$ to a space $\Phi(\boldsymbol{x}) \in R^{N'}$ of higher dimension $N' \gg N$, called *feature space*, where the discriminant surface is linear. For example, the quadratic application

$$\varphi : \boldsymbol{x} \to \Phi = \left(x_1, x_2, \ldots, x_N, x_1^2, x_1 x_2, \ldots, x_1 x_N, x_2^2, x_2 x_3, \ldots x_{N-1} x_N, x_N^2\right)$$

is an example where vector $\boldsymbol{\Phi}$ has $N' = N + N(N+1)/2$ components: the N components of $\boldsymbol{x}$ plus the $N(N+1)/2$ products of couples of components of $\boldsymbol{x}$. Training sets that are separable by a second order surface in input space $\boldsymbol{x} \in R^N$ become *linearly separable* in the space of quadratic features $R^{N'}$, and a simple perceptron in feature space can solve the problem. When the problem is linearly separable, we have already seen that there is an infinite number of separating hyperplanes. The SVM solution is the solution with *maximal stability* in feature space. If the following condition on the aligned fields in feature space is obeyed,

$$y^k \boldsymbol{\Phi}(\boldsymbol{x}^k) \cdot \boldsymbol{w} > 1,$$

the margin, which is the distance of the hyperplane to the closest example is $1/\|\boldsymbol{w}\|$. Therefore, maximizing the margin is equivalent to minimizing a cost that is the norm of the weight vector,

$$E = \frac{1}{2} \boldsymbol{w} \cdot \boldsymbol{w}$$

under the above constraints of aligned fields larger than 1 in feature space. If the latter can be satisfied, the solution is a hard margin SVM. Minimizing E with those constraints is a quadratic optimization problem in dimension N', in a convex domain. Its solution is unique, and several algorithms have been optimized to solve it (some of them are available at the URL http://kernel-machines.org). The important point is that the solution can be expressed as follows:

$$\boldsymbol{w} = \sum_{K=1}^{M} c^k y^k \boldsymbol{\Phi}(\boldsymbol{x}^k),$$

with $c^k \geq 0$. It can be shown that there are two subsets of examples: those with $c^k = 0$ and those with $c^k > 0$. The former do not contribute to the weights. If they were excluded from the training set, the solution would be the same. Those with $c^k > 0$, which actually determine the solution, are called support vectors. They verify $y^k \boldsymbol{w} \cdot \boldsymbol{\Phi}(\boldsymbol{x}^k) \equiv 1$, meaning that they are exactly at the margin. In other words, they lie at a distance $1/\|\boldsymbol{w}\|$ of the separating hyperplane.

Remark 1. If the training set is not linearly separable in the feature space determined by function $\boldsymbol{\Phi}(\boldsymbol{x})$, the algorithm does not converge, and does not give even an approximation to the solution.

Remark 2. The same algorithm of constrained quadratic minimization may be used to find the maximal margin perceptron in input space, if the training set is known to be linearly separable. Otherwise, one has to map the patterns to a feature space of dimension high enough to guarantee linear separation.

If the hard margin constraints cannot be satisfied, two possibilities arise: either another application φ is chosen, or the constraint that all examples should be correctly classified and lie outside the margin are relaxed. In the latter case, new variables ζ^k, called slack variables, are introduced. Then the function to be minimized is

$$\Gamma(C) = \frac{1}{2}\boldsymbol{w}\cdot\boldsymbol{w} + C\sum_{k=1}^{M}(\zeta^k)^n,$$

where C is a *positive* hyperparameter ($C > 0$) to be chosen, and n a positive exponent ($n > 0$), under the constraints

$$y^k\boldsymbol{\Phi}(\boldsymbol{x}^k)\cdot\boldsymbol{w} > 1 - \zeta^k$$
$$\zeta^k \geq 0.$$

The minimization of $\Gamma(C)$ under the above two constraints defines the soft margin SVM. Clearly, the slack variables allow the violation of the hard margin constraints. If $0 < \zeta^k < 1$, the distance of the examples to the hyperplane is smaller than $1/\|\boldsymbol{w}\|$, but they are correctly classified. Conversely, those with $\zeta^k > 1$ are incorrectly classified. In order to minimize the number of misclassifications, they must be penalized: that is the rationale for including the second term in the cost function $\Gamma(C)$. The larger the exponent n, the more penalized the misclassifications. However, in order to keep a quadratic minimization problem, the possible values of n are restricted to $n = 1$ or $n = 2$. In such conditions, the solution of the soft margin problem is still unique, and can be found using algorithms of quadratic minimization under constraints. The solution has the same form as the above expression of $\boldsymbol{w}$, with the examples with $\zeta^k \neq 0$ included in the sum. In principle, the coefficient C in $\Gamma(C)$ is arbitrary. Its value controls the relative importance given to the training errors with respect to the margin maximization. Theoretical studies have shown that the value of C has a large influence in the generalization properties of the soft margin SVM's. The choice of its value is a practical problem, and several heuristics have been proposed for the applications of the algorithm.

In practice there is no need to define explicitly the application $\varphi : \boldsymbol{x} \rightarrow \boldsymbol{\Phi}(\boldsymbol{x})$. In fact, from the expression of $\boldsymbol{w}$, classifying a new input $\boldsymbol{x}$ only requires the scalar products $\boldsymbol{\Phi}(\boldsymbol{x})\cdot\boldsymbol{\Phi}(\boldsymbol{x}^k)$ of the input features with those of the support vectors. If the application φ has some general (not too restrictive) properties, those scalar products may be evaluated directly in input space through

$$\Phi(\boldsymbol{x}) \cdot \Phi(\boldsymbol{y}) = K(\boldsymbol{x}, \boldsymbol{y}),$$

where $K(\boldsymbol{x}, \boldsymbol{y})$ is a convolution operator or kernel. Therefore, after training, only the support vectors in input space and the coefficients c^k must be stored in memory. They allow computing the classifier response to any input vector. Thus, if the kernel K is known, it is unnecessary to keep in memory the weight vector $\boldsymbol{w}$, that may have a huge number of components (exactly $N' + 1 \gg N$, since $\boldsymbol{w}$ is a vector in feature space). Moreover, it is not even necessary to define explicitly the application φ: the kernel is sufficient. This is why SVM's are also called kernel machines. One of the most popular kernels is the Gaussian operator

$$K(\boldsymbol{x}, \boldsymbol{y}) \propto \exp\left(-\frac{(\boldsymbol{x} - \boldsymbol{y})^2}{\sigma^2}\right),$$

generally called RBF (for radial basis function). That kernel corresponds to an infinite dimension feature space! Such operators were mentioned previously, in Chap. 1.

As explained at the end of this chapter, the generalization error is a decreasing function of the ratio of the dimension of the space where the perceptron performs the separation to the number of training patterns M. In the case of SVM's, the former is the feature space. Since the number M of training patterns remains fixed, one can wonder whether SVM's are able to generalize at all [Buhot et al. 2000]. It has been proved that the generalization error of the SVM's is bounded by the fraction of training patterns that are support vectors. Note that only values smaller than 0.5 are of interest, since a generalization error of 0.5 means that the classifier has probability 0.5 of misclassification: in other words, it has the same performance as a random classifier. The interest of this bound is that it can be determined in applications (it amounts to simply counting the number of support vectors, and to divide it by M). That problem, as well as other properties of SVM's, is subject to active theoretical research (see for example [Risau-Gusman et al. 2000a,b, 2001; Dietrich et al. 1999; Risau-Gusman et al. 2002]). The interested reader may read the recent Ph.D. thesis of [Risau-Gusman 2001].

6.6 Problems with More than two Classes

A straightforward way of discriminating patterns among more than two classes is to separate each class from all the others. A problem with K classes $y_1, y_2, \ldots, y_K$ is thus reduced to K problems of binary discrimination, with each classifier dedicated to recognize one and only one class. However, it may happen that more than one classifier recognizes the same input pattern. In that case, usually the class is decided through a vote, based on the value of the potential of the output neuron. The underlying rationale, called winner takes all (WTA), is that the larger the potential on the output neuron, the more confident we are on its classification.

We will show below that the probabilistic interpretation of the classification is based on the distance of the examples to the discriminant surfaces, that is, the absolute value of the potential divided by the norm of the weight vector. Therefore, our confidence in a classification should be based on distances and not on bare potentials, unless the weights are normalized. But a deeper problem posed by the WTA procedure is the following: the output unit only reflects the properties of the internal representations. Our confidence should depend on the distances of the input vector to the discriminant surfaces in input space, which are proportional to the potentials of the hidden neurons. It may happen that the input pattern lies so close to one discriminant surface in input space that its class is uncertain. However, its internal representation may have a large stability (see Fig. 6.18), and win in the WTA procedure against the other classifiers.

Another way of dealing with the problem of multiple classes is to construct trees of neural networks. To this end, we choose a *sequence of classes* in an arbitrary order, for example $\{K, 2, \ldots, 1\}$, and we learn the discrimination between the first class and the $K-1$ others. In our example, we may define targets $y = 1$ for the examples of the first class (in our example, y_K), and $y = -1$ for the others. Then, we restrict the training set to patterns of the classes not yet discriminated ($\{2, \ldots, 1\}$ in our example), and we learn the separation of class 2 from the others. The procedure is repeated until the two remaining classes are separated. One interest of this heuristics is that the successive training sets have decreasing sizes. The resulting network has a tree structure. In order to classify a new input, it has to be first classified by the first network. If the output is $\sigma = +1$, the class is K. Otherwise ($\sigma = -1$) the pattern is presented as input to the second network. The procedure stops as soon as one network recognizes (output $\sigma = +1$) the pattern. Since the sequence of classes selected at the beginning is arbitrary, in principle one should compare the outputs of different trees, each tree corresponding to a different sequence of classes. However, if the number of classes is large (typically for $K > 4$) this method is inapplicable. Another solution was proposed in the section "methodology" of Chap. 1: if the classes not mutually linearly separable one may resort to pairwise separation. For a problem with K classes, this requires the construction of $K(K+1)/2$ classifiers which in many practical applications turn out to be linear. Since there is no arbitrary sequence chosen a priori, there is no need to compare the outputs of $K!$ classifiers. One advantage of this solution is that one can use different descriptors for the different separations, which may simplify the problem. We have shown in Chap. 1 how to estimate the probability that a given pattern belongs to each of the possible classes, based on the results obtained in the pairwise separations.

6.7 Theoretical Questions

6.7.1 The Probabilistic Framework

Learning from examples makes sense only if there is some regularity in the data. Within the statistical formulation of training, it is generally assumed that the patterns are {input-output} pairs drawn independently at random from an unknown probability distribution $p(\boldsymbol{x}, y)$. In particular, the probability of the learning set L_M is

$$p(L_M) = \prod_{k=1}^{M} p(\boldsymbol{x}^k, y^k) = \prod_{k=1}^{M} p(\boldsymbol{x}^k) P(y^k \mid \boldsymbol{x}^k).$$

The second term above corresponds to the following process: first the input $\boldsymbol{x}^k$ is drawn at random with probability density $p(\boldsymbol{x}^k)$; given $\boldsymbol{x}^k$, the class y^k is selected with a conditional probability $P(y^k \mid \boldsymbol{x}^k)$. The case of deterministic classes considered in this chapter is just a particular case of this formulation.

Remark. The "teacher-student" paradigm, suggested in Chap. 2 for regression testing, is frequently used in classification theory. It is usually assumed that the components of the input patterns are either Gaussian variables:

$$p(x_i^k) = \frac{1}{\sqrt{2\pi}} \exp\left(-\frac{\left(x_i^k\right)^2}{2}\right),$$

or uniformly distributed variables within some interval $[-a, a] : p(x_i^k) = 1/2a$. Then, the classes of the input vectors $\boldsymbol{x}^k$ are defined by a "teacher" network of weights $\boldsymbol{w}^*$. For example, if the teacher is a deterministic perceptron, one has $P(y^k \mid \boldsymbol{x}^k) = \Theta(y^k \boldsymbol{w}^* \cdot \boldsymbol{x}^k)$. The aim of learning is to find weights $\boldsymbol{w}$ that convey good generalization properties to the "student". Besides the examples of L_M, the "student" is expected to classify correctly any pattern drawn at random with the same probability $p(\boldsymbol{x})$ as the training set.

Because the training set L_M is probabilistic, the student weights $\boldsymbol{w}$ depend on the particular realization of L_M. Therefore, $\boldsymbol{w}$ is a random variable. In this paragraph we apply the method of *Bayesian inference* to the determination of the probability distribution $p(\boldsymbol{w} \mid L_M)$. This method is based on Bayes theorem, introduced in Chap. 1, which can formally be written as follows:

$$p(\boldsymbol{w} \mid L_M) P_B(L_M) = P(L_M \mid \boldsymbol{w}) p_0(\boldsymbol{w}),$$

where $P_B(L_M)$ is defined below; $p_0(\boldsymbol{w})$ is the a priori probability of the classifier parameters (the weights in the case of neural networks) before learning, and $P(L_M \mid \boldsymbol{w})$, called evidence, is the probability of the training set L_M when the student has weights $\boldsymbol{w}$. The a posteriori probability density function for the student weights is

$$p(\boldsymbol{w} \mid L_M) = \frac{P(L_M \mid \boldsymbol{w}) p_0(\boldsymbol{w})}{P_B(L_M)},$$

where

$$P_B(L_M) = \int d\boldsymbol{w}\, P(L_M \mid \boldsymbol{w}) p_0(\boldsymbol{w})$$

guarantees the correct normalization of $p(w \mid L_M)$. It is the marginal probability of the examples within the type of students corresponding to prior p_0. Depending on the hypothesis implicit in the choices of the prior $p_0(\boldsymbol{w})$ and the evidence $P(L_M \mid \boldsymbol{w})$, different Bayesian inferences will be obtained.

Remark. The expression of the a posteriori probability density function for the student weights is Bayes formula *applied to the classifier parameters*, considered as random variables that depend on the realizations of the training set. Note that, in Chap. 1, Bayes formula was applied *to the pattern classes*, considered as random variables depending on the realizations of the vector of descriptors $\boldsymbol{x}$. Those are two different applications of Bayes formula, both within the field of patterns classification.

Usual priors for perceptrons are the Gaussian prior,

$$p_0(\boldsymbol{w}) = \frac{1}{(2\pi)^{N/2}} \exp\left(-\frac{\|\boldsymbol{w}\|^2}{2}\right),$$

and the *uniform prior* on the surface of a hypersphere of radius equal to the norm of the weight vector. For example,

$$p_0(\boldsymbol{w}) = \delta(\|\boldsymbol{w}\|^2 - 1)$$

imposes a unitary norm. In the case of a student perceptron that performs linear discriminations, the above relation is an appropriate choice, since we have already seen that only the orientation of $\boldsymbol{w}$ must be learnt. Note that the above priors do not introduce any information. In the case of the Gaussian priors, it amounts to assume that any weight vector has a non-vanishing probability, with a preference for weights of small norm. With the uniform prior, all orientations have the same probability. Any additional information about the problem should be included in the prior, through an educated choice of $p_0(\boldsymbol{w})$. The other term of the a posteriori probability density function for the student weights that must be provided is the evidence. It contains the information about the examples of the learning problem. If the examples are independent, one can write

$$P(L_M \mid \boldsymbol{w}) = \prod_{k=1}^{M} P(y^k \mid \boldsymbol{x}^k, \boldsymbol{w}) p(\boldsymbol{x}^k),$$

where $p(\boldsymbol{x}^k)$ is the probability density of the input vectors. $P(y^k \mid \boldsymbol{x}^k, \boldsymbol{w})$, the evidence for example k, is the probability that a network with weights $\boldsymbol{w}$ assigns the correct class y^k to the input $\boldsymbol{x}^k$ belonging to L_M.

Remark 1. All choices made before training, such as the network architecture (multilayered or not), the activation function (binary or real-valued), the feature space for the SVM's, correspond to different priors, and are included implicitly in $p_0(\boldsymbol{w})$.

Remark 2. Note that if the evidence is multiplicative, which is a consequence of the assumed independence of the patterns, then the expectation of any additive function of the examples is the sum of the expectations. This remark, developed in the next paragraph, justifies the use of cost functions that are sums of partial costs per example.

6.7.2 A Probabilistic Interpretation of the Perceptron Cost Functions

Within the probabilistic framework, considering a linear student perceptron corresponds to the implicit assumption that the discrimination problem is linearly separable. If we also assume that the task is deterministic, then the evidence of an example k is

$$P(y^k \mid \boldsymbol{x}^k, \boldsymbol{w}) = \Theta(z^k),$$

where $z^k = y^k \boldsymbol{x}^k \cdot \boldsymbol{w}$ is the aligned field. The expectation that a student with weights $\boldsymbol{w}$ misclassifies example k is

$$\varepsilon_t^k = 0 \cdot \Theta(z^k) + 1 \cdot \Theta(-z^k).$$

Therefore, the expected number of training errors is

$$E = \sum_{k=1}^{M} \Theta(-z^k),$$

which is equal (up to an irrelevant constant factor $1/M$) to the cost function $C(\boldsymbol{w}) = (1/M) \sum_{k=1}^{M} V(\boldsymbol{z}^k)$, with $V(z)$ given by $V(z) = \Theta(-z)$.

Remark. The previous relation shows that the weights that minimize $C(w)$ with partial cost $\Theta(-z)$ are those that minimize the expected classification errors if the classification is deterministic.

If we assume that the inputs are perturbed by an *additive noise* we have $\boldsymbol{x}^k = \bar{\boldsymbol{x}}^k + \boldsymbol{\eta}^k$, where the components of the vector $\eta_x \in R^N$ are random variables of zero mean, satisfying $\eta_i^k \ll x_i^k$. The stability of an example k is thus $\gamma^k = \bar{\gamma}^k + \delta^k$, with $\bar{\gamma}^k \equiv y^k \boldsymbol{x}^k \cdot \boldsymbol{w}/\|\boldsymbol{w}\|$. Then, $\delta^k = \boldsymbol{\eta}^k \cdot \boldsymbol{w}/\|\boldsymbol{w}\|$ is a random variable with zero mean and probability density function $p(\delta^k)$. The probability of misclassification of an example k belonging to the training set is

$$P(\bar{\gamma}^k + \delta^k < 0) = P(\delta^k < -\bar{\gamma}^k) = \int_{-\infty}^{-\bar{\gamma}^k} p(\delta^k)\, d\delta^k.$$

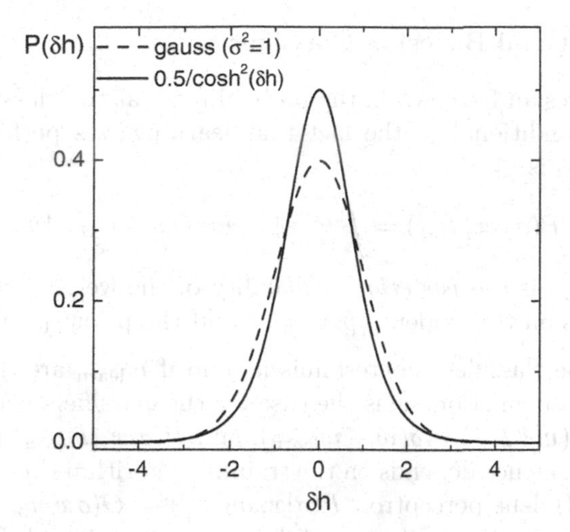

Fig. 6.20. Comparison between a Gaussian and the noise distribution proposed in this section

Depending on the specific expression of the noise term $p(\delta)$, the probability of misclassification has various expressions. Assume that $p(\delta)$ has the following bell-shaped distribution,

$$p(\delta) = \frac{\beta}{2\cosh^2(\beta\delta)},$$

which is close to a Gaussian, as may be seen on Fig. 6.20. In the latter relation, the parameter $\beta - 1$ plays the same role as the variance of the Gaussian: the larger β, the narrower the distribution. Replacing $p(\delta)$ into $P\left(\bar{\gamma}^k + \delta^k < 0\right)$, and neglecting first order terms in δ, we obtain the expected training error on example k,

$$\varepsilon_t^k = \frac{1}{2}\left[1 - \tanh(\beta\gamma^k)\right].$$

This expression is nothing but the cost function of the Minimerror algorithm.

Remark 1. Within the probabilistic formulation, the cost of Minimerror is the expected training error under additive noise on the inputs.

Remark 2. If the noise is Gaussian $p(\delta) = 1/\sigma\sqrt{2\pi}\exp(-\delta^2/2\sigma^2)$ the training error is proportional to the Error function. The latter is more difficult to treat numerically than the hyperbolic tangent. This justifies the above mentionned bell shaped noise assumed for $p(\delta)$ in the training algorithms.

6.7.3 The Optimal Bayesian Classifier

Within the Bayesian framework, the probability that the class of a new input pattern is σ, conditional to the fact that learning was performed with the training set L_M, is

$$P(\sigma \mid \boldsymbol{x}, L_M) = \int P(\sigma \mid \boldsymbol{x}, \boldsymbol{w})\, p(\boldsymbol{w} \mid L_M)\, \mathrm{d}\boldsymbol{w},$$

where $p(\boldsymbol{w} \mid L_M)$ is the *posterior* probability of the weights $p(\boldsymbol{w}|L_M)$, which in turn depends on the evidence $p(L_M|\boldsymbol{w})$ and the prior $p_0(\boldsymbol{w})$.

Remark. If the classifier is deterministic, and if $\boldsymbol{w}_{\mathrm{learn}}$ are the weights that minimize the cost function, as is the case for the classifiers considered in this chapter, then $p(\boldsymbol{w} \mid L_M) = \delta(\boldsymbol{w} - \boldsymbol{w}_{\mathrm{learn}})$, and $P(\sigma|\boldsymbol{x}, \boldsymbol{w}_{\mathrm{learn}})$ is either 0 or 1. Note that the evidence depends on the training algorithm through the weights $\boldsymbol{w}_{\mathrm{learn}}$. For a student perceptron $P(\sigma|\boldsymbol{x}, \boldsymbol{w}_{\mathrm{learn}}) = \Theta(\sigma \boldsymbol{x}.\boldsymbol{w}_{\mathrm{learn}})$. Therefore, if $\boldsymbol{x}.\boldsymbol{w}_{\mathrm{learn}} > 0$, we have $P(\sigma = +1 \mid \boldsymbol{x}, \boldsymbol{w}_{\mathrm{learn}}) = 1$ and $P(\sigma = -1 \mid \boldsymbol{x}, \boldsymbol{w}_{\mathrm{learn}}) = 0$, and symmetrically for $\boldsymbol{x}.\boldsymbol{w}_{\mathrm{learn}} < 0$. The output of a Bayesian perceptron is therefore nothing but the output of the simple perceptron with weights $\boldsymbol{w}_{\mathrm{learn}}$.

Some classifiers are not deterministic. In that case, the probability law $P(\sigma \mid \boldsymbol{x}, \boldsymbol{w})$ is different from the Heaviside function Θ assumed in this chapter. For example, if the inputs of the perceptron are subject to additive noise η, the probability that the response to pattern $\boldsymbol{x}$ is σ can be written as:

$$\begin{aligned} P(\sigma \mid \boldsymbol{x}, L_M) &= P(\sigma \boldsymbol{x} \cdot \boldsymbol{w}_{\mathrm{learn}} + \delta > 0) \\ &= P(\delta > -\sigma \boldsymbol{x} \cdot \boldsymbol{w}_{\mathrm{learn}}) \\ &= \int_{-\sigma \boldsymbol{x} \cdot \boldsymbol{w}_{\mathrm{learn}}}^{\infty} p(\delta)\, d\delta, \end{aligned}$$

where δ stands for $\sigma\eta \cdot \boldsymbol{w}_{\mathrm{learn}}$.

Another case of non-deterministic output arises when the posterior probability of the weights $p(\boldsymbol{w} \mid L_M)$, is not a delta function. For example, consider training a perceptron with the error counting cost function, from a set of linearly separable examples. That cost is highly degenerate: there is a continuum of weights that learn without errors (more generally, that continuum exists whenever the task to be learnt can be performed by the student classifier). Samples of those weights may be obtained using the perceptron algorithm, since the result depends on the weights initialization and the order of the updates. The weights that classify correctly the training patterns is a dense subset $\mathbf{w}_{\mathrm{learn}}$ of weights that occupy a finite volume Ω in weight space. Thus, the posterior probability $p(\boldsymbol{w} \mid L_M)$ is constant in that volume and vanishes outside. To guarantee correct normalization, $p(\boldsymbol{w} \mid L_M) = \Omega^{-1}$. After replacing in $P(\sigma \mid \boldsymbol{x}, L_M)$, we have

$$P(\sigma \mid \boldsymbol{x}, L_M) = \int_{\boldsymbol{w} \in \boldsymbol{w}_{\mathrm{learn}}} \Theta(\sigma \boldsymbol{x} \cdot \boldsymbol{w}) \Omega^{-1}\, \mathrm{d}\boldsymbol{w},$$

where the integral is over the volume Ω. The class assigned by the optimal Bayesian classifier to an input is such that the posterior probability of the class, $P(\sigma \mid \boldsymbol{x}, L_M)$ is maximal. In the particular case of the perceptron, all the weights in volume Ω are equiprobable, and the Bayesian prescription is equivalent to a vote: it assigns to each new input $\boldsymbol{x}$ the class σ that maximizes $P(\sigma \mid \boldsymbol{x}, L_M)$. If $P(+1 \mid \boldsymbol{x}, L_M) > P(-1 \mid \boldsymbol{x}, L_M)$, the optimal Bayesian decision is that the class of $\boldsymbol{x}$ is $\sigma = +1$, otherwise it is $\sigma = -1$. That result was mentioned in Chap. 1 under the term "Bayes decision rule". Notice that it takes this simple form only because $p(\boldsymbol{w}|L_M)$ is a constant.

Remark. In the particular case of a simple perceptron that learns a linearly separable task, the optimal Bayesian decision is to classify the new input like the *majority* of the possible error-free solutions $\boldsymbol{w}_{\text{learn}}(L_M)$. In other words, the Bayesian decision in this case is the *winner-takes-all* solution.

6.7.4 Vapnik's Statistical Learning Theory

In this paragraph we present some results of Vapnik's statistical learning theory, without proofs (see [Vapnik 1995]). The main question addressed by this theory is: are there conditions that guarantee that the minimization of the number of training errors provides the classifier with a minimal generalization error, when the distribution $p(\boldsymbol{x}, y)$ is unknown? In other words, under such conditions, the weights that minimize the training error ε_t should also minimize the generalization error ε_g, irrespective of the probability distribution of the patterns and of the particular realization of the training set. Clearly, a necessary condition that must hold is that the minimum of the training error or *empirical risk* (which is the quantity actually minimized through the learning process) tends to the generalization error, called *functional risk* (the quantity we would like to minimize) when the number of training examples increases. More specifically, the following limit must be obeyed:

$$\lim_{M \to \infty} \varepsilon_t(\boldsymbol{w}_{\text{learn}}; L_M) \to \inf_{w} \varepsilon_g(\boldsymbol{w}),$$

where $\boldsymbol{w}_{\text{learn}}$ are the classifier weights that minimize the number of training errors. If the above relation is satisfied, then the training error is a good estimator of the generalization error: minimizing the former is a good strategy to minimize the latter. Notice that when the student architecture is well adapted to the task, the right-hand term vanishes. This is indeed the case for a perceptron learning a linearly separable task. We have mentioned previously that, in such a case, there are infinitely many weights that cancel ε_t. In fact, there is a finite volume Ω of solutions in weight space. In that case, the condition is satisfied by any algorithm able to find a linear separation. However, in general the student architecture is *not* perfectly adapted to the learning problem. Then $\inf_w \varepsilon_g(\boldsymbol{w}) \neq 0$, and it is difficult to guarantee that an algorithm will find the weights that satisfy the limit, especially if the cost function presents local minima.

Since the training set is a random variable, the general conditions that guarantee that the limit condition is verified for any L_M are not trivial. Vapnik showed that the condition is satisfied if and only if the probability of the largest difference between both terms in vanishes uniformly:

$$\lim_{M\to\infty} P\left[\sup_{\boldsymbol{w},L_M} \left[\varepsilon_g(\boldsymbol{w}) - \varepsilon_t(\boldsymbol{w};L_M)\right] > \delta\right] = 0.$$

The meaning of the above equation is the following: suppose that we have all the possible training sets L_M of M examples, drawn at random with the unknown probability $p(L_M) = \prod_{k=1}^{M} p(\boldsymbol{x}^k, y^k) = \prod_{k=1}^{M} p(\boldsymbol{x}^k)P(y^k \mid \boldsymbol{x}^k)$. The argument in brackets in the above convergence condition means that we determine, for each L_M, weights such that the difference between ε_t (the fraction of misclassified patterns) and ε_g (the generalization error) is maximal. Loosely speaking, the probability P represents the fraction of training sets for which this difference is larger than δ. This means that P is the probability of the *worst case*, since it is the fraction of training sets that exhibit a training error very different from the generalization error. Now, in order to guarantee the good quality of learning, we have to be sure that ε_t and ε_g are close to each other in all cases. That is why the worst case is considered. If the condition of uniform convergence is verified, then ε_t is a good estimator of ε_g for all training algorithms and training sets L_M. That condition guarantees that weights that minimize ε_t, but that endow the network with a very poor generalization performance do not exist (in probability). Since the convergence condition is an asymptotic law, the "guarantee" will be valid only for sufficiently large M. More precisely, Vapnik proved the following inequality: for any δ,

$$\lim_{M\to\infty} P\left[\sup_{\boldsymbol{w},L_M} \left[\varepsilon_g(\boldsymbol{w}) - \varepsilon_t(\boldsymbol{w};L_M)\right] > \delta\right] \leq 4\exp\left[-(M\delta^2 - G(M))\right],$$

where $G(M)$, called growth function, is an upper bound of the *logarithm of D(M,N)*, the number of dichotomies (separations into two subsets) that the student network can perform in dimension N given a training set of M points. We will see below how that number has been computed in the case of the perceptron. $G(2M)$ is an increasing function of its argument, which is independent of the particular training set to be learnt; it depends only on the architecture of the student network: the number of hidden units, of weights, etc. Note that the second term in the above relation is a useful bound (≤ 1) only for $G(M)/M < \delta^2$. Thus, the above condtion is meaningful only if G increases slower than linearly with M.

To summarize, the condition of uniform convergence that guarantees good generalization after training from a set of examples, has been transformed into the problem of determining the student network growth function $G(2M)$ as a function of the size M of the training set. In particular, the bound proves that the generalization error can be bounded only if G increases with M slower that a linear function. If the growth function of the network is known, the bound gives the expected degree of confidence on the classification of new inputs, since from ε_t and M, which are both available in practice, one can bound ε_g.

6.7.4.1 The Vapnik-Chervonenkis Dimension

The question of bounding the generalization error of a network learning M examples in N dimensions has been reduced to that of how its growth function G varies with the size M of the training set. More precisely, exp $G[M]$ is an upper bound to the logarithm of the number $D(M,N;L_M)$ of dichotomies realizable by the student network. In other words,

$$G(M) = \ln \sup_{L_M} D(M,N;L_M).$$

Thus, we have to count the number of dichotomies of M points that the network can perform. One dichotomy of a set L_M of M points is a separation of L_M into two subsets. For example, there are 2^M possible dichotomies of M points in input space. Those correspond to all the ways of assigning classes ± 1 to the examples. If the network is able to implement all of them, then $G(M) = M \log 2 \propto M$, and the bound $\lim_{M\to\infty} P[\sup_{\boldsymbol{w},L_M}[\varepsilon_g(\boldsymbol{w}) - \varepsilon_t(\boldsymbol{w};L_M)] > \delta] \leq 4\exp[-(M\delta^2 - G(2M))]$ is useless. Clearly, if M is small enough, then a perceptron is able to perform all the dichotomies. For example, two patterns in dimension $N=2$ can always be separated by a perceptron. Even three examples are separable into all the possible dichotomies, provided the points are not aligned. In a space of larger dimension N, that non-alignment condition is called condition of *general position* (which is equivalent to asking that no subset of more than N points lies on a hyperplane). Coming back to two dimensions, it is easy to verify that beyond three points, only a fraction of all the possible dichotomies is linearly separable. If all 2^M dichotomies are realizable, the network is said to learn *by heart*; in that case $G(M) \propto M$ (where $\propto$ means proportional) and the bound is useless. That result can be understood intuitively: if the network can learn by heart, either it has too many parameters (in which case $G(M)$ is too large) or M is too small. In both cases, the information conveyed by the training set is not sufficient to compute weights with good generalization properties.

In general, whatever the network architecture, there is a maximal number of examples M_{VC}, called Vapnik-Chervonenkis (or VC) dimension that the student can learn by heart. Beyond that number, the network can only perform a subset of all the possible dichotomies. That is, only for $M > M_{VC}$, $G(M)$ grows more slowly than M and $\lim_{M\to\infty} P[\sup_{\boldsymbol{w},L_M}[\varepsilon_g(\boldsymbol{w}) - \varepsilon_t(\boldsymbol{w};L_M)] > \delta] \leq 4\exp[-(M\delta^2 - G(2M))]$ becomes a true bound. The behavior of G is the following:

$$G(M) \propto \begin{cases} M & \text{if } M < M_{VC} \\ M_{VC} \ln \dfrac{M}{M_{VC}} & \text{if } M > M_{VC}. \end{cases}$$

The training set is a sufficient constraint for the student network only if $M > M_{VC}$. Otherwise, either the network is too complex to capture the regularities of the task to be learnt, or, equivalently, the number of examples is not large enough. Therefore, it is important to know the networks' VC dimension M_{VC}. In the case of the perceptron with N inputs and one threshold (or bias), with M examples to learn, the $N+1$ components of the weight vector $\boldsymbol{w}$ must verify the M inequalities $z^k(\boldsymbol{w}) > 0$ ($k = 1, \ldots, M$). Now, the maximal number of compatible independent inequalities (that is, that have a non trivial solution) is $N+1$. If we had more than $N+1$ inequalities, the system might become incompatible. We cannot guarantee that a solution exists for arbitrary training sets if $M > N+1$. In fact, the solution exists only if the training set is linearly separable. Thus, the VC dimension of the perceptron is

$$M_{VC} = N + 1.$$

Since $M \gg M_{VC}$ is necessary to have good generalization, a lot of theoretical effort has been devoted to determining the Vapnik-Chervonenkis dimension of neural networks. However, for networks more complex than the perceptron, only some approximations of M_{VC} for particular architectures are available. For example, it has been established [Baum 1989] that the VC dimension of networks with one hidden layer of H neurons, having $N_w = (N+1)H + (H+1)$ weights (including the bias), satisfies

$$2N \left\lceil \frac{H}{2} \right\rceil \leq M_{VC} \leq 2N_W \log_2(eH),$$

where $\lceil \; \rceil$ stands for the integer part, and e is the base of the neperian logarithm. The left hand side of the above relation states that, if we have M examples, we should use a number of hidden neurons $H \ll M/N$. This result is somewhat disappointing, as it simply tells us that the network number of parameters (which is of order NH) should be smaller than the number of patterns.

6.7.5 Prediction of the Typical Behavior

A different theoretical approach allows characterizing the learning problems by their typical properties, i.e., properties that are verified with probability 1. That means that the probability that a system does not obey the predicted behavior vanishes. This is similar in spirit to the law of large numbers, which states that the average of N independent identically distributed random variables tends to the expectation of the random variable when $N \to \infty$, with probability 1. Along the same lines, Vapnik's theory states the conditions under which the typical gap between ε_t and ε_g is arbitrarily small in the asymptotic limit $M \to \infty$. Typical properties are thus asymptotic properties.

One limit that helps understanding the properties of learning is to take $N \to \infty$, $M \to \infty$, keeping $\alpha \equiv M/N$ constant. This limit is interesting because the *relative training set size* α, which is the ratio of the number of examples to the dimension of input space, is a meaningful quantity in actual (finite size) applications. Keeping the ratio of the number of training patterns to the network's *number of parameters* (instead of N) constant also provides interesting limits. Those limits are generally (although not necessarily) studied with methods of statistical mechanics [Engel 2001]. In that framework, they are known as the thermodynamic limit. Interestingly, the typical properties evaluated in this limit are still valid for M and N large but finite.

6.7.5.1 The Typical Capacity of the Perceptron

The *capacity* of a network is the largest number of patterns the network can learn to discriminate, with probability 1, irrespective of the discrimination task, i.e., whatever the labels y^k of the examples. We mentioned above that the VC-dimension of the perceptron is $N+1$: in other words, if $M < N+1$, M examples in general position can always be separated by a perceptron. However, the probability that *more* than M_{VC} points are separable *does not vanish* abruptly at $M = N+1$. The typical capacity of a perceptron was first determined by Cover in 1965 through an inductive geometrical reasoning. He counted the number of dichotomies of M points that a perceptron can perform in a space of N dimensions. Note that a perceptron performs the dichotomies by means of oriented hyperplanes; in the following, the two orientations of the hyperplane are considered as two different dichotomies. In general, we have seen that a same dichotomy of L_M (the same assignment of classes to the M points) can be performed by many different (an infinite number of) hyperplanes, but it is counted only once. In the particular case of a perceptron without threshold (nor bias) the following result is obtained: for $M < N$, the number of linearly separable dichotomies of M in dimension N is $D(M,N) = 2^M$. For $M > N$, the result is

$$D(M,N) = 2\sum_{m=0}^{N-1} \binom{M-1}{m}.$$

The above result is a geometrical property of points in an N-dimensional space, irrespective of the training algorithm.

Since the total number of possible dichotomies of M points is 2^M, the probability $P_{LS}(L_M)$ that a set of M points in N dimensions is linearly separable is

$$P_{LS}(L_M) = \frac{D(M,N)}{2^M},$$

which may be written as the sum of the $N-1$ first terms in the expansion of the binomial $(1/2+1/2)^{M-1}$. That sum is equal to $1/2$ for $N-1 = M/2$. Probability $P_{LS}(L_M)$ is shown on Fig. 6.21 for different values of M and N.

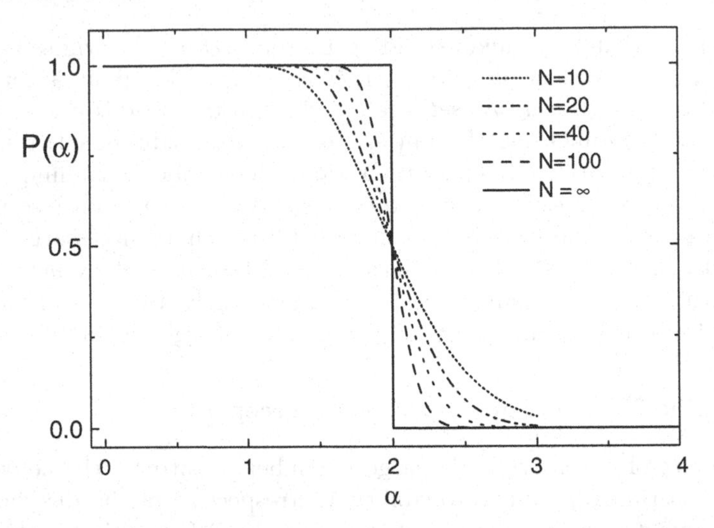

Fig. 6.21. Probability of linear separation of M points in N dimensions

For N large, the probability of linear separation is almost 1 if $M \leq 2N$, and drops abruptly to zero beyond $M/N \sim 2$. Hence, although we cannot guarantee that *any* training set with $M \leq 2N$ is linearly separable, this is highly probable; the larger N and M, the larger the probability of linear separation when $\alpha < 2$. In the thermodynamic limit $N \to \infty$, $M \to \infty$, with $\alpha \equiv M/N =$ constant, the *typical capacity* of the perceptron is $\alpha_c = 2$. Strictly speaking, α_c indicates the transition between the regime where linear separability has probability 1 and the regime where that probability vanishes, in the thermodynamic limit. It is important to emphasize that the behavior of $P_{LS}(L_M)$ is already very close to the asymptotic thermodynamic limit for values of N of the order of 100. That is why typical learning properties provide very useful hints for systems with large, but finite, dimension N.

6.8 Additional Theoretical Material

6.8.1 Bounds to the Number of Iterations of the Perceptron Algorithm

In this section, we provide the computation of the bounds used to prove the convergence of the perceptron algorithm. In order to obtain a lower bound to the norm of the weight vector, we take into account the fact that $\boldsymbol{w}_*$ is unitary:

$$\|\boldsymbol{w}(t+1)\| = \|\boldsymbol{w}(t+1)\|\|\boldsymbol{w}_*\| \geq \boldsymbol{w}(t+1) \cdot \boldsymbol{w}_*.$$

We denote by $k(t)$ the label of the examples learnt at iteration t. After iteration t, the weight vector $\boldsymbol{w}(t+1)$ can be written as

$$\begin{aligned}\boldsymbol{w}(t+1) &= \boldsymbol{w}(t) + y^{k(t)}\boldsymbol{x}^{k(t)} \\ &= \boldsymbol{w}(t-1) + y^{k(t)}\boldsymbol{x}^{k(t)} + y^{k(t-1)}\boldsymbol{x}^{k(t-1)} \\ &= \dots \\ &= \sum_{i=1}^{t} y^{k(i)}\boldsymbol{x}^{k(i)},\end{aligned}$$

where we assumed that the weights were initialized with the *tabula rasa* option: $\boldsymbol{w}(0) = 0$. Taking the scalar product of $\boldsymbol{w}(t+1)$ with the unitary vector $\boldsymbol{w}_*$, and using the above bound for $\|\boldsymbol{w}(t+1)\|$, we deduce the following lower bound,

$$\begin{aligned}\|\boldsymbol{w}(t+1)\| &\geq \sum_{i=1}^{t} \gamma^{k(i)}(\boldsymbol{w}_*) \\ &\geq t\,\gamma_{\min}(\boldsymbol{w}_*),\end{aligned}$$

where $\gamma_{\min}(\boldsymbol{w}_*)$ is the stability of the pattern with smallest stability. Since $\boldsymbol{w}_*$ is a separating hyperplane, $\gamma_{\min}(\boldsymbol{w}_*) > 0$.

An upper bound to $\|\boldsymbol{w}(t+1)\|^2$, can be obtained as follows:

$$\begin{aligned}\|\boldsymbol{w}(t+1)\|^2 &= (\boldsymbol{w}(t) + y^{k(t)}\boldsymbol{x}^{k(t)}) \cdot (\boldsymbol{w}(t) + y^{k(t)}\boldsymbol{x}^{k(t)}) \\ &= \|\boldsymbol{w}(t)\|^2 + 2y^{k(t)}\boldsymbol{x}^{k(t)} \cdot \boldsymbol{w}(t) + \|y^{k(t)}\boldsymbol{x}^{k(t)}\|^2.\end{aligned}$$

The cross-product is negative. As before, we have explicitly

$$\begin{aligned}\|\boldsymbol{w}(t+1)\|^2 &\leq \|\boldsymbol{w}(t)\|^2 + \|\boldsymbol{x}^{k(t)}\|^2 \\ &\leq \dots \\ &\leq \sum_{i=1}^{t} \|y^{k(i)}\boldsymbol{x}^{k(i)}\|^2 \\ &\leq t\|\boldsymbol{x}_{\max}\|^2,\end{aligned}$$

where we used the fact that $|y^k| = 1$. $\|\boldsymbol{x}_{\max}\|$ pertains to the example in L_M of maximal norm. Figure 6.7 illustrates the growth of the norm of $\boldsymbol{w}$ upon learning. From the above lower and upper bounds, we obtain

$$t\gamma_{\min}(\boldsymbol{w}_*) \leq \|\boldsymbol{w}(t+1)\| \leq \sqrt{t}\|\boldsymbol{x}_{\max}\|.$$

Those bounds are shown on Fig. 6.8.

6.8.2 Number of Linearly Separable Dichotomies

In this section we summarize the proof of [Cover 1965]. Consider a set L_m of m points in a space of dimension n. If no subset of $n+1$ points among the m points is linearly dependent, the m points are said to lie in general position.

Notice that if the components x_i of the input patterns are real-valued, the probability that $n+1$ points be on a same hyperplane is zero. In the case of binary input components, that probability does not vanish. In the following we restrict to points in general position.

As mentioned above, if $m \leq n$, all dichotomies of m points in general position are linearly separable. The following computation is interesting for $m > n$. We denote by $D(m, n)$ the number of dichotomies of L_m generated by the hyperplanes in dimension n. For points in general position, that number depends on m and n only. We have

$$D(m, 1) = 2; \quad D(1, n) = 2 \quad \text{for all } m, n$$

since there are two ways of separating (by assigning classes ± 1) m points in 1 dimension, and there are also two ways of assigning a class to a single point in dimension n, with a hyperplane containing the origin. Consider now a training set that includes the patterns of L_m and a new point $\boldsymbol{x}^{m+1}$, $L_{m+1} = L_m \cup \boldsymbol{x}^{m+1}$. It may happen that two hyperplanes that generated the same dichotomy in L_m assign two different classes to $\boldsymbol{x}^{m+1}$. In that case, there exists a hyperplane H^0, containing $\boldsymbol{x}^{m+1}$, such that it generates the same dichotomy on the patterns of L_m. H^0 is ambiguous with respect to $\boldsymbol{x}^{m+1}$. Then, H^0 generates a dichotomy of L_m in the space of dimension $n-1$ orthogonal to H^0. There is a one-to-one correspondence between the $D(m, n-1)$ dichotomies in the space of dimension $n-1$ and the ambiguous dichotomies of the new point in the n-dimensional space. Since there exist $D(m, n)$ dichotomies of L_m, and each one induces two dichotomies of L_{m+1}, we obtain the following recurrence [Cover 1965]:

$$D(m+1, n) = D(m, n) + D(m, n-1)$$

from which the expression of $D(M, N)$ is obtained.

References

1. Anlauf J.K., Biehl, M. [1989], The AdaTron: An adaptive perceptron algorithm, *Europhys. Lett.* 10, pp 687–692
2. Baum E.B., Haussler D. [1989], What size net gives valid generalization?, *Neural Computation*1, pp 151–160
3. Blake, C.L., Merz C.J. [1998], UCI Repository of machine learning databases, available from http://www.ics.uci.edu/mlearn/MLRepository.html
4. Buhot A., Gordon M.B. [1997], Cost function and pattern distribution of the Bayesian perceptron, *Phys. Lett. A* 228, pp 73–78
5. Buhot A., Torres Moreno J.M., Gordon M.B. [1997], Finite size scaling of the Bayesian perceptron, *Phys. Rev. E* 55, pp 7434–7440
6. Buhot A., Torres Moreno J.M., Gordon M.B. [1997], Numerical simulations of an optimal algorithm for supervised learning, *European Symposium on Artificial Neural Networks*, Proceedings, M. Verleysen éd., pp 151–156

7. Buhot A., Gordon M.B. [2000], Storage capacity of a constructive learning algorithm, *J. Phys. A* 33, pp 1713–1727
8. Cover T.M. [1965], *IEEE Trans. Elect. Comp.*, 14, pp 326–334
9. Cover T.M., Thomas J. A. [1991], *Elements of Information Theory*, John Wiley
10. Cybenko G. [1989], Approximation by superpositions of a sigmoidal function, *Mathematics of Control, Signals and Sytems* 2, pp 303–314
11. Dietrich R., Opper M., Sompolinsky H. [1999], Statistical Mechanics of Support Vectors Networks, *Phys. Rev. Lett.* 82, pp 2975–2978
12. Engel A. and Van den Broeck C. [2001], Statistical Mechanics of Learning, Cambridge University Press
13. Gardner E. [1989], J. of Physics A: Mathematical and General 22, N12, In the honour of E. Gardner
14. Godin Ch. [2000], Contributions à l'embarquabilité et à la robustesse des réseaux de neurones en environnement radiatif, thèse de l'École nationale supérieure de l'aéronautique et de l'espace, available from http://www-drfmc.cea.fr
15. Gordon M.B., Grempel D. [1995], Learning with a temperature dependant algorithm. *Europhys. Lett.* 29, pp 257–262
16. Ho E., Kashyap R.L. [1965], An algorithm for linear inequalities and its applications, *IEEE Transactions on Electronic Computers*, 14, pp 683–688
17. Hopfield J.J. [1982], *Proc. Natl. Acad. Sci. USA*, 79, p. 2554
18. Krauth W., Mézard M. [1987], Learning algorithms with optimal stability in neural networks, *J. Phys. A* 20, L745–L752
19. Risau-Gusmán S., Gordon M.B. [2000a], Understanding stepwise generalization of Support Vector Machines: a toy model, *Advances in Neural Information Processing Systems* 12, S.A. Solla, T.K. Leen, K.-R. Müller (éd.), MIT Press, pp 321–327
20. Risau-Gusmán S., Gordon M.B. [2000b], Generalization properties of finite size polynomial Support Vector Machines, *Phys Rev E* 62, pp 7092–7099
21. Risau-Gusmán S., Gordon M.B. [2001], Statistical Mechanics of Soft Margin Classifiers, *Phys. Rev. E* 64, 031907
22. Risau-Gusmán S. [2001], Étude de propriétés d'apprentissage des machines à exemples supports (SVM) par des méthodes de physique statistique, thèse de l'Université de Grenoble I—Joseph-Fourier, available from http://www.uni-bielefeld.de/ZIF/complexity/publications.html, ref. 2001/072
23. Risau-Gusmán S., Gordon M.B. [2002], Hierarchical learning in polynomial support vector machines, *Machine Learning* 46, pp 53–70
24. Rosenblatt F. [1958], The perceptron: A probabilistic model for information storage and organization in the brain, *Phys. Rev.* 65, p. 386
25. Torres Moreno J.M. [1997], Apprentissage et généralisation par des réseaux de neurones: étude de nouveaux algorithmes constructifs, thèse de l'Institut national polytechnique de Grenoble, available from http://www-drfmc.cea.fr
26. Torres Moreno J.M., Gordon M. B. [1998], Characterization of the Sonar Signals Benchmark, *Neural Processing Letters* 7, pp 1–4
27. Torres Moreno J.M. and Gordon M.B. [1998], *Efficient adaptive learning for classification tasks with binary units*, Neural Computation 10, pp. 1017–1040
28. Vapnik V. [1995], *The Nature of Statistical Learning Theory*, Springer

7

Self-Organizing Maps and Unsupervised Classification

F. Badran, M. Yacoub, and S. Thiria

This chapter is dedicated to the second group of neural networks: Topological self-organizing maps. Those models are subject to unsupervised learning, in contrast with multilayer perceptrons, which were described in previous chapters. Primarily, the purpose of those models is purely descriptive: some structure is sought in given data. There is neither precise action to perform, nor desired response to obtain. Alternatively, information compression can be considered as the purpose of unsupervised learning: a compact description of the data, with minimal distortion, is sought.

The unsupervised learning methods that are used by topological self-organizing maps stemmed from techniques that were first designed for competitive learning. Among pioneering works in the field, one may quote [Didday 1976] and [von der Malsburg 1973]. The models were made of parallel filters that analyzed the same observation. For that observation, the filters' responses were different, and the filter that generated the highest response was said to win the competition. That winner is then favored by competitive learning, and the training algorithm enhanced the response of that filter to that observation. The same operation is performed for all observations of the training set until stabilization of the parameters of the filters. At that stage, each filter has been made sensitive to features that are specific to a subset of the data set: it operates as a feature detector.

Topological maps or self-organizing maps were first introduced by T. Kohonen in 1981. The first models were designed for processing high-dimensional data. Very large data sets with high dimensional data vectors were involved in the applications under consideration. In order to process such data, the topological map visualization methodology is designed to partition available data into clusters of data that exhibit some similarity. The training process is driven by the data set. The specificity of topological maps is to provide the clusters with a neighborhood structure, which is actually a graph structure on a discrete set. Low-dimension lattices (1D, 2D or 3D grid) are most frequently considered.

The most important feature of self-organizing maps is the possibility of comparing clusters, which summarize the data. Each observation is allocated to a cluster. Each cluster is projected onto a node of the map. The comparison of projections stemming from different observations allows estimating the proximity between their respective clusters: similar observations are projected onto the same node. Otherwise, the dissimilarity increases with the distance that separates the two projections; that distance is computed on the map. Thus, the cluster space is identified to the map, so that projection enables visualizing simultaneously the cluster space and the observation space.

Unsupervised classifiers and self-organizing maps are closely related; most such methods of clustering aim at aggregating similar data. In that context, similar means close with regard to the application field and the underlying metric. The topological ordering is the specific contribution of neural networks with unsupervised learning to clustering, a key theme in data analysis [Duda et al. 1973; Jain et al. 1988].

In current decision systems, any clustering may contribute to supervised classification as well. Most applications that use self-organizing maps are classifiers. Moreover, some of them are perform regression. Several explanations help to understand that fact:

- Straightforward modifications of the basic algorithm allow its use as a supervised training algorithm [Cerkassky et al. 1991].
- Results of unsupervised training algorithms may easily be integrated into data processing systems that touch the same areas of interest as multilayer Perceptrons. Therefore, self-organizing maps are used to pre-process data: information provided by self-organizing maps may be processed by other algorithms for regression or classification.

Actually, clustering or unsupervised classification turns out to be complementary to discrimination or supervised classification (as described in Chap. 6 of this book). It can be considered in a sense, that any application project uses supervised information to some extent. Any system needs to be validated before use: therefore, available expert knowledge must be used, since an expert has processed some available data so that the associated desired response is known and may be used to tune the automatic system. In particular, this knowledge may be used to improve unsupervised models. If expert knowledge is widely available, then it is possible to take advantage of it from the beginning of the analysis, using supervised forms of self-organizing maps. Conversely, if it is scarce, it can be only used to interpret results of the unsupervised analysis: expert knowledge will be used after achieving clustering tasks. Thus, the approach is sequential: first, a partition of the data set is sought; the recognition itself is subsequently performed.

Self-organizing maps and their theoretical foundations are presented in this chapter. Those algorithms are described under a unified formalism, in order to connect them with data analysis methods from which they actually stemmed: self-organizing map algorithms may be viewed as extensions of well-known algorithms of pattern recognition and clustering. The formalism that we use here is slightly different from the original Kohonen formalism. We will discuss all the necessary links between the various versions of the basic algorithm. Then a section will show in detail how expert knowledge can be used after performing unsupervised training.

This chapter is also application-oriented to a large extent. Two detailed studies of real-world applications are presented. Numerous self-organizing map based concrete projects were carried out in various application fields. Some recent books describe some of those applications [Oja et al. 1999; Kohonen 2001]. A review paper provides a fairly complete bibliography of all papers published between 1981 and 1997 ([Kaski et al. 1998] www.soe.ucsc.edu/NCS). The Helsinki University Web site (http://www.cis.hut.fi/ research/som-research/) addresses a large variety of topics: computer vision, image analysis, image compression, medical imagery, handwriting recognition, speech recognition, signal analysis, music analysis, process control, robotics, Web searching and so on.

The first application that is described in the present chapter deals with remote sensing. By analyzing the details of the modeling that was performed, we will help understand how self-organizing maps are used to perform data analysis. Kohonen's research group performed the second application: the Websom system, which is aimed at document searching on the Web. This application is interesting because the relevant data exhibit very large dimensionality. It is a striking example demonstrating the expected computational power of self-organizing maps.

7.1 Notations and Definitions

This section defines the notations that will be used throughout the present chapter. The set D denotes the observation space. We assume that the observations are real-valued and multidimensional; therefore, D is a subset of the n-dimensional vector space $\mathbb{R}^n$. Each vector belonging to D is associated to a particular encoding of an individual observation, which is taken from the given population. N observation vectors are assumed to be available: they are associated to N individuals. They form the subset $A = \{z_i; i = 1, \ldots, N\}$. Actually, A is included in D. Naturally, it is assumed that A is a representative sample of the considered population. According to that assumption, A is the training set that allows parametric estimation.

All the methods that will be described aim, in a first step, at reducing the information that is present in D. They do so

- by building a finite subset $W = \{\boldsymbol{w}_c; c = 1, \ldots, p\}$ of D; those n-dimensional vectors will be called reference vectors or simply reference throughout this chapter;

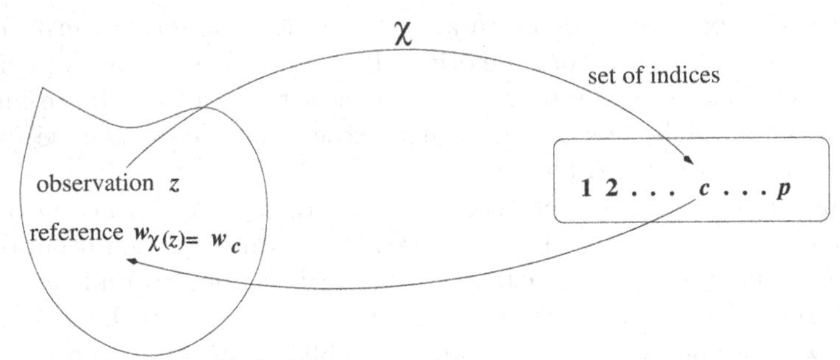

Fig. 7.1. General diagram of the modeling process: one observation z is associated to an index c that is selected among p indices using a function χ; that index allows defining the associated reference w_c

- by defining an allocation function χ from D into the index set $\{1, \ldots, p\}$; that function performs a partition $P = \{P_1, \ldots, P_c, \ldots, P_p\}$ of D into p subsets, $P_c = \{z \in D/\chi(z) = c\}$.

Figure 7.1 describes graphically the modeling process: one observation z is associated to an index c that is selected among p indices using a function χ; that index allows defining the associated reference w_c. Thus, the reference vector w_c is the representative example of the set P_c. It summarizes all the information contained in P_c. In the following, reference w_c or its associated index c will be used, depending on the context, for representing the observation subset P_c. We will estimate the model parameters from the observations of the training set A. Therefore, we denote by n_c the number of elements of P_c.

The knowledge of the reference vector set W and of the allocation function χ generates what is called a *vector quantization*. All known methods to determine W and χ can be derived from a variational principle and amount to a cost function minimization. Each method has a specific cost function. The latter incorporates the specific properties of the associated quantization. The vector quantization permits the allocation of a reference $w_{\chi(z)}$ to any observation $z \in D$. That reference index is $\chi(z)$. Furthermore, the knowledge of the allocation function χ completely determines the partition of the set D into p subsets.

Although the cost functions are different for different methods, all methods that will be described share common features. In the following, the formalism of dynamic clustering will be used. That approach is iterative. Each iteration consists in two steps: a minimization step computes the reference vectors, and an allocation step changes the allocation function χ. Under some assumptions, the cost function decreases at each step and eventually converges towards a local minimum. That minimum depends strongly on the choice of the reference vector set that was selected to initialize the algorithm.

The k-means algorithm is a traditional unsupervised classification algorithm. It is the ancestor of self-organizing maps. In the next section, we describe both the most classical form of that algorithm, and its variants that give insight into the connections with self-organized maps.

For all methods, we first describe the standard version of each algorithm. Then we describe its most popular variants (stochastic or probabilistic versions).

7.2 The k-Means Algorithm

7.2.1 Outline of the k-Means Algorithm

The most known vector quantization method is the k-means algorithm. That method finds the set of reference vectors W and the allocation function χ by minimizing the cost function:

$$I(W, \chi) = \sum_{\boldsymbol{z}_i \in A} \left\| \boldsymbol{z}_i - \boldsymbol{w}_{\chi(\boldsymbol{z}_i)} \right\|^2 = \sum_{c} \sum_{\boldsymbol{z}_i \in P_c \bigcap A} \left\| \boldsymbol{z}_i - \boldsymbol{w}_c \right\|^2.$$

The quantity

$$I_c = \sum_{\boldsymbol{z}_i \in P_c \bigcap A} \left\| \boldsymbol{z}_i - \boldsymbol{w}_c \right\|^2$$

is the local inertia, with respect to the reference vector $\boldsymbol{w}_c$, of the observations of the learning set A that are allocated to that reference vector. Therefore, those observations belong to the subset P_c. That inertia is the squared quantization error performed when the observations of the subset P_c are replaced by the reference vector $\boldsymbol{w}_c$ that represents them. The total cost $I(W, \chi)$, which is to be minimized, is the sum of the local inertias I_c. In order to minimize $I(W, \chi)$, one must define the allocation function χ. The quantity to minimize becomes

$$I(W, \chi) = \sum_{c} I_c = \sum_{c} \sum_{\substack{\boldsymbol{z}_i \in A \\ \chi(\boldsymbol{z}_i) = c}} \left\| \boldsymbol{z}_i - \boldsymbol{w}_c \right\|^2.$$

The algorithm is implemented sequentially. An iteration is split into two phases. The first sep consists in minimizing $I(W, \chi)$: assuming that the reference vectors are kept fixed, it computes the activation function χ. In the second step, the value of the allocation function takes on the value that was just computed: the cost function is then minimized with respect to the parameters W of the reference set. In that two-phase iterative process, the value of the cost function $I(W, \chi)$ decreases at each step.

Thus, an iteration can be summarized as follows:

- *Allocation phase:* $I(W,\chi)$ is minimized with respect to the χ; during that phase, the reference vectors retain their previous values (or the initial values for the fist iteration). The minimization is performed when each observation $\boldsymbol{z}_i$ is allocated to the reference $\boldsymbol{w}_c$ by the allocation function χ:
$$\chi(\boldsymbol{z}) = \arg\min_r \|\boldsymbol{z} - \boldsymbol{w}_r\|^2 . \tag{1}$$
In that relation, r varies from 1 to p (the number of reference vectors). By allocating the closest reference vector (in the sense of Euclidean distance) $\boldsymbol{w}_c$ to each observation $\boldsymbol{z}_i$, the cost function $I(W,\chi)$ is minimized. The new allocation function χ defines a new partition P of the set D (the closest reference vector has to be understood according to the Euclidean distance). In the following, n_c is the cardinal of the set $A \cap P_c$.
- *Minimization phase:* $I(W,\chi)$ is minimized with respect to the reference set W; the allocation function χ that was computed at the previous step is kept constant. The cost function $I(W,\chi)$ is then a convex quadratic function with respect to W. Its global minimum is reached for
$$\frac{\partial I}{\partial W} = \left[\frac{\partial I}{\partial \boldsymbol{w}_1}, \frac{\partial I}{\partial \boldsymbol{w}_2}, \ldots, \frac{\partial I}{\partial \boldsymbol{w}_p}\right]^T = 0.$$

The computation of the gradient that is associated to each reference vector $\boldsymbol{w}_c$ provides a new set of vector equations
$$2 \sum_{\substack{\boldsymbol{z}_i \in A \\ \chi(\boldsymbol{z}_i)=c}} (\boldsymbol{z}_i - \boldsymbol{w}_c) = 0,$$
which define the new reference vectors
$$\boldsymbol{w}_c = \frac{1}{n_c} \sum_{\boldsymbol{z}_i \in P_c \bigcap A} \boldsymbol{z}_i . \tag{2}$$

That algorithm can be proved to converge. If the allocation function that was computed in the first phase is applied, the class of an observation $\boldsymbol{z}$ changes only if its contribution to the global inertia that is computed with respect to the reference set W decreases. Therefore, that global inertia is smaller than the current value of $I(W,\chi)$. The second phase consists in updating the reference set W. Each reference vector $\boldsymbol{w}_c$ defines the center of inertia of the observation set $P_c \cap A$. That requires that $I(W,\chi)$ decrease, since it is the inertia with respect to the center of inertia of partition P. When the two phases are alternatively iterated, the cost function $I(W,\chi)$ decreases. $I(W,\chi)$ is expressed as a function of the trace of partition P on the data set A. That trace is a partition of A. Since the number of partitions of set A is finite, the iterative process converges to a local minimum of the cost function $I(W,\chi)$ with respect to the reference set and to the allocation function.

The implementation of the k-means algorithm can be summarized as follows:

k-Means Algorithm

1. Initialization phase: $t = 0$, choose the initial p reference vectors (randomly, in general), choose the maximal number of iterations N_{iter}.
2. Iteration: at iteration t, the reference set W^{t-1} is known from the previous step:
   Allocation phase: update the allocation function χ^t that is associated to the reference set W^{t-1}: a reference vector is allocated to each observation $\boldsymbol{z}$, as given in (1).
   Minimization phase: compute the new reference vectors W^t, as given in (2).
3. Iterate until the specified maximum number of iterations is reached, or until I stabilizes.

Note that the k-means algorithm may be considered as belonging to the family of dynamic clustering algorithms [Didday 1976]. It is a general method that provides a local minimum of a cost function. That method is based on using two entities: the set of partitions of the original data set into p subsets, and the space W of the representation (which may be different from the data set). Then, a subset P_k of the partition will be represented by an element $\boldsymbol{w}_k$, which will be its associated element of W. The discrepancy between an element $\boldsymbol{x}$ of the data set and its associated element $\boldsymbol{w}_k$ will be assessed by a positive dissimilarity function d such that the smaller $d(\boldsymbol{x}, \boldsymbol{w}_k)$, the better $\boldsymbol{x}$ agrees with $\boldsymbol{w}_k$. Thus, it is necessary to define a partition $P = \{P_k/k = 1 \dots p\}$ into p data subsets and jointly a set $W = \{\boldsymbol{w}_k/k = 1 \dots p\}$ of p representative elements such that they minimize a cost function. The latter will be defined from the training set by

$$H(P, W) = \sum_{k=1}^{p} \sum_{x_i \in P_k \bigcap A} d(\boldsymbol{x}, \boldsymbol{w}_k).$$

The dynamic clustering algorithm minimizes that function iteratively way. First, p representative elements are selected to initialize the process. Then, the general iteration consists of two phases: first an allocation phase that minimizes the cost function with respect to the partition, given the representative elements. Then, during the subsequent phase, the criterion is minimized with respect to the p representative elements, retaining the previous allocation function. In the particular case of k-means algorithm, the reference vectors are the representative elements and the dissimilarity function d is the Euclidean distance.

7.2.2 Stochastic Version of k-Means

The previous algorithm has all the shortcomings of deterministic optimization algorithms. Generally, those algorithms depend strongly on initial conditions, and converge to a local minimum. The optimization mechanism does not allow exploring all the local minima of the cost function. As shown in Chap. 2, optimization can be improved simply by running several optimization processes from various initial conditions, and selecting the best local minimum. In the case of unsupervised learning, the best reference vector set and the best partition will be selected, i.e., those which generate the smallest value of the cost $I(W, \chi)$.

At each iteration, during the minimization phase, the reference set that minimizes the cost function $I(W, \chi)$ for a given allocation function χ is determined. Yet, it is not necessary to complete the move towards the global minimum of the cost function to guarantee that it decreases. At time t, given the allocation function χ^t, finding a reference vector set W^t such that

$$I\left(W^t, \chi^t\right) \leq I\left(W^{t-1}, \chi^t\right)$$

is sufficient.

One may implement a simple gradient descent algorithm, which guarantees the decrease of $I(W, \chi)$ at each step. The computation of the gradient requires the computation of the partial derivatives of $I(W, \chi^t)$ with respect to all the components of each reference vector $\boldsymbol{w}_c$,

$$\frac{\partial I}{\partial \boldsymbol{w}_c} = \sum_{\substack{\boldsymbol{z}_i \in A \\ \chi^t(\boldsymbol{z}_i)=c}} 2\left(\boldsymbol{w}_c - \boldsymbol{z}_i\right).$$

The computation of the reference vectors that was performed by relation $\boldsymbol{w}_c = 1/n_c \sum_{\boldsymbol{z}_i \in P_c \bigcap A} \boldsymbol{z}_i$ at each step is replaced by

$$\boldsymbol{w}_c^t = \boldsymbol{w}_c^{t-1} - \mu_t \frac{\partial I}{\partial \boldsymbol{w}_c} = \boldsymbol{w}_c^{t-1} - 2\mu_t \sum_{\substack{\boldsymbol{z}_i \in A \\ \chi^t(\boldsymbol{z}_i)=c}} \left(\boldsymbol{w}_c^{t-1} - \boldsymbol{z}_i\right).$$

That is the simple gradient descent optimization method that was described in Chap. 2. The allocation function χ^t that appears in the expression of the gradient is defined in the allocation phase of iteration t, the quantity μ^t is the training rate at iteration t, and the reference vector $\boldsymbol{w}_c^{t-1}$ was computed during previous iteration. That algorithm is not adaptive, since it minimizes the global cost function $I(W, \chi)$. To implement any change, the whole data set $\boldsymbol{A}$ has to be used.

The adaptive or stochastic version of the k-means algorithm is obtained by the following modification of the basic optimization procedure. The minimization is now performed stochastically: the terms of the sum in relation $I(W, \chi)$ are considered separately. At each iteration, a single observation $\boldsymbol{z}_i$ of the data set is presented. It leads to the update of the closest reference vector $\boldsymbol{w}_{\chi(\boldsymbol{z}_i)}$. It amounts to decrease only the single term $\|\boldsymbol{z}_i - \boldsymbol{w}^t_{\chi(\boldsymbol{z}_i)}\|^2$ of $I(W, \chi)$ by gradient descent.

Then the partial gradient $2(\boldsymbol{w}^t_{\chi(\boldsymbol{z}_i)} - \boldsymbol{z}_i)$ is used to update the reference vector $\boldsymbol{w}_{\chi(\boldsymbol{z}_i)}$ as follows:

$$\boldsymbol{w}^t_{\chi^t(\boldsymbol{z}_i)} = \boldsymbol{w}^{t-1}_{\chi^t(\boldsymbol{z}_i)} - 2\mu_t \left(\boldsymbol{w}^{t-1}_{\chi^t(\boldsymbol{z}_i)} - \boldsymbol{z}_i\right).$$

A good minimum is obtained by presenting each observation of the data set $\boldsymbol{A}$ repeatedly (N_{iter} must be large enough). When updating the reference vectors, the gradient step μ decreases. When training starts, the value of μ is relatively large, and the decrease of the cost function is not strictly guaranteed. As training proceeds, μ^t becomes small enough, each reference vector update is small, and several updates must be performed to produce a significant change in the cost function. In that case, there is no major difference between the total gradient and the addition of several steps of the partial gradient. Then the stochastic gradient algorithm behaves as the classical version of the k-means algorithm does. The stochastic algorithm shows that k-means may be considered as a competitive algorithm, where each observation of the data set attracts the closest reference vector. Repeatedly presenting each observation while the gradient step μ decreases allows finding a satisfactory partition P such that each reference vector is the center of inertia of each subset of the partition.

The following summary of stochastic k-means may be useful for algorithm implementation:

Stochastic k-Means

1. Initialization: $t = 0$,
   Choose the initial p reference vectors (randomly, in general),
   choose the maximal number of iterations N_{iter} and the law of decrease of the gradient step μ^t.
2. Iteration t: keeping the reference set W^{t-1} constant, as computed at the previous iteration, choose randomly or sequentially an observation $\boldsymbol{z}_i$, and compute the gradient step (or learning rate) μ^t.
   Allocation phase: given W^{t-1}, $\boldsymbol{z}_i$ is assigned assign to the closest reference element of W^{t-1}, which defines a new allocation function χ^t.
   Minimization phase: the new reference vector $\boldsymbol{w}^t_{\chi(\boldsymbol{z}_i)}$ is computed (2).
3. Iterate until the specified maximum number of iterations is reached, or until I stabilizes.

The learning rate must decrease as the number of iterations t increases. It may be piecewise constant, equal to $1/\sqrt{t}$, or have any other ap propriate form.

The three experiments that are shown on Fig. 7.2 allow understanding the evolution of the k-means algorithm, classical and stochastic. They demonstrate the sensitivity of the solution to the number of reference vectors and to their initialization. For those experiments, the observations were generated randomly from spherical Gaussian laws with standard deviation $\sigma = 0.1$. Those laws are called the Gaussian modes. The first experiment is seeks a two-class partition; it shows the evolution of the set of reference vectors that capture observations from the four modes. During training, the two reference vectors are attracted by the two blocks made of the observations of the left-hand and right-hand sides. They stabilize at the centers of the two observation blocks. The second experiment makes use of the same observation data set, but seeks a four-class partition, with two different initializations of the reference vectors: at the center in the first experiment, and at the bottom right in the second experiment. In the first case, the position, which is symmetric with respect to the problem, allows finding the four classes produced by the four Gaussian modes. With the second initialization, three reference vectors are assigned to the data generated by the two right-hand Gaussian modes, and the last one is assigned to the data generated by the other two modes.

7.2.3 Probabilistic Interpretation of k-Means

The k-means is minimizes the cost function $I(W, \chi)$, which is the sum of the local inertias I_c. We defined that cost function by following geometric and kinetic intuition. It is possible to follow another approach. Actually, the cost function has a natural probabilistic interpretation. In order to get insight into that, a probabilistic model of data generation must be defined: we assume that the observations of the training set are an i.i.d. sample of a mixture of p Gaussian modes,

$$p(\boldsymbol{z}) = \sum_{c=1}^{p} \alpha_c f_c(\boldsymbol{z}), \quad \text{with} \sum_{c=1}^{p} \alpha_c = 1.$$

Each Gaussian mode has density f_c with expectation $\boldsymbol{w}_c$ and covariance matrix equal to Σ_c. Therefore, this density is given by

$$f_c(\boldsymbol{z}) = \frac{1}{(2\pi)^{n/2} \det(\Sigma_c)^{1/2}} \exp\left[-\frac{1}{2}(\boldsymbol{z} - \boldsymbol{w}_c)^T \Sigma_c^{-1} (\boldsymbol{z} - \boldsymbol{w}_c)\right].$$

It is well known that a Gaussian mixture model is a general formalism, which can be used for modeling complex probability distributions [Duda et al. 1973]. The mixture assumption states that each observation is a realization of one of the hidden random variables with normal density f_c. The mode is

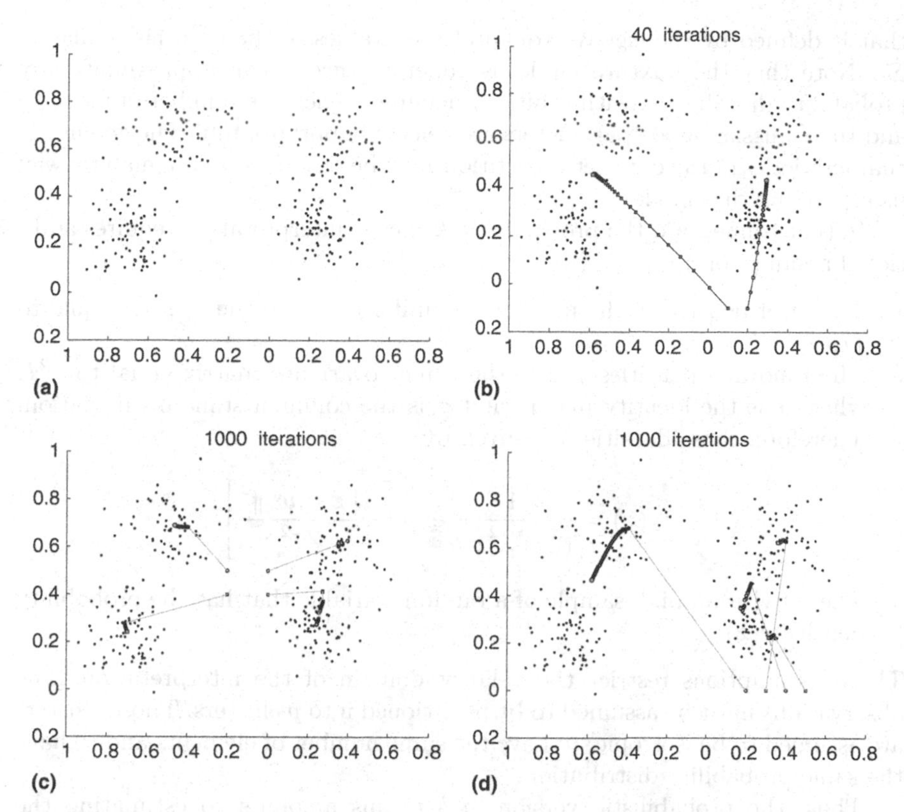

Fig. 7.2. Examples of operation of the k-means algorithm: sensitivity to initial conditions and to the number of reference vectors. Observations and reference vectors are shown on the same picture. (**a**) Representation of the learning set A: the data are generated from four Gaussian modes. (**b**) Evolution of the two reference vectors that were initialized at bottom right of the picture: each reference vector is assigned the observations that are generated from two Gaussian modes. Pictures (**c**) and (**d**) show the evolution of four reference vectors that were generated in two different ways. (**c**) The reference vectors are initialized at the center of the picture; each of them is assigned observations coming from one a Gaussian mode (**d**) The reference vectors are initialized at the bottom right of the picture: three reference vectors share the observations generated by two Gaussian modes; the last reference vector collects the observations generated by the other two modes

selected among p Gaussian modes with the prior distribution α_c. Equivalently, to generate the data, one must first choose randomly the mode according to the discrete probability α_c, and then to generate the observation from the probability law of the selected mode. Thus, that model generates a data set, which is partitioned by construction into p subsets. The subset that is labeled by index c contains about $N\alpha_c$ observations. Those observations are split around the reference vector $\boldsymbol{w}_c$. The subset has an ellipsoidal shape that is defined by the eigenvectors and eigenvalues of the covariance matrix Σ_c. Note that the mixture model is general, since it can approximate any probability distribution with arbitrary accuracy when the number of modes p and the Gaussian mode parameters are selected appropriately. The geometric characteristics of the data set repartition may be described in an analytic way using the mixture model.

In that framework, the probabilistic k-means interpretation requires additional assumptions:

- The prior density on the mode set is uniform, i.e. all the α_c's are equal to $1/p$.
- The p normal densities f_c has the same covariance matrix equal to $\sigma^2 I$, where I is the identity matrix and σ is the common standard deviation. Therefore, those densities are given by

$$f_c(\boldsymbol{z}) = \frac{1}{(2\pi)^{n/2}\sigma^n} \exp\left[-\frac{\|\boldsymbol{z}-\boldsymbol{w}_c\|^2}{2\sigma^2}\right].$$

- The set A is an i.i.d. sample of a random variable that has the probability density $p(\boldsymbol{z})$.

Those assumptions restrict the validity domain of the interpretation. The observations must be assumed to be partitioned into p clusters. Those clusters are assumed to be isotropic, to have the same number of elements and to have the same probability distribution.

Thus, the probabilistic version of k-means amounts to estimating the reference vectors and the common standard deviation by maximizing the likelihood of the data set A. That estimation is performed by maximizing $p(\boldsymbol{z}_1, \boldsymbol{z}_2, \ldots, \boldsymbol{z}_N)$ where $\boldsymbol{z}_1, \boldsymbol{z}_2, \ldots, \boldsymbol{z}_N$ are the observations. Under the independence assumption one has

$$p(\boldsymbol{z}_1,, \boldsymbol{z}_2, \ldots, \boldsymbol{z}_N) = \prod_{i=1}^{N} p(\boldsymbol{z}_i).$$

As in the previous section, the allocation function χ is supposed to assign to each observation $\boldsymbol{z}_i$ its generating mode. The random generating modes are the mixture components. Therefore the allocation function χ defines a partition of the training set A into p subsets. If the classifying likelihood is defined by

$$p(\boldsymbol{z}_1, \boldsymbol{z}_2, \ldots \boldsymbol{z}_N \mid \chi) = \prod_{i=1}^{N} \alpha_{\chi(\boldsymbol{z}_i)} f_{\chi(\boldsymbol{z}_i)}(\boldsymbol{z}_i) = \left(\frac{1}{p}\right)^N \prod_{i=1}^{N} f_{\chi(\boldsymbol{z}_i)}(\boldsymbol{z}_i),$$

then maximizing the classifying likelihood amounts to minimizing

$$V(W, \sigma) = -\ln p(\boldsymbol{z}_1, \boldsymbol{z}_2, \ldots, \boldsymbol{z}_N)$$

$$= \frac{1}{2\sigma^2} \sum_{i=1}^{N} \left\| \boldsymbol{z}_i - \boldsymbol{w}_{\chi(\boldsymbol{z}_i)} \right\|^2 + Nn \ln \sigma + \text{cte}$$

$$= \frac{1}{2\sigma^2} I(W, \chi) + Nn \ln \sigma + \text{cte}.$$

The minimization of $V(W, \sigma, \chi)$ may be performed in two steps:

- In the first step, the cost function $I(W, \chi)$ that appears in the expression of $V(W, \sigma)$ is minimized. One recognizes the global inertia term discussed in the previous section. The k-means algorithm is implemented (also in two steps as described above). That step leads to a local minimum of $I(W, \chi)$, denoted as $I_{\min}$.
- In the second step the quantity

$$\frac{1}{2\sigma^2} I_{\min} + Nn \ln \sigma$$

is minimized with respect to σ. That expression is minimum when its derivative is equal to zero. Therefore, one has

$$\sigma = \sqrt{\frac{I_{\min}}{Nn}}.$$

Thus, the k-means algorithm can be interpreted in a probabilistic framework. The minimization of the cost function $I(W, \chi)$ amounts to the parametric estimation of a probabilistic mixture model under very restrictive assumptions. The assumption of the isotropic identical distribution of the components of the mixture, with the single covariance matrix $\sigma^2 I$, should be emphasized. From a geometric point of view, that algorithm assumes that the data are split into p equally weighted spherical clusters with the same radius. Such is not always the case, so that the assumption is a severe limitation to the application range of the original k-means algorithm.

The following simulation gives insight into the behavior of k-means when the true data distribution does not comply with the assumptions of the probabilistic model. The observations that are shown on Fig. 7.3 are significantly non-isotropic and do not comply with the assumption of equal standard deviations. Therefore, the implementation of k-means in that case favors a solution, which is associated to a partition into two subsets that are as spherical as possible. Thus, the reconstructed partition is far from the original one (Fig. 7.3b).

In order to circumvent the problem, it may be efficient to display a larger number of reference vectors: Fig. 7.4 shows the reference vectors and the associated partition if five reference vectors are used. In that case, four reference vectors are allocated to the left mode while the last reference vector represents the other mode. Then, the problem of clustering the modes to reconstruct the two original classes must be solved. Alternative data analysis methods, such as hierarchical classification, can be taken advantage of. That methodology will be demonstrated in the section "Classification and topological maps," where the introduction of expert knowledge is addressed.

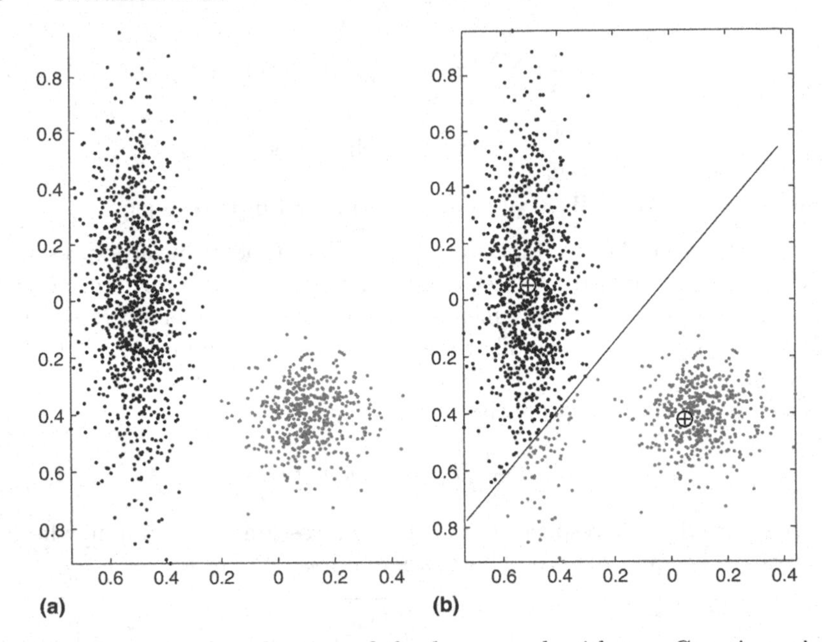

Fig. 7.3. Example of application of the k-means algorithm: a Gaussian mixture model of two modes with different anisotropic covariance matrixes generates the data. Crosses denote the positions of the reference vectors. (**a**) Training set A. (**b**) The reference vectors and the associated partition after stabilization of the algorithm; an oblique line separates the two sets. The algorithm has not reconstructed the true partition

In order to reconstruct the original distributions, the assumption of isotropic covariance must be relaxed. That is possible if the covariance matrices Σ_c of the Gaussians are not supposed to be identical (yet semi-positive definite). Then, the $n(n-1)/2$ elements of the matrices must be estimated, in addition to the reference vectors $\boldsymbol{w}_c$. The model is more complex since it has a larger number of parameters. A maximum likelihood methodology may perform the estimation, using the EM (Expectation-Maximization) algorithm [Dempster et al. 1977].

7.3 Self-Organizing Topological Maps

7.3.1 Self-Organizing Maps

In the early 1980's, Kohonen described a self-organization algorithm that defines a projection of the data space D onto a discrete, low-dimensionality space. That space has a non-oriented graph structure that is a generally a 1-, 2- or 3-dimensional mesh; that graph will be hereinafter termed the map. Actually, the set C is made of interconnected neurons: the connections between

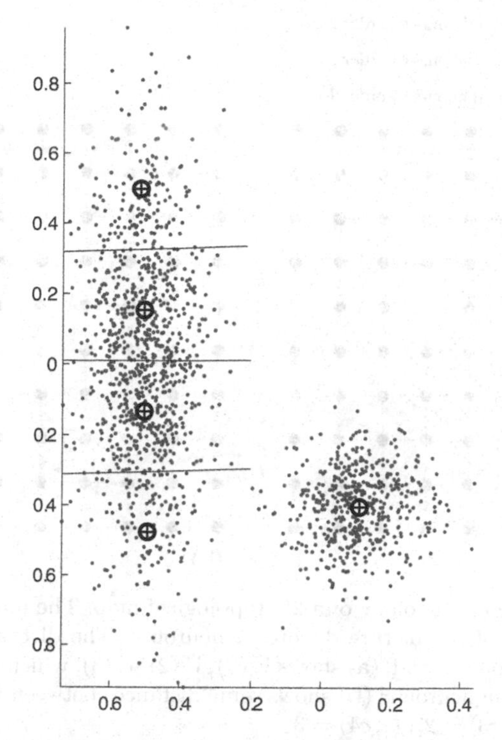

Fig. 7.4. Application of k-means with 5 reference vectors on the same data distribution as in Fig. 7.3; observations are distributed according two Gaussian modes with different anisotropic covariance matrices. Crosses denote the positions of the reference vectors. Four reference vectors are allocated to the first Gaussian mode. The last reference vector is allocated to the second Gaussian mode

neurons are the edges of the graph. For simplicity, we denote the whole graph and the set of its nodes with the same letter C. The graph structure allows the definition of an integer distance δ on C as follows: the length of a path on the graph is the number of edges of that path. For all the couple of neurons (c, r) of the map, $\delta(c, r)$ is the length of the shortest path on C between c and r. For any neuron c, that integer distance leads to defining the neighborhood of c of order d,

$$V_c(d) = \{r \in C, \delta(c, r) \leq d\} .$$

As mentioned above, the maps that are currently used are regular lattices. Therefore, the distance and the neighborhoods are quite easy to visualize, and they define the discrete topology on the map in a straightforward way. Examples of distance and neighborhoods are shown on Fig. 7.5 for a 2D grid.

For self-organizing maps, an association is sought between neurons of C and reference vectors in data space D, similarly to k-means. Training enables the set of reference vectors to sample the underlying probability distribution on the data set as faithfully as possible. In the case of topological maps,

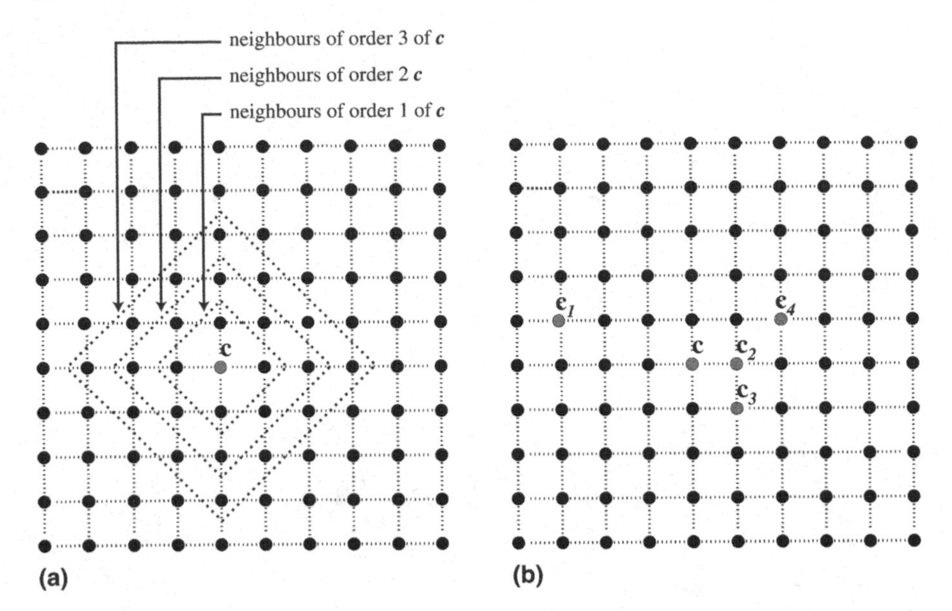

Fig. 7.5. The discrete topology of a 2D-topological map. The map features 10×10 neurons; each dot of the picture denotes a neuron c. The distance δ between two neurons is defined on the grid. (**a**) shows $V_c(1), V_c(2), V_c(3)$, which are neighborhoods of order 1, 2 and 3 of neuron c; (**b**) shows some distances between neurons: $\delta(c, c1) = 4, \delta(c, c2) = 1, \delta(c, c3) = 2, \delta(c, c4) = 3$

an additional constraint is imposed to retain the topology of the map: two neighboring neurons r and c are associated to reference vectors $\boldsymbol{w}_c$ and $\boldsymbol{w}_r$ that are close for the Euclidean distance in data space D.

That cursory description shows clearly that the self-organizing map algorithm is an extension of k-means. We will further show that it minimizes an appropriate cost function, which takes into account the inertia of the partition of the data set, and which guarantees that the topology of C is retained. In order to design such a cost function, the inertia function of the k-means algorithm will be generalized, by adding specific terms that take into account the topology of the map, through the distance δ and the associated neighborhoods.

The concept of neighborhood is taken into account through kernel functions K which are positive and such that $\lim_{|x| \to \infty} K(x) = 0$. Those kernels define influence regions around each neuron c. The distances $\delta(c, r)$ between neurons c and r of the map allow the definition of the relative influence of the neurons on elements of the data set. The quantity $K(\delta(c, r))$ quantifies that influence.

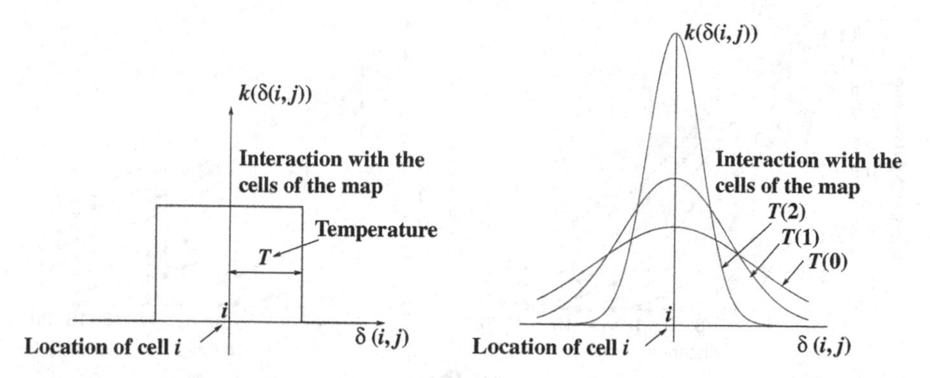

Fig. 7.6. Threshold neighborhood kernel (*left picture*) and Gaussian neighborhood kernel (*right picture*). In the case of the threshold kernel, neurons either belong to the neighborhood and share the same influence, or do not belong to the neighborhood, hence have no influence at all. In the case of Gaussian kernels, the influence between two neurons depends on their mutual distance

To take advantage of the size of the neighborhood, the family of kernels K^T that is parameterized by T (where T stands for temperature) will be used:

$$K^T(\delta) = K(\delta/T)$$

Figure 7.6 shows kernel functions that are commonly used in applications:

- $K(\delta) = \begin{cases} 1 & \text{if } \delta < 1 \\ 0 & \text{otherwise} \end{cases}$ hence $K^T(\delta) = \begin{cases} 1 & \text{if } \delta < T \\ 0 & \text{otherwise} \end{cases}$
- $K(\delta) = \exp(-|\delta|)$ hence $K^T(\delta) = \exp\left(-\frac{|\delta|}{T}\right)$
- $K(\delta) = \exp(-\delta^2)$ hence $K^T(\delta) = \exp\left(-\frac{\delta^2}{T^2}\right)$.

Figure 7.7 shows graphs of various kernels, for different values of parameter T. If we choose a level α such that the influence of a neuron that is below α is considered negligible ($K^T(\delta) < \alpha$), the radius of the effective neighborhood of a neuron can be computed for each value of T. For neuron c, that influence zone is exactly the ball $V_c^T = \{r \in C / K^T(\delta(c, r)) > \alpha\}$. Figure 7.7 shows that the size of the neighborhood decreases with T: the smaller T, the fewer the neurons that belong to the neighborhood V_c^T. The self-organizing map training algorithms minimize a cost function. When the minimum is reached, one gets a partition that is made of sets that are compact enough, and, in addition, it is possible to define an order that stems from the topology of the map. That cost function will be hereinafter noted as J_{som}^T. It plays the role of the cost function I of the k-means algorithm that was described in the previous section. We will now consider the most popular function J_{som}^T, which is

$$J_{\text{som}}^T(\chi, W) = \sum_{z_i \in A} \sum_{c \in C} K^T(\delta(c, \chi(z_i))) \|z_i - w_c\|^2.$$

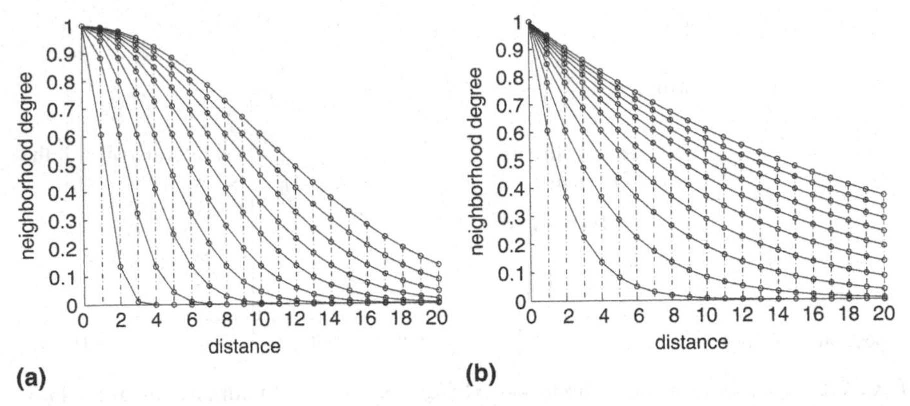

Fig. 7.7. Families of kernel functions that are used to control the neighborhood on the map; x-axis: distance on the map (length of the shortest path between two neurons). The curves show the kernels for different values of T; from top to bottom, T takes on values from 10 to 1 (**a**) $K^T(\delta(c_1,c_2)) = \exp(-0.5\ \delta(c_1,c_2)/T)$ (**b**) $K^T(\delta(c_1,c_2)) = \exp(-0.5\ \delta^2(c_1,c_2)/T^2)$

In that relation, χ is an allocation function, and W is the set of the p reference vectors of the map. $\chi(\boldsymbol{z}_i)$ stands for the neuron of the map C that is associated to the observation $\boldsymbol{z}_i$, and $\delta(c,\chi(\boldsymbol{z}_i))$ is the distance on the map C between a neuron c and the neuron that is allocated to observation $\boldsymbol{z}_i$. As for the k-means algorithm, it is possible to view the links between the map and the data space. Actually, the basic principles of those two algorithms are very similar, as shown on Fig. 7.8. The difference stems from the fact that the set of labels, shown on Fig. 7.1, is replaced by the label *graph* of the map. The cost function J^T_{som} is a mere extension of the k-means cost function $I(W,\chi) = \sum_{\boldsymbol{z}_i\in A}\|\boldsymbol{z}_i - \boldsymbol{w}_{\chi(\boldsymbol{z}_i)}\|^2$, where the Euclidean distance between an observation $\boldsymbol{z}_i$ and its associated reference vector is replaced by a generalized distance, denoted d^T, which takes into account all the neurons of the map

$$d^T\left(\boldsymbol{z}_i, \boldsymbol{w}_{\chi(\boldsymbol{z}_i)}\right) = \sum_{c\in C} K^T\left(\delta\left(c,\chi\left(\boldsymbol{z}_i\right)\right)\right)\|\boldsymbol{z}_i - \boldsymbol{w}_c\|^2 .$$

Note that the distance between $\boldsymbol{z}$ and $\boldsymbol{w}_{\chi(\boldsymbol{z})}$, as expressed by the distance function d^T, is a weighted sum of the Euclidean distances between $\boldsymbol{z}$ and all the reference vectors of the neighborhood of the neuron $\chi(\boldsymbol{z})$. Function J^T_{som} is equal to the function $I(W,\chi)$ if parameter T is small enough. In that case, the distance d^T is identical to the Euclidean distance.

The minimization of the cost function $J^T_{\text{som}}(\chi, W)$ is performed in different ways, depending on whether an adaptive or a batch optimization is desired. In addition, a probabilistic formalism leads to a third version, which explicitly estimates probability densities. Those three versions of the topological map training algorithm are presented in the next sections.

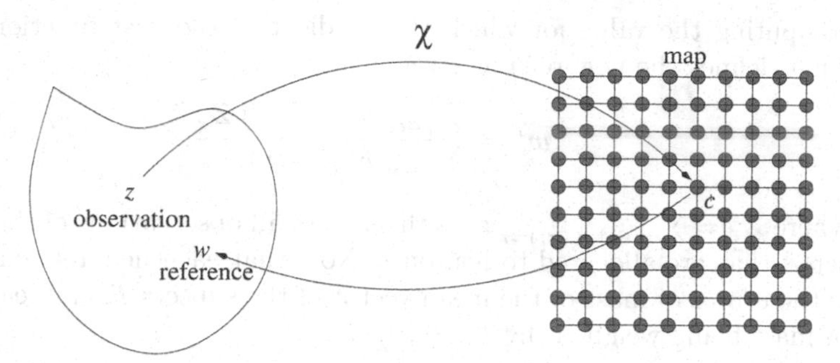

Fig. 7.8. Basic principle of self-organizing map modeling of data space. A label c, which is selected among P neurons of map C, is associated to any observation $\boldsymbol{z}$ of the data set D, using the allocation function $\chi(\chi(\boldsymbol{z}_i = c))$; that label allows the definition of the reference vector $\boldsymbol{w}_c$

7.3.2 The Batch Optimization Algorithm for Topological Maps

In the present section, we describe the minimization of the cost function $J^T_{\text{som}}(\chi, W)$. The only difference between the k-means and the self-organizing map algorithm is the difference between the two cost functions. When T is kept constant, the minimization of J^T_{som} may be written in the dynamic clustering formalism (see the section that is devoted to k-means). Here, just as in the previous section, that formalism provides a proof of convergence of the algorithm to a local minimum of the cost function.

When T is kept fixed, the minimization of J^T_{som} is performed iteratively. Each iteration has two phases. The first phase is an allocation phase and the second phase is a minimization phase where the cost function that is associated to the current partition is minimized:

- *Allocation phase.* $J^T_{\text{som}}(\chi, W)$ is minimized with respect to the allocation function χ. The set W of reference vectors is kept fixed during that phase. The expression of $J^T_{\text{som}}(\chi, W)$ and of $d^T\left(\boldsymbol{z}_i, \boldsymbol{w}_{\chi(\boldsymbol{z}_i)}\right)$ show that the best allocation function is defined for each observation $\boldsymbol{z}$ by

$$\chi^T(\boldsymbol{z}) = \arg\max_{r \in C} \sum_{c \in C} K_T\left(\delta\left(c, r\right)\right) \left\|\boldsymbol{z} - \boldsymbol{w}_c\right\|^2 = \arg\max_{r \in C} d^T\left(\boldsymbol{z}, \boldsymbol{w}_r\right).$$

  That phase allows defining an allocation function χ and the associated partition of data space D. Then the closest reference vector with respect to the weighted distance d^T is allocated to each observation.
- *Minimization phase.* $J^T_{\text{som}}(\chi, W)$ is minimized with respect to the reference vector set W. That minimization is performed while freezing the allocation function χ that was previously computed. Since J^T_{som} is convex with respect to the parameters from W, the minimization can be performed by

computing the value for which the gradient of the cost function is zero. That defines the new reference vector set

$$\boldsymbol{w}_c^T = \frac{\sum_{r \in C} K(\delta(c, r)) \boldsymbol{Z}_r}{\sum_{r \in C} K(\delta(c, r)) n_r},$$

where $\boldsymbol{Z}_r = \sum_{\boldsymbol{z}_i \in A, \chi(\boldsymbol{z}_i)=r} \boldsymbol{z}_i$ is the sum of all observations of the training set A that are allocated to neuron r. Note that each new reference vector is the center of mass of the mean vector of the subsets $P_r \cap A$, each center of mass being weighted by $K(\delta(c, r))n_r$.

To summarize, we get the following algorithm:

Batch Algorithm of Topological Maps: T Fixed

1. *Initialization*: $t = 0$. Select the p reference vectors (randomly, in general), the structure of the map and its size, the maximum number of iterations N_{iter}.
2. *Iteration* t. The reference vector set W^{t-1} is known from the previous step,
   Allocation phase: update the allocation function χ^t that is associated to W^{t-1}. Then each observation $\boldsymbol{z}_i$ is allocated to a reference vector according to

   $$\chi^T(\boldsymbol{z}) = \arg\max_{r \in C} \sum_{c \in C} K_T(\delta(c, r)) \|\boldsymbol{z} - \boldsymbol{w}_c\|^2 = \arg\max_{r \in C} d^T(\boldsymbol{z}, \boldsymbol{w}_r);$$

   Minimization phase: apply relation

   $$\boldsymbol{w}_c^T = \sum_{r \in C} K(\delta(c, r)) \boldsymbol{Z}_r / \sum_{r \in C} K(\delta(c, r)) n_r$$

   to compute the new set W^t of reference vectors.
3. Iterate until the maximum iteration is reached, or until J_{som}^T stabilizes in a local minimum according to a stopping criterion.

As for k-means, a close look at the behavior of self-organizing maps for simple examples gives insight into the implementation problems that may arise. The following numerical experiment illustrates the role of the temperature parameter T in the minimization. The data are the same as on Fig. 7.2 in the section on k-means. As mentioned before, the data are a sample of a uniform mixture of four Gaussian modes with partial pairwise overlap. On Fig. 7.9, the results (topological graph and quantization) are displayed in data space. Kohonen's representations are used. The observations and the reference vectors are shown on the same diagram. The map-induced topology of neighboring neurons is shown as well. Reference vectors that are relative to neighboring neurons on the graph are connected by edges on the picture. At initialization, reference vectors were selected randomly around the center of the observation

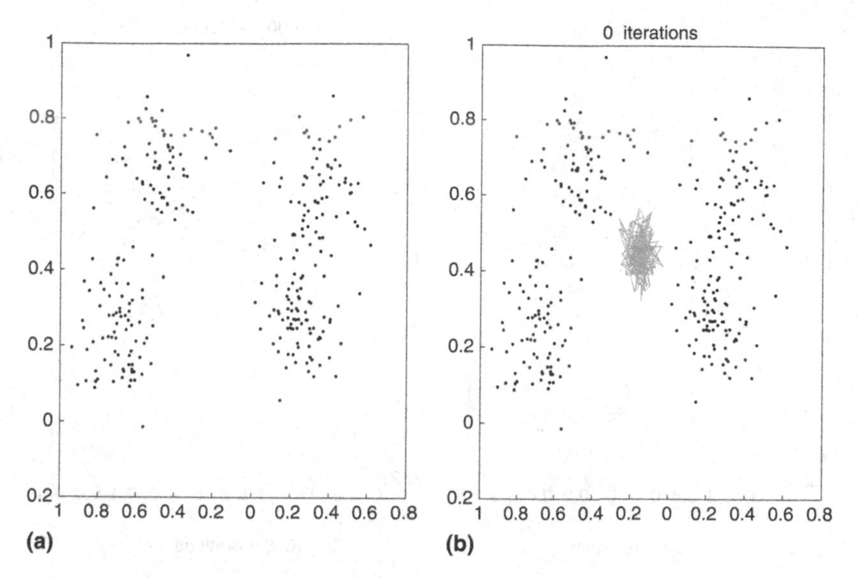

Fig. 7.9. Observation set and initial order generated on the map by random selection of the reference vectors

cloud according to a peaked Gaussian law (its standard deviation is equal to 0.01). Initially, no ordering between the positions of the reference vectors can be observed.

Figure 7.10 shows the maps that are obtained for four distinct values of $T : T = 10$, $T = 5$, $T = 3$ and $T = 1$.

For large values of T, the reference vectors are gathered around the center of mass of the observation cloud. For small values of T, the neighborhood interaction is weaker and the map is unfolded from the same initialization.

The above procedure, for a fixed value of the temperature parameter T, finds a local minimum of the cost function J^T_{som} with respect to χ and W. Actually, Kohonen originally suggested to repeat that minimization a number of times, with a monotonous decrease of T. In that approach, the process performs successive steps of Fig. 7.10. The reference vectors are randomly initialized and order appears when T value is still large: the map then unfolds until it covers the whole space of the observation distribution. The performance of the model on completion of training, and the associated partition, depend on the parameters of the minimization algorithm. The most important parameters are:

- the temperature variation interval $[T^{\min}, T^{\max}]$ of the temperature parameter T, i.e. the initial value of $T(T^{\max})$ and its terminal value $(T^{\min})$;
- the number of times N_{iter} the iterative step is repeated;
- the cooling schedule, i.e. how T decreases in time when it spans the temperature interval $[T^{\min}, T^{\max}]$.

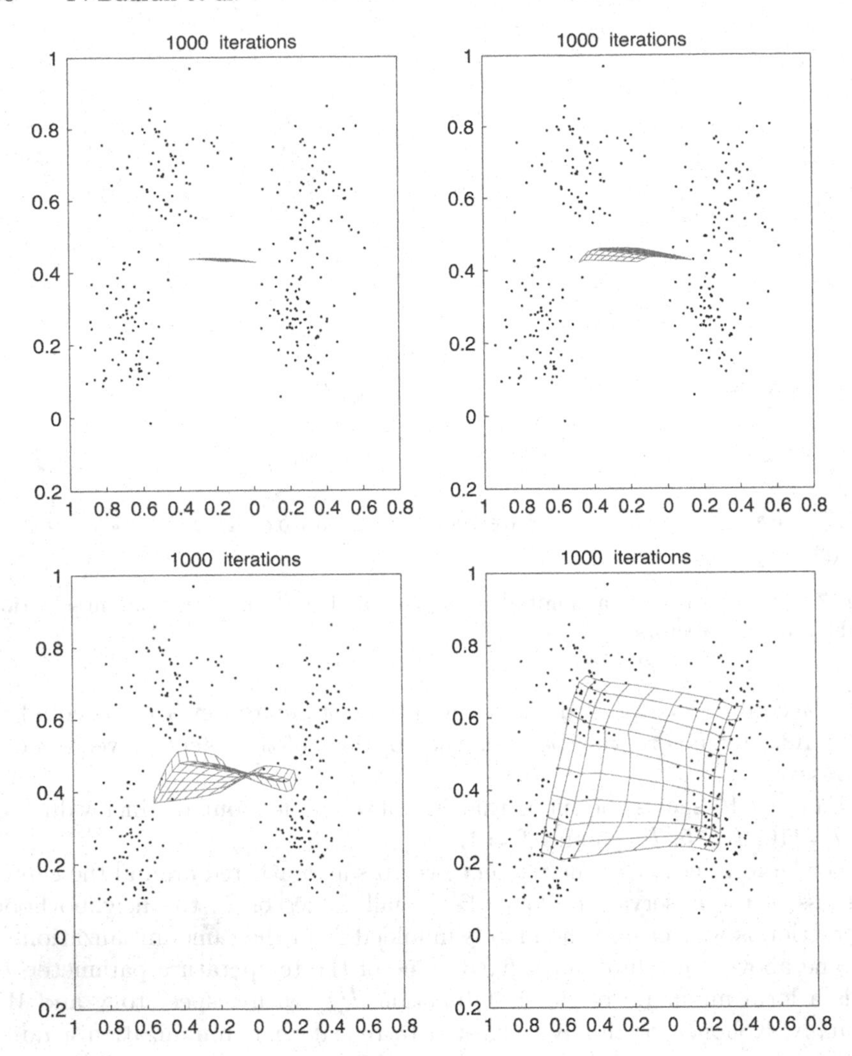

Fig. 7.10. Performing batch algorithm with T fixed (*from top to bottom and from left to right*: $T = 10$, $T = 5$, $T = 3$ and $T = 1$)

Figure 7.11 shows that the cooling schedule may be crucial. On that figure, several final results are displayed for the same training data set, the same temperature variation interval and different cooling schedules. If temperature decreases too fast (quenching), the self-organization is not efficient and the induced neighborhood relation among reference vectors from the data space is not a faithful representation of the graph-induced topology. Using neuronal analogy, one may say that the receptive fields of neurons that are close in cortical maps are not close in the perceptual space. Note that topological self-organization is highly sensitive to the whole parameter set of the algorithm. No

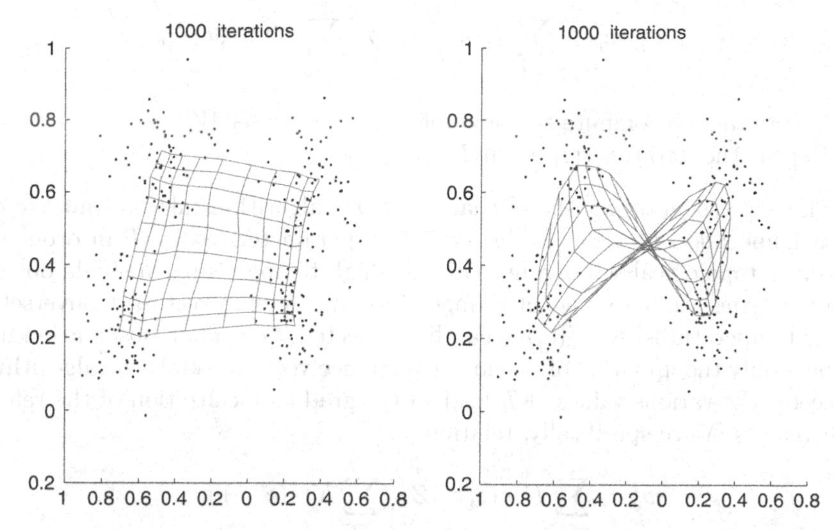

Fig. 7.11. Representation of the order that is induced by the topological map. The maps are obtained for different cooling schedules of T, the same random initialization at the center of the observation cloud and the same variation interval for T

general rule guarantees an appropriate self-organization of the reference vector set. Therefore, it is advisable to test that the algorithm has performed well and that the final set of reference vectors exhibits appropriate self-organization. A relevant real-life example is described in detail below.

For implementation purpose, the batch algorithm of topological maps for a usual cooling schedule can be summarized as follows:

Batch Optimization of Topological Maps (T Decreasing)

1. *Initialization:* perform the dynamic clustering SOM (Self-organizing map) algorithm for $T = T_{\max}$, set $t = 0$
2. *Iteration t.* The reference vector set W^{t-1} is known from previous step. Compute the new temperature value according to the cooling schedule:

$$T = T_{\max}\left(\frac{T_{\min}}{T_{\max}}\right)^{\frac{t}{N_{\text{iter}}-1}}$$

For that temperature T, perform sequentially the two following phases:

- *Allocation:* update the allocation function that is associated to W^{t-1}. A reference vector is allocated to each observation belonging to the training data from relation:

$$\chi^T(\boldsymbol{z}) = \arg\max_{r\in C} \sum_{c\in C} K_T\left(\delta\left(c,r\right)\right)\left\|\boldsymbol{z}-\boldsymbol{w}_c\right\|^2 = \arg\max_{r\in C} d^T\left(\boldsymbol{z},\boldsymbol{w}_r\right).$$

- *Minimization:* apply relation

$$\boldsymbol{w}_c^T = \sum_{r \in C} K(\delta(c,r))\boldsymbol{Z}_r / \sum_{r \in C} K(\delta(c,r))n_r$$

in order to compute the new reference vector set W^t.

3. Repeat the iterative step until $T = T_{\min}$.

The expression of $\boldsymbol{w}_c^T$ shows that the SOM algorithm takes advantage of the neighborhood function $K^T(\delta)$, which is parameterized by T in order to introduce topological self-organization. At high temperatures, a single observation $\boldsymbol{z}_i$ generates a significant change of many reference vectors. Conversely, at low temperatures, $K^T(\delta(c,r))$ can be neglected if $c \neq r$: an observation influences only the updating of its closest reference vectors. While the algorithm proceeds, the various values of T lead to the gradual localization of the reference vectors. More specifically, relation

$$\boldsymbol{w}_c^T = \sum_{r \in C} K(\delta(c,r))\boldsymbol{Z}_r / \sum_{r \in C} K(\delta(c,r))n_r$$

shows that, for a given temperature, the update of a reference vector $\boldsymbol{w}_c$ depends on the observations of the training set that belong to the subset P_c of the partition, and also of the observations of P_r if r is located in a significant neighborhood of c, $r \in V_G^T = \{r | K^T(\delta(c,r)) \leq \alpha\}$.

The smaller T, the smaller the number of neurons belonging to a neighborhood V_c^T, and the smaller number of observations that have an influence on the updating of a reference vector. For T small enough, V_c^T is restricted to the single neuron c, and J_{som}^T strictly amounts to relation the k-means cost function; in that case, SOM is strictly identical to k-means.

Since Kohonen's original SOM training algorithm includes a cooling schedule within bounds that are defined by the temperature variation interval $[T^{\min}, T^{\max}]$, the convergence to a solution occur in two steps. The first step takes place for large values of T: repeated iterations of the dynamic clustering SOM algorithm (with T fixed) tends to guarantee a topological similarity between the reference vector set and the map. The second step takes place at low temperature T: the algorithm tends to get more similar to k-means until complete similarity when T is very small and when $K(\delta(c,r)) \equiv 0$ for any couple (c, r) of distinct neurons. Thus, the first step may be considered as the initialization step of the k-means, using initial reference vectors that retain the topological structure of the map.

The following experiments show how the maps unfold and finally cover the manifold of the training observations. Figure 7.12 describes the progress of training for two different topologies (1D chain and 2D grid) for the four-gaussian-mixture example. The 1-D chain features 50 neurons. The 2-D grid is a 10×10 square mesh. When the reference vectors are initialized randomly around the center of the training data set, the following behavior is observed:

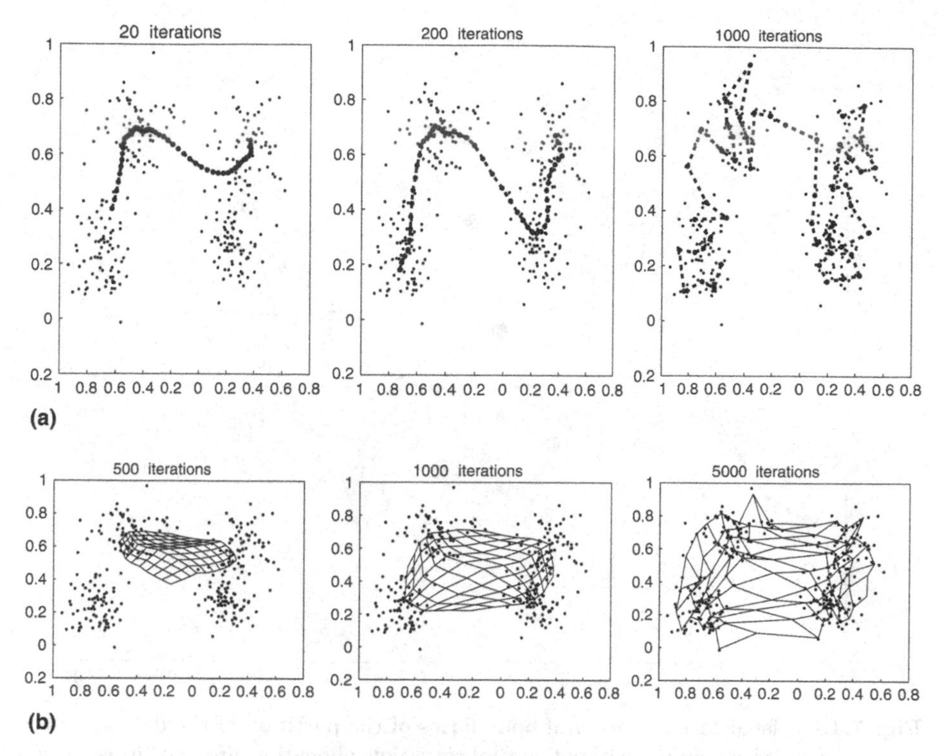

Fig. 7.12. Evolution of the batch training algorithm for the four Gaussian mixture example (pictures a and b) for two different topologies: 1-D with 50 neurons et 2-D with 10×10 neurons. Top pictures display the 1-D map after 20, 200, 2,000 iterations. The same experiment was performed for the 2-D map model; the bottom pictures show the evolution after 50, 500, 5,000 iterations. In the two cases, when convergence is reached, the map covers the whole support of the observation density

- During the first phase, when T is large, the map collapses onto the center of mass, and topological self-organization appears. Then, as T decreases, the map is organized in order to minimize the total inertia of the partition that is associated to the reference vector set. At the end of the algorithm, some reference vectors are positioned at the heart of the observation cloud. Others are trapped in void or low-density regions.
- A close look at the resulting partition provides an interpretation of the hidden structure of observations. Figure 7.13 displays the map. The neurons that have not captured any observation are shown as black dots. Thus, it is possible to separate the data set into two distinct clusters: the algorithm detects natural boundaries.

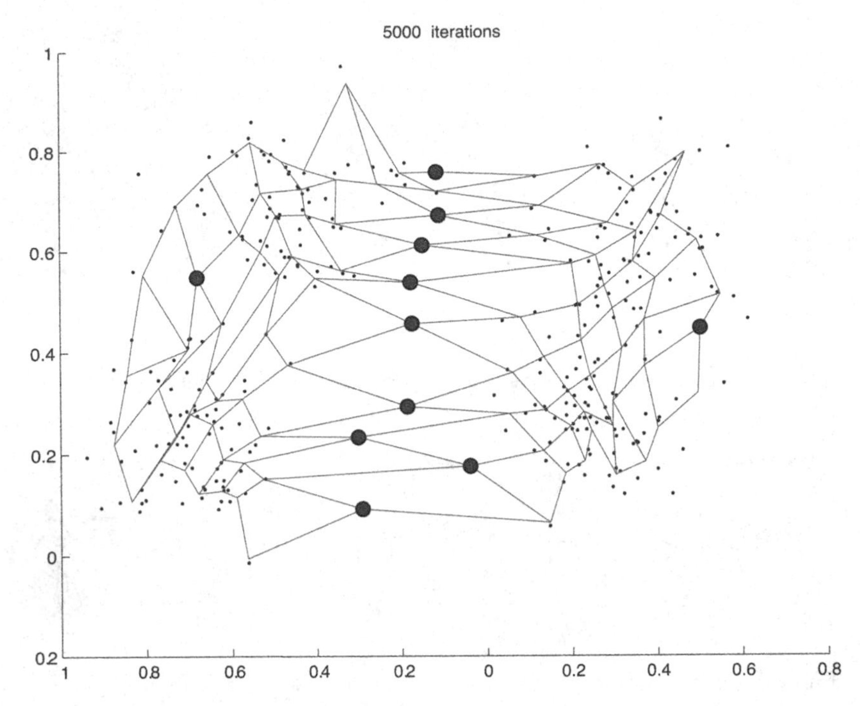

Fig. 7.13. Visualization of natural boundaries of the partition of the data set into two subsets. The neurons without any observation allocation are shown as black dots

7.3.3 Kohonen's Algorithm

The original SOM algorithm, as suggested by T. Kohonen, stems from the dynamic clustering version that was described above. We will now describe its specific features. As for k-means, a stochastic version of SOM is available. It suffices to observe that, during the minimization phase, it is not necessary to terminate the minimization process and compute the global minimum of $J^T_{\text{som}}(\chi, W)$ for a given allocation function χ: one just has to make it decrease. Therefore, relation $\boldsymbol{w}^T_c = \sum_{r \in C} K(\delta(c,r))\boldsymbol{Z}_r / \sum_{r \in C} K(\delta(c,r))n_r$ may be replaced by a simple gradient descent step. Thus, at iteration t and for neuron c, one has

$$\boldsymbol{w}^t_c = \boldsymbol{w}^{t-1}_c - \mu^t \frac{\partial J^T_{\text{som}}}{\partial \boldsymbol{w}^{t-1}_c},$$

where μ^t is the gradient step at iteration t,

$$\frac{\partial J^T_{\text{som}}}{\partial \boldsymbol{w}_c} = 2 \sum_{\boldsymbol{z}_i \in A} K^T \left(\delta\left(c, \chi\left(\boldsymbol{z}_i\right)\right)\right) \left(\boldsymbol{z}_i - \boldsymbol{w}_c\right).$$

That batch algorithm requires that the whole data training set A be available. The contribution of the single observation $\boldsymbol{z}_i$ to parameter $\boldsymbol{w}_c$ to the update is $2K^T(\delta(c,\chi(\boldsymbol{z}_i)))(\boldsymbol{z}_i - \boldsymbol{w}_c^{t-1})$. Alternatively, one may use the stochastic gradient algorithm that computes the reference set once again, at each presentation of an observation $\boldsymbol{z}_i$. That adaptive version is closer to training processes in natural systems. It was the initial version that was suggested by Kohonen. It differs from the batch version that was presented above in two respects: first, the data flow is used instead of the stored data; second the allocation function χ is not the same; Kohonen's algorithm uses the same as in k-means: $\chi(\boldsymbol{z}_i) = \arg\min_c \|\boldsymbol{z}_i - \boldsymbol{w}_c\|^2$.

Therefore, at each presentation of an observation, the new reference vectors are computed for all the neurons of the map C, depending on the selected neuron,

$$\boldsymbol{w}_c^t = \boldsymbol{w}_c^{t-1} - \mu^t K^T\left(\delta\left(c, \chi_t\left(\boldsymbol{z}_i\right)\right)\right)\left(\boldsymbol{w}_c^{t-1} - \boldsymbol{z}_i\right).$$

Thus, Kohonen's algorithm may be summarized as follows:

Kohonen's Algorithm

1. Initialization
   - select the structure and size of the map;
   - choose the initial position of the p reference vectors (usually, this choice is random);
   - choose $T_{\max}, T_{\min}$ and the maximum number of iterations N_{iter};
   - initialize $t = 0$.
2. Iteration t: with the reference vector set W^{t-1}, as computed at the previous iteration:
   - take the current observation $\boldsymbol{z}_i$ (or select randomly an observation from the training set);
   - compute the new value of T according to the cooling schedule:

$$T = T_{\max}\left(\frac{T_{\min}}{T_{\max}}\right)^{\frac{t}{N_{\text{iter}}-1}}$$

   - For that value of T, the following two phases must be performed:
     - *Allocation phase*: W^{t-1} being known, neuron $\chi^t(\boldsymbol{z}_i)$ is assigned to the current observation $\boldsymbol{z}_i$ by the allocation function $\chi(\boldsymbol{z}) = \arg\min_r \|\boldsymbol{z} - \boldsymbol{w}_r\|^2$;
     - *Minimization phase*: the new set of reference set W^t is computed; the reference vectors are updated according to

$$\boldsymbol{w}_c^t = \boldsymbol{w}_c^{t-1} - \mu^t K^T(\delta(c, \chi_t(\boldsymbol{z}_i)))(\boldsymbol{w}_c^{t-1} - \boldsymbol{z}_i),$$

       depending on their distance to the neuron that was selected during the allocation phase.
3. Iterate with decreasing temperature T, until the maximum number of iterations N_{iter} is reached.

7.3.4 Discussion

An in-depth analysis of Kohonen's algorithm unravels its salient features.

- In the update rule for reference vectors, the gradient step μ^t decreases as the number of iterations increases. When the algorithm starts, μ^t is large, and J^T_{som} is not guaranteed to decrease. Later, when the gradient step becomes small enough, the reference vector updates are small for each iteration. In that situation, Kohonen's SOM algorithm behaves in a way similar to the dynamic clustering SOM version.
- If we assume that $K^T(\delta)$ becomes negligible when distance δ exceeds a given threshold d^T, then $K^T(\delta(c,r))$ is significant only for neurons that belong to a given neighborhood of neuron c, whose size is tuned by d^T. That neighborhood will be denoted as $V_c(d^T)$. Thus, when an observation $\boldsymbol{z}_i$ is taken into account, the reference vector $\chi(\boldsymbol{z}_i)$ will be updated, together with the reference vectors of all neurons of the neighborhood $V_{\chi(\boldsymbol{z}_i)}(d^T)$.
- From the point of view of the neuronal representation, the operation of Kohonen's maps can be understood by taking into account the lateral connections between neurons of the map: each neuron c is connected to neighboring neurons r, and any modification of the reference vector $\boldsymbol{w}_c$ generates updates for all reference vectors that are associated to neurons belonging to $V_c(d^T)$ with intensity $K^T(\delta(c,r))$, which decreases with increasing distance $\delta(c,r)$.
- If $K^T(\delta)$ is chosen as a threshold function (see Fig. 7.6), it is constant on the interval $[-d^T, d^T]$ and equal to zero elsewhere, the difference between Kohonen's SOM and k-means is clear. The weight update is the same for the two algorithms; however, in Kohonen SOM, not only is the closest reference vector r updated: the reference vectors associated to neurons of the neighborhood $V_c(d^T)$ are updated as well. Thus, topological self-organization arises: neurons that are close on the map represent observations that are close in data space.
- When temperature T is small, updates according to relation $\boldsymbol{w}_c^t = \boldsymbol{w}_c^{t-1} - \mu^t K^T(\delta(c, \chi_t(\boldsymbol{z}_i)))(\boldsymbol{w}_c^{t-1} - \boldsymbol{z}_i)$ are performed for a subset of all neurons, and, when $d^T < 1$, Kohonen's SOM algorithm is identical to stochastic k-means. Actually, in that case, the only neuron to be updated is the winner of the competition selected by the allocation function χ.

The fact that self-organizing maps are considered as belonging to the family of neural methods stems from the fact that the neural interpretation allows a crisp understanding of the training process. In the following section, we elaborate on that point.

7.3.5 Neural Architecture and Topological Maps

The training algorithms that were described in the previous section allow the determination of the reference vector set $W = \{\boldsymbol{w}_c; c \in C\}$ of a self-organizing

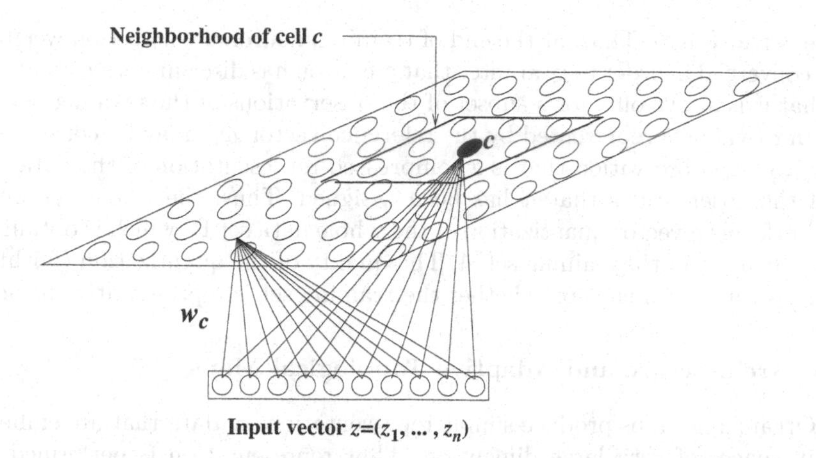

Fig. 7.14. 2-D topological map. The network has two layers: an input layer contains the observations, and a representation layer, for which a topology must be defined (distance δ between neurons and neighborhood function). Each neuron c stands for a reference vector $\boldsymbol{w}_c$; it is fully connected to the input layer. The connection weight vector of each neuron c is the reference vector $\boldsymbol{w}_c$ associated to neuron c

map. A 2-layer neural network provides a joint representation of the map and of the reference vectors (see Fig. 7.14):

- Observations are present in the input layer of the network. The state of each unit is a component of one observation. Therefore, the number of neurons of that layer is equal to the dimension of input space.
- The second layer is the neuronal map. The structure of the map may be decided a priori. In more flexible versions, the structure can evolve during training. The neurons simply compute a distance. Each neuron c is connected to all input units. The reference vector that is associated to the current neuron c of the second layer is actually the vector of connection weights afferent to neuron c. Each neuron has n afferent connections since it is connected to all units of the input layer. When an observation $\boldsymbol{z}$ is presented to the input layer, the output of neuron c of the map is $\|\boldsymbol{z} - \boldsymbol{w}_c\|^2$.

During training, the network connection weights change using various updating rules. Thus, the neurons of the map compute their distances to the current observation in parallel. The main feature of the self-organization process is to focus the adaptation process on the most active area of the map. Kohonen's original algorithm, which is the simplest one, considers that the active zone is the neighborhood of the neuron c that is closest to the observation under consideration, i.e. whose output $\|\boldsymbol{z} - \boldsymbol{w}_c\|^2$ is smallest. That neighborhood generates topological constraints that lead to self-organization.

As indicated in the previous section, it models in a simple way, the lateral coupling between an active neuron and its neighbors on the connection graph of the cortical map. Thus, at the end of training, neuronal connection weights have converged in order to guarantee that a neuron has discriminative abilities, i.e. that it is active only for a subset of the observations of the training set. A neuron c, which is represented by the reference vector $\boldsymbol{w}_c$, may be considered as an average observation that is a compressed representation of the data set P_c of the observations that it has been assigned. Thus, the whole neuronal map performs a vector quantization of the whole data set D, which is obtained by the analysis of the training set A. The quality of the quantization (faithful or not) strongly depends on whether the training set is representative or not.

7.3.6 Architecture and Adaptive Topological Maps

Self-Organizing Maps produce simple representations of data that are embedded in spaces of very large dimension. That representation is performed in a low-dimension discrete set C with a graph structure. The problem of the choice of architecture consists in selecting a suitable graph structure for the map, i.e. a structure that is appropriate for the specific problem of interest. Therefore, one must define a measure of the adequacy of a map to the problem of interest. The data set D and the map C are related in two ways: the embedding of C into D that maps each neuron c of C onto a reference vector $\boldsymbol{w}_c$ of C, and the allocation function χ of D into C, which associates to each observation vector in D a neuron c of the map. Those two mappings have to be topologically consistent in the following sense:

- Two neurons that are neighbors in the map C must be represented by two reference vectors that are close in D.
- Reciprocally, data that are approximately similar must be allocated by χ to the same neuron or to neighboring neurons.

If the dimension of the map does not fit with the underlying dimension of the data cloud (dimension of the manifold that is generated by the observations), two observations that are close in data space D may be allocated to distant neurons in the map. Yet, the topological consistency is an interesting property because it allows reducing the dimension of the data while retaining similarities. In previous sections, it was assumed that the graph structure of the map was given a priori. That choice was not data-driven, which has shortcomings: it does not guarantee the adequacy between the structure of the map and the internal structure of the data distribution.

Usually, in applications, the dimension of the data space may be very large if the number of features describing the data is large, but the observations are not distributed uniformly in the data set. They are located in specific regions with various concentrations. Reference vectors must be located in high-density regions, and one must avoid wasting reference vectors by locating them in void regions. The choice of the graph structure of the map is very important because, when it is appropriate, it guarantees the topological consistency of the map and a good representation of the underlying data probability distribution.

In order to solve that problem, one may consider an oversize map (with respect to the underlying dimension of the problem) and apply Kohonen SOM algorithm to it. After training, neurons that capture no observation of the training set are discarded. Kohonen's algorithm is performed on the resulting new structure. That *pruning* process is iterated as long as necessary.

Another technique consists in defining the map (size and dimension) during training while updating the reference vectors. Thus, the map is built incrementally, allowing the addition of some neurons and the deletion of others. Several methods have been proposed in the literature. They can be classified into two categories:

- In the first category, the dimension k of the map is decided a priori, and the map is built adaptively by addition and deletion of neurons. In order to perform those operations in a systematic way, simple graphical structures such as segments for $k = 1$, triangles for $k = 2$, tetrahedrons for $k = 3$ and so on, are processed. [Oja et al. 1999].
- A second category allows the data to drive the selection of the dimension of the map, which may vary from one region to another. The neural gas algorithm [Oja et al. 1999; Fritzke 1995] is building the graph by introducing the connection links in the data space itself. In that method, whenever an observation is presented, the two closest reference vectors are selected; if they are connected, the connection is activated, otherwise the corresponding connection link is created. The connection links that are inactive during a fixed number of iterations are deleted.

7.3.7 Interpretation of Topological Self-Organization

The structure of the cost function J^T_{som} gives insight into the topological self-organization during training. The subsets $P_r \cap A$ generate a partition of the training set A, so that J^T_{som} can be written as follows:

$$J^T_{\text{som}} = \sum_r \sum_{z_i \in P_r \bigcap A} \sum_c K^T(\delta(c,r)) \|z_i - w_c\|^2$$

$$= \sum_c \sum_r \sum_{z_i \in P_r} K^T(\delta(c,r)) \|z_i - w_c\|^2.$$

Decomposing that relation shows that the cost function J^T_{som} generates a vector quantization and guarantees topological consistency

$$J^T_{\text{som}} = \left[\sum_c \sum_{r \neq c} \sum_{z_i \in P_r} K^T(\delta(c,r)) \|z_i - w_c\|^2 \right] + K^T(\delta(c,c)) \sum_c \sum_{z_i \in P_r} \|z_i - w_c\|^2$$

$$= \frac{1}{2} \sum_{c} \sum_{r \neq c} K^T \left(\delta\left(c,r\right)\right) \left[\sum_{z_i \in P_r} \left\| \boldsymbol{z}_i - \boldsymbol{w}_c \right\|^2 + \sum_{z_i \in P_c} \left\| \boldsymbol{z}_i - \boldsymbol{w}_r \right\|^2 \right]$$

$$+ K^T \left(\delta\left(c,c\right)\right) \left[\sum_{c} \sum_{z_i \in P_c} \left\| \boldsymbol{z}_i - \boldsymbol{w}_c \right\|^2 \right].$$

This decomposition gives two terms, the sum of which must be minimized:

- The second term is I of k-means, weighted by $K^T(\delta(c,c)) = K(0)$. Its influence is controlled by the temperature parameter T: the smaller the temperature, the more influential that term during minimization. It tends to build a partition into compact subsets, and the reference vectors tend to be the centers of mass of the partition subsets.
- The first term enforces the topological consistency constraint: if two neurons r and c are close on the map, $K^T(\delta(c,r))$ is large, because $\delta(c,r)$ is small. Minimizing that term decreases the distance between the subsets P_c and P_r that are allocated to c and r. Thus, proximity on the map enforces proximity in the data set.

The above form of J^T_{som} also gives insight into the presentation of the algorithm as consisting in two different steps that depend on the temperature T (see above the section on batch optimization algorithm of topological maps). The first step occurs when T is large: the first term is dominant, and the task of the algorithm is mainly to guarantee the topological consistency of the map. The second step occurs at lower temperature. In that case, the second term becomes dominant and the algorithm essentially minimizes the inertia of the partition. The temperature allows performing the appropriate tradeoff between the two terms of J^T_{som}. Since the topological self-organization occurs during the first part of training, then the minimization is useful to obtain subsets that are as compact as possible. It is the k-means phase of the algorithm that consists in approximating locally the data distribution. Thus, the algorithm may be cursorily described as a version of the k-means algorithm subject to the constraint of topological consistency of the reference vectors with the map.

The following experiment gives insight into the difference between SOM and k-means. We consider again the example that was displayed on Fig. 7.2[d] to illustrate k-means. In that case, a topological map with a 1D chain structure is used with four neurons, and the parameters of the map are estimated from the training set, generated from a mixture of four Gaussians.

The four reference vectors were initialized at the bottom right of the figure just as for the previous k-means experiment. The two solutions that are obtained by k-means and SOM are shown on Fig. 7.15. The map topology constraint allows locating the four neurons at the centers of the four Gaussian modes. Thus, the SOM algorithm was able to determine the solution of the k-means problem under the topological consistency constraint (Fig. 7.15 [b]);

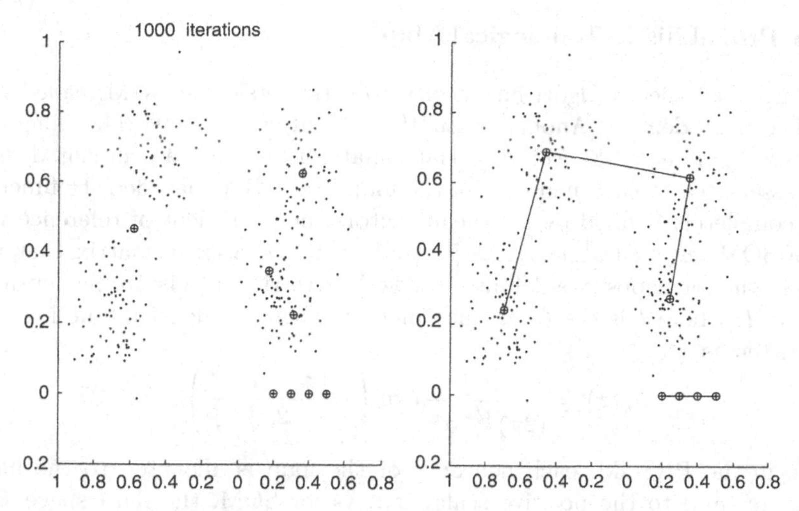

Fig. 7.15. Comparison k-means (**a**) and SOM (**b**) for the same initialization. The reference vectors are initialized in the right bottom of the picture

that solution is different from the solution that was found by the straightforward implementation of k-means (Fig. 7.15 [a]). To summarize, the map provided a better representation of the training set.

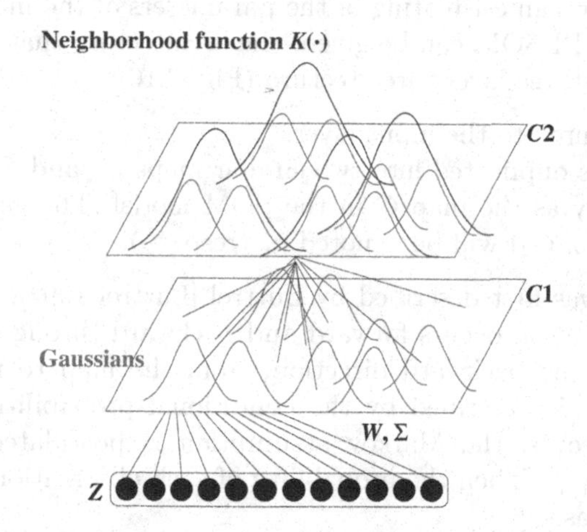

Fig. 7.16. Modeling of the SOM according to a probability density mixture. The map is shown in the neural formalism: 3 layer architecture: an input layer and two layers that are maps with similar size and similar topology. A neuron of C_1 represents a gaussian with expectation vector $\boldsymbol{w}_c$ and scalar covariance matrix $\sigma_c I$; a neuron of C_2 represents a Gaussian mixture, whose density is given by $p(\boldsymbol{z}) = \sum_{c_2} p(c_2)p_{c_2}(\boldsymbol{z})$ where $p_{c_2}(\boldsymbol{z}) = \sum_{c_1} p(c_1 \mid c_2)p(\boldsymbol{z} \mid c_1)$

7.3.8 Probabilistic Topological Map

Similarly to k-means algorithm, a probabilistic version of SOM, called PRSOM, can be defined [Anouar et al. 1997; Gaul et al. 2000]. The difference between SOM and PRSOM is essentially that, for PRSOM, a Gaussian density f_c is associated to each neuron c of the map. Each Gaussian density function f_c is completely defined by the mean vector (the equivalent of reference vector of SOM) $\boldsymbol{w}_c = (w_c^1, w_c^2, \ldots, w_c^n)$, and by its covariance matrix that is a square symmetric positive-definite matrix Σ_c, restricted to isotropic densities: $\Sigma_c = \sigma_c^2 I$, where I is the (n, n) unit matrix. Then the density functions can be written as

$$f_c(\boldsymbol{z}) = \frac{1}{(2\pi)^{n/2}\,\sigma_c^n} \exp\left(-\frac{\|\boldsymbol{z} - \boldsymbol{w}_c\|^2}{2\sigma_c^2}\right).$$

Thus, in the PRSOM, each neuron c of the map is allocated to the mean vector $\boldsymbol{w}_c$ and to the positive scalar σ_c. As for SOM, the data space D is partitioned into subsets of the family $\{P_c / c \in C\}$. The subset P_c is described by the density function f_c : $\boldsymbol{w}_c$ represents its associated reference vector, and σ_c estimates the standard deviation of the observation of $P_c \cap A$ around $\boldsymbol{w}_c$. The two parameter sets $W = \{\boldsymbol{w}_c; c \in C\}$ and $\sigma = \{\sigma_c; c \in C\}$ define completely the PRSOM. Their values must be estimated during training from the training set A.

If we assume that the data underlying distribution is a Gaussian mixture, the PRSOM allows an estimating of the parameters of the mixture. A neural interpretation of PRSOM can be given: the architecture that is associated to the PRSOM has three layers architecture (Fig. 7.16):

- Data is presented to the input layer.
- The map C is duplicated into two similar maps C_1 and C_2 that have the same topology as the map C in the SOM model. The generic neuron of maps C_1 (resp. C_2) will be denoted c_1 (resp. c_2).

That approach was first described by Luttrel [Luttrel 1994]; it assumes that a random propagation occurs forward and backward through the 3 layers of the network. In the backward direction, from the map to the data space, that propagation is described by the conditional probabilities $p(c_1|c_2)$ and $p(\boldsymbol{z}|c_1, c_2)$. Moreover, the Markov assumption is postulated, namely that $p(\boldsymbol{z}|c_1, c_2) = p(\boldsymbol{z}|c_1)$. Then the probability of each observation $\boldsymbol{z}$ can be computed explicitly as

$$p(\boldsymbol{z}) = \sum_{c_2} p(c_2) p_{c_2}(\boldsymbol{z})$$

with

$$p_{c_2}(\boldsymbol{z}) = \sum_{c_1} p(c_1 \mid c_2) p(\boldsymbol{z} \mid c_1).$$

The probability density is fully determined by the network architecture, which provides an expression of the conditional probability $p(c_1|c_2)$ using the neighborhood relation on the map and the conditional density of the observation $p(\boldsymbol{z}\,|c_1) = f_{c_1}(\boldsymbol{z}, W_{c_1}, \sigma_{c_1})$. If we assume that the neighborhood relationships permit the definition

$$p(c_1|c_2) = \frac{1}{T_{c_2}} K^T(\delta(c_1, c_2)), \quad \text{with } T_{c_2} = \sum_r K^T(\delta(c_2, r));$$

then the posterior probability densities of the observations may be expressed as a function of the Gaussian distributions of the neurons:

$$p_{c_2}(\boldsymbol{z}) = \frac{1}{T_{c_2}} \sum_{r \in C_1} K^T(\delta(c_2, r))\, f_r(\boldsymbol{z}, \boldsymbol{w}_r, \sigma_r).$$

Thus, $p_{c_2}(\boldsymbol{z})$ can be interpreted as a local mixture of Gaussian densities that are associated to each neuron of the map. The set of average vectors $W = \{\boldsymbol{w}_c; c \in C\}$ and the set of scalar standard deviations $\sigma = \{\sigma_c; c \in C\}$ are the parameters to be estimated by training. The probabilistic formalism makes it possible now to maximize the likelihood of the observation set just as for the probabilistic version of k-means. If the observations of the training set A are assumed to be the independent, and that each observation $\boldsymbol{z}_i$ is generated by the Gaussian mode $p_{\chi(\boldsymbol{z}_i)}$ that is associated to neuron $\chi(\boldsymbol{z}_i)$, and if it is further assumed that neurons c_2 of C_2 have similar prior probabilities, the classifying likelihood can be written as

$$p(\boldsymbol{z}_1, \boldsymbol{z}_2, \ldots, \boldsymbol{z}_N \mid W, \sigma, \chi) = \prod_{i=1}^{N} p_{\chi(\boldsymbol{z}_i)}(\boldsymbol{z}_i),$$

which must be maximized with respect to the parameters of the model W, σ and the allocation function χ. According to the usual strategy, it is performed by a minimization process

$$E(W, \sigma, \chi) = -\sum_{i=1}^{N} \ln \sum_{r \in C} K^T(\delta(\chi(\boldsymbol{z}_i), r))\, f_r(\boldsymbol{z}_i, \boldsymbol{w}_r, \sigma_r)$$

by using the dynamic clustering formalism. The phases of allocation and minimization are sequentially and alternatively iterated until convergence:

- *Allocation phase.* Assume that the parameters $\{W, \sigma\}$ have the values computed at the previous iteration or at initialization. Then E must be minimized with respect to the allocation function χ. A new allocation function must be found that assigns each observation $\boldsymbol{z}$ to a neuron. That step generates a new partition of the training data space D. It can easily be seen that the optimal allocation function associates to a given observation $\boldsymbol{z}_i$ the most probable neuron c according to the density p_{c_2}:

$$\chi(\boldsymbol{z}) = \arg\max_{c_2} p_{c_2}(\boldsymbol{z})$$

- Minimization phase. During that phase, the allocation function is kept constant, and $E(W, \sigma, \chi)$ is minimized with respect to W and σ.

The parameters W and σ are updated as in the batch version of the SOM algorithm by canceling the partial derivatives of the cost function $E(W^t, \sigma^t, \chi^t)$. To solve the equation, an iterative procedure is used as in [Duda et al. 1973], assuming that for ith iteration the initial values of the parameters are close to the optimal values. The update relations are the following:

$$\boldsymbol{w}_r^t = \frac{\sum_{i=1}^{N} \boldsymbol{z}_i K\left(\delta\left(r, \chi^{t-1}\left(\boldsymbol{z}_i\right)\right)\right) \dfrac{f_r\left(\boldsymbol{z}_i, \boldsymbol{w}_r^{t-1}, \sigma_r^{t-1}\right)}{P_{\chi^{t-1}(\boldsymbol{z}_i)}\left(\boldsymbol{z}_i\right)}}{\sum_{i=1}^{N} K\left(\delta\left(r, \chi^{t-1}\left(\boldsymbol{z}_i\right)\right)\right) \dfrac{f_r\left(\boldsymbol{z}_i, \boldsymbol{w}_r^{t-1}, \sigma_r^{t-1}\right)}{P_{\chi^{t-1}(\boldsymbol{z}_i)}\left(\boldsymbol{z}_i\right)}}$$

$$\left(\sigma_r^t\right)^2 = \frac{\sum_{i=1}^{N} \left\|\boldsymbol{w}_r^{t-1} - \boldsymbol{z}_i\right\|^2 K\left(\delta\left(r, \chi^{t-1}\left(\boldsymbol{z}_i\right)\right)\right) \dfrac{f_r\left(\boldsymbol{z}_i, \boldsymbol{w}_r^{t-1}, \sigma_r^{t-1}\right)}{P_{\chi^{t-1}(\boldsymbol{z}_i)}\left(\boldsymbol{z}_i\right)}}{n \sum_{i=1}^{N} K\left(\delta\left(r, \chi^{t-1}\left(\boldsymbol{z}_i\right)\right)\right) \dfrac{f_r\left(\boldsymbol{z}_i, \boldsymbol{w}_r^{t-1}, \sigma_r^{t-1}\right)}{P_{\chi^{t-1}(\boldsymbol{z}_i)}\left(\boldsymbol{z}_i\right)}}.$$

In both above relations, the parameters at iteration t are expressed as functions of the parameters at iteration $t-1$.

Since the model is complex, an appropriate initialization is desirable. Since PRSOM can be considered as extensions of SOM, one can first perform a SOM estimation of the reference vector set W in order to initialize the mean vector set of PRSOM.

Thus, the PRSOM training algorithm can be summarized as follows:

PRSOM Algorithm with Constant Temperature T

1. *Initialization*: $t = 0$. The initial values W^0 of the references are computed using a SOM training algorithm, the σ_r^0 is computed by the mean of the local inertia I_r (Sect. 7.2.1). The initial allocation function χ^0 is derived from the update relation

$$\boldsymbol{w}_r^t = \frac{\sum_{i=1}^{N} \boldsymbol{z}_i K\left(\delta\left(r, \chi^{t-1}\left(\boldsymbol{z}_i\right)\right)\right) \dfrac{f_r\left(\boldsymbol{z}_i, \boldsymbol{w}_r^{t-1}, \sigma_r^{t-1}\right)}{P_{\chi^{t-1}(\boldsymbol{z}_i)}\left(\boldsymbol{z}_i\right)}}{\sum_{i=1}^{N} K\left(\delta\left(r, \chi^{t-1}\left(\boldsymbol{z}_i\right)\right)\right) \dfrac{f_r\left(\boldsymbol{z}_i, \boldsymbol{w}_r^{t-1}, \sigma_r^{t-1}\right)}{P_{\chi^{t-1}(\boldsymbol{z}_i)}\left(\boldsymbol{z}_i\right)}},$$

$$\left(\sigma_r^t\right)^2 = \frac{\sum_{i=1}^{N} \left\|\boldsymbol{w}_r^{t-1} - \boldsymbol{z}_i\right\|^2 K\left(\delta\left(r, \chi^{t-1}\left(\boldsymbol{z}_i\right)\right)\right) \dfrac{f_r\left(\boldsymbol{z}_i, \boldsymbol{w}_r^{t-1}, \sigma_r^{t-1}\right)}{P_{\chi^{t-1}(\boldsymbol{z}_i)}\left(\boldsymbol{z}_i\right)}}{n \sum_{i=1}^{N} K\left(\delta\left(r, \chi^{t-1}\left(\boldsymbol{z}_i\right)\right)\right) \dfrac{f_r\left(\boldsymbol{z}_i, \boldsymbol{w}_r^{t-1}, \sigma_r^{t-1}\right)}{P_{\chi^{t-1}(\boldsymbol{z}_i)}\left(\boldsymbol{z}_i\right)}}.$$

The maximal number of iterations N_{iter} is chosen.

2. *Iteration* $t : W^{t-1}$ and σ^{t-1} were computed at the previous iteration.
   - *Minimization phase*: computation of the new parameters W^t and σ^t;
   - *Allocation phase*: update of the allocation function χ^t that is associated to W^t and σ^t from relation $\chi(\boldsymbol{z}) = \arg\max_{c2} p_{c2}(\boldsymbol{z})$.
2. Iterate until $t > N_{\text{iter}}$ or until stabilization of the cost function $E(W, \sigma, \chi)$.

As for SOM training, PRSOM uses a neighborhood whose size is controlled by the temperature parameter T. During training, the size of the neighborhood decreases according to the cooling schedule. At the end of training, the map provides an organized structure of the average vector set, and the partition associated to the map is defined by the final allocation function $\chi^{N_{\text{iter}}}$. As for other versions of SOM, the data space D is divided into M subsets: each neuron c of the map represents a data subset $P_c = \{\boldsymbol{z}/\chi^{N_{\text{iter}}}_{(\boldsymbol{z})} = c\}$. That map and that partition were determined from probability distributions, whereas SOM just uses Euclidean distances. The probability density estimation gives access to additional information that may be useful for application purposes. Actually, that information is crucial as far as classification problems are concerned. No stochastic version of PRSOM is available: a large sample of the data is necessary to estimate the initial variance before updating the parameters.

PRSOM provides a lot of additional information about the training data (tracking outliers, computing probabilities, etc.). However, that model can be used only if the training observation set is large enough to allow an accurate estimation of the standard deviations of the Gaussian modes of the mixture in the initialization phase. Remote sensing, where a tremendous amount of data is available, is ideally suited to applications of SOM. The detection of ocean color is described in the next section.

7.4 Classification and Topological Maps

Among the various applications of SOM, many of them are classification tasks. As stated above, classification is not a straightforward application of self-organization: unsupervised learning provides an allocation function that assigns any observation to a cluster of a partition of the training set, irrespective of the semantics of the data. In such problems, it is assumed that a lot of noise-corrupted observations are available with not knowledge of their class. The partition that is obtained depends on the probability density underlying the training set. Regions that contain a high density of data will be covered by a fine partition; low-density regions will be covered by a coarse partition. The large amount of data available in high-density regions provides accurate information on those regions. On the other hand, the geometry of the partition depends on the nature of the encoding of the observations. Thus, for a given problem and a given data set, several different encodings may generate several partitions of the data space. With the SOM algorithm, the selection of the code provides information about the problem of interest. The basic principle of the algorithm is to favor the emergence of clusters (the partition subsets) that are appropriate for the application under consideration. If the application involves a classification task into S classes, each partition subset must be included in one class as completely as possible. Then, one can assign one of the S classes to a whole cluster. Since each subset is assigned to one neuron of the map, the classification problem amounts to labeling each neuron of the map. The label set is the set of the S classes of the problem. Labeling can be performed in two different ways. Since each reference vector represents a subset of the partition P, and since the reference vector may be interpreted as an average experiment, it is possible to use expert knowledge to recognize the class of the reference vector on the basis of its characteristics:

1. by asking an expert of the domain to classify some data extracted from the training set,
2. by first aggregating the neurons on a statistical basis and then use the expert knowledge to label the clusters.

7.4.1 Labeling the Map Using Expert Data

Assume that a S-class classification task must be performed, and that the labels of those classes must belong to a label set $L = \{l_i, i = 1, \ldots, S\}$. At the end of SOM training, when all parameters of the map are estimated, each observation $\boldsymbol{z}$ is assigned to a neuron $c = \chi(\boldsymbol{z})$, so that the label l_c of that neuron can be assigned to the observation. Therefore, the problem is: how to label the neurons of the map with the labels of L?

Labeling the neurons of the map is the first step in the design of a classifier from a SOM. If the amount of data classified by the expert is very large, labeling may be performed by majority voting (see hereafter Fig. 7.17):

- Assign the expert-classified data to the various neurons of the map using the allocation function provided by the SOM training.
- For every neuron c, select the label l_i that is the most commonly used label for the expert classified-data assigned to neuron c.
- All the data belonging to the subset that is represented by neuron c are now labeled by label l_i.

At the end of the labeling phase, the set of neurons c that have the same label l can be used to approximate the probability distribution of the data of class l. The larger the amount of expert-classified data, the better the classifier. Of course, neurons that represent data lying on the boundaries of the classes may get the wrong label. Another source of error is the lack of expert-classified data in some subset represented by a given neuron: the corresponding region of the data space is thus poorly identified.

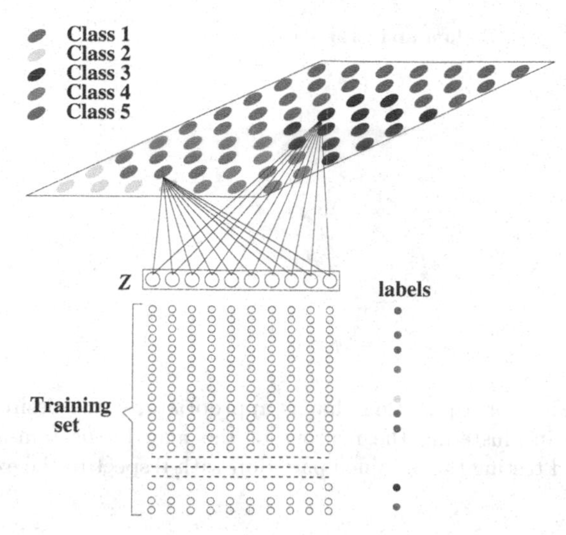

Fig. 7.17. Map labeling using expert-classified data. Classified data are assigned to the relevant neurons of the map. Then each neuron is labeled using majority voting among classified data that are allocated to that neuron

7.4.2 Searching a Partition that Is Appropriate to the Classes

If the amount of expert-labeled data is too small, the above labeling method is inappropriate. Majority voting result has a large variance and may generate classification error with significant probability. The presence of a single wrongly labeled observation may lead to assigning a wrong label to the associated neuron. Thus, a whole region of the data space may be wrongly classified. Furthermore, due to the small amount of labeled data, a significant number of subsets of the partition may include no labeled data, so that the algorithm is not able to provide them with any label.

In that case, it is possible to take into account an additional phase, in which the various observation subsets are clustered as appropriately as possible. A coarser partition is sought, and labeling will be performed after that additional clustering phase. When one fuses several partition subsets, more expert-classified data are available to label a larger subset. Of course, as before, the whole process is valid only if the original clustering is consistent with the classification, so that majority voting can select the right label.

If the map and the partition that are provided by SOM are assumed to be relevant, then the two following additional assumptions are taken into consideration:

- The data quantization is correct, so that each reference vector is a good representative of its allocated data.
- Topology is relevant: two subsets that are represented by neighboring neurons on the map contain observations that are close in data space.

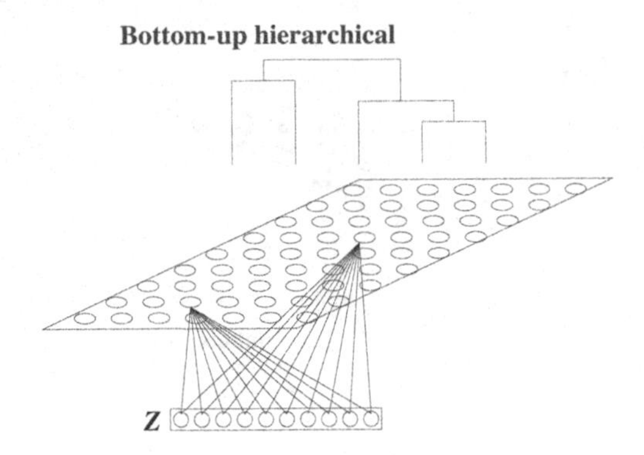

Fig. 7.18. Looking for a partition that is appropriate to the desired classes. The method consists in clustering the neurons of the map by bottom-up hierarchical classification, and testing the obtained partition with respect to the expert-classified data

Underlying the second assumption is the hypothesis that there exists a structure on the dataset that fits the classification problem, and that it is possible to exhibit that structure with the topological self-organization of the map. Thus, two subsets that are represented by neighboring neurons have a strong probability of representing observations that belong to the same class.

Of course, those assumptions are very strong. It is implicitly supposed that a right data encoding is already known to perform the classification. Therefore, that point must be solved in a preliminary analysis, providing an appropriate data representation, stemming from an appropriate variable selection and the design of a relevant coding. The effect of the coding process on the classification result will be shown in the section devoted to applications.

Bottom-up hierarchical classification [Jain et al. 1988] may perform the second stage of the process by appropriately clustering the neurons (see Figs. 7.18 and 7.19).

This method computes a partition hierarchy. The various partitions of the hierarchy are found iteratively. The initial partition is the finest one. It is made of all the singletons of the map. From that initial partition, two subsets of the current partition are clustered at each iteration. To select the two subsets that are going to be clustered, a measure of the similarity between two subsets is defined. Among all the possible subset pair, the pair that is made of the most similar subsets, with respect to the chosen similarity criterion, is selected.

Summary of the hierarchical classification algorithm:

Hierarchical Classification Algorithm

3. *Initialization.* Consider the finest partition that is made by all the singletons; each neuron is allocated to a distinct subset. Choose the desired number of subsets K.

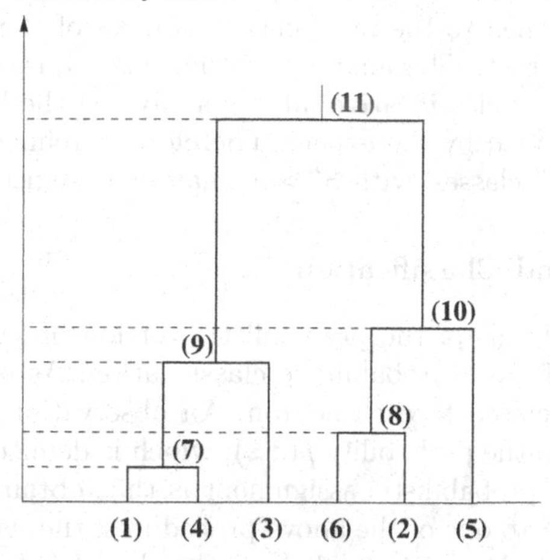

Fig. 7.19. Clustering the neurons of the map using bottom-up hierarchical classification: the leaves of the tree represent the neurons (here there are six neurons); for each cluster, the vertical axis provides the clustering index for the selected similarity

4. For a given partition, find the closest subsets in the sense of the selected similarity criterion, and cluster them together in order to get a single subset.
5. If the number of clusters of the current partition is larger than K, go to step (2), otherwise terminate the algorithm.

Several similarity measures are proposed in the literature [Jain et al. 1988]. The most popular one is due to Ward. It consists in aggregating clusters in such a way that the sum of the cluster inertia is as small as possible. That is a way of favoring clusters that are as compact as possible in the (Euclidean) data space. If that criterion is selected to cluster the neurons of the map, the working space is the data space and the associate reference vectors are the neurons. Conversely, since the neurons are distributed on the map with its awn discrete graph topology, one may choose to favor aggregation in a way that takes into consideration that discrete structure. Those clusters will be made of neurons that tend to form connected sets on the map [Murtagh 1985; Yacoub et al. 2001]. The choice one of those two strategies, or the use of a hybrid strategy that combines both of them, may have a crucial influence on the results.

Hierarchical classification allows generating an arbitrary number of subsets, since clustering may be stopped at any iteration. For a given similarity measure, the number of partition elements depends on the number S of classes that are sought. That number also depends on the agreement between the unsupervised statistical partitioning and the partition into S classes that is determined by the application. That number may be higher than S if a class is not homogeneous from a statistical point of view. Then it is inferred that the expert has assigned to the same class instances of the observation space that are quite different. The analysis of the most consistent partition, obtained by hierarchical classification, allows analyzing the homogeneity of the classification performed by the expert. Therefore, a refinement of the expert classification into S' classes, with $S' > S$, may be designed.

7.4.3 Labeling and Classification

After labeling of the map, the probabilistic version of self-organizing maps (PRSOM) can perform a probabilistic classification. As mentioned above, a normal law is associated to each neuron. An observation $\boldsymbol{z}$ is assigned to a neuron according to the probability $p(c|\boldsymbol{z})$, which is defined by Bayes relation as shown below. A probabilistic assignment is thus obtained. Since the map is labeled according to one of the above procedures, the *posterior* probability that the considered observation $\boldsymbol{z}$ belongs to class l can be estimated. The PRSOM stem from a probabilistic modeling where it is assumed that the observations are generated according to the mixture distribution:

$$p(\boldsymbol{z}) = \sum_{c} p(c)\, p_c(\boldsymbol{z})$$

where $p_c(\boldsymbol{z})$ is also a normal law local mixture

$$p_c(\boldsymbol{z}) = \frac{1}{T_c} \sum_{c} K^T(\delta(c,r))\, f_r(\boldsymbol{z}, \boldsymbol{w}, \sigma_r)$$

where $T_c = \sum_c K^T(\delta(c,r))$ and f is a normal law with mean $\boldsymbol{w}_r$ and scalar covariance matrix $\sigma_r^2 I$. The quantities $p_c(\boldsymbol{z})$ are computed from the neurons of the map and the quantities $p(c)$ are computed from the partition that has been proposed by PRSOM. If N stands for the observation number of the training set A and n_c is the number of observations that are assigned to neuron c by the allocation rule $\chi(\boldsymbol{z}) = \arg\max_c p(\boldsymbol{z} \mid c)$, the prior probability $p(c)$ of neuron c can be estimated as n_c/N. Then Bayes rule allows computing the posterior probability of neuron c given observation $\boldsymbol{z}$:

$$p(c|\boldsymbol{z}) = \frac{p(c)\, p_c(\boldsymbol{z})}{p(\boldsymbol{z})} = \frac{n_c p_c(\boldsymbol{z})}{\sum_{r \in C} n_r p_r(\boldsymbol{z})}.$$

After training, the topological map that is proposed by PRSOM determines the parameters of the normal laws that characterize the various neurons. For any observation $\boldsymbol{z}$, it is then possible, applying the above relation , to compute the posterior probability of an obervation being assigned to a given neuron. Since a class is represented by a subset of neurons, the posterior probability that the observation $\boldsymbol{z}$ belongs to the class l_i is derived from the neurons that are labeled by l_i. If the subset of those neurons is denoted C_i, one gets

$$p(l_i \mid z) = \sum_{c \in C_i} p(c \mid z) = \frac{\sum_{c \in C_i} n_c p(z)}{\sum_{r \in C} n_r p_r(z)}$$

where

$$p_c(z) = \frac{1}{T_c} \sum_c K^T(\delta(c,r)) f_r(z, w, \sigma_r).$$

Note that this probability relies on the labeling of the map. That step is crucial for the computation of the posterior probabilities. Their consistency depends on the quality of the map. Thus, the classifier performances depend jointly on the amount of expert data, on the accuracy of the approximation of the observation density, and on the topological order that is built by the self-organization process.

The knowledge of posterior probabilities leads to a classification rule that is based on probabilistic estimation. Using those relations, the vector of class membership probabilities can be computed for each observation z. Finally, the assignment of the observation to a class is performed by the application of Bayes rule: choose the class for which the membership probability is highest.

7.5 Applications

Self-organizing maps gave rise to a large number of applications. Specific developments were required for some of them, but they are in actual operation. At the moment, the most important research center for those topics is located at University of Technology of Helsinki (UTH). The major part of the research that is developed in its computer science laboratory (Laboratory of Computer and Information Science) is performed by the Neural Network Research Center, created by T. Kohonen in 1994, and now headed by E. Oja. The description of a large number of applications is now available on the Web site of NNR (http://www.cis.hut.fi/research/). The main research axis and the current applications are generally focused on self-organizing maps. Companies now exploit many applications. They arose from original, multidisciplinary research, and several research groups specialized such fields as bioinformatics, speech and writing analysis and recognition, and image analysis.

Actually, the implementation of self-organizing maps into larger systems widely uses the specific features of the application domain. The coding of the information, the organization of data bases, the analysis and visualization of the data, require specific, multidisciplinary research whose results are crucial for the performance of the self-organizing maps.

In the following, two applications will be described in detail. They were selected as representative of the domains to which self-organizing maps are relevant. The target of this presentation is twofold:

- The first section describes a satellite remote sensing application. It is a field of growing importance, and a lot of statistical problems must be solved by physicists and research engineers who are in charge of designing models. Considering that a very large amount of data is now available, this field is particularly suited to neuronal modeling. That application will fully illustrate the methodology that was described in previous sections. It uses the probabilistic model of self-organizing maps (PRSOM).
- The second section gives a brief account of one of the most popular applications that was developed at UTH: the WEBSOM system. That is devoted to information research on the Web. The earlier version was implemented in 1995. The salient feature of that application is the high dimensionality of the data. Dimensioning the topological map with a very large set of neurons and tuning the algorithm (regarding computing time and convergence accuracy) were the basic issues that were successfully faced at UTH. The development of WEBSOM spurred research oriented towards shortening the training phase for the design, and towards shortening the document research time during the exploitation phase.

7.5.1 A Satellite Remote Sensing Application

A lot of data is generated by the observation of earth with on-board sensors, and handed to geophysicists. All the neural methods that are presented in this book are helpful to process those data because they solve multidimensional statistical problems. Among those methods, unsupervised training is especially useful, because it allows extracting information even when expert information is scarce. Gathering expert information often requires costly analyses (ground mission, sophisticated biological and chemical analyses). That explains why expert-appraised data is scarce as compared to the amount of available satellite data.

Self-organizing bring valuable contributions to satellite data analysis, because estimating the observation probability density, and designing representative data partitioning, can be performed in a relatively straightforward way. Such information provides new insights into the physical phenomena of interest:

1. PRSOM estimate the variance and local uncertainty of the observations.
2. The partitions that are obtained are useful to the expert of the various application fields (physicists, chemists ...) because they may serve as an accurate summary of the observation set. Investigating such a summary may be crucial for understanding the phenomena of interest.
3. In all the fields that are concerned by experiments, heavy and expensive experimental campaigns are carried out regularly. With respect to the amount of satellite data, the expert-assessed observations are scarce, but they contain extremely valuable information. A few expert-labeled observations allow the identification of subsets of the partition from the

topological map. The classification methodology that is presented below in the section «Classification and PRSOM» allow that identification.

4. In order to demonstrate the various capabilities of self-organizing maps, the presentation of the next application is organized as follows:
   - Description of the application field, of the problems of interest, and of the available data;
   - Description of experiments that allow the understanding of the influence of data encoding on the partition and on the topological order that is obtained;
   - Description of experiments that allow the assessment of the impact of expert knowledge

7.5.1.1 The Color of the Ocean

The biological activity of the ocean is crucial for the natural economy of the earth, for it is strongly connected to the fishing resource, and is part of the biochemical cycles with strong climate consequences. During the past few years, several multi-spectral sensors dedicated to measure the color of the ocean were launched on different satellites ((MOS, POLDER-1, OCTS, SEAWIFS, MODIS) or will be in the future (MERIS, POLDER-2, GLI). They are expected to estimate the chlorophyll contents of the upper layer of the ocean, and to assess its space and time variability.

Two steps are needed to recover the pigment fields from the ocean color satellite data. The first step deals with atmospheric correction (indeed, the atmosphere contributes more than 80% of the measured signal); the second step is dedicated to ocean interaction (see Fig. 7.20). The atmospheric correction algorithm currently computes sea-level reflectance by canceling atmospheric effects (aerosols and air molecules). The second step aims at inverting that reflectance to provide the chlorophyll pigment concentration. This is a tricky issue, because one has to take into account both uncertainties that arise from atmospheric corrections, and uncertainties that are intrinsically related to the variation of biological population. Thus, the knowledge of the atmospheric aerosol composition and of the water constituent concentration is crucial. The following experiments aim at assessing various classes of aerosols and waters from the top of atmosphere spectra, which are provided by the sensors.

7.5.1.2 The Data

The data that are used hereafter were provided by US radiometer SeaWifs, which was located on the SeaStar satellite. That sensor has eight spectral bands in the visible and the near infrared spectrum (see Table 7.1).

For each measurement location at the ocean surface, the observation vector is 8-dimensional. Its components are the eight radiances that are measured on the top of atmosphere. The following results are a representative example

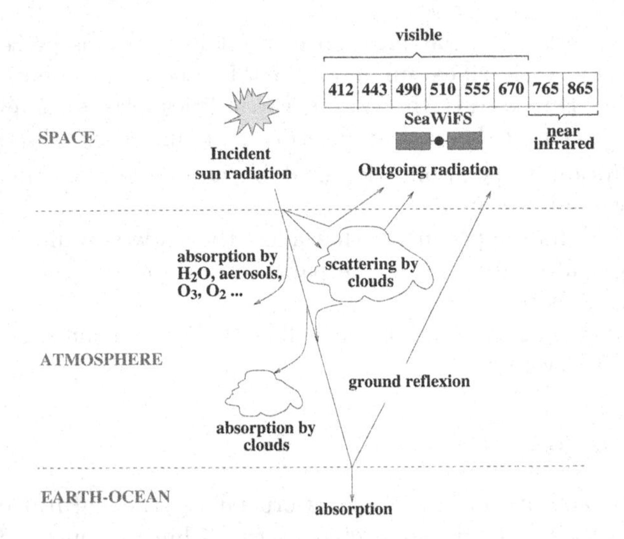

Fig. 7.20. Outline of the physical phenomena that are relevant to explain the observations. The observations consist in the set of reflectance spectra that are measured by the SeaWifs radar. That radar uses eight wavelengths for analyzing the surface

of ocean color data processing. They come from a SeaWifs image. This image was over West Africa and the Canarie islands on January 5th, 1999. Two pictures of that zone, which were obtained on the same day with different resolutions, are available: a LAC (*Local Area Coverage*) image, with $2141 \times 793 = 1,697,813$ pixels (Fig. 7.21) and lower resolution, GAC (*Global Area Coverage*) image with $536 \times 199 = 106,664$ pixels. The topological map design was performed from a sampled LAC image. Sampling was performed by line decimation of the original LAC image. Thus the training set had $238 \times 793 = 188,734$ pixels. The quality of the resulting vector quantization, was assessed from the full LAC picture. Since 90% of the pixels had no contribution to

Table 7.1. Spectral bands of SeaWifs

Bands K	Wavelengths (nanometers) λ_k
1	412
2	443
3	490
4	510
5	555
6	670
7	765
8	865

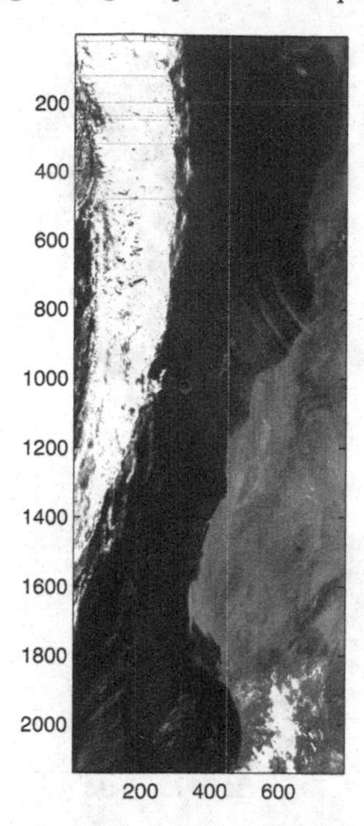

Fig. 7.21. ThiS SeaWifs image has been taken over West Africa and the Canarie islands on January 5th, 1999. LAC picture 2141 × 793

the training phase, and considering the large amount of available data, the performances thus estimated may be considered as fully significant.

The expertise was provided through the GAC image. That image had additional information, from two sources:

- Information that was provided by SeaWifs was available, such as land masking, presence of clouds.
- A pixel classification of the GAC image, using different optical models provided from atmosphere experts, was also available. Figure 7.22 shows the expert-processed GAC picture. On that image, five classes were identified by the expert: aerosols that arise from the desert, so-called case 2 waters, which are waters with high organic matter contents, sea aerosols, clouds and land. Actually the pixels that are assigned the land label are residual pixels for which no label was provided by the expert
- Note that both types of information may contain errors, just as any expert-based classification of highly complex phenomena. For instance, the expert sought five classes. Therefore, different aerosols may have been clustered in the same class if he had no appropriate physical model to discriminate between them.

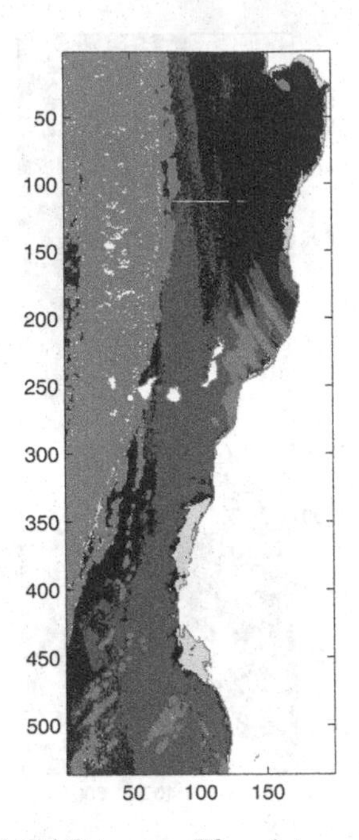

Fig. 7.22. Expert-processed GAC image. The picture represented five zones whose boundaries had been defined by the expert: the aerosols coming from the desert (*black*), the case-2 waters (*light grey*), the marine aerosols (*dark grey*), the clouds (*medium grey*) and the earth (*white*)

7.5.1.3 The Role of Encoding

Numerical experiments were performed using different encodings of SeaWifs spectra.

The first encoding uses directly the reflectance at the top of the atmosphere. In order to normalize the influence of the wavelengths, the reflectance spectrum values were normalized to the $[-1, +1]$ interval. The reflectance for wavelength λ_k, is denoted $\rho(\lambda_k)$, the normalization has been computed from the learning set for each wavelength, $(k = 1, \ldots, 8)$. Thus, each observation was encoded by an 8-dimensional vector, each component of that vector being the normalized reflectance for a given wavelength. Since the reflectance values $\rho(\lambda_k)$ are numbers between 0 and 1, they were normalized between -1 and 1, using to $2\rho(\lambda_k) - 1$.

The sampled LAC image that was used for training (actually one picture line out of ten), after coding as described above, will be denoted by $\mathrm{App}_{\mathrm{cod1}}$. A second coding was performed, in order to highlight the shape of the spectra of interest. To that effect, the slopes of spectra were taken into account. They were computed for each wavelength. Thus, the kth component of the new coding vector was computed from the reflectance as

$$\Delta\rho(\lambda_k) = \frac{\rho(\lambda_{k+1}) - \rho(\lambda_k)}{\lambda_{k+1} - \lambda_k}.$$

Seven slopes were thus computed for k from 1 to 7. In order to keep information about the spectrum intensity, a component of the observation vector was the norm $\|\rho\|$ of the spectrum. Thus, each observation was encoded in an 8-dimensional vector

$$[\Delta\rho(\lambda_1), \ldots, \Delta\rho(\lambda_7), \|\rho\|]^{\mathrm{T}}.$$

The sampled LAC image that was used for training was encoded according to that second scheme; the result will be denoted $\mathrm{App}_{\mathrm{cod2}}$. Normalization between –1 and 1 was performed as previously. Since the slopes and the norm do not have the same order of magnitude, the normalization was implemented separately for each component as

$$2\frac{x - \min}{\max - \min} - 1,$$

where x is a derivative (namely $(\Delta\rho(\lambda_k)k = 1, \ldots, 7)$, min (resp. max) are the minimum (resp. the maximum) over the set of all the derivatives in $\mathrm{App}_{\mathrm{cod2}}$.

For all test data, the same encodings were used. The following numerical experiments are an illustration of the methodology, which has been described in the section "Classification and PRSOM." They use quantizations, which are followed by classifications. The quantizations are obtained from the probabilistic maps, and clustering is performed by hierarchical classification. All self-organizing maps have the same architecture:

- The input layer has 8 units
- The map is 2-dimensional, with 10×10 neurons. The neighborhoods are defined from the exponential kernel family $K(\delta) = \exp(-\delta^2)$.

7.5.1.4 Quantization Using PRSOM

In the first part of the study, PRSOM was used for determining the patterns that are representative summaries of the set of all observed spectra. In that case, a fine quantization of the training set is sought. Actually, the result is a summary of the training set; if it is statistically representative, it is also a summary of all observations. Otherwise, the generalization may be poor since a subset of the set of observations was overlooked. The two encoding schemes that were described above (normalized radiance vales for the first one, slopes + norm for the second one), resulted in different maps. Those maps will illustrate the importance of the encoding process for quantization and topological order. Each map quantizes the observation set intro 100 subsets. Figure 7.23 shows the map that was obtained with $\mathrm{App}_{\mathrm{cod1}}$; on that figure, the number that is located above the neuron indicates how many pixels from the training set are allocated to that neuron. Figure 7.24 shows the

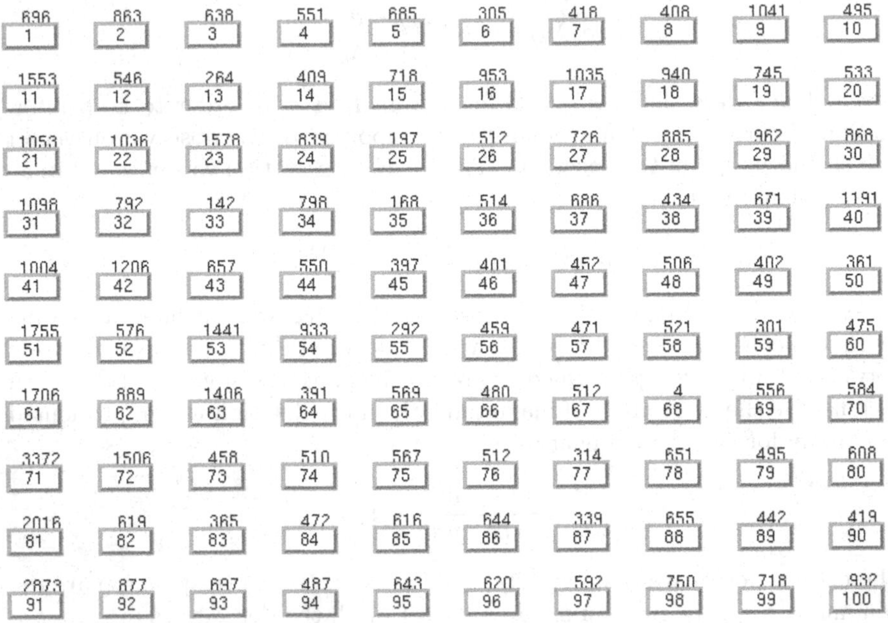

696 863 638 551 685 305 418 408 1041 495
1 2 3 4 5 6 7 8 9 10
1553 546 264 409 718 953 1035 940 745 533
11 12 13 14 15 16 17 18 19 20
1053 1036 1578 839 197 512 726 885 962 868
21 22 23 24 25 26 27 28 29 30
1098 792 142 798 168 514 686 434 671 1191
31 32 33 34 35 36 37 38 39 40
1004 1206 657 550 397 401 452 506 402 361
41 42 43 44 45 46 47 48 49 50
1755 576 1441 933 292 459 471 521 301 475
51 52 53 54 55 56 57 58 59 60
1706 889 1406 391 569 480 512 4 556 584
61 62 63 64 65 66 67 68 69 70
3372 1506 458 510 567 512 314 651 495 608
71 72 73 74 75 76 77 78 79 80
2016 619 365 472 616 644 339 655 442 419
81 82 83 84 85 86 87 88 89 90
2873 877 697 487 643 620 592 750 718 932
91 92 93 94 95 96 97 98 99 100

Fig. 7.23. PRSOM (10×10) map obtained using App_{cod1} as the training set. The map representation displays the 2D topological ordering. Each square contains a number that identifies the neuron, and the figure above each square is the number of pixels of the training set that are assigned to that neuron

same map, and, for each neuron, the associated variance. It is clear that scattered values of the reflectance are represented at the upper right corner of the map, whereas the lower left corner contains representative of similar spectra. Physical considerations provide an interpretation of the various zones of the map:

- The spectra are more stable if the sky is clear and if the signal permits an analysis of the ocean
- High variability zones may stem from atmospheric influence that is due to aerosols or to cloud reflection

The hundred patterns associated to the neurons are characterized by spectra that are of the same kind as the observations (vectors of $\mathbb{R}^8$). Figure 7.25 is presenting for the first coding, the set of patterns and their topological connections through the map. It is the same map than in Figs. 7.23 and 7.24 but each neuron is now attached to its encoded pattern. Each encoded pattern is an average spectrum over the allocated reflectance observations. The patterns are organized according an order, which is visualized through the map. The patterns that are associated to weak variance local densities have similar shapes: the observation space is sampled there in a very fine way. Patterns that belong to the high variance regions are sampling the observation space

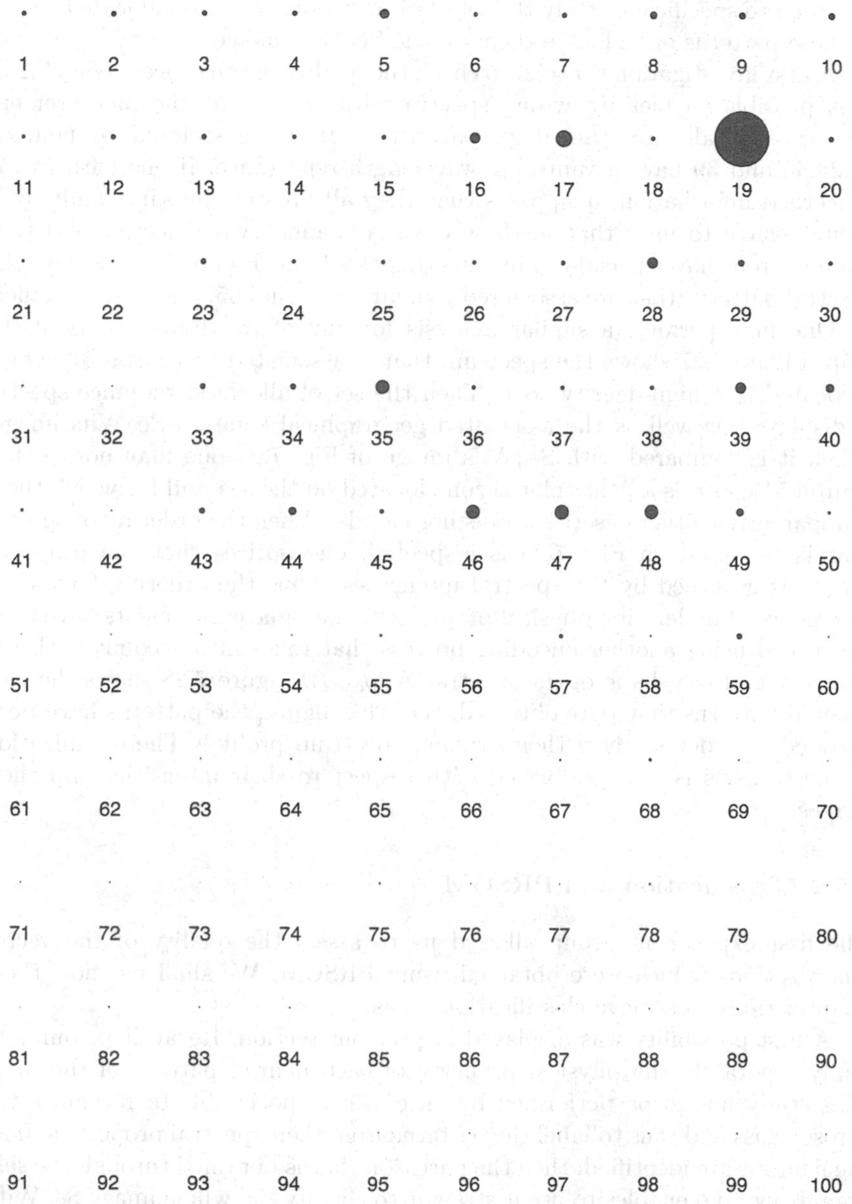

1 2 3 4 5 6 7 8 9 10

11 12 13 14 15 16 17 18 19 20

21 22 23 24 25 26 27 28 29 30

31 32 33 34 35 36 37 38 39 40

41 42 43 44 45 46 47 48 49 50

51 52 53 54 55 56 57 58 59 60

61 62 63 64 65 66 67 68 69 70

71 72 73 74 75 76 77 78 79 80

81 82 83 84 85 86 87 88 89 90

91 92 93 94 95 96 97 98 99 100

Fig. 7.24. Estimated variance for PRSOM. The map is the (10×10) map of the previous figure. The area of the disk above the neuron is proportional to the estimated variance of the associated Gaussian distribution

in a coarser way. The visualization technique enables to select some patterns and then to specifically study the set of observations that are allocated to one of these patterns or to locate them on the SeaWifs image.

A first investigation is useful to check the quality of the process (Fig. 7.25): it is possible to identify wrong spectra with respect to the measurement process. Actually, on this map, patterns that are associated to neurons 17,28,35 and 39 have a vanishing wavelength reflectance. If one gathers the concerned information, it appears that they all present the same fault. It is then possible to infer that in these cases a channel was defective and that some neurons have specialized in detecting this fault. Figure 7.26 displays the spectral patterns that are associated to neurons 17 and 35. and their variance.

One may perform a similar analysis for any of the 100 neurons of the map. Figure 7.27 shows the spectrum that is associated to neuron 51, which is located in a high-density zone. Then the set of allocated radiance spectra is displayed as well as the associated geographical zone on SeaWifs image. When it is compared with SeaWifs image of Fig. 7.21 one may notice that neuron 51 controls a light colored zone located on the sea and for which there is apparently neither desert aerosols nor clouds. When the ordering of spectra that is proposed on Fig. 7.25 is inspected, one notices that the proposed coding is governed by the spectral intensities. Thus the ordering favors the emergence of underlying physical properties. The same experiments have been performed using another encoding process that takes into account both the intensity and the shape of the spectra (App_{cod2}). Figure 7.28 shows the new ordered patterns that were obtained. (On that figure, the patterns have been decoded in order to show their original spectrum profile). The organization of the neurons is now performed with respect to their intensities and their shapes.

7.5.2 Classification and PRSOM

The first experiment group allowed us to assess the quality of the vector quantizations, which were obtained using PRSOM. We shall use now these quantizations to achieve classification tasks.

A first possibility was displayed in previous section. Recall it amounts to study separately the physical property of each neuron pattern of the map. This study has to be performed by an expert, who is able to recognize the aerosol class and thus to label the patterns from their spectral properties. If all the neurons are identified, then the partition that is obtained through the self-organizing map enables to use it straight to classify the whole image SeaWifs. Moreover, if the learning set is representative of the physical problem, it may be used to label other SeaWifs images that share the same physical properties.

If this identification process is not possible, i.e. if the expert is not able to label accurately every neuron of the map, it is possible to cluster the neurons according to an unsupervised way. One can proceed as it was demonstrated in the previous section “Classification and topological map,” by aggregating the

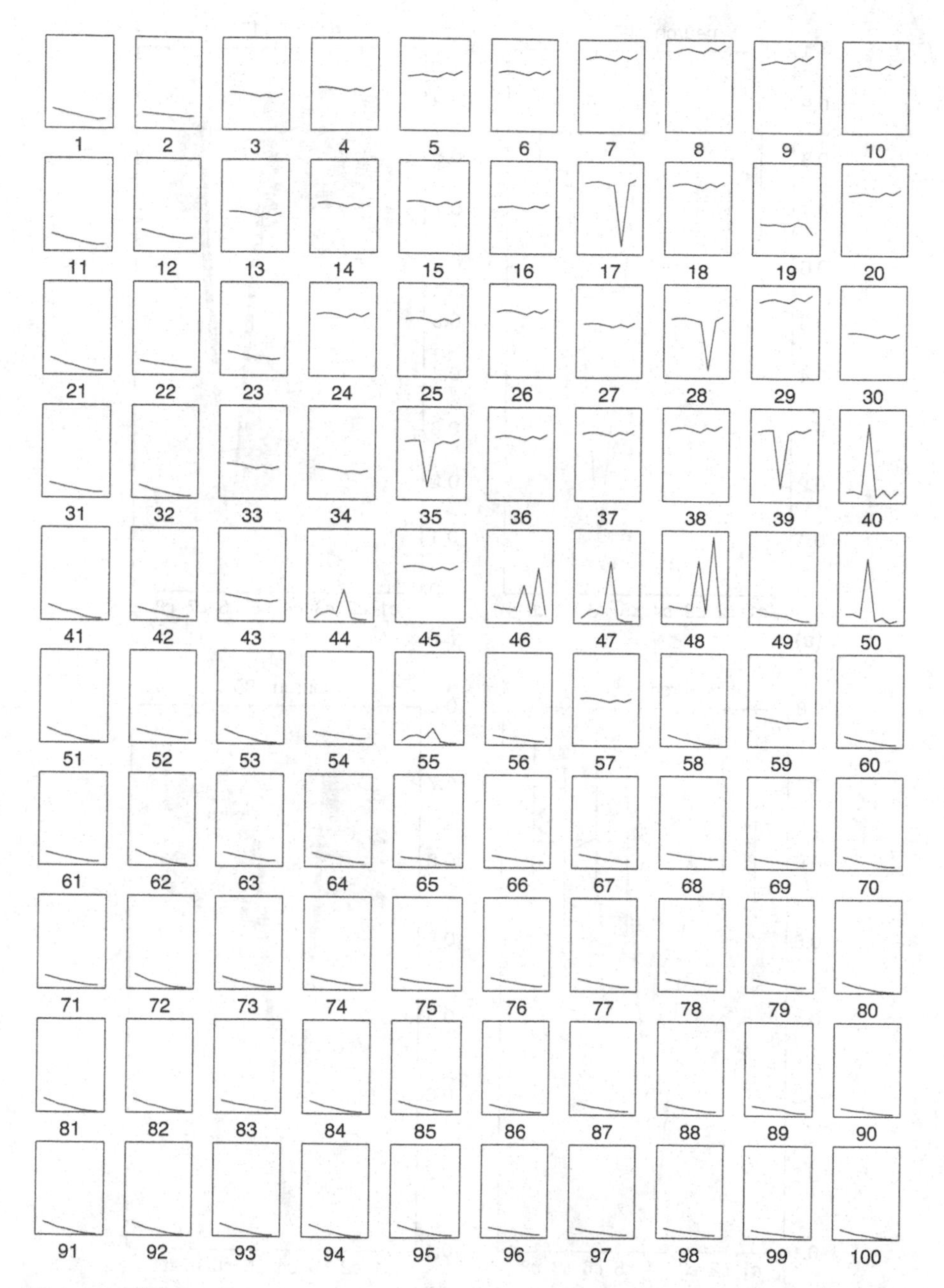

Fig. 7.25. Vector quantization from the previous PRSOM (10 × 10) map from the training set App_{cod1}. The number that identifies the neuron is above the square, the associated spectral pattern is drawn inside each frame

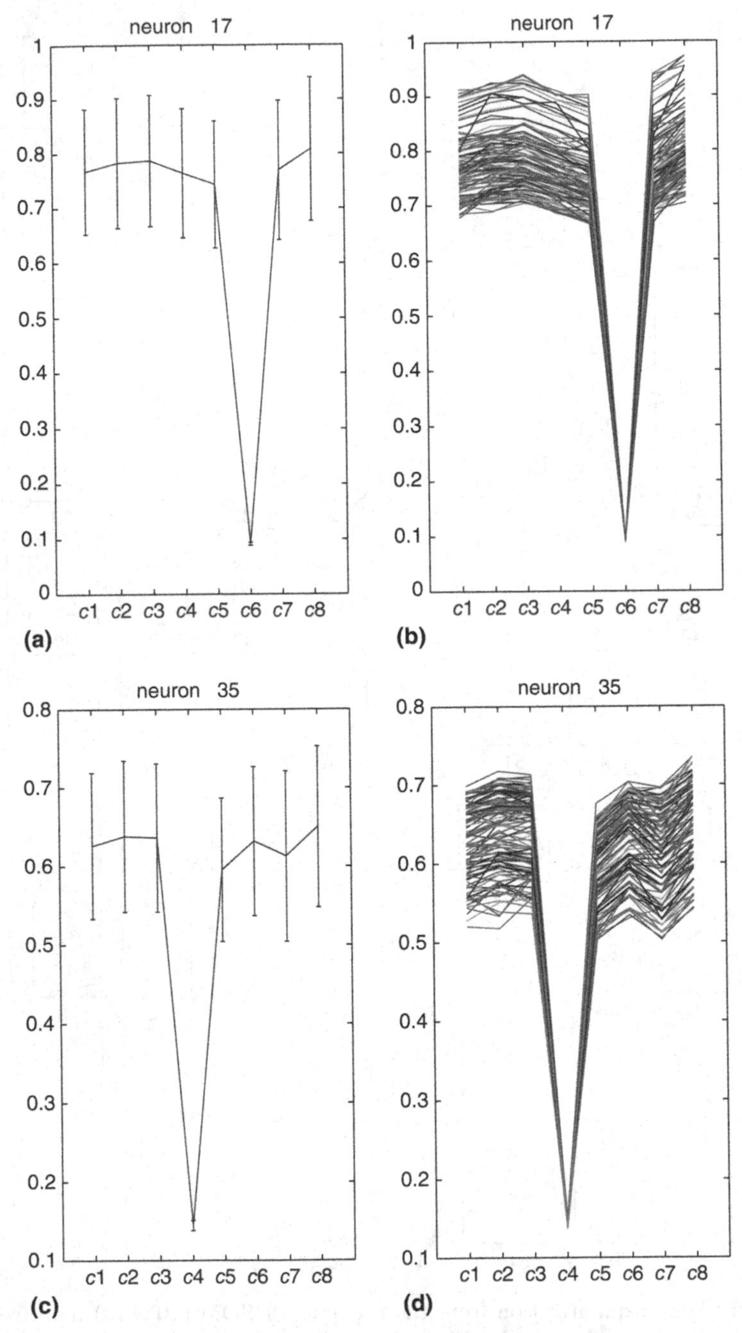

Fig. 7.26. Figures (**a**) and (**c**) represent the spectral patterns that are associated to neurons 17 and 35. Vertical bars represent the variance for each wavelength. Figures (**b**) and (**d**) represent the radiance subset that are respectively represented by neurons 17 and 35. (PRSOM 10×10 map obtained from App_{cod1})

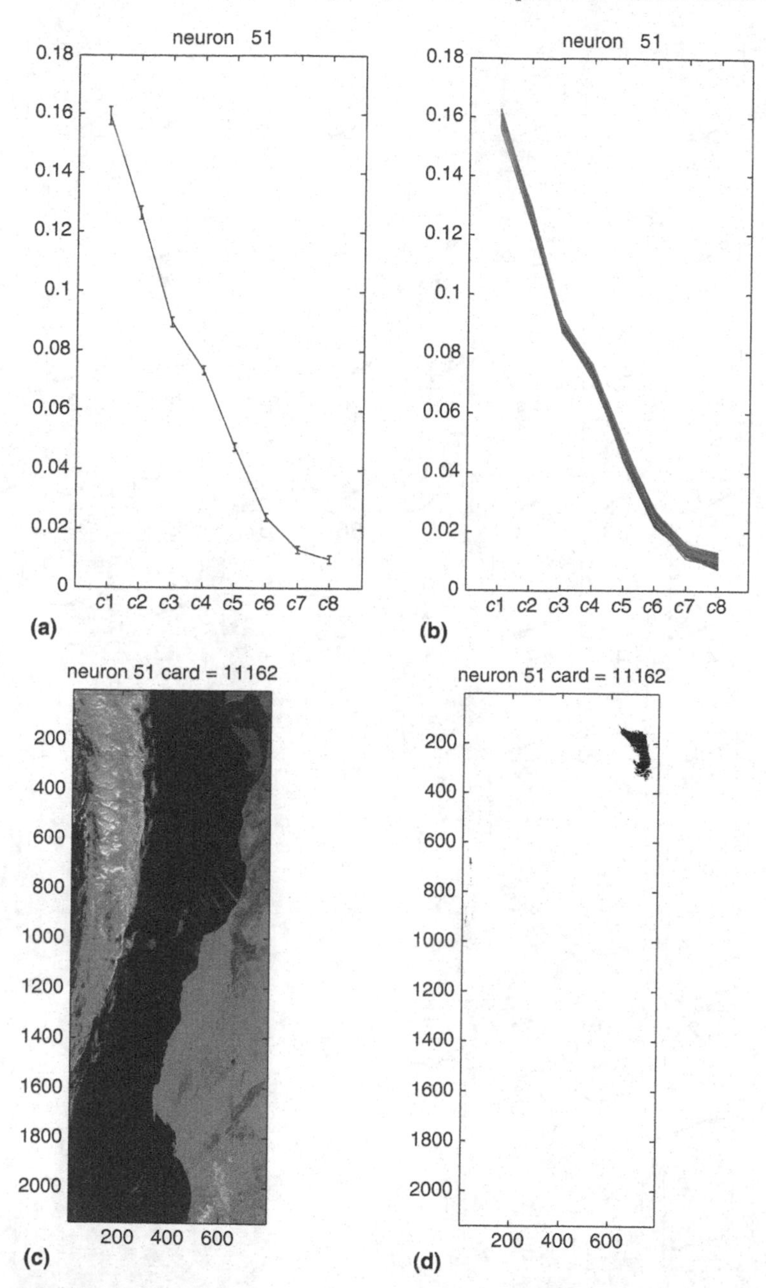

Fig. 7.27. Representation of neuron 51 spectral characteristics: (**a**) representation of the spectral observation vectors of the learning set App_{cod1} that are allocated to neuron 51 (**b**) representation of the spectral observation vectors of the geographical zone, which is allocated to neuron 51, (**c**) representation of the geographical zone in the total image, (**d**) representation in black of this geographical zone. (PRSOM 10×10 map obtained from App_{cod1})

Fig. 7.28. Representation of vector quantization, which is associated to PRSOM map that is obtained from App_{cod2}. The identification number of the neuron is above the frame. The associate pattern spectrum is depicted inside the frame

neurons into classes and trying to label the classes that were obtained through hierarchical classification. In order to illustrate as best as possible the quality of results when PRSOM and BUHC (Bottom-Up Hierarchical Classification) are sequentially processed, two experiments of different complexity are presented:

- The first experiment is relative to the determination of a mask that is able to detect the thick clouds and to discriminate them from other spectra. It is known that clouds are strongly reflecting the signal: the top of atmosphere signal that are registered by the satellite sensors are presenting stronger and more variable intensities than when sea or aerosols are concerned. The discrimination between thick clouds and other constituents amounts to build a binary classifier. Since clearly distinct properties in the observation set physically characterize this problem, the two classes that are searched have to be fully separated.
- In the second experiment one tries to recover the five classes, which have been identified by the expert; these classes have been determined by comparing the data with aerosol physical models. Actually, the number of classes is higher and the expert has possibly introduced a lot of mistakes, so this problem is far more complex.

The two vector quantizations that have been obtained using PRSOM will be used to recognize the expert-identified classes. Class determination will be performed through bottom-up hierarchical classification using the Ward index that has been previously defined in the paragraph ≪Looking for a partition that suits the classes of interest≫.

In the first experiment, bottom-up hierarchical classification is performed on the PRSOM 10×10 map obtained from App_{cod1}. Since the searched classification is a binary classification to select thick clouds, clustering has been pursued up to the obtention of two classes. Figures 7.29 and 7.30 show the classifications that were obtained on the topological map and on the image. The visualization of the map enables to observe the neurons of each class. Clearly, the associated zones are well connected on the map. This classification has been compared with the expert-based classification by computing the confusion matrix. Here, SeaWifs provided the expert-based classification since the cloud mask is available. The confusion matrix is represented on Table 7.2; it

Table 7.2. Confusion matrix that compares the SeaWifs labeled classification and the classification that was obtained from PRSOM + BUHC. PRSOM was obtained from Appcod1 and BUHC uses Ward index

	PRSOM + BUHC	
	Clouds	Apparent sea
SeaWifs-labeled clouds	0.91	0.09

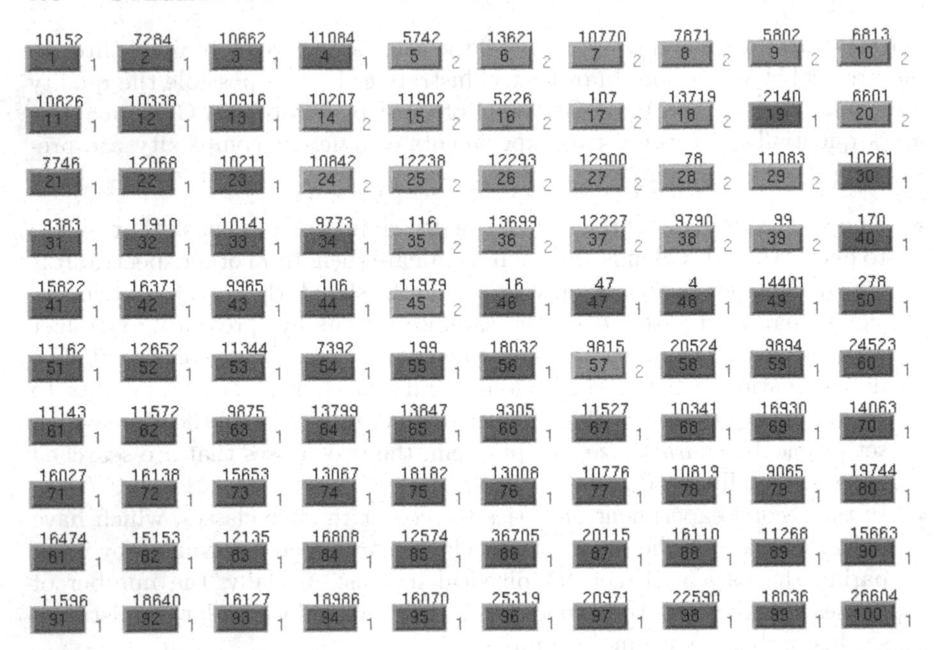

Fig. 7.29. Presentation on the map of the two classes that are proposed by PRSOM + BUHC: PRSOM was determined from $\mathrm{App}_{\mathrm{cod1}}$, and BUHC uses the Ward index. The number at the right of the neuron frame represents the class that is obtained through hierarchical classification. The set of dark grey neurons constitutes class 1 and the set of light grey neurons constitutes class 2

allows to compare the two classifications. The division between two geographical zones that is displayed on Fig. 7.30 shows a good adequacy to the division that comes straight from the satellite data. It shows that the two classes have been correctly demarcated by using PRSOM algorithm and then bottom-up hierarchical classification onto the raw physical measurements without any additional expert information. The nice consistency of hierarchical classification results at the top level of the classification tree allows expecting that clustering is consistent at any level of the hierarchy.

The second experiment is searching a zone that is already provided by expert knowledge. Actually this class is case 2 water. This experiment confirms the quality of vector quantization. In Fig. 7.22, this zone, which is endowed with particular optical properties is colored in light grey. Investigation of the various geographical zone that are associated to the 100 neurons of the map enables to select three neurons. The subset of their associated spectra allows to find a geographical zone that can be superimposed to the expert-based region. Figure 7.31 shows the three patterns (w_{33}, w_{82}, w_{93}) and the associated geographical zones.

If the full hierarchy that arises from BUHC is investigated, one notices that the three neurons of interest (namely 33, 82 and 93 constitute a subset that is

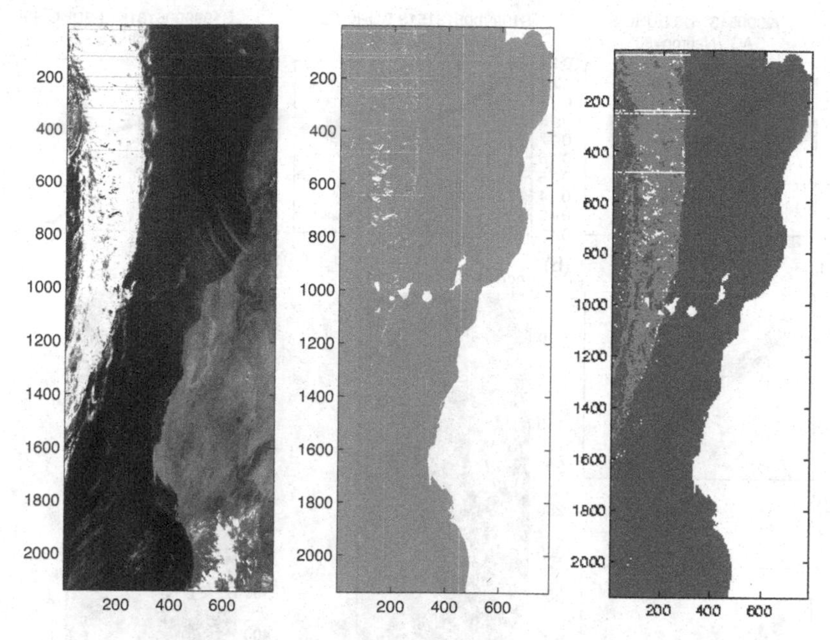

Fig. 7.30. Presentation of the two classes that are proposed by PRSOM + BUHC: the *left* image is representing the original SeaWifs image the *center* image represents the land mask that is provided by SeaWifs (in *white* on the image) and the *right* image presents the binary classification that is provided by PRSOM + BUHC (the light zone is similar to the class 2 of Fig. 7.29 and represent thick clouds, the dark zone includes various aerosols); PRSOM was achieved using App_{cod1} as learning set and BUHC uses the Ward index

clustered at the level 35 of the hierarchy. Thus, it is possible to propose from this result a classifier that enables to automatically identify the case 2 water labeled pixels. The neurons 33, 82 and 93 will be labeled case 2, all the other neurons will be labeled by a negative label that points out they do not belong to case 2. Then, it appears that in that case, using the hierarchical clustering without additional expert knowledge may allow retrieving information which comes from the physics of measure.

The last experiment is using straight expert knowledge. Knowledge is used to label neurons according to the methodology that was presented previously in the section about classification and topology. The test set is the expert-labeled GAC image. Each neuron is then mastering pixels of GAC image that are allocated to that neuron. These pixels constitute a subset of labeled elements. The label of the associate neuron is determined by a majority voting. The two classifications are compared on Table 7.3 where the confusion matrix is represented. It is clear that the neurons correctly provide the expert knowledge.

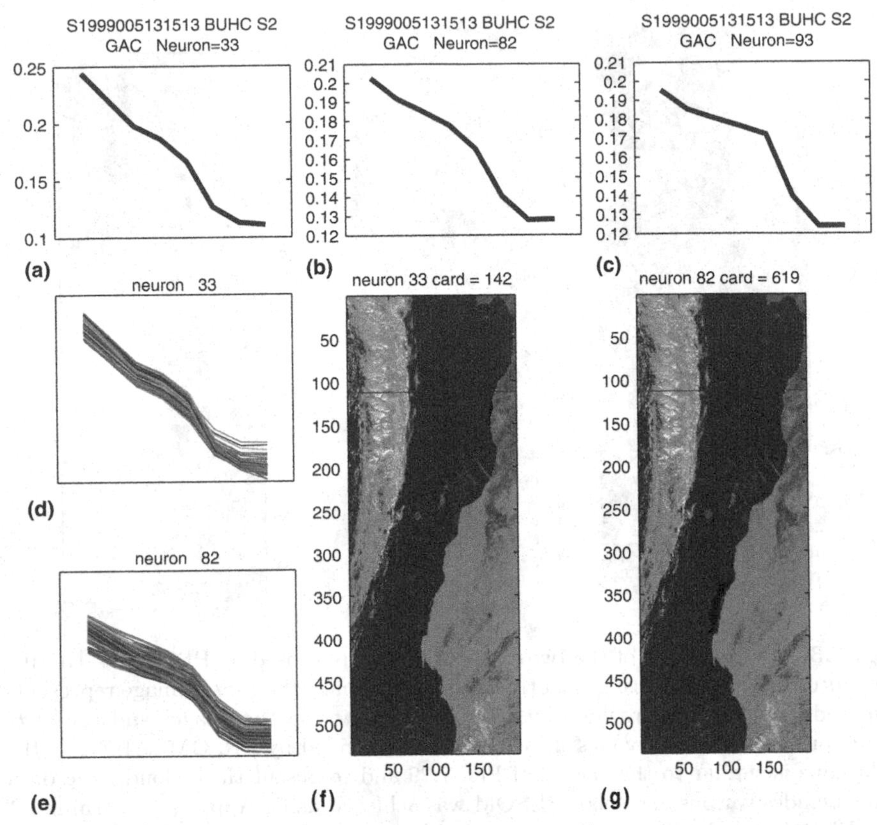

Fig. 7.31. Representation of the patterns, of the observation subsets and of the geographical zones that are allocated to neurons 33, 82 and 93 (PRSOM map 10×10 using App_{cod2} as learning set). Figures (**a**), (**b**), (**c**) represent the neuron spectra, Figures (**d**) and (**e**) represent the associated learning set spectra for neurons 33 and 82. The zones that are associated to neurons 33 and 93 have been colored in black in figure (**f**) and (**g**)

Table 7.3. Confusion matrix allowing the comparison between the expert-based classification (GAC image) and the PRSOM-based classification. On this map, the hundred neurons have been labeled using majority vote from expert data

		PRSOM + majority vote			
		Class 1	Class 2	Class 3	Class 4
Expert	Marine aerosols	0.8	0.04	0	0.16
	Clouds	0.03	0.91	0.01	0.05
	Case 2	0.03	0.22	0.71	0.03
	Desert aerosols	0.1	0.04	0	0.86

The bunch of results that were obtained in that application demonstrates that topological maps are correctly operating when they are used to process numerical data.

The following application is due to T.Kohonen. It shows that the algorithm is performing well when it is used for textual processing.

7.5.3 Topological Map and Documentary Research

This last paragraph presents another real-world application in a field that is completely different: documentary research. The general objective of the Websom system that was created by Kohonen and his collaborators is to create a content-based labeling of a set of texts. The current working version allows organizing 7,000,000 texts in a single documentary data basis. Thus documents with close-by semantics are endowed with neighbor label. A visual inspection of the basis representation provides a global hint about the content of the documents that are stored in a particular zone of the basis. Looking for the keywords that are associated to the zone and considering the topics of the different documents allow document searching in an original way. Considering his short description of Websom system's main characteristics, one feels how self-organizing maps are used: semantically close observations (texts) are allocated to neighbor neurons on the map. In order for the application to be operational, several additional properties have to be checked:

- As for the remote-sensing satellite data application, the quality of the system depends closely on the semantics of the texts of interest.
- Documentary research is useful only if the number of stored texts is large enough and if the visualization is fine enough. Thus the dimension of the map has to be very high.
- The system is supposed to be operated on-line, thus it has to work fast.
- The basis algorithms has to be changed to allow
  1. introducing a linguistic knowledge that enables textual manipulation,
  2. training high-dimensional maps to be able to process as many documents as possible,
  3. using a friendly interface which really helps the user to perform document research,
  4. reducing the duration of an average research session.

7.5.3.1 Information Coding

When a text is preprocessed significant information is extracted that depends on the specificities of the general field of the research. Of course, the encoding has to be made according the specifications of the topological maps: Kohonen's algorithm is processing numerical multidimensional data. Thus any text has to be represented by an n-dimensional numerical vector. The current version of Websom system is processing a corpus that contains 6,840,568 English summaries. The average length of such abstract is 132 words. To process also number and symbols, it has been decided to cancel the words that are too scarce (less that 50 occurrences) and also to cancel 1355 words that are semantically poor. Eventually a set of 43,222 words has been considered for the whole corpus.

Several versions of the system exist. The earlier was coding straight the text histogram by a vector, the length of which was the word number of the corpus. According to that coding, each component of the text representative vector represents the weighted occurrence frequency of the associated word in the text. The weights were fixed according to the influence of the word on the document global meaning. This dimension was too large to allow further processing. Several data compression methods were proposed to cope with this high dimension problem: projection reduction (principal component analysis) or random projection. Eventually a random projection method was implemented. A 500-dimensional vector represents each text. Such a vector is a text summary coming from a statistical analysis of the text vocabulary. The coding complexity is O(NL)+(n), where N is the document number, L is the average number of distinct words in a document and n is the initial histogram dimension. To appreciate the reduction, it is interesting to point out that the simpler projection compression method has a complexity equivalent to *Nld.* Thus, the reduction is quite important and enables to extend Websom over the whole corpus.

7.5.3.2 Specific Features of Learning Process

A visual representation of the corpus organization is possible through the two-dimensional map. That is a great help for documentary research. At the end of the learning phase, the allocation phase that associates a neuron to a document enables to locate a given document with respect to the global corpus: texts with similar meanings are supposed to be located in close zones on the map. In Websom's last version, the corpus is divided into 21 sections (agriculture, transportation, chemistry, electricity, etc.). To extract this information, each neuron is endowed with one of the section labels and a set of key words. These keywords are extracted from the subset of texts that are allocated to this neuron. More precisely, the type is determined through a majority voting over the text subset and the keywords are selected by building the intersection of the key-words set of every text of the text subset.

When Websom is used, texts with close meaning are projected in closed regions of the two-dimensional map. So, projecting the text onto the map enables to locate its meaning with respect to the whole set of texts of the learning basis, actually the whole corpus. Using the map labeling enables to interpret a new text through an automatic process. The neighbor neurons provide subsidiary information that allows a finer understanding.

Considering the very large number of documents that lie in the basis, a large amount of neurons are required in order to perform a fine enough document analysis. The more important change that was introduced enables to train fast high-dimensional maps. The topological map that is used for Websom is composed by 1,002,240 neurons. It is impossible to train this map because the connection number is too large: 1,002,240 × 500. The new idea relies on the simple fact that a good initialization considerably increases the convergence speed. This good initialization is found through a hierarchical procedure that enables to guide the training from one step to the next one. In the Websom implementation, the parameters are tuned using a first rectangular map of 435 neurons. This first map is extracted form the learning basis. Then a second map that uses a finer sampling is initialized using the results of the first one: the initial values of a parameter of the second map are obtained through an interpolation of the values of the three closest neighbors extracted from the 435 neurons of the first map. In that way, the number of neurons increases from step to step up to 1,002,240 neurons. For each step, there is a new learning phase of the whole corpus. The initial learning phase (for the 435 neurons of the first map) requires 300,000 iterations; every further learning phase only requires five iterations of the "dynamical clouds" versions of the algorithms. In such a way, it is possible to train very large maps. Moreover, the hierarchical order that was found in the previous steps is used to find the closest neighbors during the successive learning steps.

7.5.3.3 Discussing Websom Performances

The various improvements, which were implemented in Websom are quite efficient with respect to the time complexity of the computation. The methodology that was previously detailed allows reducing the number of operations from $O(dN^2)$ for the original Kohonen algorithm to $O(dN^2)$ $O(dM^2)+$ $O(dN)$ $+O(M^2)$ for Websom. In that expression, N is the number of neurons in the actual map, M is the number of neurons in the initial map and d is the dimension of the input layer ($d = 500$ for Websom). The comparisons that were achieved with the original Kohonen methodology show that the last version of the implementation has the same performances than the original algorithm with respect to the quantization error and the classification error. The final version of the map was obtained through a six-week learning phase that was performed on a six-processor computer (SGI O2000). The performances over the seven millions text basis go up to 64% of correct classification. As it is generally the case for data mining applications, the interface was very carefully designed : the map is presented as a sequence of HTML pages. It is easy to explore it by using the mouse. A simple click enables to reach the documents and then to visualize and to read them.

References

1. Anouar F., Badran F., Thiria S. [1997], Self Organized Map, A Probabilistic Approach, *Proceedings of the Workshop on Self-Organized Maps*, Helsinki University of Technology, Espoo, Finlande, 4–6 juin 1997

2. Bock H.H. [1996], Probabilistic Models in Data Analysis, Computational Statistics and Data Analysis, *23*, pp 5–28
3. Bock H.H. [1998], Clustering and neural networks, *in* Rizzi et al. (éd.), *Advances in data science and classification*, Springer verlag, pp 265–278
4. Cerkassky Y., Larmnajafih [1991], Constrained topological mapping for non parametric regression analysis, *Neural Network*, vol. 4, pp 27–40
5. Dempster A.P., Laird N. M., Rubin D. [1977], Maximum Likelihood from incomplete data via the E.M. algorithm (with discussion), *Journal of the Royal Statistical Society*, series B 39, pp 1–38
6. Didday E., Simon J.C. [1976], Clustering Analysis, *in Digital Pattern Recognition*, K.S. Fu, Springer verlag
7. Duda R.O., Hart P.E. [1973], *Pattern Classification and Scene Analysis*, John Wiley
8. Fritzke B. [1995], *A growing Neural Gaz Network learns topology*, D.S. Touretzky and T.K. Leen (editors), Advanced in Neural Information Processing Systems 7, MIT Press, Cambridge MA
9. Gaul W., Opitz O., Schader M. (éd.) [2000], *Data Analysis Scientific Modeling and Practical Application*, Springer
10. Jain A.K., Dubes R.C. [1988], *Algorithms for Clustering Data*, Prentice Hall
11. Kaski S., Honkela T., Lagus K., Kohonen T. [1998], WEBSOM-self-organizing maps of document collections, *Neurocomputing*, vol. 21, pp 101–117
12. Kaski S., Kangas J., Kohonen T. [1998], Bibliography of self organizing map (SOM) papers 1981–1997, Neural Computing Survey, vol. 1, pp 102–350. On peut trouver cet article à l'adresse: http://www.icsi.berkeley.edu/~JAGOTA/ncs/
13. Kohonen T. [1984], Self organization and associative memory, Springer Series *in Information Sciences*, 8, Springer Verlag, Berlin (2nd éd. 1988)
14. Kohonen T., Kaski S., Lagus K., Salojrvi J., Honkela J., Paatero V., Saarela A. [2000], Self organization of a massive document collection, *IEEE transaction on neural networks*, vol. 11, no 3
15. Kohonen T. [2001], *Self Organizing Maps*, Springer, 3e édition
16. Luttrel S.P. [1994], A bayesian analysis of self-organizing maps, *Neural Comput*, 6
17. Murtagh F. [1985], A survey of algorithms for contiguity-constrained clustering and related problems, *The Computer Journal*, vol. 28, pp 82–88
18. Oja E., Kaski S. [1999], *Kohonen Maps*, Elsevier
19. Vichi M., Bock H.H. [1998], Advances in Data Science and Classification, Springer, Heidelberg, pp 397–402
20. Von der Malsburg C. [1973], Kybernetik 14, 85
21. Yacoub M., Badran F., Thiria S. [2001], *Topological Hierarchical Clustering : Application to Ocean Color Classification*, ICANN'2001, Springer 2001, Proceedings, pp 492–499

8

Neural Networks without Training for Optimization

L. Hérault

Previous chapters have shown neural networks as powerful tools for modelling, control, discrimination and automatic classification. In those fields, non-linearity and training properties are used: a parameterised static or dynamic function is used, and the parameters are estimated through training. This chapter will focus on another way of taking advantage of neural networks: non-linearity and dynamics properties are also used, but the parameters of those networks are naturally derived from the application, without training. That approach is particularly well suited to solving optimisation problems.

What decision should to be taken? How to minimize the production costs through an optimised tasks scheduling and an optimal management of flows and resources? How to increase productivity? How to make the best use of available resources to fulfil a request at a minimum cost? An optimisation problem is the core issue behind all these questions. In fact, the optimisation task deals with the choice between several alternatives. This choice is governed by the desire to make the best decision, which is often expressed as the selection of a solution satisfying the problem requirements with a minimum realization cost.

8.1 Modelling an Optimisation Problem

When facing an optimisation problem, the first step consists in reformulating it in a mathematical way. That modelling is a crucial step and is sometimes critical, since quantifying the quality of a solution is not that simple, and the mathematical formulation (sometimes named coding) of a problem has an influence on the choice of the methodology to solve it. That phase requires a close cooperation between the optimisation experts and the application experts, who are looking for a solution. The result of that step is a mathematical model, generally defined by

- some variables;
- an objective, which is a function of the variables, and is generally expressed as the minimization of a cost, or a set of costs, of the solution;
- some constraints to be satisfied by the solution. Some are essential; then they are named strict constraints. Other ones express some preferences; then they are named relaxable constraints.

The problem is then to find, in a limited time, a set of variable values that reach the objective while satisfying the constraints. Depending on the applications, the number of optimisation variables ranges from a few to hundreds of thousands, and the expected response time from a few microseconds to several hours. Moreover, it is sometimes desirable to find a set of solutions among which the user will make a choice.

The variables of an optimisation problem can be

- continuous variables,
- discrete variables, e.g. binary variables (the problem is said to be combinatorial),
- mixed variables, i.e., some variables are continuous while others are discrete (those problems are said to be mixed).

A problem, besides, may not contain constraints, or cost function to be optimised. In that last case, the question is then related to the existence of a solution satisfying all the constraints; such a problem is named constraint satisfaction problem. When all constraints cannot be satisfied, i.e., when the problem has no solution, it is sometimes interesting to find a satisfactory trade-off that violates a few relaxable constraints.

The cost functions to be optimised can be expressed with various formulations, more or less complex, e.g. as linear or quadratic functions of the variables.

Sometimes, the data necessary to solve the optimisation problem are not immediately available, but may become available gradually: the methods must then update dynamically the solutions.

Finally, some problems are distributed: some locally connected decision centres make partial decisions according to the available local data, and that set of local decisions must be a good solution of the global problem.

8.1.1 Examples

There is a very large variety of combinatorial optimisation problems. They are commonly encountered in many industrial applications. Among them, the author has been directly involved in the following typical areas:

- Military: resource allocation problems (weapon allocation on moving targets [Hérault 1995a,b,c]), target tracking [Hérault 1997a].

- Computer aided design (CAD): design of complex materials, e.g. optical filters or composite materials [Boudet et al. 1996]; shape and structure optimization.
- Telecommunications: deflection routing of packets in ultra high speed networks [Hérault 1997b], code optimization in broadband wireless systems (CDMA), and optimization of the radiation pattern of antenna arrays.
- Nuclear: stock management of perishable material [Privault et al. 1998a,b].
- Spatial: scheduling; mission planning, resource allocation, etc.
- Human resources: management optimization, allocation of persons to positions, optimization of timetables.
- Signal and image processing: particle tracking [Dérou et al. 1996], pattern recognition [Hérault et al. 1993].
- Industry: job shop scheduling, packaging and routing problems.

Many other applications are described in [Takefuji 1992; Cichocki et al. 1993; Dagli 1994; Takefuji et al. 1996].

Those problems can often be encoded as mathematical problems of graph theory, such as graph colouring, graph partitioning, graph matching, as well as extraction of sub-graphs with specific properties like cliques, paths, cycles, etc. [Gondran et al. 1995]. As a consequence, it is important to study, jointly with the applications, how to solve those families of generic problems. The travelling salesman problem is one of those archetypal hard combinatorial problems; we describe it in the following section.

8.1.2 The Travelling Salesman Problem (TSP)

Throughout this chapter, the graph theory problem named travelling salesman problem, which is a reference among the combinatorial problems, will be frequently used as a problem example.

Reminder of the TSP

This is a tour optimisation problem: a travelling salesman has to visit some towns with known geographic coordinates. To reduce its costs, he looks for a tour that is as short as possible in terms of distance, and which goes through each town exactly once.

Many algorithms have been proposed in the literature to solve this problem [Reinelt 1994; Gondran et al. 1995].

An example of problem with 101 towns is illustrated by Figs. 8.1 and 8.2. Many applications can be modelled as this type of combinatorial problem: such is the case, for instance, for the problem of finding the shortest path of an industrial tool that has to machine some parts of an object (e.g., drill holes in a printed circuit board).

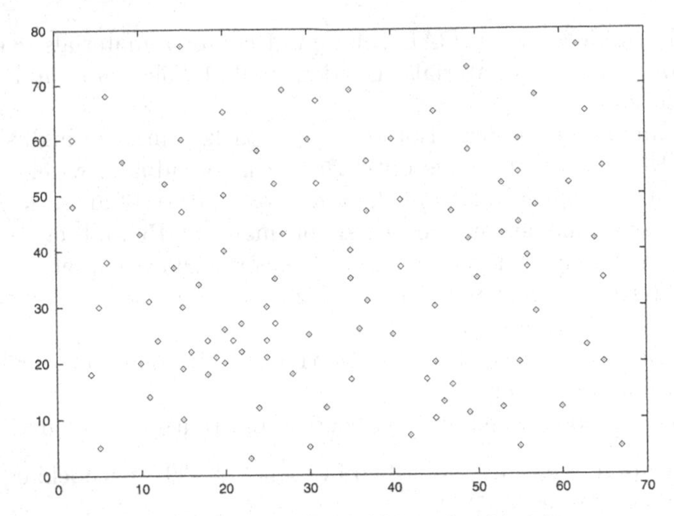

Fig. 8.1. Example of TSP with 101 cities

8.2 Complexity of an Optimization Problem

Optimisation problems can be classified as a function of their intrinsic complexity.

In industrial applications, an optimisation problem can often be modelled with a cost function and a set of constraints that are linear functions of the optimisation variables (real numbers). In that case, the problem is a linear programming problem and an exact solution can be found in a reasonable time: the optimal solution can be found in a number of iterations that varies

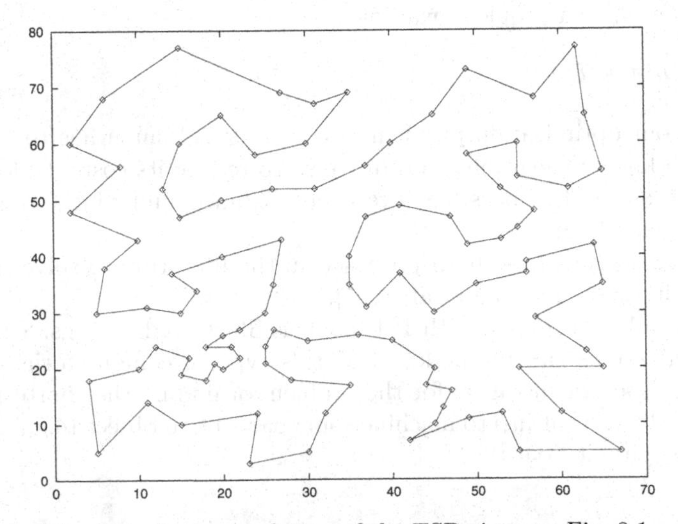

Fig. 8.2. Optimal solution of the TSP given on Fig. 8.1

polynomially with the number of variables. A number of methods were found to solve efficiently this type of problems [Schrijver 1986], the most famous one being the simplex (which has already been mentioned in the theoretical and algorithmic addenda of Chap. 2, to put the dynamic models under a canonical form). Nevertheless, when the number of variables becomes very high, the simplex is very slow; other methods, such as the interior points methods, avoid that drawback [Gonzaga 1992].

Practically, it is sometimes necessary to linearise some problems which are non-linear. Nevertheless, the resulting modelling error can be large: it is then necessary to use other methods, which can attack directly non-linear problems. As described in the previous chapters, neural networks have that specific property.

Many constrained optimisation problems encountered in industrial applications are combinatorial problems. Underlying graph theory problems are often NP-complete [Garey et al. 1979]: in other words, the number of possible solutions undergoes combinatorial explosion, i.e., it grows exponentially with the number of variables. It is then possible to use heuristics, which are ad-hoc techniques. Although unable to find exact solutions, they are able to find good solutions in a reasonable computational time, but with no guarantee on their optimality.

8.2.1 Example

The above-described TSP is combinatorial and belongs to the family of NP-complete problems [Reinelt 1994]. The number of potential solutions, i.e., of possible tours, for a TSP with N cities, is $N!$. Thus, for the example given Fig. 8.1, the number of potential solutions is greater than 10^{159}. Obviously, an exhaustive enumeration of all those possible solutions cannot possibly be performed. Even an approach consisting in picking at random many potential solutions and keeping the best one would not provide good results. In fact, those solutions would represent a tiny part of the set of possible solutions. Therefore, using more elaborate techniques for providing good solutions in a reasonable search time is mandatory.

8.3 Classical Approaches to Combinatorial Problems

To solve the linear programs with integer variables, which are combinatorial problems frequently encountered in applications, some heuristics have been proposed, such as cutting planes methods, which add constraints to reduce the convex envelop of the solutions [Schrijver 1986].

Branch-and-bound methods are exact methods, which try to enumerate, in an educated way, the feasible solutions, in such a way that good solutions are found rapidly [Prins 1994]. They split the solution space into subsets that are smaller and smaller, most of them being eliminated after bounds calculation before being constructed explicitly: that is the reason why they are called implicit enumeration methods. When applied to combinatorial problems, those approaches require a computational time that increases exponentially with the number of variables, but their complexity is, however, smaller than an exhaustive enumeration. Nevertheless, those approaches are generally inefficient to solve large-size problems with hundreds of variables, because they are too computationally demanding.

Dynamic programming is an enumerative approach based on the idea that solutions to sub-problems of the problem can help to guide the search of the optimal solution of the global problem. Here again, that approach has limitations for large size problems.

Local search methods, used as early as the seventies, are generally used when the previous approaches cannot be used or require a too long resolution time. From an initial feasible solution, they try to improve it through an exploration of a small neighbourhood. If a better solution is found, it becomes the current solution and the search is iterated from that solution. The algorithms stops when no solution in the neighbourhood improves the current solution. Metaheuristics such as simulated annealing or tabu search, which will be described below, are refinements of those techniques, which avoid the local minima of the cost function. To that end, from a feasible solution, a small neighbourhood is explored as previously, but a move in the solution space is authorized according to an ad hoc criterion. This criterion may authorize moves towards solutions that may be worse that the current one, so as to escape from local minima.

Constraint programming methods couple a reduction of the search space with efficient branch and bound methods. Their efficiency has been ascertained to solve some large-size problems, where constraint satisfaction is more important than cost minimisation.

8.4 Introduction to Metaheuristics

At the beginning of the 1980's, methods known under the generic name of metaheuristics were proposed in order to attack problems of high complexity [Reeves 1995; Osman et al. 1996; Aarts et al. 1997; Teghem et al. 2001]. Those approaches are iterative: they converge towards attractors that encode good solutions of the problem. Therefore, the underlying theory is related to the theory of dynamic systems, discrete or continuous, multi-dimensional and generally non-linear. Such systems have also been analysed in Chap. 2 and 4. Those approaches are derived from the mathematical modelling of some natural process:

- The modelling of physics process, by some methods of statistical physics, has given rise to new methods such as simulated annealing.

- The modelling of processes with memory gave rise to the so-called tabu search. The specificity of that approach stems from the fact that it stores in memory the recent past of the search, and possibly parts of the faraway past.
- The modelling of the nervous system gave rise to neural algorithms. Being massively parallel and allowing hardware implementations (analog, digital, optical), they are particularly attractive to solve problems requiring very short resolution time.
- The modelling of genetics initiated the development of genetic or evolutionary algorithms, in which potential solutions are considered as individuals evolving inside a population.
- The modelling of the learning process gave rise to reinforcement learning methods (see also Chap. 5). The corresponding algorithms can be distributed over a network of calculators, each of them learning how to react optimally (in a stationary environment), from the knowledge of an evaluation of its decisions by its neighbors. Those approaches are particularly suitable for solving dynamic problems in non-stationary environments.

Finally, it is important to emphasize that there exist hybrid methods that couple some metaheuristics, or metaheuristics and conventional approaches.

Before considering neural networks to solve optimisation problems, we present the most closely related metaheuristics: simulated annealing.

8.5 Techniques Derived from Statistical Physics

It is possible to devise an analogy between combinatorial optimisation problems and the modelling of complex systems by statistical physics methods. A complex physical system has a multitude of possible states; among them, an equilibrium state is a state for which a quantity that depends on the system state (e.g., its free energy) is minimum. The search for an equilibrium state of a system simulated on a computer amounts to searching the minimum of a function that depends on the system state, which might be defined by a huge number of variables (such as the positions of the particles of the system). Therefore, the simulation of macroscopic properties of a physical system amounts to the search for a minimum of thermodynamical quantities such as the free energy. By establishing an analogy between thermodynamical quantities and the cost functions of an optimisation problem, it becomes possible to find a minimum of the cost function, hence a solution to the optimisation problem, by taking advantage of simulation techniques derived from statistical physics.

Remark. To this end, the optimisation problem is assimilated to a system of interacting particles, whose states code for the values of the variables. The optimal solution is then considered as a fundamental state (state with minimum energy) of this physical system. In order to get close to those fundamental states, some results of statistical physics are used.

Two hypotheses on the physical system gave rise to two families of optimisation methods: the canonical analysis and the microcanonical analysis.

8.5.1 Canonical Analysis

Canonical analysis assumes that the physical system under investigation is not isolated: it can exchange heat with its environment. As a consequence, we will include in the methods a temperature parameter, denoted by T.

At a given temperature, the thermodynamical equilibrium of such a system is a state such that the free energy of the system is minimum. This free energy is defined as the difference between the internal energy and the temperature-entropy product. Moreover, it is well known that the entropy is zero at zero temperature. Those results can be used to solve an optimisation problem by associating the cost function of the problem with the internal energy of the system.

During the search of an optimum, it is important to explore a large territory in the solution space: to this end, thermal noise is used.

Finally, statistical physics teaches us that, at thermodynamical equilibrium, the system states are distributed according to a Boltzmann law: the probability that the system be in a state of energy E_0 is given by

$$P(E_0) = \frac{\exp\left(-\dfrac{E_0}{kT}\right)}{Z(T)}.$$

In this equation, k is the Boltzmann constant and $Z(T)$ is a normalization function ensuring that the sum of the probabilities of all the accessible energies is 1. From this property, it is clear that the most probable states at thermodynamical equilibrium are those of minimal energy. This property is of great interest to solve optimisation problems.

Remark. The number of possible states is *finite* since we are considering a combinatorial problem: therefore the energy states are *discrete*. As a consequence, the notion of probability can be used; for a problem with continuous variables, where the energies have continuous values, the notion of probability density function, introduced in Chap. 2, should be used.

8.5.1.1 Simulated Annealing

Moreover, metallurgy teaches us that a good way to reach low energy states of a solid consists in heating the material, and then letting it cool slowly. That process, called annealing, forces the system evolution towards low energy states; a slow cooling process (as opposed to a fast cooling, called quenching) avoids the system being trapped in metastable states corresponding to high-energy local minima.

The principle of simulated annealing, defined independently by Kirkpatrick in 1983 [Kirkpatrick et al. 1983], Siarry in 1984 [Siarry et al. 1984] and Cerny in 1985 [Cerny 1985], consists in implementing those concepts as a numerical algorithm. The idea is the following: at decreasing temperature steps, the algorithm uses an iterative procedure, proposed by Metropolis in 1953, to reach a thermodynamical quasi-equilibrium state. That procedure allows escaping from local minima with a probability that increases with temperature. When the algorithm reaches the very low temperatures, the most probable states are excellent solutions to the optimisation problem.

The Metropolis Algorithm

In 1953, Metropolis devised an iterative algorithm that allows finding the thermodynamical equilibrium state of a simulated system at a given temperature T [Metropolis et al. 1953]. It consists in iterating the two following steps:

- Evaluation of the energy variation associated with a random elementary transition from the current state i, of energy E_i, to a new state j, of energy E_j: $\Delta E_{ij} = E_j - E_i$.
- Acceptance of the transition to that new state with probability A_{ij} defined as:

$$A_{ij}(T) = \begin{cases} 1 & \text{if } \Delta E_{ij} = 0, \\ \exp\left(-\dfrac{\Delta E_{ij}}{T}\right) & \text{otherwise.} \end{cases}$$

Simulated Annealing

The simulated annealing algorithm consists in decreasing the temperature in a systematic way, from a high initial temperature, within the Metropolis algorithm. In practice, several cooling schedules can be used. One is a geometric decrease, in which the temperature at step k is given by

$$T(k) = \alpha T(k-1) \,,$$

where α is a constant strictly smaller than 1, but close to 1.

Two types of simulated annealing were defined, depending on the cooling schedule:

- in homogeneous simulated annealing, the temperature parameter is decreased only when the thermodynamical equilibrium is reached at the current temperature; that algorithm assumes that the Metropolis procedure has been iterated an infinite number of times, and therefore has only a theoretical interest;
- in practice, the temperature parameter is decreased after a finite number of evaluations of transitions at a given temperature: then the algorithm is called in-homogeneous.

Coding of the Optimization Problems

The practical performances of the simulated annealing are closely related to the coding of the problem, and in particular to the choice of

- the variables;
- the elementary transitions, which define the topology of the solution space: in that space, the distance between two states is the number of elementary transitions necessary to go from one state to the other;
- the functions coding the cost and the constraints. Relaxable constraints can be combined with the cost function; strict constraints can, for instance, be automatically satisfied by an ad-hoc choice of the elementary transitions.

Some Theoretical Results

The theoretical behaviour of the simulated annealing has been investigated in great detail, through a Markov chain modelling (see Chap. 5). A good overview of those results is given in [Aarts et al. 1989]. We summarize here the most important results.

The Metropolis algorithm at a given temperature converges asymptotically towards the thermodynamical equilibrium at that temperature, which is characterized by a stationary distribution of the states.

The homogeneous simulated annealing, which assumes that a stationary distribution was reached by the Metropolis algorithm at each temperature step, converges towards the optimal solutions of the problem, irrespective of the cooling schedule.

As far as inhomogeneous simulated annealing (the only one used in practice) is concerned, Hajek found a necessary and sufficient condition on the cooling schedule between two (or more) elementary transitions [Hajek 1988]: the temperature at the k-th transition, or at the k-th temperature step, must satisfy

$$T(k) \geq \frac{C}{\log(1+k)},$$

where C is a constant equal to the maximum depth of the local minima.

Pros and Cons

That technique is very successful for two main reasons. First, the values of the parameters of the method can be easily determined and a black-box operation is often possible for real applications. Secondly, theoretical results show that simulated annealing can reach solutions as close as expected of the optimal solutions at a higher speed than an exhaustive exploration of the solution space. Nevertheless, it is at the cost of the user's patience. In practice it is generally possible, depending on the amount of time available to solve the problem, to automatically adjust the internal parameters of the method in order to make the best use of the allotted time, and find a solution that is a good trade-off between the quality of the solution and the time required to find it. Finally, this algorithm can be generalized to continuous optimisation problems.

The main drawback of simulated annealing is in his reputation of slowness. Depending on the requirements of the user on the solution quality, it may need many computations associated with the transition evaluations. This can be redhibitory in applications that require very short response times.

Besides, the parallelisation of the algorithm is difficult, particularly if one wishes to keep the properties of theoretical convergence towards the optimal solutions. Finally, the Metropolis algorithm, within the simulated annealing, may appear costly: indeed, it requires a good numerical accuracy. Moreover, the acceptance criterion of an elementary transition requests a random pick of a real number and the computation of an exponential.

Many faster variants were devised [Dowsland 1995]. Some exploit a priori knowledge on the solution space, when it is available. Others use various decreasing laws of the temperature. Finally, the parallelization of the algorithm was investigated in depth. Nevertheless, the benefit in terms of convergence speed is at the cost of a loss of the theoretical properties of asymptotic convergence towards the optimal solutions.

Nevertheless, it is important to emphasize the existence of fast methods which are simple to implement, and which keep these theoretical properties. A good one consists in a crafty parallelisation of the simulated annealing by using several computers in parallel [Roussel-Ragot et al. 1990]. The other one consists in using the rescaled simulated annealing or the microcanonical simulated annealing, which is described in the next section.

8.5.1.2 Rescaled Simulated Annealing

This algorithm speeds up the convergence of the simulated annealing in many practical problems, while keeping the theoretical properties of asymptotic convergence towards the optimal solutions [Hérault 2000].

It is derived from the fact that the Metropolis criterion is impatient, leading to some waste of time. In fact, The Metropolis criterion makes the current state dive into low energy states as soon as possible. However, on high temperature steps, all the local minima visited are of poor quality. As a consequence, the acceptance criterion requires a lot of transitions to escape from high-energy local minima, and thus waste elementary transitions and time.

To overcome that drawback, at each temperature step, the elementary transitions are allowed only in an energy slice centred on a target energy E_{target}, which is a decreasing function of the temperature. In order to enforce that behaviour, it is sufficient to modify the energies of the states inside the Metropolis algorithm. The computation of the energy variation associated with an elementary transition from state i to state j becomes

$$\Delta E_{ij} = \left(\sqrt{E_j} - \sqrt{E_{\text{target}}}\right)^2 - \left(\sqrt{E_i} - \sqrt{E_{\text{target}}}\right)^2 .$$

This modification is a generalization of the Metropolis algorithm: the latter is retrieved by imposing $E_{\text{target}} = 0$.

It can be shown that the following decreasing law of the target energy guarantees satisfactory asymptotic convergence to the optimal solutions,

$$E_{\text{target}} = \alpha T^2,$$

where α is a positive real number.

That generalization of simulated annealing can be viewed as modifying the energy landscape during the convergence of the algorithm. The rescaled simulated annealing starts from a flattened energy landscape, which is then progressively unfolded in an ad-hoc manner, at each temperature step. It is important to note that that unfolding does not affect the location of the extrema of the original function. To illustrate this behaviour, consider the example of a mono-dimensional function to be minimized, given on Fig. 8.3. Figure 8.4 shows the evolution of that function as a function of the target energy. When the latter is zero, the recomputed function is the original one. When the target energy is high, the minima of the recomputed function correspond to the maxima of the original function. Therefore, if the target energy is high at the beginning of the search, the most probable states derived from the Metropolis criterion will be the crests of the original function. Once the target energy is smaller than the absolute minimum of the original function, the recalculated function and the original one have their minima at the same locations, with the same relative depths. Moreover, when the target energy decreases, the minima of the recomputed function converge to those of the original function. As in simulated annealing, homogeneous and inhomogeneous algorithms can be defined. In terms of coding, no modification compared to the simulated annealing is necessary.

The properties of asymptotic convergence to optimal solutions have been proved; they compare favourably with those of simulated annealing.

The Metropolis algorithm minimizes the free energy of the system with the corrected energies. In other respects, it can be shown that the asymptotic convergence to thermodynamical equilibrium (stationary distribution of the states) is estimated faster than with the Metropolis algorithm. Therefore, at each temperature step, the distance to stationarity after a finite number of elementary transitions is smaller than with the original metropolis algorithm.

As with simulated annealing, it can be shown that the homogeneous algorithm converges asymptotically to the optimal solutions.

As far as the inhomogeneous algorithm is concerned, a sufficient condition on the cooling schedule between two elementary transitions is the following:

$$T(k) \geq \frac{C_1}{\log(1+k) + C_2},$$

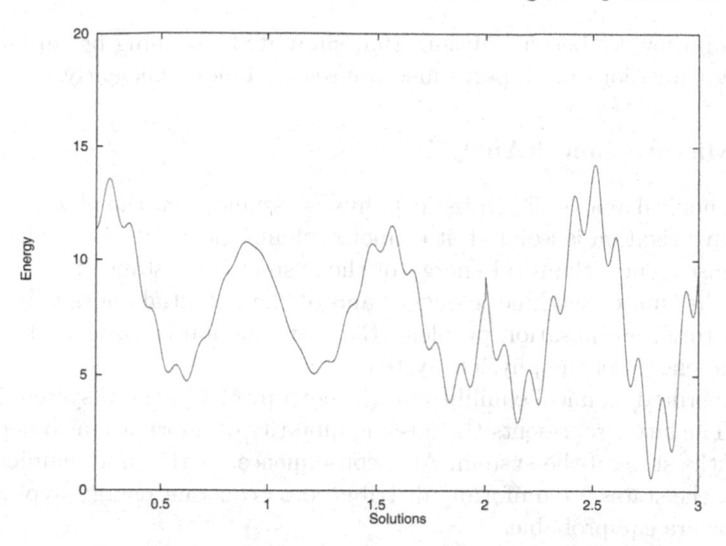

Fig. 8.3. Example of 1D function to be minimized

where C_1 and C_2 are real positive numbers.

This cooling schedule is faster than the law given by Mitra et al. in 1986 for the simulated annealing [Mitra et al. 1986].

Pros and Cons

In practice, the qualities of the simulated annealing are preserved with the rescaling of the energies. Moreover, when the allotted time is limited, that

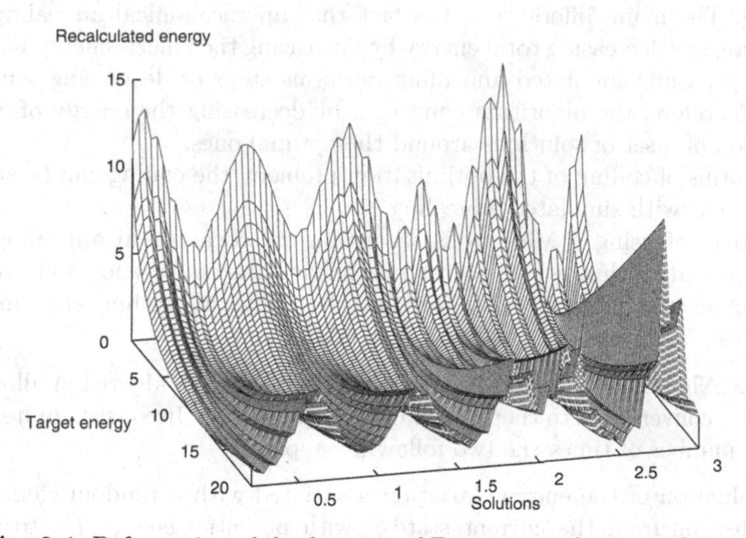

Fig. 8.4. Deformation of the function of Fig. 8.3 as a function of E_{target}

algorithm provides better solutions that simulated annealing because the elementary transitions to be performed are selected more efficiently.

8.5.2 Microcanonical Analysis

Microcanonical analysis in statistical physics assumes that the physical system under investigation is isolated: it cannot exchange heat with its environment. As a consequence, the total energy of the system is constant. That total energy is the sum of the kinetic energy and of the potential energy. To solve a combinatorial optimisation problem, the cost function is associated with the potential energy of the physical system.

At thermodynamical equilibrium, the entropy of a physical system is maximum. The latter represents the missing quantity of information to determine exactly the state of the system. As a consequence, at thermodynamical equilibrium, the states are uniformly distributed on constant energy hypersphere, and thus are equiprobable.

Thermal noise is used in canonical analysis for exploring a wide area in solution space; similarly, some kinetic noise is used in microcanonical analysis.

When the number of particles in interaction is very large, statistical physics shows that the canonical analysis and the microcanonical analysis are equivalent. Nevertheless, applied to combinatorial optimisation, microcanonical analysis has some advantages in terms of implementation simplicity and convergence speed.

8.5.2.1 Microcanonical Annealing

The principle of microcanonical annealing is similar to that of simulated annealing. The main difference is the fact that microcanonical annealing performs steps of decreasing total energy by decreasing the kinetic energy between two steps, while simulated annealing performs steps of decreasing temperature. Therefore, the algorithm converges by decreasing the energy of energy reduction of a set of solutions around the optimal ones.

In terms of coding of the optimisation problem, the coding can be strictly the same as with simulated annealing.

Instead of using a Metropolis algorithm, microcanonical annealing uses the Creutz algorithm, which allows maximizing the entropy for a given total constant energy [Creutz 1983]. As the Metropolis algorithm, that method evaluates a sequence of elementary transitions.

Creutz Algorithm. For a total energy E_t, an iterative algorithm allows asymptotic convergence to thermodynamical equilibrium. It consists in iterating a large number of times the two following steps:

- Evaluation of the energy variation associated with a random elementary transition from the current state i, with potential energy E_i, to a new state j, of energy: $\Delta E_{ij} = E_j - E_i$.

- Acceptance of the transition to that new state if $\Delta E_{ij} \leq E_t - E_i$.

Here, $E_t - E_i$ is the kinetic energy of the system when it is in state i. Note that transitions to states with higher potential energy are allowed, under the condition that enough kinetic energy is available to compensate the increase of the potential energy, and therefore keep a constant total energy.

Microcanonical Annealing

The microcanonical annealing algorithm consists in reductions of the total energy, from an initial high total energy, within the Creutz algorithm. Several decreasing laws for the total energy can be used in practice, similarly to the cooling schedules in simulated annealing.

Pros and Cons

The Creutz algorithm is much simpler than the Metropolis algorithm, and requires fewer computations. Moreover, it does not require a good numerical accuracy. If the problem is coded in integer numbers, it is not necessary to make calculations with floating point numbers. Moreover, it is neither necessary to compute an exponential, nor to pick at random a real number to compare it with the result of the exponential function. Therefore, the computations used are extremely simple.

For large-size problems, microcanonical annealing provides solutions whose quality is comparable to that of the simulated annealing, while being particularly cheap in terms of required calculations. Moreover, a parallelization of this approach is possible.

Note that microcanonical annealing generates families of solutions with comparable qualities, because all the solutions with the same level of total energy are equiprobables. When the total energy is small, all accessible states have a smaller energy (since the kinetic energy is a positive quantity) and are good solutions.

However, at present, there is no proof of convergence, in contrast to simulated annealing. One of the reason is the existence of energetic barriers that cannot be jumped with this algorithm, contrary to the simulated annealing: in the Creutz criterion, the transition probability from a state i to a state j can be zero when E_t becomes small. More efforts are still needed to prove the convergence properties.

8.5.3 Example: Travelling Salesman Problem

Let us consider the following coding for the travelling salesman problem. It is one of the simplest coding schemes [Reinelt 1994].

Initially, the algorithm starts from a random feasible tour. A tour is represented by a vector of integer numbers, where the i-th component indicates the position of city i in the tour.

The transition from one state to another one is defined as the exchange of the respective positions of two cities, chosen randomly in the tour. That type of transition has two advantages. Firstly, it is simple to implement, thus quite cheap. Moreover, it allows visiting the space of feasible tours. A transition between two states is made of two steps:

- random selection of two cities,
- exchange of their positions in the tour.

8.5.3.1 Examples of Annealing Algorithms

In the following, we give some practical examples of annealing algorithms: simulated annealing, rescaled simulated annealing and microcanonical annealing. More specifically, we present quasi-homogeneous algorithms, i.e., algorithms that perform, at each step, a large number of elementary transitions bringing close to thermodynamical equilibrium.

Remarks. For comparison purposes, note that

- Minor modifications are requested to upgrade a standard simulated annealing algorithm to a rescaled simulated annealing; only a few lines of code must be added. In practice, the performances of those two approaches can be compared at a small software development cost. In a fixed resolution time, rescaled simulated annealing will perform a smaller number of elementary transitions than simulated annealing; their evaluation is slightly more complex, but they are more efficient: the smaller the allotted resolution time, the larger the gain in efficiency.
- Microcanonical annealing is simpler to implement than simulated annealing, and can produce as good results. But if the user has a limited time to solve the problem, it might appear less efficient, because it builds during the search some impassable energetic barriers, which might trap it into areas of solution space where no good solution exists.

8.5.3.2 Simulated Annealing

Initialization of the Algorithm

Define the minimal percentage p of accepted transitions on the first step.
Determine the initial temperature $T = T_0$ such that $p\%$ of the tested transitions are accepted.
Generate randomly an acceptable solution and compute its energy E.
Select the maximal number of tested transitions on each step: Nb_{maxtest}.
Select the parameter for the temperature decrease between two steps: *dec*.

Simulated Annealing Algorithm

Set $Nb_{\text{accepted}} = 1$.
While Nb_{accepted} is non-zero
  $Nb_{\text{tested}} = Nb_{\text{accepted}} = 0$
  While $Nb_{\text{tested}} < Nb_{\text{maxtest}}$ /* Metropolis algorithm */
    Increment Nb_{tested}.
    Pick at random a valid transition.
    Calculate the energy variation ΔE.
    If $\Delta E < 0$ then /* Accepted transition */
      Perform the transition.
      Update the energy E: $E := E + \Delta E$.
      Increment Nb_{accepted}.
      Compare the new state with the best state found from the beginning of the search, and store it in memory if it is better.
    Otherwise
      Pick at random a real number *rand* in $[0, 1]$.
      If $rand < \exp(-\Delta E/T)$ then /* Accepted transition */
        Perform the transition.
        Update the energy E: $E := E + \Delta E$.
        Increment Nb_{accepted}.
    Decrease the temperature: $T := dec\, T$ /* Annealing */
  Return the best state encountered during the search.

8.5.3.3 Rescaled Simulated Annealing

Initialisation of the Algorithm

Define the minimal percentage p of accepted transitions on the first step.

Determine the initial temperature $T = T_0$ such that $p\%$ of the tested transitions are accepted.

Initialize α, for instance with a value close to $\sqrt{\hat{E}}/T_0$, where $\hat{E}$ is the mean energy of the solutions.

Generate randomly a valid solution and calculate its energy E.

Select the maximal number of tested transitions on each step: Nb_{maxtest}.

Select the parameter for the temperature decrease between two steps: *dec*.

Rescaled Simulated Annealing Algorithm

Set $Nb_{\text{accepted}} = 1$.
While Nb_{accepted} is non-zero
  $Nb_{\text{tested}} = Nb_{\text{accepted}} = 0$
  While $Nb_{\text{tested}} < Nb_{\text{maxtest}}$ /* Generalized Metropolis algorithm */
    Increment Nb_{tested}.
    Pick at random a valid transition.
    Calculate the energy variation ΔE.
    Modify the energy variation by subtracting $-2\alpha T(\sqrt{E + \Delta E} - \sqrt{E})$ from it
    If $\Delta E < 0$ then /* Accepted transition */
      Perform the transition.
      Update the energy E: $E := E + \Delta E$.
      Increment Nb_{accepted}.
      Compare the new state with the best state found from the beginning of the search, and store it in memory if it is better.
    Otherwise
      Pick at random a real number *rand* in $[0, 1]$.
      If *rand* $< \exp(-\Delta E/T)$ then /* Accepted transition */
        Perform the transition.
        Update the energy E: $E := E + \Delta E$.
        Increment Nb_{accepted}.
  Decrease the temperature: $T := dec\,T$ /* Annealing */
Return the best state encountered during the search.

8.5.3.4 Microcanonical Annealing

Initialisation of the Algorithm

Define the minimal percentage p of accepted transitions on the first step.
Choose the initial total energy E_t such that $p\%$ of the tested transitions are accepted.
Generate randomly a valid solution and calculate its energy E.
Select the maximal number of tested transitions on each step: Nb_{maxtest}.
Select the parameter for the temperature decrease between two steps: *dec*.

Microcanonical Annealing Algorithm

Set $Nb_{\text{accepted}} = 1$.
While Nb_{accepted} is non-zero
  $Nb_{\text{tested}} = Nb_{\text{accepted}} = 0$
  While $Nb_{\text{tested}} < Nb_{\text{maxtest}}$/* Creutz algorithm */
    Increment Nb_{tested}.
    Pick at random a valid transition.
    Calculate the energy variation ΔE.
    If $\Delta E = E_t - E$ is positive, then /* Accepted transition */
      Perform the transition.
      Update the energy E: $E := E + \Delta E$.
      Increment Nb_{accepted}.
      If $\Delta E < 0$, compare the new state with the best state found from the beginning of the search, and memorize it if it is better.
  Decrease the total energy: $E_t := dec\, E_t$ /* Annealing */
Return the best state encountered during the search.

To come close to optimal solutions of the problem illustrated on Fig. 8.1, the following values of the parameters can be used:

- $p = 90\%$
- $dec = 0,99$
- $Nb_{\text{maxtest}} = 500000$.

With those parameter values, Figs. 8.5 and 8.6 show, for the standard simulated annealing and the rescaled simulated annealing, the decrease of the mean energy of the states visited during the convergence, as well as the length of the best tour, found as the search converges. The mean energy of the solutions is computed as

$$\langle E\rangle = \frac{\sum_i E_i \exp\left(-\frac{\left(\sqrt{E_i} - \alpha T\right)^2}{T}\right)}{\sum_i \exp\left(-\frac{\left(\sqrt{E_i} - \alpha T\right)^2}{T}\right)}.$$

Those curves were obtained by computing averages over 100 runs with different initializations of the random number generator. With identical parameters in the two algorithms, one observes that the energy rescaling speeds

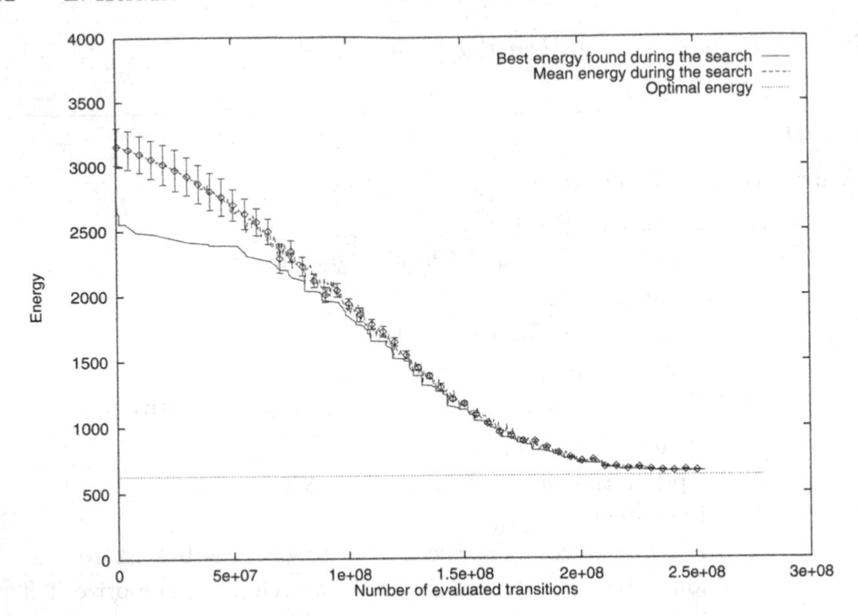

Fig. 8.5. Simulated annealing for the travelling salesman problem of Fig. 8.1: evolution of the mean energy of the visited solutions, and of the best solution found during the convergence

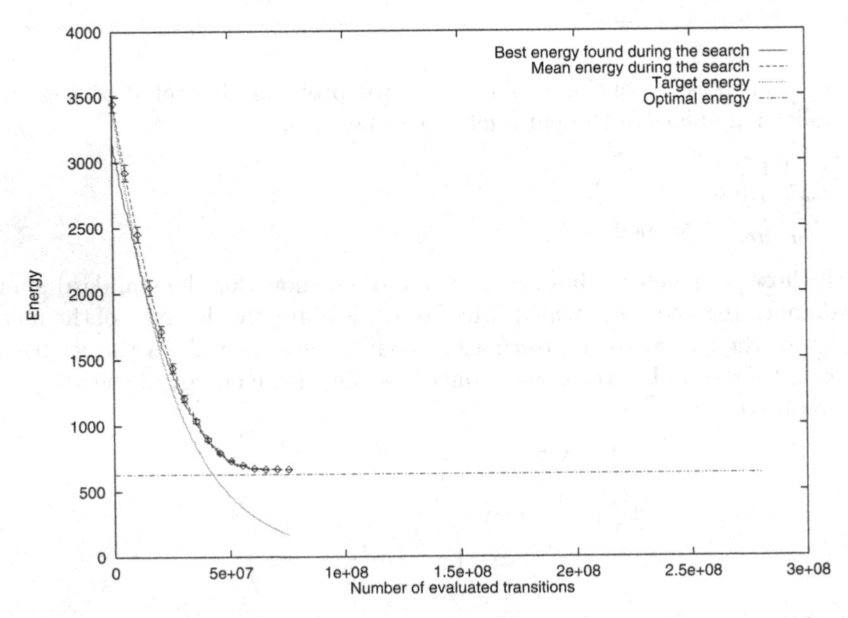

Fig. 8.6. Rescaled simulated annealing for the travelling salesman problem of Fig. 8.1: evolution of the mean energy of the visited solutions, and of the best solution found during the convergence. They follow the target energy as long as they are greater than the minimal reachable energy

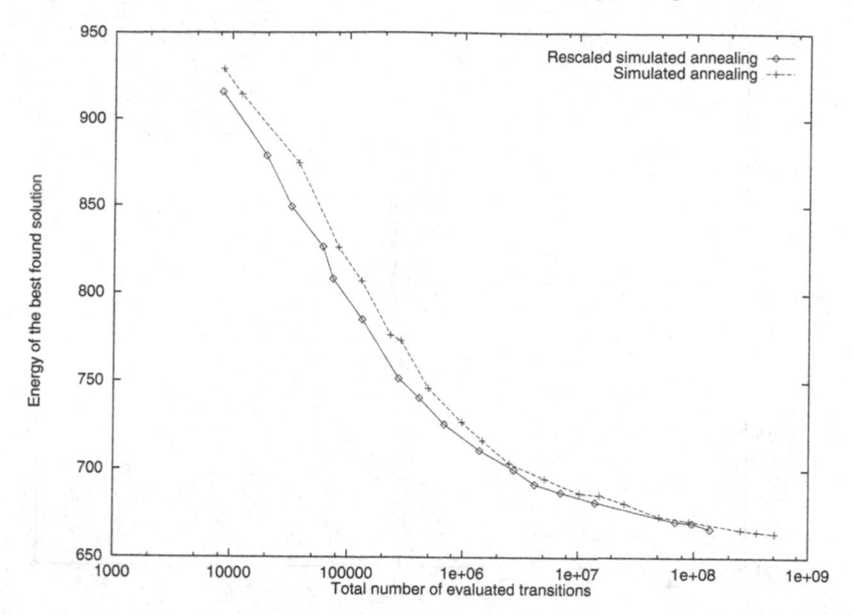

Fig. 8.7. Comparison of the performances of the standard and rescaled simulated annealing as a function of the total number of transitions tested by the algorithm. Each value on the curves is computed over 100 runs of the algorithms

up the convergence. Figure 8.7 compares the average quality of the solutions obtained with simulated annealing and with microcanonical annealing, as a function of the total number of tested transitions. As expected, the smaller the allotted resolution time, the larger the gain in terms of performance.

8.6 Neural Approaches

At the beginning of the 1980s, recurrent neural networks (defined in Chaps. 2 and 4) were shown to be able to solve optimisation problems. For that purpose, they have two major advantages: first, neural algorithms often solve very efficiently some optimisation problems; in addition, neural algorithms can give rise to high-speed numerical, analog, and even optical electronic implementations, taking advantage of parallelism.

As compared to other metaheuristics, recurrent neural networks are particularly well suited to problems that require extremely short response times, and possibly a hardware implementation.

8.6.1 Formal Neural Networks for Optimization

In recurrent neural networks used in optimization, neurons are either binary (their activation function is a hard limiter between -1 and $+1$), or with a

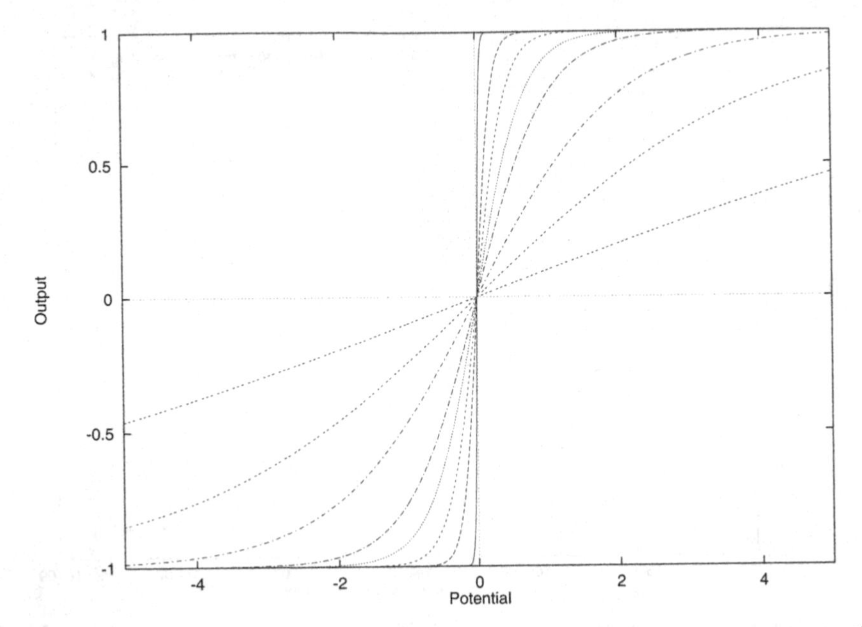

Fig. 8.8. Sigmoïd function for different values of γ ($\gamma = 50, 10, 5, 3, 2, 1, 0.5, 0.2$)

sigmoidal activation function: in the latter case, the output y_i of neuron i is given by: $y_i = \tanh(\gamma v_i)$, where γ is the slope at the origin of the sigmoïd, and where v_i is the potential of neuron i, defined, as in the previous chapters, for a network of N neurons mutually connected, as

$$v_i = \sum_{j=1}^{N} w_{ij} y_j + I_i \ ,$$

where I_i is the constant input (bias) of neuron i.

Remark. In contrast to what was done in the previous chapters, for the neural networks dedicated to optimisation, we will distinguish explicitly the bias from the other inputs of the neurons.

The only difference with neurons used in Chaps. 2 to 4 is therefore in the fact that the slope γ might be different from 1. Note that the sigmoid approximates the hard limiter when γ increases (Fig. 8.8); that is the reason why the inverse of the slope can be considered as a temperature, by analogy with the algorithms described in the previous sections.

It is sometimes preferable to use neurons with continuous outputs between 0 and 1. They can be obtained directly from the previous formula by the change of variable $(v_i + 1)/2$.

When the outputs must be binary after convergence of the network, instead of using the previous sigmoid function, for which 0 (or -1) and 1 are

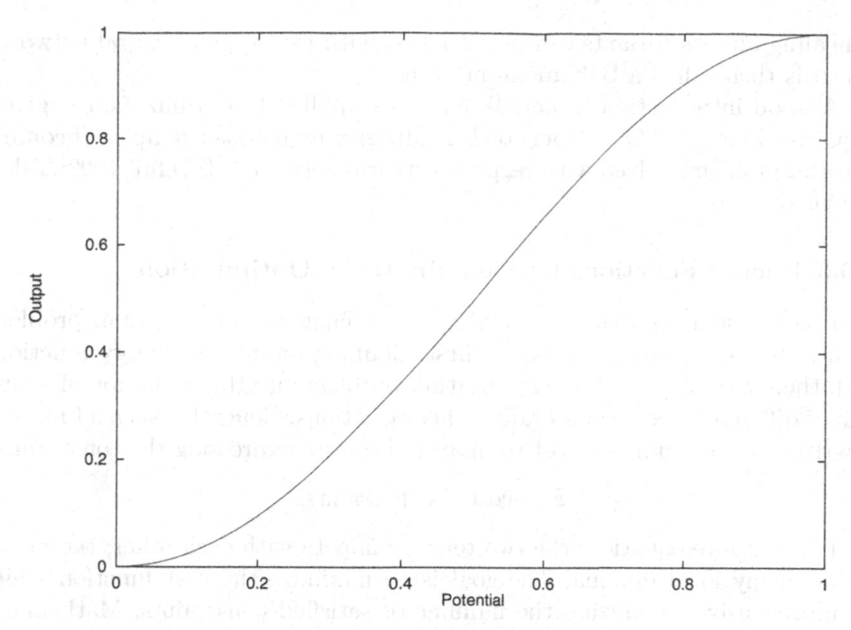

Fig. 8.9. Activation function for $\rho = 1$

asymptotic values, one can also use the following activation function, which is continuous on [0,1]:

$$y_i = \begin{cases} 0 & \text{if } v_i = 0 \\ \frac{1}{2}\left[1 + \sin\left(\pi\left(\frac{v_i}{\rho} - \frac{1}{2}\right)\right)\right] & \text{if } 0 < v_i < \rho \\ 1 & \text{if } v_i = \rho. \end{cases}$$

Figure 8.9 shows the shape of this activation function when $\rho = 1$. In that activation function, ρ is a strictly positive real number that controls the maximal slope of the function; the latter is continuous, differentiable and monotonic increasing. When ρ tends to 0, the activation function tends to a step.

8.6.2 Architectures of Neural Networks for Optimisation

Recurrent neural networks are the neural techniques that are the most frequently used for solving optimization problems. As explained in Chap. 2, the graph of the connections of those networks has at least a cycle. For optimization, those networks have no control input: they evolve with their own dynamics, from an initial state (often random), to an *attractor* that encodes a solution of the optimization problem. We will show later that simulated annealing and its variants can be modelled with that type of neural network, which is then called a Boltzmann machine.

A good introduction to neural networks applied to optimization is given in [Cichocki et al. 1993]. Papers of Takefuji give numerous examples of combinatorial problems solved with recurrent neural networks [Takefuji 1992; Takefuji et al. 1996].

8.6.3 Energy Functions for Combinatorial Optimisation

The most usual approach to solving a combinatorial optimisation problem with a neural network consists in, first, defining an ad hoc energy function, and, then, to transform the minimisation problem into the resolution of a system of differential equations or difference equations. Generally, such a function is written as the sum of a cost term and of a term expressing the constraints:

$$E = \text{cost} + \text{constraints}.$$

In the above equation, the two terms compete with each other; therefore, E has many local minima. The goal is to minimize the cost function while simultaneously maximizing the number of satisfied constraints. Mathematically, the minimization of the cost function often amounts to minimizing a function $E(\boldsymbol{x})$ over a *finite* set of point X, generally chosen as a *hypercube* in a space of dimension N, $\{-1, 1\}^N$ or $\{0, 1\}^N$. Vector $\boldsymbol{x}$ contains the variables of the optimization problem: $\boldsymbol{x} = [x_1, x_2, \ldots, x_N]^T$. The function $E(\boldsymbol{x})$ is defined as

$$E(\boldsymbol{x}) = E_c(\boldsymbol{x}) + \sum_k a_k E_k(\boldsymbol{x}),$$

where $E_c(\boldsymbol{x})$ is the cost function, the $E_k(\boldsymbol{x})$ are penalty terms associated with violations of constraint, and the a_k are weights to guarantee a satisfactory balance between the minimization of the cost and the satisfaction of the constraints.

For many optimization problems, the function E can be expressed as a quadratic form of the type

$$E(\boldsymbol{x}) = -\frac{1}{2} \sum_{i=1}^{N} \sum_{j=1}^{N} T_{ij} x_i x_j - \sum_{i=1}^{N} I_i x_i.$$

In the above relation, the quantities T_{ij} and I_i are fully determined by the energy function to minimize. For a given energy function $E(\boldsymbol{x})$, the T_{ij} express a local curvature,

$$T_{ij} = -\frac{\partial^2 E}{\partial x_i \partial x_j},$$

and the quantity I_i is given by the first derivative of the energy,

$$I_i = -\left.\frac{\partial E}{\partial x_i}\right|_{\boldsymbol{x}=0}.$$

We will show below that a recurrent neural network, such as the Hopfield network can minimize the energy function.

8.6.4 Recurrent Hopfield Neural Networks

A Hopfield neural network [Hopfield 1982, 1984], as defined in Chap. 4, has one layer of fully connected neurons; a delay of one time unit is associated to each connection; the state vector being the vector of neuron outputs, the order of the network is equal to the number of neurons.

Principle of Hopfield Neural Networks for Optimization

When applied to optimization, the network is used as follows: from an initial state, the network evolves freely towards an attractor, which is generally a time-independent state (a fixed point of the dynamics). Then the network is said to have converged: the outputs of neurons no longer evolve any longer.

More details on the convergence properties can be found in the papers of Goles [Goles 1995].

The dynamics of the network is generally *asynchronous*: between two instants of time, a single neuron, randomly selected, is updated; in other words, its potential is computed, and its output is appropriately updated.

When those networks are used to solve optimization problems, the weights of the connections are found analytically from the formulation of the optimization problem; generally, they are directly derived from the energy function associated to the problem, as will be exemplified below. Moreover, in the attractor where the network has converged, the outputs of the neurons code for a solution of the optimization problem.

8.6.4.1 Binary Hopfield Neural Networks

The neural network initially proposed by Hopfield was a discrete-time recurrent neural network, with a symmetrical connection matrix (matrix of w_{ij} coefficients) with a null diagonal [Hopfield 1982]. It has been presented in Chap. 4.

Since each connection has a delay of one time unit, the potential of neuron i at time k is the weighted sum of the activities of the other neurons at time $k-1$: $v_i(k) = \sum_{j \neq i} w_{ij} y_j(k-1) + I_i$, where $y_i(k)$ is the output of neuron i at time k, w_{ij} is the weight of the connection between neuron j and neuron i, and I_i is the bias (constant input) of neuron i.

The attractors on which the network converges are the minima of a function, called network energy, defined by

$$E(\boldsymbol{y}) = -\frac{1}{2} \sum_{i=1}^{N} \sum_{j=1}^{N} w_{ij} y_i y_j - \sum_{i=1}^{N} I_i y_i ,$$

where $\boldsymbol{y}$ is the vector of neuron outputs, i.e., the state vector of the system: $\boldsymbol{y} = [y_1, y_2, \dots, y_N]^{\mathrm{T}}$.

That function is an N-variable function that generally has a large number of local minima.

A natural link between such a function and the energy function of a combinatorial problem can be found. That is the reason why recurrent neural networks are interesting for solving optimization problems.

8.6.4.2 Analog Hopfield Neural Networks

Hopfield described a continuous (called analog) version of the above binary recurrent neural network [Hopfield 1984]. In that case, the associated energy function is defined as

$$E(\boldsymbol{y}) = -\frac{1}{2}\sum_{i=1}^{N}\sum_{j=1}^{N} w_{ij} y_i y_j - \sum_{i=1}^{N} I_i y_i + \frac{\alpha}{\gamma}\sum_{i=1}^{N}\int_0^{y_i} \Psi^{-1}(\boldsymbol{y})\,\mathrm{d}\boldsymbol{y}\ ,$$

where α is a positive real number, and Ψ is the activation function of the neurons.

Generally, the last term in that energy function is negligible with respect to the previous ones, when the slope γ is large, or when α is small.

Hopfield and Tank first applied that type of neural network to combinatorial optimization [Hopfield et al. 1985].

A potentially interesting feature of that type of network is the fact that they can give rise to the hardware implementation of analog ASICs, by interconnecting a set of resistors, some non-linear amplifiers with symmetric outputs, external current sources and some capacitors [Newcomb et al. 1995].

The equations that govern the evolution of a continuous neuron i is the following:

$$\begin{cases} \dfrac{\mathrm{d}v_i}{\mathrm{d}t} = \mu_i \left[-\alpha_i v_i - \dfrac{\partial E(\boldsymbol{y})}{\partial y_i} \right] \\ y_i = \tanh\left(\dfrac{\mu_i}{T}\right), \end{cases}$$

where $\mu_i = 1/\tau_i$ is a positive real number which parameterizes the convergence speed, α_i is a positive real number, T is the *temperature* (inverse of the slope at the origin of the neuron's activation function) and $E(\boldsymbol{y})$ is the energy function of the problem, which is not necessarily quadratic.

The derivative of the energy function E versus time can be written, from the above equations, as

$$\frac{\mathrm{d}E}{\mathrm{d}t} = \sum_{i=1}^{N} \frac{\partial E}{\partial y_i}\frac{\mathrm{d}y_i}{\mathrm{d}t} = -\sum_{i=1}^{N} \tau_i \frac{\mathrm{d}y_i}{\mathrm{d}v_i}\left(\frac{\mathrm{d}v_i}{\mathrm{d}t}\right)^2 - \sum_{i=1}^{N} \alpha_i v_i \frac{\mathrm{d}y_i}{\mathrm{d}t}.$$

Clearly, the first term in that equation is always negative or zero because $\tau_I > 0$ and $\mathrm{d}y_i/\mathrm{d}v_i > 0$. As for the second term, it can be positive, negative or zero; as a consequence, the derivative of the energy function with respect to time can be positive [Takefuji 1992]. In order to avoid such a situation, a slightly different motion equation is often used. It is given by

$$\frac{\mathrm{d}y_i}{\mathrm{d}t} = -\mu_i \frac{\partial E(\boldsymbol{y})}{\partial y_i}, \quad i = 1, \ldots, N.$$

As for the neuron outputs, the following equations are obtained:

$$\frac{\mathrm{d}y_i}{\mathrm{d}t} = -\frac{\mu_i}{T}(1 - y_i^2)\frac{\partial E(\boldsymbol{y})}{\partial y_i}, \quad i = 1, \ldots, N.$$

Thus, the dynamic properties of an analog Hopfield neural network are governed by a system of non-linear differential equations.

With the above modification, it can be easily proved that any change of state (i.e., of output) of a neuron decreases, or keeps constant, the network energy,

$$\frac{\mathrm{d}E}{\mathrm{d}t} = \sum_{i=1}^{N} \frac{\partial E}{\partial y_i}\frac{\mathrm{d}y_i}{\mathrm{d}t} = \sum_{i=1}^{N}\left[-\tau_i \frac{\mathrm{d}v_i}{\mathrm{d}t}\right]\frac{\mathrm{d}y_i}{\mathrm{d}t} = -\sum_{i=1}^{N} \tau_i \frac{\mathrm{d}y_i}{\mathrm{d}v_i}\left(\frac{\mathrm{d}v_i}{\mathrm{d}t}\right)^2$$
$$\frac{\mathrm{d}E}{\mathrm{d}t} \leq 0.$$

In other words, the system of equations constrains the energy function E to monotonically decrease to a local minimum. More precisely, the system has the following property: from an initial point $\boldsymbol{x}$ inside the hypercube $\{-1, 1\}^N$, the dynamic system converges to a local minimum of the energy function $E(\boldsymbol{y})$, located either on a vertex or on the surface of the hypercube. When the attractor is on the surface, any vertex of that surface can be chosen as a solution since the energy is constant on that surface. Therefore, any solution where the network converges is locally optimal.

In practice, with a dedicated hardware implementation of such a network, the convergence times are of the order of a few nanoseconds or microseconds. That allows generally several resolutions of the problem, starting from different initial points. That strategy can sometimes be efficient, but generally the attractors are not satisfactory.

To solve efficiently combinatorial optimization problems, it is often preferable to use continuous neurons with a variable slope. Thus, by evolving inside the hypercube of solutions, and not only on its vertices as a binary neural network does, the high-energy local minima are better avoided during convergence. Moreover, it has been shown that a continuous neural network is much faster and reliable than a binary neural network with an asynchronous update of the neurons [Ayer et al. 1990; Lee et al. 1991a]. Minimizing $E(\boldsymbol{y})$ with a continuous $\boldsymbol{y}$ reduces significantly the probability of being trapped in a local minimum, because the valleys of the solution space are wider than in the binary case. Unfortunately, analog neurons evolving inside the hypercube rarely converge towards a vertex of the latter. In order to force the analog neurons to finally have binary values that code for a solution of the problem, several possibilities exist. One of them consists in increasing the slope at the origin of the activation function of the analog neurons during convergence. As a further refinement, one may add to the energy function a penalty term of the form

$$E_y = \sum_{j=1}^{N} (1 - y_j^2).$$

That term is zero if the neuron outputs are discrete, with values in $\{-1, 1\}$.

8.6.4.3 Application of Hopfield Neural Networks to Optimization

To summarize: from the above-described convergence properties, it ensues that Hopfield neural networks can be applied to optimization problems with the following methodology, featuring four steps:

1. Find an appropriate encoding of the problem. The question is, first, to find a representation of the problem such that a solution, i.e., an instantiation of the problem variables, is represented by the neuron outputs after convergence.
2. Express the cost function and the constraints under the form of the energy of a Hopfield network. Try to express the cost function as a quadratic function. As for the constraints to be satisfied, they are of two types: those defined by the optimization problem, and those resulting from the problem encoding selected in the previous step. Sometimes, it is not possible to find a quadratic energy. It is then necessary to try to express the problem as the minimization of a function F, which can be differentiated with respect to neuron outputs.
3. Find the equations of the neurons.
4. Randomly start the network. Generally, when no a priori knowledge on the location of an optimal solution is available, the neuron states are randomly initialized. Then, the equations of the neurons will drive the network towards a local minimum of the energy.

That type of methodology has been applied to solve a wide variety of combinatorial problems, which are generally amenable to problems of graph theory.

Table 8.1. Valid solution for a TSP with 5 cities

	1	2	3	4	5
City A	0	0	1	0	0
City B	1	0	0	0	0
City C	0	1	0	0	0
City D	0	0	0	1	0
City E	0	0	0	0	1

8.6.4.4 Example of Hopfield Neural Network to Solve the TSP

To illustrate the previous methodology with a concrete example, we will detail the resolution of a traveling salesman problem, proposed by Hopfield and Tank in 1985 [Hopfield et al. 1985]. This resolution is not the most efficient one, as we will se later, but it has an educational interest.

Step 1: Encoding the Problem

The first step consists in mathematically reformulating the problem, in order to solve it with a neural network.

A problem solution, i.e., a tour of the different cities, can be encoded as a permutation matrix, i.e., a square matrix with binary elements, which contains exactly one 1 in each row and in each column. In other words, given N cities, a tour is represented by a (N, N) having N^2 elements. In that matrix, a row represents a city, and a column the position of a city in the tour. Therefore, a feasible tour is represented by a permutation matrix that has exactly N 1s and $N^2 - N$ 0s. Table 8.1 shows an example of a matrix encoding a feasible solution to a problem with 5 cities.

The corresponding solution is a tour in which the cities A, B, C, D and E are visited in the following order: B-C-A-D-E-B.

A neuron is associated to each matrix coefficient. Therefore, that encoding makes use of N^2 neurons and N^4 connections between the neurons. We will denote by $y_{i,j}$ the output of the neuron (i, j) corresponding to the i-th row and to the j-th column, and $v_{i,j}$ its potential. Moreover, the distances between any pair of cities are known a priori: d_{ij} represents the distance between city i and city j.

Step 2: Defining the Network Energy

8.6.4.5 Cost Function

The cost function to be minimized is the length of the journey performed during the round. It can be expressed as a function of the outputs of the N neurons that are equal to 1 in the permutation matrix. Mathematically, it is given by

$$F_c = \frac{1}{2}\sum_{i=1}^{N}\sum_{j=1}^{N}\sum_{\substack{k=1\\k\neq i}}^{N} d_{i,k}y_{i,j}(y_{k,j-1}+y_{k,j+1}),$$

where, by definition, $y_{k,0} = y_{k,N}$ and $y_{k,N+1} = y_{k,1}$.

It can be written in a simpler way as follows:

$$F_c = \frac{1}{2}\sum_{i=1}^{N}\sum_{j=1}^{N}\sum_{\substack{k=1\\k\neq i}}^{N} d_{i,k}y_{i,j\oplus 1}.$$

In that equation, $d_{i,k}$ is the distance between cities i and k. The operator $\oplus$ is defined as follows:

$$\text{For} \quad j < N, j \oplus 1 = j+1, \quad \text{and} \quad N \oplus 1 = 1.$$

The term in the multiple sums is non-zero when city i is position j, and that term is equal to the distance of the journey between that city and the next one on the tour. The output of the neuron (i, j) is therefore multiplied by the outputs of the neurons of column $j \oplus 1$. Therefore, each neuron is connected to $2N$ neurons in the network. As a consequence, the encoding of that cost function does not require N^4 connections, but N^3.

8.6.4.6 Constraints

To guarantee that, after convergence, the matrix of neuron outputs is a permutation matrix that guarantees the validity of the solution, some additional constraints are defined. A penalty function is associated to their violation.

For a tour to be valid, each city must be visited exactly once. That requires that, in each row, no two neurons have their outputs equal to 1. Therefore, the following function is defined, such that it is non-zero if at least two neurons have an output equal to 1 in a row:

$$F_1 = \frac{1}{2}\sum_{i=1}^{N}\sum_{j=1}^{N}\sum_{\substack{l=1\\l\neq j}}^{N} y_{i,j}y_{i,l}.$$

Similarly, at each step on the tour, the traveling salesman cannot be in more than one city. As a consequence, in each column, there no two neurons have their outputs equal to 1 after convergence of the network. Therefore, the following function is defined, such that it is non-zero if at least two neurons have an output equal to 1 in a column:

$$F_2 = \frac{1}{2}\sum_{i=1}^{N}\sum_{j=1}^{N}\sum_{\substack{k=1\\k\neq i}}^{N} y_{i,j}y_{k,j}.$$

The above two constraints are not sufficient to guarantee the validity of a solution. In fact, without any new constraint, the minimization process would naturally drive the network in a state where all the neuron outputs are 0. It does not make sense. The validity of a solution requires that exactly N neurons have an output equal to 1 after convergence. Therefore, a third constraint function is defined,

$$F_3 = \frac{1}{2}\left[\sum_{i=1}^{N}\sum_{j=1}^{N} y_{i,j} - N\right]^2.$$

8.6.4.7 Energy of the Neural Network

The total energy of the Hopfield network is the weighted sum of the above functions,

$$E = F_c + a_1 F_1 + a_2 F_2 + a_3 F_3.$$

Constants a_1, a_2 and a_3 must be adjusted according to the relative weights of the various constraints.

Step 3: Finding the Equations of the Neurons

The energy function of the problem can be expressed as a quadratic function, which is the energy of a Hopfield neural network,

$$E(y) = -\frac{1}{2}\sum_{i=1}^{N}\sum_{j=1}^{N}\sum_{k=1}^{N}\sum_{l=1}^{N} w_{ij,kl}\, y_{i,j}\, y_{k,l} - \sum_{i=1}^{N}\sum_{j=1}^{N} I_{i,j}\, y_{i,j}.$$

In that equation, the weights $w_{ij,kl}$ are determined from the analytical form of the cost and of the constraints. Considering the first constraint F_1, its contribution to the synaptic coefficients is given by

$$-\delta_{i,k}(1-\delta_{j,l}) \quad \text{where } \delta_{x,y} = 1 \text{ if and only if } x = y.$$

Similarly, the contribution of the constraint F_2 to the synaptic coefficients is

$$-\delta_{j,l}(1-\delta_{i,k}).$$

The contribution of the constraint F_3 is -1.

Finally, the contribution of the cost function is given by

$$-d_{i,k}(\delta_{l,j+1} + \delta_{l,j-1}).$$

Therefore, the final expression of the weights is

$$w_{ij,kl} = -2 d_{i,k}\delta_{1,j\oplus l} - a_1\delta_{i,k}(1-\delta_{j,l}) - a_2\delta_{j,l}(1-\delta_{i,k}) - a_3.$$

The external input on the neuron (i,j) is: $I_{i,j} = a_3 N$.

The equation of neuron (i, j) is easily derived, for instance in the case of binary neurons. At time $t + 1$, one gets

$$\Delta v_{i,j}(t+1) = \sum_{\substack{k=1 \\ k \neq i}}^{N} \sum_{\substack{l=1 \\ l \neq j}}^{N} w_{ij,kl} y_{k,l}(t) + I_{i,j}.$$

In the case of analog neurons, the motion equations are given by

$$\frac{\mathrm{d}v_{i,j}}{\mathrm{d}t} = \mu_{i,j} \left(-\alpha_{i,j} v_{i,j} + \sum_{\substack{k=1 \\ k \neq i}}^{N} \sum_{\substack{l=1 \\ l \neq j}}^{N} w_{ij,kl} y_{k,l} + I_{i,j} \right),$$

where $\mu_{i,j}$ and $\alpha_{i,j}$ are real positive numbers.

Step 4: Starting the Dynamics

From a valid random solution, the initial values of the neuron inputs and outputs can be defined. Then, a random asynchronous dynamics, based on the motion equations of the neurons, makes the network converge to a local minimum of the energy, in which the inputs and the outputs of the neurons do not evolve any longer. After convergence, reading out the values of the neuron outputs provides the encoding of a possible solution to the problem. The validity of the solution must nevertheless be checked. If the solution is not valid (i.e., if one or more constraints are violated), the network can be started again from another random initial state, and/or the weighting constants can be adjusted, for instance by increasing the constant associated to the violated constraint.

8.6.4.8 Limitations of Hopfield Neural Networks

Solving optimization problems with Hopfield neural networks raises some problems.

The main difficulty stems from the fact that the dynamics drives often the network into a local minimum of the energy, which is not necessarily close to the optimum, or which does not correspond to a valid solution. That is due to the fact that the constraints are combined with the cost function in the network energy.

Moreover, the Hopfield neural network makes no difference between the strict constraints and the preference constraints, except by different weights in the energy function.

Finally, the values of the parameters that weight the different terms of the network energy have an influence on the number of iterations to convergence.

Those difficulties have given rise to numerous investigations aiming at overcoming those limitations.

8.6.5 Improvements of Hopfield Neural Networks

Numerous studies are in progress to overcome the limitations of the recurrent neural networks without training for optimization. All of them will not be mentioned in this chapter, but a few tracks recently explored will be detailed in the following.

8.6.5.1 Improvements of the Encoding of the Energies to Minimize

To avoid being trapped in high-energy local minima corresponding to unacceptable solutions, it is often fruitful to develop alternative encoding schemes for the problem. Given a problem, different energy functions can be defined, each of them being characterized by a more or less complex solution space to explore. Along those lines, in the case of the TSP, Brandt suggested that, in order to avoid being trapped in a high-energy local minimum during the convergence, the following energy should be minimized [Brandt et al. 1988]:

$$E = F_c + \frac{\gamma}{2}\left[\sum_{i=1}^{N}\left[1-\sum_{j=1}^{N} y_{i,j}\right]^2 + \sum_{j=1}^{N}\left[1-\sum_{i=1}^{N} y_{i,j}\right]^2\right].$$

With the same goal, Szu, in 1988 [Szu 1988], proposed another energy function,

$$E = F_c + a_1 F_1 + a_2 F_2 + a_3\left[\sum_{i=1}^{N}\left[1-\sum_{j=1}^{N} y_{i,j}\right]^2 + \sum_{j=1}^{N}\left[1-\sum_{i=1}^{N} y_{i,j}\right]^2\right].$$

8.6.5.2 Analog Hopfield Networks with Annealing

As explained briefly in previous sections, a first solution consists in varying the temperature τ_i (the inverse of the slope at the origin of the activation function of neuron i) in the analog Hopfield networks. As with simulated annealing, it is initialized to a high value, and it decreases during convergence. At high temperatures, the system behaves like a quasi-linear system, because the activation functions are quasi-linear over a wide range of potential values. Therefore, neuron outputs vary between -1 and $+1$. By decreasing the temperature, neuron outputs tend to the values -1 or $+1$, which code for a solution. During convergence, a critical temperature T_c can be observed; below that temperature, the system starts to freeze, i.e., the neuron outputs significantly evolve towards $+1$ or -1 [Hérault et al. 1989]. That temperature can be estimated theoretically for some optimization problems [Peterson et al. 1989]. When it is the case, it is not necessary decrease the temperature regularly: it is sufficient to let the system converge towards an equilibrium at the critical temperature, and then to *quench* the system by brining it to a temperature close to 0. At the end of this second step, all neurons are almost binary.

8.6.5.3 Probabilistic Networks

To avoid getting trapped in a high-energy local minimum, another approach consists in defining probabilistic equations, by adding a noise term in the motion equations of analog neurons. That noise term is such that its influence decreases during the convergence of the network [Asai 1995]. The modified equation of an analog neuron thus becomes

$$\frac{\mathrm{d}y_i}{\mathrm{d}t} = -\mu_i \left[(1 - y_i)^2 \frac{\partial E(\boldsymbol{y})}{\partial y_i} + c(t) N_i \right], \quad i = 1 \ldots, N,$$

where the N_i are non-correlated sources of white noise with zero mean, and $c(t)$ provides the decreasing law of the noise. Typically, $c(t)$ has the following form:

$$c(t) = c_0 \exp\left(-\frac{t}{\tau}\right).$$

In practice, it is necessary to add an uncorrelated noise source to each neuron, and to decrease gradually its power during the convergence. The benefit, as with simulated annealing, is to provide the network with the capacity of escaping from high-energy local minima, and thus to converge to much better solutions. That is the reason why that technique is sometimes called hardware simulated annealing.

8.6.5.4 Boltzmann Machine

The Boltzmann machine was first described in 1984 and 1985 [Hinton et al. 1984; Ackley et al. 1985]. It can be considered as a combination of the principles of simulated annealing with those of binary Hopfield neural networks. Its architecture is similar to that of a binary Hopfield neural network.

The energy of a Boltzmann network can be expressed under the same form as the energy of a binary Hopfield neural network,

$$E(\boldsymbol{y}) = -\frac{1}{2} \sum_{i=1}^{N} \sum_{j=1}^{N} w_{ij} y_i y_j.$$

The first step of the updating of a neuron i consists in computing the energy variation generated by its change of state. The energy variation associated with the change from 0 to 1 of a neuron output is given by

$$\Delta E_i = E_{y_i=1} - E_{y_i=0} = -\sum_{\substack{j=1 \\ j \neq i}}^{N} w_{ij} v_j.$$

Then, the state of the neuron changes with the probability given by the Metropolis criterion used in simulated annealing. During the convergence of the network, as with simulated annealing, the temperature parameter decreases progressively.

Considering for instance the TSP, the same encoding as the one used by Hopfield networks, and based on permutation matrices, can be performed. The binary outputs of the neurons code a solution of the problem. Energies associated with the cost function and the constraints are exactly the same as those defined for a binary Hopfield network. The neuron outputs are randomly initialized with the values 0 or 1. Therefore the initial state does not necessarily code for a valid solution of the problem.

To present, we have proceeded as with Hopfield networks. But the updating of the neurons is different. To make the network evolve, a neuron i is picked at random. Independently of the value of its output, the probability p_i of setting the output of that neuron to 1 is determined. To this end, the difference ΔE_i between the network energy value when p_i is equal to 0 and when p_i is equal to 1 is computed. These energy values depend on the current state of the other neurons of the network. The probability to set the output of neuron i to 1 is computed from ΔE_i and from the temperature according to the formula used in simulated annealing,

$$p_i = \begin{cases} 1 & \text{if } \Delta E_i \leq 0, \\ \exp\left(-\dfrac{\Delta E_i}{T}\right) & \text{otherwise.} \end{cases}$$

If the probability is equal to 1, the output of neuron i is set to 1. Otherwise, a random number between 0 and 1 is generated. If it is lower than the computed probability, then the output of neuron i is set to 1; otherwise, it is set to 0. That updating procedure is iterated until the system is brought close to thermodynamical equilibrium at temperature T.

The above-described procedure is applied on decreasing temperature steps, as in simulated annealing.

The network has converged when the number of state changes minimal.

The above procedure shows the close relation between simulated annealing, Hopfield recurrent neural networks and Boltzmann machines. That is the reason why the algorithms derived from statistical physics are often considered as neural techniques. The convergence properties of Boltzmann machines are considered in detail in [Aarts et al. 1989].

8.6.5.5 Mean Field Annealing

Another approach, whose objectives are similar to those of the previous technique, is the *mean field theory*, or *mean field annealing*. The principle is to manipulate statistical means of the visited states in the Metropolis procedure used in simulated annealing. Unfortunately, it is necessary to make the so-called mean field approximation handle those means mathematically. That approximation consists in replacing very complex functions by their Taylor development truncated around a saddle point (see below): that is the main difference between simulated annealing and mean field annealing.

The performances of mean field annealing are sometimes improved by performing some normalization on groups of neurons [Hérault et al. 1989; Peterson et al. 1989]. Some constraints of the type one output out of N equal to 1 are then naturally taken into account, without resorting to additional penalty terms in the energy function. Similarly to microcanonical annealing [Hérault et al. 1993], it is possible to build a microcanonical mean field annealing algorithm [Lee et al. 1991b].

The equations of the mean field approximation, giving the neuron outputs between -1 and $+1$, are defined as follows:

$$y_i = \tanh\left(-\frac{1}{T}\frac{\partial E(\boldsymbol{y})}{\partial y_i}\right), \quad i = 1, \ldots, N.$$

In mean field theory, the temperature is fixed, and it is chosen around the critical temperature T_c, which is generally very difficult to estimate. That is the reason why it is preferable to use the mean field annealing with a decreasing temperature according to a given schedule during convergence [van den Bout et al. 1989, 1990; Peterson et al. 1988, 1989; Peterson 1990; Hérault et al. 1989, 1991]. Contrary to simulated annealing, mean field annealing is a deterministic and intrinsically parallel method, described by the following system of differential equations:

$$\frac{\mathrm{d}y_i}{\mathrm{d}t} = -\mu_i\left(y_i - \tanh\left(-\frac{1}{T}\frac{\partial E(\boldsymbol{y})}{\partial y_i}\right)\right), \quad i = 1, \ldots, N.$$

8.6.5.6 Pulsed Neural Networks

Pulsed neural networks [Hérault 1995c] do not suffer the limitations of the Hopfield networks in terms of constraint violation. They are defined with models of binary neurons that are more general than those used in the Hopfield neural networks. In a Hopfield network, the non-linearity of a neuron is only in its activation function: its potential is a linear function of inputs, and its output is a non-linear function of its potential. In a pulsed neural network, the potential of a neuron is a function, possibly non-linear of its inputs, and the motion equation has a more general form,

$$y_i = \boldsymbol{\Phi}[v_i(k-1), y_1(k-1), \ldots, y_N(k-1)].$$

The dynamics associated with that type of networks alternates several constraint satisfaction phases, during which the network converges towards valid solutions, and pulsation phases, during which the network tries to escape from local minima to go to better minima. In practice, the network proposes regularly some valid solutions (if they exist) of satisfactory quality. The selected solution will then be chosen among the set of proposed solutions. That type of network has been used successfully to solve problems where the solutions are required in very short response times, such as resource allocation problems (for instance, problems of allocation of weapons on targets [Hérault 1995a,b,c], and problems of extraction of maximum cliques encountered in fluid mechanics [Dérou et al. 1996]).

8.6.5.7 High-Order Neural Networks

In order to better satisfy the constraints during convergence, some neural networks with continuous outputs have been defined; their particularity is that they use penalty terms of order greater than 2 [Sun 1993]. That gives rise to equations of neurons that are based on Newton methods, which have a convergence speed of order 2, such as, for instance,

$$\frac{\mathrm{d}v_i}{\mathrm{d}t} = -\frac{\dfrac{\partial E}{\partial y_i}}{\dfrac{\partial^2 E}{\partial y_i^2}}.$$

That type of network has been successfully applied to combinatorial problems, by using a particular dynamics, which consists in selecting the neuron to be updated by the following procedure:

- Compute in parallel all the Δv_i associated with the penalty energy of the constraints, and select those which have the smallest Δv_i.
- Compute the gradient of the cost function for the neurons selected in the previous step, and keep the neuron degrading the least the cost value.

Example of the TSP

We consider again the TSP, with the same encoding as for a Hopfield network (permutation matrix encoded by the neuron outputs); in order to make use of the above equations, the penalty terms associated with the constraint violation must be of high order. They are derived from the following conditions, which are required for a solution to be valid:

- For each city i: $\prod_{j=1}^{N}(1-y_{i,j})=0$ and $\forall\, j,l\; y_{i,j}y_{i,l}=0$.
- For each position j on the tour: $\prod_{i=1}^{N}(1-y_{i,j})=0$ and $\forall\, i,k\; y_{i,j}y_{k,j}=0$.

An energy associated with the constraint violations is then derived,

$$E_c = \sum_{i=1}^{N}\prod_{j=1}^{N}(1-y_{i,j})^2 + \sum_{j=1}^{N}\prod_{i=1}^{N}(1-y_{i,j})^2 + \sum_{i=1}^{N}\sum_{j=1}^{N}\sum_{\substack{l=1\\ l\neq j}}^{N} y_{i,j}^2 y_{i,l}^2 + \sum_{j=1}^{N}\sum_{i=1}^{N}\sum_{\substack{k=1\\ k\neq i}}^{N} y_{i,j}^2 y_{k,j}^2.$$

8.6.5.8 Lagragian Neural Networks

Kuhn and Tucker showed that the solutions of a continuous optimization problem with constraints are the saddle points of an associated function, called Lagrangian [Kuhn et al. 1951]. Therefore, those problems can be solved by gradient methods applied to their Lagrangian.

Lagrangian neural networks are recurrent neural networks with continuous outputs, which have the specific property of converging to the saddle points of a Lyapunov function, instead of converging to local minima. They can be used to solve combinatorial optimization problems [Maa et al. 1992; Mjolsness et al. 1990, 1991; Zhang 1992; Rangarajan et al. 1996].

Example of the TSP

An augmented lagrangian can be defined for the TSP encoded as for the Hopfield network,

$$\begin{aligned} L(\boldsymbol{y}, \lambda, \mu) = Cost &+ \sum_{i=1}^{N} \lambda_i^0 \left[\sum_{j=1}^{N} y_{i,j} - 1 \right] + \sum_{j=1}^{N} \lambda_j^1 \left[\sum_{i=1}^{N} y_{i,j} - 1 \right] \\ &+ \frac{1}{2} \sum_{j=1}^{N} \mu_i^0 \left[\sum_{j=1}^{N} y_{i,j} - 1 \right]^2 + \frac{1}{2} \sum_{j=1}^{N} \mu_i^1 \left[\sum_{i=1}^{N} y_{i,j} - 1 \right]^2 . \end{aligned}$$

In that equation, parameters $\{\lambda_i^0\}$, $\{\lambda_i^1\}$ are the Lagrange multipliers, while parameters $\{\mu_i^0\}$, $\{\mu_i^1\}$ are the penalty weights.

Continuous equations for the neuron outputs and the Lagrange multipliers are defined as follows:

$$\begin{aligned} \frac{\mathrm{d}y_{i,j}}{\mathrm{d}t} &= -\mu \left[\sum_{\substack{k=1 \\ k \neq i}}^{N} d_{i,k} y_{k,j} + \lambda_i^0 + \lambda_j^1 + \mu_i^0 \left[\sum_{j=1}^{N} y_{i,j} - 1 \right] + \mu_j^1 \left[\sum_{i=1}^{N} y_{i,j} - 1 \right] \right] \\ \frac{\mathrm{d}\lambda_i^0}{\mathrm{d}t} &= -\rho \left[\sum_{j=1}^{N} y_{i,j} - 1 \right] \\ \frac{\mathrm{d}\lambda_j^1}{\mathrm{d}t} &= -\rho \left[\sum_{i=1}^{N} y_{i,j} - 1 \right] . \end{aligned}$$

In the above equations, μ and ρ regulate the convergence speed of the network.

That approach is far less sensitive than Hopfield networks to the penalty parameters associated with the constraint violations. However, convergence times can appear to be much longer.

8.6.5.9 Potts Neural Networks

To solve many combinatorial optimization problems, it is often interesting to encode the neurons in a vector form, by generalizing the neuron structure with K activation potentials, and K coupled outputs, whose sum is constant. By analogy with the Potts spin glasses in statistical physics, those neurons are called Potts neurons.

That form allows a natural encoding of constraints of the type 1-out-of-K or n-out-of-K. In other words, among the K outputs of a vector neuron, exactly n must be at +1 in a solution, all the others being −1. That type of constraint could be expressed under the form of an energy to minimize. But that strict constraint might not be satisfied at the convergence on a local minimum.

Making use of Potts neurons consists in replacing sigmoid activation functions by multidimensional activation functions. For a constraint of type 1-out-of-K, that amounts to defining binary neurons (i, a), with potential v_{ia} and output y_{ia} defined by

$$y_{ia} = \frac{\exp(v_{ia})}{\sum_{b=1}^{K} \exp(v_{ib})}, \quad i = 1, \ldots, N, \quad a = 1, \ldots, K.$$

Clearly, for each Potts neuron i, we have $\sum_{a=1}^{K} y_{ia} = 1$.

In other words, at the end of the convergence process, the constraint is automatically satisfied.

The mean field equations previously mentioned can also be defined with analog Potts neurons [Peterson et al. 1989; Peterson 1990; Hérault et al. 1989, 1991].

The use of those neurons has the following benefits:

- The energy function contains a smaller number of competing terms.
- It is not necessary to tune a parameter which weight an energy function associated with the violation of this type of constraint.
- Results are generally much better than those provided by scalar neurons.

That type of neural network has been applied to many problems, such as graph partitioning problems [Hérault et al. 1989], knapsack problems [Ohlsson et al. 1993], or scheduling problems [Lagerholm et al. 1997].

8.6.5.10 Rangarajan Neural Networks

When the solutions of a problem are expressed under the form of a permutation matrix, the neural network proposed by Rangarajan can be used. It has the advantage of not requiring the addition of penalty terms to the energy function in order to satisfy this constraint [Rangarajan et al. 1995, 1999]. It combines the Hopfield network with a projection algorithm on the space of doubly stochastic matrices, derived from the two following theorems [Sinkhorn 1964].

A *doubly stochastic matrix* is a matrix with real positive coefficients, such that the sum of the coefficients of each row and of each column is equal to 1.

Theorem 1. *To a* strictly *positive $N \times N$ matrix corresponds exactly a doubly stochastic matrix T_A which can be expressed under the form $T_A = D_1 A\, D_2$, where D_1 and D_2 are positive diagonal matrices. Matrices D_1 and D_2 are unique, within a scalar factor.*

Theorem 2. *The alternative normalization of rows and columns of a strictly positive $N \times N$ matrix converges to a strictly positive doubly stochastic matrix.*

Remark 1. That normalization consists, when considering an element of the matrix, in dividing it successively by the sum of the row coefficients, and then by the sum of the column coefficients.

Therefore, theorem 2 is a projection algorithm on the space of doubly stochastic matrices. Combined with an analog annealed recurrent neural network, it provides a powerful algorithm that guarantees the convergence of the network to a valid solution (coded by a permutation matrix) of good quality.

Example of the TSP

Let us consider the TSP with the same coding as the one used by a Hopfield network (permutation matrix coded by the neuron outputs).

The energy function is defined as follows:

$$E = \sum_{i=1}^{N} \sum_{j=1}^{N} \sum_{\substack{k=1 \\ k \neq i}}^{N} \mathrm{d}_{i,k} y_{i,j} y_{k,j\oplus 1} - \frac{\gamma}{2} \sum_{i=1}^{N} \sum_{j=1}^{N} y_{i,j}^2,$$

where γ is a strictly positive real number.

The second term in this equation is called *auto-amplification* term; it is useful to avoid converging to a null matrix.

The motion equation of an analog neuron $(i,\, j)$ is defined by

$$\frac{\mathrm{d}v_{i,j}}{\mathrm{d}t} = -\frac{\mathrm{d}E}{\mathrm{d}y_{i,j}} = -\sum_{\substack{k=1 \\ k \neq i}}^{N} \mathrm{d}_{i,k} y_{k,j\oplus 1} + \gamma y_{i,j}.$$

To enforce the strict positivity of the neuron outputs, the authors use an exponential activation function instead of a sigmoid function,

$$y_{i,j} = \exp(\beta v_{i,j}),$$

where β is a positive real number.

The dynamics of that analog neural network can be either synchronous, or random asynchronous. After an update of the neurons, the outputs are projected on a close doubly stochastic matrix by an alternative normalization of rows and columns:

- normalization of column j: $y_{i,j} := (y_{i,j})/(\sum_{i=1}^{N} y_{i,j})$,
- normalization of row i: $y_{i,j} := (y_{i,j})/(\sum_{j=1}^{N} y_{i,j})$.

During convergence, an annealing is performed by increasing the parameter β in order to force the neuron outputs to converge to binary values.

That type of approach is very efficient when the neuron outputs must be under the form of a permutation matrix.

Very good results were obtained on other combinatorial problems [Rangarajan et al. 1999].

8.6.5.11 Mixed-Penalty Neural Networks

Recurrent neural networks with *mixed penalties* can be used to solve 0/1 linear programming problems, which are combinatorial problems. Those networks can be viewed as annealed analog Hopfield neural networks, in which the energy function includes additional terms to help the convergence. Those terms are directly inspired from energy functions used in *interior point methods* for linear programming problems [Gonzaga 1992].

Consider the constraint satisfaction problem consisting in the search for Q binary variables that simultaneously satisfy M linear inequalities. That problem is NP-complete. It can be defined as follows:

$$g_k(\boldsymbol{y}) \leq 0, \quad k = 1, \ldots, M, \ \boldsymbol{y} \in \{0,1\}^Q.$$

One associates those variables to analog neuron outputs y_i.

The energy of the neural network is given by [Privault et al. 1998a]

$$E = \sum_{i=1}^{Q} y_i \left(1 - y_i\right) + \sum_{k=1}^{M} \left[\frac{\gamma}{2}\delta_k \left(g_k \left(\boldsymbol{y}\right)\right)^2 - \frac{1}{\alpha}\left(1 - \delta_k\right) \ln \left|g_k \left(\boldsymbol{y}\right)\right|\right],$$

where γ and α are positive real numbers, and where δ_k is equal to 1 when constraint k is violated, and to 0 otherwise.

In that energy, the first term penalizes the fact that neuron outputs are not binary.

The second term corresponds to the constraints. For the kth term in the sum

- The first part penalizes the violation of the k-th constraint: it is an exterior penalty term.
- The second part is an interior penalty term, which prevents the system form being attracted to bad local minima. It is not associated to a constraint violation. On the contrary, it is applied to the k-th constraint only when that constraint is satisfied; nevertheless, its goal is to keep the current state of the network far from the constraint boundaries of equations $g_k(\boldsymbol{y}) = 0$, and therefore to stay close to the *analytic centre* of the problem, defined by

$$\arg\min_{\boldsymbol{y}} \left[-\ln \left[\prod_{k=1}^{M} -g_k(\boldsymbol{y}) \right] \right].$$

During convergence, an annealing is performed by increasing the coefficient α, so that states close to the constraint boundaries are visited.

That network was assessed on numerous practical examples of an industrial problem containing up to 30,000 binary variables and 1,500 inequalities. In each example, a solution has been found in less than 500 complete updates of the network.

8.7 Tabu Search

Tabu search algorithms are iterative algorithms, whose main feature is that it stores in memory the near past during the search (and possibly scraps of the faraway past) in order to avoid retracing one's steps. Moreover, some mechanisms allow intensifying the search in a part of space where some solutions appear interesting. On the contrary, some other mechanisms allow to diversify the search, in other words to leave for new parts of the solution space when the part currently explored does not seam to contain interesting solutions.

Glover described that metaheuristics in 1986, but similar ideas were proposed at that time by Hansen. At present, an increasing number of publications are devoted to Tabu search. That metaheuristics was applied to many applications, and provided excellent results [Glover et al. 1997]. Among those applications, one finds many combinatorial problems such as scheduling, vehicle routing, resource allocation, graph coloring, graph partitioning problems.

Similarly to simulated annealing and its variants, that metaheuristics may require a long resolution time. A notable difference with the methods derived from statistical physics is that far less theoretical results exist. As a consequence, the tuning of the parameters is often more empirical and difficult. Nevertheless, when faced to a practical optimization problem to be solved in a limited time, it is difficult to anticipate whether simulated annealing or tabu search would provide the best solution.

8.8 Genetic Algorithms

Genetic algorithms were first described by Holland in 1975. Initially, they were not designed for optimizing functions, but for modeling adaptive behaviors. In fact, genetic algorithms model an evolution process of species, drawing inspiration from the evolution theory of Darwin.

In a genetic algorithm applied to optimization, a potential solution is considered as an individual in a population [Goldberg 1989]. The value of the cost function associated with a solution measures the fitness of the associated individual to its environment. A genetic algorithm simulates the evolution during several generations of an initial population whose individuals are poorly fitted, by means of genetic operators of *mutation* and *crossover*. After some generations, the population is made of well-fitted individuals, in other words of supposedly good solutions to the problem. The main difference with simulated annealing and tabu search is that genetic algorithms manipulate populations of solutions, instead of manipulating a single solution, which is improved statistically in an iterative way. Genetic algorithms can be considered as generalized local search algorithms.

At present, genetic algorithms have important limitations, mainly due to a very difficult tuning (coding of the solutions, types of genetic operators, size of the initial population, required number of generations, percentage of mutations, of crossovers, etc.). Moreover those algorithms are slow and can require large memory storage for the individuals of several generations. In terms of theoretical results, at the present time, there are far less solid theoretical results than with other metaheuristics such as methods derived from statistical physics.

8.9 Towards Hybrid Approaches

At present, in order to solve hard combinatorial problems encountered in real applications, an tendency emerges, aimed at building complex methods incorporating knowledge and techniques coming from various horizons (linear programming, branch and bound, simulated annealing, tabu search, neural networks, etc.).

For instance, to solve a 0/1 linear programming problem with tens of thousands variables and constraints, associated with a real stock management problem, a metaheuristics was developed around a core based on simulated annealing, with the following specific features [Privault et al. 1998b]:

- The initial solution is derived from a binarization of the solution of the continuous problem, obtained with the simplex algorithm.
- During the search, some intensification and diversification mechanisms derived from tabu search are used.
- Since it is difficult to find valid solutions, as soon as a new linear constraint is satisfied, that constraint is no more violated afterwards.

Another way to combine efficiently different concepts encountered in the metaheuristics is presented for the TSP in [Charon et al. 1996].

8.10 Conclusion

8.10.1 The Choice of a Technique

The points developed in the previous sections allow making some choices as a function of the particularities of the problem to solve:

- Analyze the complexity of the problem:
  1. theoretical complexity, in order to determine whether a classical approach is sufficient: intrinsic complexity, number of variables, number of constraints, type of costs to minimize;
  2. practical complexity: computation time required for evaluating a candidate solution, constraints on the global resolution time, requirements on the quality of the solution.
- Define how the method is to be used (automatic, semi-automatic with tunable parameters, etc.) and assess the degree of skill of the users.
- Assess precisely the requirements on the quality of the solution; for instance, if the data of the problem to solve are corrupted by noise, it is not necessary to go as close as possible of the optimum.
- Assess the available development time.

If the requirement on the quality of the solution is demanding, and if automatic operation is required, annealing algorithms are powerful. Tabu search requires a generally longer development time, but can provide in some cases better results than simulated annealing. Genetic algorithms require a very long development time and a new tuning of the internal parameters as a function of the problem data. Recurrent neural networks are more adapted to mean size problems, where the resolution time is more important than the requirements in terms of quality of the solution.

To reduce the resolution time while producing good solutions, a hybrid approach is often the best choice, but at the cost of a development time that may be important.

To summarize, when facing a combinatorial problem, the comparison of the performances of the different metaheuristics is difficult and must be performed accurately and rigorously. A presentation the frequently encountered pitfalls, and a sound evaluation methodology, are provided in [Barr et al. 1995; Hooker 1995; Rardin et al. 2001].

References

1. Aarts E., Korst J. [1989], *Simulated Annealing and Boltzmann Machines – a Stochastic Approach to Combinatorial Optimization and Neural Computing*, John Wiley & Sons Ed., 1989
2. Aarts E., Lenstra J.K. [1997], *Local Search in Combinatorial Optimization*, John Wiley & Sons Ed., 1997
3. Ackley D.H., Hinton G.E., Sejnowski T.J. [1985], A learning algorithm for Boltzmann machines, *Cognitive Science*, 9, pp 147–169, 1985
4. Asai H., Onodera K., Kamio T., Ninomiya H. [1995], A study of Hopfield neural networks with external noise, *1995 IEEE International Conference on Neural Networks Proceedings*, New York, États-Unis, vol. 4, pp 1584–1589
5. Ayer S.V.B. et al. [1990], A theoretical investigation into the performance of the Hopfield model, *IEEE Transactions on Neural Networks*, vol. 1, pp 204–215, June 1990

6. Barr R.S., Golden B.L., Kelly J.P., Resende M.G.C., Stewart W.R. [1995], Designing and reporting on computational experiments with heuristic methods, *Journal of Heuristics*, vol. 1, no 1, pp 9–32, 1995
7. Boudet T., Chaton P., Hérault L., Gonon G., Jouanet L., Keller P. [1996], Thin film designs by simulated annealing, *Applied Optics*, vol. 35, no 31, pp 6219–6226, Nov. 1996
8. Brandt R.D., Wang Y., Laub A.J., Mitra S.K. [1988], Alternative networks for solving the TSP and the list-matching problem, *Proceedings of the International Joint Conference on Neural Networks*, San Diego, II, pp 333–340, 1988
9. Cerny V. [1985], A thermodynamical approach to the travelling salesman problem: an efficient simulated algorithm, *Journal of Optimization Theory and Applications*, n 45, pp 41–51, 1985
10. Charon I., Hudry O. [1996], Mixing different components of metaheuristics, Chap. 35 de [Osman 1996]
11. Cichocki A., Unbehauen R. [1993], *Neural Networks for Optimization and Signal Processing*, John Wiley & Sons Ed., 1993
12. Creutz M. [1983], Microcanonical Monte Carlo simulations, *Physic Review Letters*, vol. 50, no 19, pp 411–1414, 1983
13. Dagli C. [1994], *Artificial Neural Networks for Intelligent Manufacturing*, Chapman & Hall, 1994
14. Dérou D., Hérault L. [1996], A new paradigm for particle tracking velocimetry, based on graph-theory and pulsed neural networks, *Developments in Laser Techniques and Applications to Fluid Mechanics*, pp 438–462, Springer-Verlag Ed., 1996
15. Dowsland K.A. [1995], Simulated annealing, Chap. 2 of [Reeves 1995]
16. Garey M.R., Johnson D.S. [1979], *Computers and intractability. A guide to the theory of NP-completeness*, W.H. Freeman and company Ed., 1979
17. Glover F., Laguna M. [1997], *Tabu search*, Kluwer Academic Publishers, 1997
18. Goldberg D.E. [1989], *Genetic Algorithms in Search, Optimization and Machine Learning*, Addison Wesley, 1989
19. Goles E. [1995], Energy functions for neural networks, *The Handbook of Brain Theory and Neural Networks*, The MIT Press, pp 363–367, 1995
20. Gondran M., Minoux M. [1995], *Graphes et algorithmes*, Éditions Eyrolles, 1995
21. Gonzaga C.C. [1992], Path-following methods for linear programming, *SIAM Review 34(2)*, pp 167–224, 1992
22. Hajek B. [1988], Cooling schedules for optimal annealing, *Mathematics of operations research*, vol. 13, no 2, pp 311–329, 1988
23. Hérault L., Niez J.J. [1989], Neural networks & graph K-partitioning, *Complex Systems*, vol. 3, no 6, pp 531–576, 1989
24. Hérault L., Niez J.J. [1991], Neural networks & combinatorial optimization: a study of NP-complete graph problems, *Neural Networks: Advances and Applications*, pp 165–213, Elsevier Science Publishers B.V. (North-Holland), 1991
25. Hérault L., Horaud R. [1993], Figure-ground discrimination: a combinatorial optimization approach, *I.E.E.E. Transactions on Pattern Analysis and Machine Intelligence*, vol. 15, no 9, pp 899–914, 1993
26. Hérault L. [1995a], Pulsed recursive neural networks & resource allocation – Part 1: static allocation, *Proceedings of the 1995 SPIE's International Symposium on Aerospace/Defense Sensing and Dual-Use Photonics*, Orlando, Florida, USA, pp 229–240, April 1995

27. Hérault L. [1995b], Pulsed recursive neural networks & resource allocation – Part 2: static allocation, *Proceedings of the 1995 SPIE's International Symposium on Aerospace/Defense Sensing and Dual-Use Photonics*, Orlando, Florida, USA, pp 241–252, April 1995
28. Hérault L. [1995c], Réseaux de neurones récursifs pulsés pour l'allocation de ressources, *Revue Automatique—Productique—Informatique industrielle (APII)*, vol. 29, numbers 4–5, pp 471–506, 1995
29. Hérault L. [1997a], A new multitarget tracking algorithm based on cinematic grouping, *Proceedings of the 11th SPIE's International Symposium on Aerospace/Defense Sensing, Simulation and Controls*, vol. 3086, pp 296–307, Orlando, Florida, États-Unis, avril 1997
30. Hérault L., Dérou D., Gordon M. [1997b], New Q-routing approaches to adaptive traffic control, *Proceedings of the International Workshop on Applications of Neural Networks to Telecommunications 3*, pp 274–281, Lawrence Erlbaum Associates Ed., 1997
31. Hérault L. [2000], Rescaled Simulated Annealing – Accelerating convergence of Simulated Annealing by rescaling the states energies, *Journal of Heuristics*, pp 215–252, vol. 6, Kluwer Academic Publishers, 2000
32. Hinton G.E., Sejnowski T.J., Ackley D.H. [1984], Boltzmann machines: constraint satisfaction network that learn, *Carnegie Mellon University technical report*, CMU-CS-84–119, États-Unis, 1984
33. Hooker J.N. [1995], Testing heuristics: we have it all wrong, *Journal of Heuristics*, vol. 1, no 1, pp 33–42, 1995
34. Hopfield J. [1982], Neural Networks and Physical Systems with emergent collective computational abilities, *Proceedings of National Academy of Sciences of USA*, vol. 79, pp 2554–2558, 1982
35. Hopfield J. [1984], Neurons with graded response have collective computational properties like those of two-state neurons, *Proceedings of National Academy of Sciences of USA*, vol. 81, pp 3088–3092, 1984
36. Hopfield J., Tank D. [1985], Neural computation of decisions in optimization problems, *Biological Cybernetics*, vol. 52, pp 141–152, 1985
37. Kirkpatrick S., Gelatt C.D., Vecchi M.P. [1983], Optimization by simulated annealing, *Science*, vol. 220, pp 671–680, 1983
38. Kuhn H.W., Tucker A.W. [1951], Non-linear programming, *Proceedings of the Second Berkeley Symposium on Mathematical Statistics and Probability*, pp 481–492, University of California Press, 1951
39. Lagerholm M., Peterson C., Söderberg B. [1997], Airline crew scheduling with Potts neurons, *Neural Computation*, vol. 9, no 7, pp 1589–1599, 1997
40. Lee B.W., Shen B.J. [1991a], Hardware annealing in electronic neural networks, *IEEE Transactions on Circuits and Systems*, vol. 38, pp 134–137, 1991
41. Lee H.J., Louri A. [1991b], Microcanonical mean field annealing: a new algorithm for increasing the convergence speed of mean field annealing, *Proceedings of the International Joint Conference on Neural Networks*, Singapore, pp 943–946, 1991
42. Maa C.Y., Shanblatt M.A. [1992], A two-phase optimization neural netwok, *IEEE Transactions on Neural Networks*, vol. 3, no 6, pp 1003–1009, 1992
43. Metropolis N., Rosenbluth A., Rosenbluth M., Teller A., Teller E. [1953], Equation of state calculations by fast computing machines, *Journal of Chemical Physics*, vol. 21, pp 1087–1092, 1953

44. Mitra D., Romeo F., Sangiovanni-Vincentelli A. [1986], Convergence and finite-time behavior of simulated annealing, *Adv. Appl. Prob.*, vol. 18, pp 747–771, 1986
45. Mjolsness E., Garrett C. [1990], Algebraic transformations of objective functions, *Neural Networks*, no 3, pp 651–669, 1990
46. Mjolsness E., Miranker W.L. [1991], A Lagrangian approach to fixed points, *Advances in Neural Information Processing Systems 3*, pp 77–83, Morgan Kaufman Pub., 1991
47. Newcomb R.W., Lohn J.D. [1995], Analog VLSI for neural networks, *The Handbook of Brain Theory and Neural Networks*, The MIT Press, pp 86–90, 1995
48. Ohlsson M., Peterson C., Söderberg B. [1993], Neural networks for optimization problems with inequality constraints – the knapsack problem, *Neural Computation*, vol. 5, no 2, pp 331–339, 1993
49. Osman I, Kelly J.P. [1996], *Meta-heuristics: theory and applications*, Kluwer Academic Publishers, 1996
50. Peterson C., Anderson J.R. [1988], Neural networks and NP-complete optimization problems: a performance study on the graph bisection problem, *Complex Systems*, vol. 2, pp 59–89, 1988
51. Peterson C., Söderberg B. [1989], A new method for mapping optimization problems onto neural networks, *International Journal on Neural Systems*, vol. 1, pp 3–22, 1989
52. Peterson C. [1990], Parallel distributed approaches to combinatorial optimization: benchmark studies on travelling salesman problem, *Neural Computation*, vol. 2, pp 261–269, 1990
53. Prins C. [1994], *Algorithmes de graphes*, Éditions Eyrolles, 1994
54. Privault C., Hérault L. [1998a], Constraints satisfaction through recursive neural networks with mixed penalties: a case study, *Neural Processing Letters*, Kluwer Academic Publishers, vol. 8, no 1, pp 15–26, 1998
55. Privault C., Hérault L. [1998b], Solving a real world assignment problem with a metaheuristic, *Journal of Heuristics*, vol. 4, pp 383–398, Kluwer Academic Publishers, 1998
56. Rangarajan A., Gold S. [1995], Softmax to softassign: neural network algorithms for combinatorial optimization, *Journal of Artificial Neural Networks*, vol. 2, no 4, pp 381–399, 1995
57. Rangarajan A., Mjolsness E.D. [1996], A Lagrangian relaxation network for graph matching, *IEEE Transactions on Neural Networks*, vol. 7, no 6, pp 1365–81, 1996
58. Rangarajan A., Yuille A., Mjolsness E.D. [1999], Convergence properties of the softassign quadratic assignment algorithm, *Neural Computation*, vol. 11, no 6, pp 1455–1474, 1999
59. Rardin R.L., Uzsoy R. [2001], Experimental evaluation of heuristic optimisation algorithms: a tutorial, *Journal of Heuristics*, vol. 7, no 3, pp 261–304, 2001
60. Reeves C.R. [1995], *Modern Heuristic Techniques for Combinatorial Problems*, McGraw-Hill, 1995
61. Reinelt G. [1994], *The travelling salesman. Computational solutions for TSP applications*, note de lecture *in Computer Science* 840, Springer Verlag, 1994
62. Roussel-Ragot P., Dreyfus G. [1990], A Problem-Independent Parallel Implementation of Simulated Annealing: Models and Experiments, *IEEE Transactions on Computer-Aided Design*, vol. 9, p. 827, 1990

63. Schrijver A. [1986], *Theory of Linear and Integer Programming*, John Wiley & Sons, 1986
64. Siarry P., Dreyfus G. [1984], Application of Physical Methods to the Computer-Aided Design of Electronic Circuits, *J. Phys. Lett.* 45, L 39, 1984
65. Sinkhorn R. [1964], A relationship between arbitrary positive matrices and doubly stochastic matrices, *The annals of mathematical statistics*, vol. 35, no 1, pp 141–152, 1964
66. Szu H. [1988], Fast TSP algorithm based on binary neuron output and analog neuron input using the zero-diagonal interconnect matrix and necessary and sufficient constraints on the permutation matrix, *Proceedings of the International Joint Conference on Neural Networks*, San Diego, II, pp 259–266, 1988
67. Sun K.T., Fu H.C. [1993], A hybrid neural network model for solving optimisation problems, *IEEE Transactions on Computers*, vol. 42, no 2, 1993
68. Takefuji Y. [1992], *Neural Network Parallel Computing*, Kluwer Academic Publishers, 1992
69. Takefuji Y., Wang J. [1996], *Neural Computing for Optimization and Combinatorics*, World Scientific, 1996
70. Teghem J., Pirlot M. [2001], *Métaheuristiques et outils nouveaux en recherche opérationnelle. Tome I: Méthodes. Tome II: Implémentations et Applications*, Hermès Editions, 2002
71. van den Bout D.E., Miller T.K. [1989], Improving the performance of the Hopfield-Tank neural network through normalization and annealing, *Biological Cybernetics*, vol. 62, pp 129–139, 1989
72. van den Bout D.E., Miller T.K. [1990], Graph partitioning using annealing neural networks, *IEEE Transactions on Neural Networks*, vol. 1, pp 192–203, 1990
73. Zhang S., Constantinides A.G. [1992], Lagrange programming neural networks, *IEEE Transactions on Circuits and Systems II: Analog and Digital Processing*, vol. 39, no 7, pp 441–452, 1992

About the Authors

Fouad Badran is a professor at CNAM (Conservatoire National des Arts et Métiers), where he teaches neural networks.

Gérard Dreyfus is a professor at the École supérieure de physique et de chimie industrielles de la ville de Paris (ESPCI), head of the Electronic Engineering Laboratory. He teaches machine-learning techniques at ESPCI and in various universities. The research activities of the laboratory are fully devoted to machine learning, with applications to data processing and to the modeling of biological neural systems.

Mirta B. Gordon, is a physicist, chief researcher at CNRS (Centre National de la Recherche Scientifique); she carries out research on neural networks and training algorithms. Formerly with the Theoretical Group of the Département de recherche fondamentale du CEA (Grenoble), she moved to the Leibnitz laboratory at Institut de mathématiques appliquées de Grenoble (IMAG).

Laurent Hérault is a project leader at the CEA–LETI, where he manages research on neural networks and combinatorial optimization applied to industrial problems. He has been a senior expert at CEA since 1998; his present research activities focus on data processing for wireless telecommunications.

Jean-Marc Martinez is a research scientist at Centre d'Études de Saclay, where he carries out research on simulation and supervision methods. He teaches statistical machine-learning methods at INSTN (Institut national supérieur des techniques nucléaires).

Manuel Samuelides is a professor at the École nationale supérieure de l'aéronautique et de l'espace (Supaéro) and head of the department of Applied Mathematics of that institute. He teaches probability theory, optimization and statistical techniques in machine learning, and pattern recognition. He carries out research on neural-network applications at the Département de traitement de l'information et modélisation of ONERA (Office national d'études et de recherches aérospatiales).

Sylvie Thiria is a professor at the University of Versailles Saint-Quentin-en-Yvelines and is a researcher at LODYC (Laboratoire d'Océanographie DYnamique et de Climatologie). Her research is centered on neural modeling and its applications to such areas as geophysics.

Méziane Yacoub is an assistant professor at CNAM.

Index